Biotechnological Innovation and Sustainable Developmental Goals

This reference book examines the role of biotechnological innovations in achieving global health targets, addressing disease burdens, and fostering a sustainable healthcare ecosystem for promoting good health and wellbeing. It covers a wide range of topics, including the evolution of healthcare technologies; epidemiological approaches in predictive medicine; management of diseases across different organs; emerging technologies in drug development, design, and delivery systems; vaccination strategies; technologies for tackling pandemics; and biotechnological applications in regenerative medicine and tissue engineering. Further, the book discusses different biotechnological advances in geriatric medicine and innovative therapies for age-related diseases. It also explores the intersection of biotechnology with maternal and child health and emphasizes the One Health approach in preventing and controlling zoonotic diseases. This book is intended for researchers, academics, and professionals in biotechnology, healthcare, and public health.

Biotechnological Innovation and Sustainable Developmental Goals

Fostering Good Health and Wellbeing

Edited by RC Sobti

CRC Press
Taylor & Francis Group
Boca Raton London New York

CRC Press is an imprint of the
Taylor & Francis Group, an **informa** business

Designed cover image: Shutterstock

First edition published 2026
by CRC Press
2385 NW Executive Center Drive, Suite 320, Boca Raton FL 33431

and by CRC Press
4 Park Square, Milton Park, Abingdon, Oxon, OX14 4RN

CRC Press is an imprint of Taylor and Francis Group, LLC

ISBN: 9781032644646 (hbk)
ISBN: 9781032660578 (pbk)
ISBN: 9781032660592 (ebk)

DOI: 10.1201/9781032660592

Typeset in Times
by Apex CoVantage, LLC

Contents

Preface

The planet's future has been uncertain due to the escalating challenges of climate change, resource depletion, environmental degradation, overpopulation, deforestation, loss of biodiversity, pollution, and the overexploitation of natural resources. These interconnected issues threaten ecosystems, disrupt food systems, and endanger human health and wellbeing. If these activities continue unchecked, the foreseeable future of the planet shall be in jeopardy. However, technological advances, particularly in biotechnology, offer hope. Biotechnology can potentially make previously unsustainable activities sustainable and pivotal in achieving the UN Sustainable Development Goals (SDGs) and underpinning human wellbeing and livelihoods.

In the medical and health sector, biotechnology enables the development of novel drugs and vaccines, addressing SDG 3: Good Health and Wellbeing. There have been remarkable conceptual and practical advances in molecular biology and biomedical sciences in recent decades, giving rise to emerging fields such as molecular medicine and personalized drug therapies. A comprehensive understanding of the mechanisms underlying the development and functioning of the human body in health and disease has emerged. Most of this knowledge has been acquired through *in vitro* and *in vivo* studies using microorganisms, cell lines, and laboratory animals, including primates and marine organisms. Insights into physiological and pathophysiological mechanisms are helping to delineate health and disease processes. Researchers are using molecular, computational, and structural biology and imaging technologies to understand the integrated functions of human body organs and other organisms.

The aftermath of the Human Genome Project and molecular developments, including next-generation sequencing and gene editing technologies, has provided new knowledge for understanding disease mechanisms and developing strategies for preventing acute and chronic diseases. Biotechnological innovations are facilitating the development of targeted medicines for cancer patients and improving longevity and quality of life. Novel molecular techniques have enabled the production of vaccines and diagnostic tools that prevent and contain epidemics while allowing for early and accurate detection of diseases.

For instance, Herceptin (trastuzumab), a monoclonal antibody-based treatment, targets the HER2 protein to treat certain breast and gastric cancers. At the same time, CAR-T cell therapy modifies a patient's T-cells to combat blood cancers like leukaemia and lymphoma, exemplifying how biotechnology enables precise, personalized treatments that improve survival rates and quality of life. In addition, mRNA vaccines like those from Pfizer-BioNTech and Moderna were developed to combat COVID-19, showcasing rapid production and effective immune response while paving the way for vaccines against other diseases, including cancer. Vaccines such as Gardasil have significantly reduced HPV-related cancers; and PCR and CRISPR-based diagnostic tools enable early, precise detection of infections like COVID-19, HIV, and Zika. Advances in biotechnology have also improved the production of recombinant insulin, enabled gene therapies like Luxturna for restoring vision, and developed stem cell therapies for regenerating tissue, enhancing longevity and quality of life.

Today, parasitic infectious diseases such as malaria, AIDS, and tuberculosis can be diagnosed rapidly and at a relatively low cost using molecular diagnostic tools, including polymerase chain reaction, recombinant antigens, and monoclonal antibodies. Modern diagnostic kits for rickettsial, bacterial, and viral diseases and radiolabeled biological treatments for imaging and analysis have been developed. Vaccines have nearly eradicated deadly diseases such as smallpox, polio, and cholera over the past century. Advances in recombinant vaccine technology also hold the potential to combat noncommunicable diseases (NCDs) such as cancer. Naked DNA, viral vectors, and plant-derived vaccines have proven effective against various bacterial and viral diseases. Therapeutic proteins, polyunsaturated fatty acids, and dietary interventions significantly prevent NCDs, which account for over 60% of deaths in developing countries. Immunotherapy, which harnesses the body's immune system, has enabled the development of vaccines to manage disease symptoms effectively. Biotechnology therapies and vaccines have reduced the spread of infectious diseases and helped contain epidemics and pandemics.

Antiretroviral therapies have extended the longevity of HIV/AIDS patients. For instance, in the "London Patient" case, targeting the C-C chemokine receptor 5 gene mutation through stem cell transplants successfully cured the disease. Biotechnology companies are working to deliver high-impact health technologies worldwide, reducing healthcare inequities, facilitating point-of-care treatment, and developing biopharmaceutical products that lower healthcare costs and promote healthier workforces and economic development.

Transgenic bacteria, yeasts, medicinal plants, and experimental animals have been used to produce recombinant therapeutic proteins and other products. Gene therapies are being introduced into these systems to produce therapeutic proteins in large quantities for purification and therapeutic use. Key recombinant proteins include erythropoietin for treating anaemia, interferon-alpha for leukaemia and viral diseases, and insulin for type 1 diabetes mellitus. Other therapeutic agents include growth hormones, cytokines, recombinant blood products, monoclonal antibodies, gene therapy products, molecular pharmaceutical agents, and bioengineered tissue products. Xenografts, bone grafts, collagen, and heart valves have also been successfully engineered.

New technologies now offer affordable drug and vaccine delivery systems, eliminating the risk of blood-borne infections caused by reusing needles. Drugs can be administered efficiently using gas jets, nasal sprays, or nanoparticle solutions. Advances in biotechnology and nanotechnology have reduced treatment dosages, improving efficacy and minimizing side effects. Although it is still in its early stages, stem cell therapy shows promising potential for therapeutic applications.

Genetically modified (GM) crops address micronutrient deficiencies essential for physiological functions, aiding children's physical growth and cognitive development. GM crops are enriched by introducing desired genes to meet global nutritional needs; for instance, iron-rich rice helps prevent anaemia, while Golden Rice, rich in vitamin A, combats childhood blindness in developing countries. Potatoes and tomatoes enhanced with vitamins and minerals have also effectively met global nutritional demands. Integrating GM foods, including flax seed, corn, maize, soybeans, papaya, squash, sugar beets, canola, and cottonseed, into staple diets can improve nutrition and quality of life. Biotechnology has also produced plant-based fungicides, herbicides, and insecticides to control pests effectively.

While knowledge in these fields has expanded rapidly, translating this information into practical applications remains challenging. A significant gap exists between the knowledge gained and its application for human benefit, and addressing these challenges requires multidisciplinary therapeutic approaches. Synthesizing this knowledge shall accelerate our understanding of disease mechanisms and facilitate the development of new strategies for prevention, diagnosis, and treatment.

This book aims to bridge this knowledge gap by providing updated insights into biomedical sciences and molecular medicine for personalized patient care. The editor expresses gratitude to all the contributors to the chapters included in this book. He also acknowledges the support of colleagues and friends who assisted in compiling this series of books on biotechnological innovations for achieving sustainable development goals. Special thanks go to Dr. Vipin Sobti, Er. Aditi, Er. Vineet, and Er. Ankit and Irene for their invaluable support throughout this endeavour.

RC Sobti

About the Editor

Professor Ranbir Chander Sobti is an Honorary Scientist at the Indian National Science Academy and Emeritus Professor at Panjab University, Chandigarh. He is a former Vice-Chancellor of Panjab University, Chandigarh and Babasaheb Bhimrao Ambedkar University (Central University), Lucknow. He started his career as a cytogeneticist and then moved on to molecular biology, including genomics, to understand the susceptibility and disease process of cancer, COPD, AIDS, metabolic syndrome, and kidney diseases. He has also used stem cells and nanoparticles to understand the process of tissue organ development through a designed de-cellularization protocol. He has published more than 350 research articles in journals of international repute and more than 60 books by international publishers such as Elsevier, Springer, CRC Press, and others. He is a Fellow of the Third World Academy of Sciences, the National Academy of Sciences India, the Indian National Science Academy, the National Academy of Medical Sciences, the National Academy of Agricultural Sciences, the Canadian Academy of Cardiovascular Diseases, the Royal Society of Biology and International Association of Cancer Research, and a few others. He was the General President of the Indian Science Congress for the 102nd session held at the University of Jammu in 2013. Dr Sobti is the recipient of many prestigious awards such as the INSA Young Scientist Medal, UGC Career Award, Punjab Rattan Award, JC Bose Oration and Sriram Oration Awards, and Life Time Achievement Awards of the Punjab Academy of Sciences, Zoological Society of India, and the Environment Academy of India. He received the most prestigious civil award, Padma Shri, from the Government of India.

1 Biotechnological Innovations for Sustainable Good Health

RC Sobti,[1] *Ranjana Bhandari,*[2]* *Ritika Patial,*[3] *and Aastha Sobti*[4]

[1] Department of Biotechnology, Panjab University, Chandigarh, India
[2] University Institute of Pharmaceutical Sciences, Panjab University Chandigarh
Hillman Comprehensive Cancer Center, Pittsburgh, Pennsylvania, USA
[3] Centre for Bioinformatics, Panjab University, Chandigarh, India
[4] Department of Immunotechnology, Lund University, Lund, Sweden

1.1 INTRODUCTION

Biotechnology is an interdisciplinary field that integrates biology, technology, engineering, and other scientific disciplines to control biological systems for diverse purposes. It entails creating products that utilize living organisms or biological procedures such as genetic recombination, cloning, and genetic alteration. The techniques and knowledge for developing and utilizing biological systems for numerous purposes have experienced significant advancements in recent years. These advancements, fueled by the development and innovation of recombinant DNA, molecular biology, bioinformatics, and cell culture techniques, have greatly extended the possibilities and potential of biotechnology.

The impacts of biotechnology on daily life are evident through its advancements in various sectors, such as healthcare, agriculture, environmental restoration, and national security (1, 2). This field has led to the development of groundbreaking vaccines, antibiotics, genetically modified crops, and bioengineered materials, making significant contributions to sectors such as pharmaceuticals and agriculture. One of the major areas of innovations with profound impacts is the field of medical biotechnology. Medical biotechnology has revolutionized medicine by providing quick and effective methods for disease diagnosis, generating new drugs and vaccines, and introducing completely novel treatment approaches. This has led to rapid growth and expansion of the biotechnology industry, particularly in countries such as the USA and Western nations. Asian countries such as China, India, South Korea, Taiwan, and Singapore have also emerged as major players in the biotechnology industry because of significant funding and regular government support in its development. Advancements in biotechnology not only have transformed the way diseases are diagnosed and treated but also have the potential to revolutionize the delivery of healthcare.

1.2 ROLE OF BIOTECHNOLOGY IN HEALTHCARE

By utilizing omics-based technologies, such as genomics, proteomics, metabolomics, and transcriptomics, biotechnology plays a vital role in the healthcare sector in gaining insights into human diseases, identifying biomarkers, and devising personalized treatment approaches (1) (Figure 1.1). Advances in immunology and genetics have helped reduce global health disparities by providing means for diagnoses, vaccines, and treatments (2). The impact of healthcare biotechnology on human wellbeing is substantial, particularly in terms of disease prevention, detection, and management, resulting in better health outcomes and financial gains (3). Additionally, the field ensures sustainable innovations in healthcare by offering prospects for biopharmaceuticals, vaccines, botanical drugs, data analysis, genomic research, and nanomedicine (4). Moreover, biotechnology contributes to the production of personalized medications and healthier food choices, demonstrating its capability to improve human health and enhance the quality of life.

1.3 HISTORICAL PERSPECTIVES ON BIOTECHNOLOGY IN MEDICINE

The development of biotechnology in medicine begins its journey in ancient eras. Biotechnology has been closely associated with medical practices since early times, such as the utilization of desiccated smallpox scabs for immunization (5). Over the years, the use of living organisms to produce therapeutic substances has transformed pharmacotherapy, resulting in the development of recombinant human therapeutic proteins and monoclonal antibodies via genetic manipulation (6). Moreover, the culmination of the Human Genome Project in 2003 made a substantial contribution to uncovering the mechanisms through which genetic abnormalities lead to diseases, establishing the profound influence of genetics and biotechnology on medical progress. Moreover, genomic technologies such as DNA cloning, sequencing, and polymerase chain reaction have revolutionized medicine by enabling the manipulation and analysis of genetic information, leading to advancements in diagnostics and treatments (7, 8). These historical perspectives highlight the ongoing evolution of biotechnology in medicine, from ancient immunization methods to contemporary genetic treatments and stem cell technologies.

* Corresponding author: akb10in@yahoo.co.uk

DOI: 10.1201/9781032660592-1

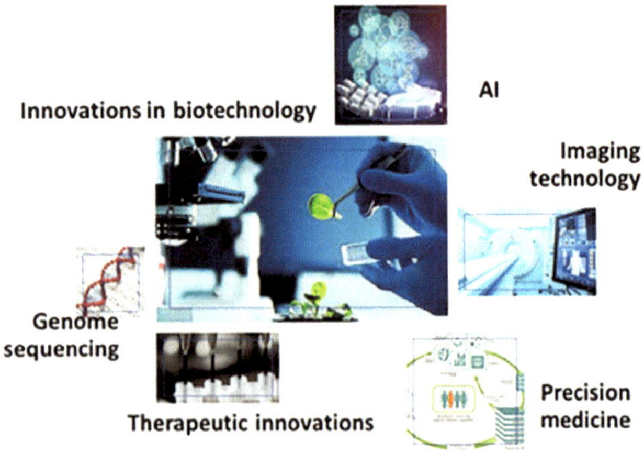

FIGURE 1.1 Medical applications of biotechnology.

1.4 DIAGNOSTIC ADVANCEMENTS

1.4.1 Genomic Sequencing and Personalized Medicine

Recent advancements in genomic sequencing technology, especially next-generation sequencing (NGS), have brought about a significant transformation to genomic research.

The swift sequencing of DNA fragments by NGS offers valuable insights into genome structure, genetic variations, and gene expression patterns, which are essential for developing personalized medicine as shown in Figure 1.2 (9). The fusion of NGS with CRISPR/Cas9 has improved the detection and targeting of genetic abnormalities in cancer, thereby promoting the advancement of precision medicine (10). Furthermore, developments in molecular point-of-care tests have simplified genetic testing at the point of care, supporting decisions of personalized treatment and dosage adjustments (11). And the genomic profiling of diseases such as lung cancer has paved the way for tailored therapies based on specific genetic changes, revolutionizing treatment outcomes and illustrating precision medicine in clinical oncology.

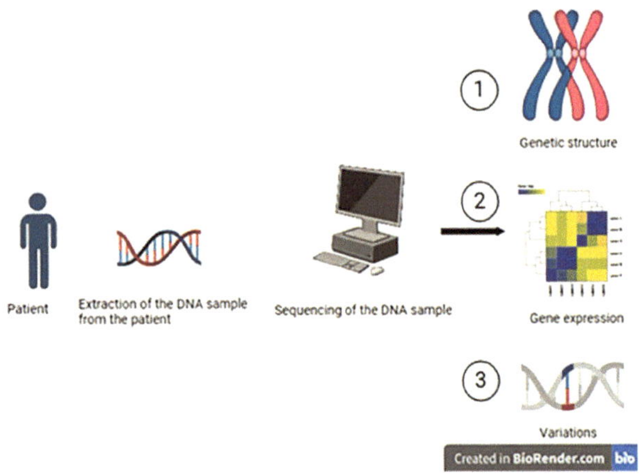

FIGURE 1.2 Genome sequencing.

1.4.2 Rapid Diagnostic Tools and Imaging Technology

Biotechnology has helped the field of medicine to undergo revolutionary trends via the development of quick diagnostic instruments and imaging technology. The speed, accuracy, and efficiency of disease diagnosis and monitoring have greatly increased because of these developments, enabling early disease detection, accurate localization of abnormalities, and focused treatment planning. For instance, supported by the US National Institutes of Health, the Rapid Acceleration of Diagnostics (RADx) Tech project significantly accelerated the commercialization of novel COVID-19 diagnostic tests (12). Molecular diagnostic tools such as nucleotide sequences and nucleic acid amplification are being used to detect various diseases including cancer and microbial infections (13). Additionally, diagnostic imaging technologies such as CT and MRI have advanced beyond traditional approaches by emphasizing increased sensitivity through updated modalities and contrast agents (14).

1.4.2.1 The Role of Artificial Intelligence in Diagnostics

Artificial intelligence (AI) has revolutionized many industries including healthcare and medicine. In the field of diagnostics, AI has significantly improved accuracy and efficiency, significantly transforming healthcare (15) for instance by using AI algorithms to analyze clinical images such as X-rays and CT scans and enhance disease detection and diagnosis (16). Applying AI to process extensive data and identify intricate patterns has improved medical decision-making, accelerated diagnoses, optimized treatment plans, and decreased human errors (Figure 1.3) (17). The use of machine learning and deep learning in handling medical has reduced errors and also enhanced the accuracy of disease diagnosis (18). However, despite its potential to aid clinical diagnoses, AI poses certain obstacles in ensuring data precision for accuracy. Nevertheless, the integration of AI with the medical sector in handling patient data such as medical images and vital signs is providing valuable assistance to healthcare professionals in the quick and accurate diagnosis of diseases (19).

1.4.3 Therapeutic Innovations

Biotechnology has significantly advanced therapeutic innovations by offering unique insights into disease pathogenesis and the development of new therapies to revolutionize disease management (20, 21), and entrepreneurship in biotechnology has produced practical applications in various sectors such as health, agriculture, and industry.

One of its key contributions to therapeutic innovations is its ability to provide unique insights into disease pathogenesis (22). By identifying underlying molecular mechanisms of diseases, biotechnology has helped researchers gain a deeper understanding of disease development and progression. This understanding has made it possible to identify new

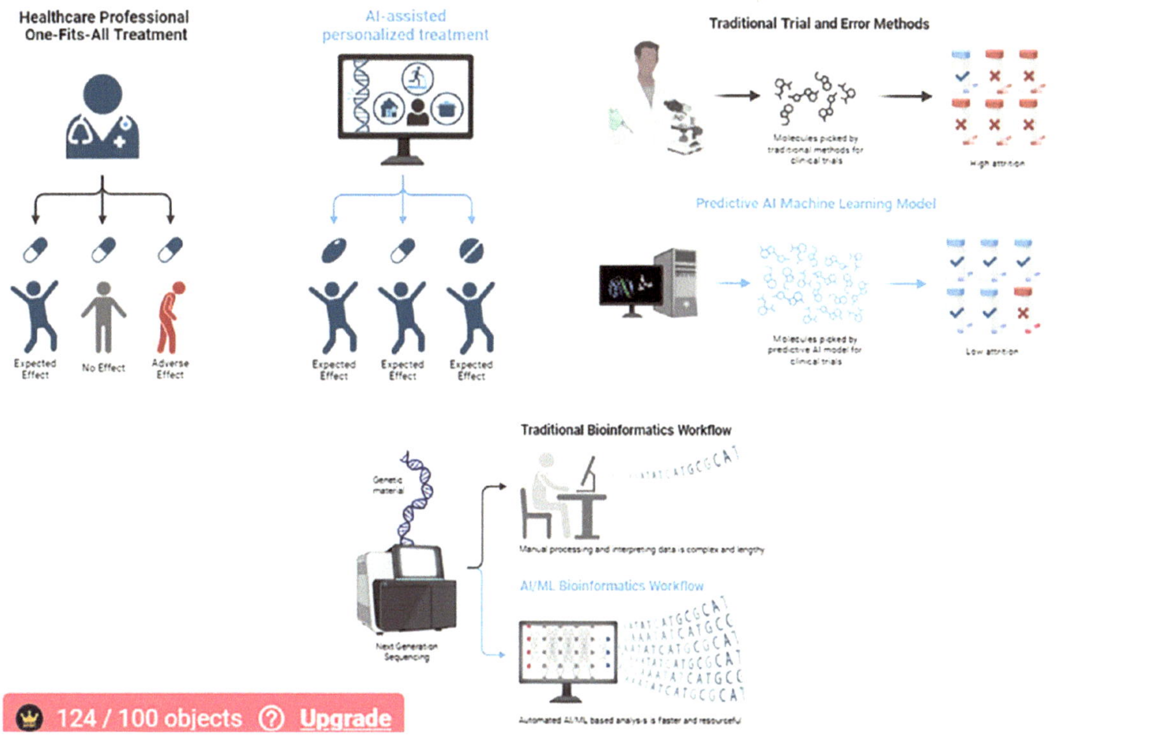

FIGURE 1.3 Artificial intelligence in healthcare has reduced errors, improved automated result analysis, and assisted in the development of precision medications.

therapeutic intervention targets and to design more focused and effective treatment strategies. Another contribution of biotechnology to therapeutic innovation is its role in expediting the development of new treatments. Through biotechnological techniques such as genetic engineering, researchers can precisely and effectively generate therapeutic proteins, antibodies, and vaccines. These advancements have shortened the drug development timeline and allowed for the rapid translation of scientific discoveries into tangible therapies.

1.4.4 Advances in Gene Editing

Advances in the field of genetic manipulation have been significantly propelled by innovations in biotechnology, particularly in gene editing technologies such as CRISPR/Cas9 (23). The precision, simplicity, and adaptability of CRISPR/Cas9 tools have transformed genome editing, allowing precise alterations in a variety of organisms. These tools have been effectively employed in the treatment of muscular dystrophies such as Duchenne muscular dystrophy, myotonic dystrophy, and congenital muscular dystrophy, demonstrating promising therapeutic capabilities (24). The versatility and efficiency of CRISPR/Cas9 are evident in its capacity to simultaneously modify multiple loci without the need for plasmids, thus simplifying gene editing in microorganisms, plants, animals, and humans (25). Moreover, improvements in CRISPR delivery methods, including nonviral strategies such as nanoparticles and cell-derived membrane vesicles, have increased the accuracy and effectiveness of gene editing (26).

1.4.5 Stem Cell Therapy and Tissue Engineering

Advances in biotechnology have had significant impacts on the development of stem cell therapies and tissue engineering. Stem cells, in combination with scaffolds, are employed for tissue repair and meat production in the food industry (27). Engineered hydrogels play a vital role in controlling the fate of stem cells, thereby improving their clinical use in tissue regeneration (28). Techniques such as 3D culture, bioreactors, and the integration of cytokines and biomaterials mimic natural microenvironments, encouraging tissue specialization (29). Tissue engineering has progressed with the advent of technologies such as 3D bioprinting, organ-on-a-chip systems, and induced pluripotent stem cells, opening new possibilities for regenerative medicine and disease modeling (30). Although stem cell engineering has benefits in tissue transplants, challenges remain in the treatment of diseases. However, these advancements collectively have propelled the field forward, tackling key obstacles and broadening the scope of therapeutic applications.

1.4.6 Regenerative Medicine

Advances in biotechnology have significantly advanced regenerative medicine. Innovations such as genetic manipulation, stem cell investigations, tissue engineering, and gene therapy have brought about transformations in this field (31). Genetic manipulation and nuclear reprogramming have enabled the development of personalized therapies for patients, thereby improving the efficacy of gene and stem cell treatments. Gene therapy, particularly that utilizing nonviral carriers, has demonstrated potential in regulating the function of genes introduced *in vivo*, a critical aspect of regenerative medicine (32). Moreover, creating organoids by cultivating stem cells in a 3D setting has introduced a new approach to tissue regeneration and personalized medicine, presenting potential solutions for organ transplantation and disease studies (33). Furthermore, advancements such as 3D printing and organ-on-a-chip technologies are pushing the boundaries of tissue engineering by producing physiologically functional organs, aiding the comprehension of disease mechanisms and pharmaceutical exploration (34).

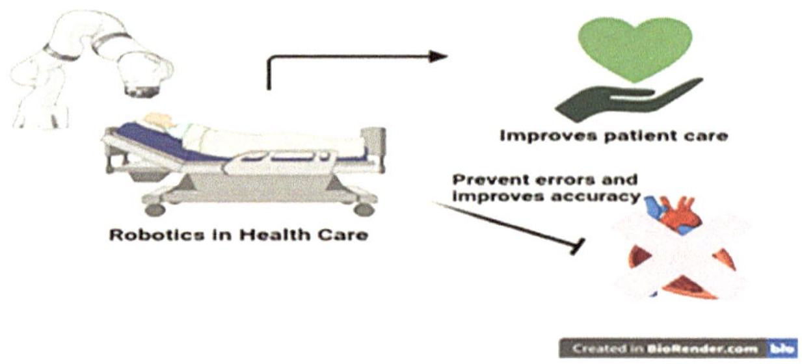

FIGURE 1.4 The integration of robotics in the healthcare sector has supported medical staff and enhanced patient care by automating responsibilities, reducing risks of exposure, and allowing healthcare providers to focus on patients' welfare.

1.4.7 Precision Medicine

Precision medicine, fueled by innovations in biotechnology, has gained the potential to revolutionize healthcare practices and improve patient outcomes. With the utilization of genomic information and personalized treatment approaches, precision medicine can tailor medical interventions according to the specific needs of individual patients. This approach allows for more accurate diagnoses, targeted therapies, and better disease management.

Integrating techniques in nucleic acid manipulation and analysis, such as PCR and sequencing, has facilitated the identification of biomarkers, mapping of cellular signaling pathways, and customization of therapeutic interventions (35). Moreover, the discovery of targetable gene mutations and the utilization of predictive biomarkers have bolstered innovative treatment strategies in oncology (36). The emergence of technologies such as AI, machine learning, and big data analysis have significantly advanced personalized healthcare, leading to more accurate and efficient therapies (37). Furthermore, improvements in magnetic resonance imaging (MRI) technology have enhanced diagnostic capabilities, thus advancing precision medicine in the management of diseases (38).

1.4.8 Synthetic Biology

Innovation in biotechnology has profoundly impacted the field of synthetic biology by introducing novel biological components and systems and modifying microbes for health purposes (39). This has allowed researchers to develop new tools and techniques for engineering biological systems, thereby enabling the design of more complex and effective synthetic organisms. Synthetic biology employs light regulation to precisely control molecular systems while addressing safety concerns through regulatory actions (40). Furthermore, it plays a critical role in addressing antibiotic resistance by developing new antimicrobial treatments (41). In the healthcare sector, synthetic biology presents a promising approach that utilizes engineered microbes as therapeutic agents, facilitating the creation of therapeutic substances based on the body's physiological conditions (42). Moreover, the design of genetic systems compatible with different hosts improves the manipulation of biological functions to enhance cellular processes (43). This integration of biotechnology and synthetic biology holds great promise for advancements in biomedical engineering and human health, leading to improved diagnostics, more effective therapies, and a better understanding of complex biological systems

1.4.9 Robotic Assistance in Healthcare

The progress in robotics in the healthcare sector signifies a notable breakthrough in the field of biotechnology. These advancements have resulted in the creation of automated systems, AI bots, and robotic devices that aid in patient care, surgical procedures, and medical treatment, particularly in challenging scenarios such as the COVID-19 crisis (44). The utilization of robotics in healthcare is geared towards diminishing interpersonal contact, improving sanitation and hygiene, and providing support while enhancing the quality of healthcare, reducing medical mistakes, and delivering tailored therapy (45). The incorporation of robotics in healthcare not only assists healthcare professionals but also improves patient care by automating tasks, minimizing exposure risks, and enabling healthcare providers to concentrate on the wellbeing of patients (46). In essence, robotics in healthcare marks a transformative change toward more effective, accurate, and supportive medical approaches, highlighting the innovative capabilities of biotechnology in enhancing healthcare results (47).

1.5 FUTURE PROSPECTS AND ETHICAL CONSIDERATIONS

1.5.1 Potential Impacts of Biotechnological Innovations on Society

The potential impacts of biotechnological innovation on society are immense. Biotechnological advancements in sectors such as healthcare and agriculture have brought advantages such as improved efficiency and effectiveness, productivity, and product quality (4). Within the healthcare industry, biotechnology has facilitated advanced therapies such as biopharmaceuticals and vaccines, thereby enhancing global health outcomes (48). In the field of agriculture, biotechnology can bolster crop protection, adaptation to climate change, and forest yield, providing potential solutions to societal and environmental issues (49). Nonetheless, apprehensions persist regarding unforeseen long-term impacts on human health and the environment alongside ethical concerns surrounding genetic manipulation and the domination of biotechnology.

The emergence and implementation of biotechnological advancements present notable ethical considerations and regulatory hurdles such as the need for clear ethical, legal, and social implications (ELSI); bioethical aspects such as informed consent and privacy rights (50); the ongoing discourse involving religious fact, ethicists, and biotechnologists on the merits and drawbacks of biotechnology products; and the absence of well-established ethics and religious viewpoints among scientists in specific areas (51). Neglecting to tackle these issues may impede the public perception and acceptance of biotechnological progress. It is essential to proactively recognize and tackle ELSI concerns; set up thorough policy frameworks; and engage interdisciplinary committees to navigate the ethical, legal, and social dimensions of biotechnological innovations to improve public confidence and acceptance (52).

1.6 ROLE OF GENETICALLY MODIFIED FOODS IN HEALTHCARE

Genetic engineering presents a promising avenue for addressing nutritional deficiencies and bolstering public health outcomes, particularly in regions where access to diverse diets or vitamin supplements is limited (53).

1.6.1 Golden Rice

Golden Rice represents a breakthrough in genetic engineering, designed to produce beta-carotene, a precursor of vitamin A. This innovation targets the pervasive issue of vitamin A deficiency, particularly in regions where rice is a dietary staple. Enriching rice with this crucial nutrient presents a targeted approach to addressing childhood blindness and enhancing immune resilience against infections (54).

1.6.2 Biofortified Maize

Genetic modification has enabled the development of maize varieties with elevated levels of nutrients such as vitamin A, iron, and zinc. These biofortified crops address widespread micronutrient deficiencies that can lead to serious health complications. By incorporating essential nutrients directly into maize, this innovation offers a sustainable strategy for improving public health (55).

1.6.3 High-Protein Soybeans

Genetically modified soybean varieties with increased protein content play a crucial role in meeting the rising demand for plant-based protein sources globally. By providing a nutrient-rich alternative for both human consumption and animal feed, high-protein soybeans address protein deficiencies and promote healthier dietary patterns (56).

1.6.4 Reducing Antinutrients

Genetic modification also enables the reduction or elimination of antinutritional compounds in crops, thereby enhancing their nutritional value and digestibility. By improving nutrient absorption and alleviating digestive discomfort associated with certain foods, this innovation supports better health outcomes and addresses challenges related to malnutrition and dietary intolerances (57).

1.7 CONCLUSION

Advancements in biotechnology have significantly transformed the healthcare sector on a global scale by facilitating early disease detection, tailored therapies, and enhanced health results. They also offer a multifaceted approach to enhancing the nutritional quality of crops and addressing malnutrition. These innovations hold significant promise for improving public health, particularly in regions where nutritional deficiencies are prevalent. Developments in genomics, proteomics, metabolomics, and transcriptomics have resulted in the identification of novel biomarkers that can aid in the early identification of diseases.

This field has brought forth pioneering methodologies including gene therapy, stem cell therapy, and CRISPR/Cas9 technology that have made notable contributions to the field of medical science. Moreover, the fusion of big data analytics and genomic research is improving the accuracy of personalized medicine and clinical procedures, and the emergence of AI is revolutionizing existing treatment methods. These innovative approaches, in conjunction with technological progress such as wearable devices and remote monitoring tools have the potential to mitigate the long-term complications of diseases such as cardiovascular disorders, ultimately enhancing the sustainability of healthcare systems and decreasing mortality rates. In the field of biotechnology, the emergence of new ideas and methods has brought forth both obstacles and opportunities for progress. The ongoing advancements in biotechnology continue to drive innovation in the medical field, offering hope for more effective and personalized approaches to diagnosis and treatment.

REFERENCES

1. Tanveer A, Yadav K, Dwivedi S, Gupta S, Yadav PK, Yadav D (2022). Overview of healthcare biotechnology, biotechnology in healthcare, Academic Press, 1–26, ISBN 9780323898379, https://doi.org/10.1016/B978-0-323-89837-9.00003-6.
2. Dwivedi S, Agrawal R, Misra R, Anjum S, Kushwaha SS, Joshi HS, Misra S, Kishore S (2022). Biotechnology in primary healthcare and hospital management, Biotechnology in Healthcare 2; 3–19, ISBN 9780323900423, https://doi.org/10.1016/B978-0-323-90042-3.00012-8.

3. Kumar B (2022). Government initiatives and policies in healthcare biotechnology, Biotechnology in Healthcare 2; 303–320, ISBN 9780323900423, https://doi.org/10.1016/B978-0-323-90042-3.00017-7.
4. Bhatti A, Rehman A, John P (2022). Challenges and opportunities in healthcare biotechnology, Biotechnology in Healthcare 2(1); 321–342, ISBN 9780323900423, https://doi.org/10.1016/B978-0-323-90042-3.00014-1.
5. Silva JM, Massabni AC (2019). The importance of biotechnology in the health area: Historical overview and future perspectives, International Journal of Advances in Medical Biotechnology – IJAMB 2(2), https://doi.org/10.25061/2595–3931/IJAMB/2019.V2I2.42.
6. Massabni AC, Silva JM (2019). The importance of biotechnology in the health area: Historical overview and future perspectives, International Journal of Advances in Medical Biotechnology – IJAMB 2; 69–74.
7. Galas DJ, McCormack SJ (2003). An historical perspective on genomic technologies, Current Issues in Molecular Biology 5(4); 123–127, PMID: 12921227.
8. Goyal PA, Bankar NJ, Mishra VH, Borkar SK, Makade JG (2023). Revolutionizing medical microbiology: How molecular and genomic approaches are changing diagnostic techniques, Cureus 15(10); 16, https://doi.org/10.7759/cureus.47106.
9. Satam H, Joshi K, Mangrolia U, Waghoo S, Zaidi G, Rawool S, Thakare RP, Banday S, Mishra AK, Das G, Malonia SK (2023). Next-generation sequencing technology: Current trends and advancements, Biology (Basel) 12(7); 997, https://doi.org/10.3390/biology12070997.
10. Selvakumar SC, Preethi KA, Ross K, Tusubira D, Khan MWA, Mani P, Rao TN, Sekar D (2022). CRISPR/Cas9 and next generation sequencing in the personalized treatment of cancer, Molecular Cancer 21(1); 83, https://doi.org/10.1186/s12943-022-01565-1.
11. De Olazarra AS, Wang SX (2023). Advances in point-of-care genetic testing for personalized medicine applications, Biomicrofluidics 3;17(3); 031501, https://doi.org/10.1063/5.0143311.
12. Schachter SC, Tessier P (2022). The role rapid acceleration of diagnostics tech continues to play in the Covid-19 pandemic and next steps, Expert Review of Molecular Diagnostics 26; 1–3, https://doi.org/10.1080/14737159.2022.2136990.
13. Safhi AY (2022). Three-dimensional (3D) printing in cancer therapy and diagnostics: Current status and future perspectives, Pharmaceuticals 27;15(6); 678, https://doi.org/10.3390/ph15060678.
14. Sim S, Wong NK (2021). Nanotechnology and its use in imaging and drug delivery (review), Biomedical Reports 14(5); 42, https://doi.org/10.3892/br.2021.1418.
15. Thomas LB, Mastorides SM, Viswanadhan NA, Jakey CE, Borkowski AA (2021). Artificial intelligence: Review of current and future applications in medicine, Federal Practitioner 38(11); 527–538, https://doi.org/10.12788/fp.0174.
16. Bajwa J, Munir U, Nori A, Williams B (2021). Artificial intelligence in healthcare: Transforming the practice of medicine, Future Healthcare Journal, https://doi.org/10.7861/fhj.2021–0095.
17. Schilling V, Beyerlein P, Chien J (2023). A bioinformatics analysis of ovarian cancer data using machine learning, Algorithms 16(7); 330, https://doi.org/10.3390/a16070330.
18. Kulkarni PA, Singh H (2023). Artificial intelligence in clinical diagnosis: Opportunities, challenges, and hype, JAMA 330(4); 317–318, https://doi.org/10.1001/jama.2023.11440, PMID: 37410477.
19. Chopra H, Shin DK, Munjal K, Dhama K, Emran TB (2023). Revolutionizing clinical trials: The role of AI in accelerating medical breakthroughs, International Journal on Surgery 109(12); 4211–4220, https://doi.org/10.1097/JS9.0000000000000705.
20. Wakchaure N, Chawla R, Arora JS (2022). Entrepreneurship in healthcare biotechnology, Biotechnology in Healthcare 2; 343–353, ISBN 9780323900423, https://doi.org/10.1016/B978-0-323-90042-3.00006-2.
21. Abinaya RV, Viswanathan P (2021). Biotechnology-based therapeutics, Translational Biotechnology, 27–52, ISBN 9780128219720, https://doi.org/10.1016/B978-0-12-821972-0.00019-8.
22. Amalu UC (2004). Plant biotechnology and food crop development in Sub-Saharan Africa, Technology in Society 26(4); 537–550, ISSN 0160–791X, https://doi.org/10.1016/j.techsoc.2003.10.003.
23. Zhou L, Yao S (2023). Recent advances in therapeutic CRISPR-Cas9 genome editing: mechanisms and applications, Molecular Biomedicine 4(1); 10, https://doi.org/10.1186/s43556-023-00115-5.
24. Fatehi S, Marks RM, Rok MJ, Perillat L, Ivakine EA, Cohn RD (2023). Advances in CRISPR/Cas9 genome editing for the treatment of muscular dystrophies, Human Gene Therapy 9–10; 388–403, https://doi.org/10.1089/hum.2023.059, PMID: 37119122.
25. Ding X, Pan Z, Yang L, Luo X, Jiang N, Zhu M, Wu C, Lan G, Li P (2022). Advances of CRISPR/Cas9 activation system, Sheng Wu Gong Cheng Xue Bao 38(8); 2713–2724, Chinese, https://doi.org/10.13345/j.cjb.220040, PMID: 36002405.
26. Zhu Y (2022). Advances in CRISPR/Cas9, Biomedicine Research International 2022; 9978571, https://doi.org/10.1155/2022/9978571.
27. Liu G, Zhou Y, Zhang X, Guo S (2022). Advances in hydrogels for stem cell therapy: Regulation mechanisms and tissue engineering applications, Journal of Materials Chemistry B 10(29); 5520–5536, https://doi.org/10.1039/d2tb01044e, PMID: 35819152.
28. Lu M, Zhao J, Wang X, Zhang J, Shan F, Jiang D (2022). Research advances in prevention and treatment of burn wound deepening in early stage, Frontiers in Surgery 9; 1015411, https://doi.org/10.3389/fsurg.2022.1015411, PMID: 36338639, PMCID: PMC9634408.
29. Benayahu D (2022). Mesenchymal stem cell differentiation and usage for biotechnology applications: Tissue engineering and food manufacturing, Biomaterials in Translational 28; 3(1); 17–23, https://doi.org/10.12336/biomatertransl.2022.01.003.
30. Ashammakhi N, GhavamiNejad A, Tutar R, Fricker A, Roy I, Chatzistavrou X, Hoque Apu E, Nguyen KL, Ahsan T, Pountos I, Caterson EJ (2022). Highlights on advancing frontiers in tissue engineering, Tissue Engineering Reviews (Part B) 28(3); 633–664, https://doi.org/10.1089/ten.TEB.2021.0012.
31. Dhanjal DS, Singh R, Sharma V, Nepovimova E, Adam V, Kuca K, Chopra C (2024). Advances in genetic reprogramming: Prospects from developmental biology to regenerative medicine, Current Medicinal Chemistry 31(13); 1646–1690, https://doi.org/10.2174/0929867330666230503144619.
32. Hosseinkhani H, Domb AJ, Sharifzadeh G, Nahum V (2023). Gene therapy for regenerative medicine, Pharmaceutics 15(3); 856, https://doi.org/10.3390/pharmaceutics15030856.
33. Arjmand B, Rabbani Z, Soveyzi F, Tayanloo-Beik A, Rezaei-Tavirani M, Biglar M, Adibi H, Larijani B (2023). Advancement of organoid technology in regenerative

medicine, Regenerative Engineering and Translational Medicine 9(1); 83–96, https://doi.org/10.1007/s40883-022-00271-0. Epub 2022 August 8.
34. Jehan Z (2022). Biotechnology applications in regenerative medicine, Biotechnology in Healthcare 2; 131–142, ISBN 9780323900423, https://doi.org/10.1016/B978-0-323-90042-3.00011-6.
35. Pereira GC (2023). Latest trends in nucleic acids' engineering techniques applied to precision medicine, Methods Molecular Biology 2575; 25–38, https://doi.org/10.1007/978-1-0716-2716-7_2.
36. Murciano-Goroff YR, Suehnholz SP, Drilon A, Chakravarty D (2023). Precision oncology in review, Cancer Discovery 12;13(12); 2525–2531, https://doi.org/10.1158/2159–8290.CD-23-1194.
37. Katarína S, Marek K, Miroslava R, Mária M (2022). Innovative technologies in precision healthcare, Biotechnology in Healthcare; 83–102, ISBN 9780323898379, https://doi.org/10.1016/B978-0-323-89837-9.00016-4.
38. Friebe M, Illanes A (2023). Advancements in medical imaging and image-guided procedures: A potential – Or rather likely – Paradigm shift in diagnosis and therapy: Understand disruption and take advantage of it! Applied Sciences 13(16); 9218, https://doi.org/10.3390/app13169218.
39. Stano P, Altamura E, Mavelli F, Singh V, Jones DD (2023). *Fiat lux*! Light-driven and light-controlled synthetic biological parts, devices, systems and processes, Frontiers in Bioengineering and Biotechnology 11; 1201962, https://doi.org/10.3389/fbioe.2023.1201962.
40. Li J, Zhao H, Zheng L, An W (2021). Advances in synthetic biology and biosafety governance, Frontiers in Bioengineering and Biotechnology 9; 598087, https://doi.org/10.3389/fbioe.2021.598087.
41. León-Buitimea A, Balderas-Cisneros FJ, Garza-Cárdenas CR, Garza-Cervantes JA, Morones-Ramírez JR (2022). Synthetic biology tools for engineering microbial cells to fight superbugs, Frontiers in Bioengineering and Biotechnology 10; 869206, https://doi.org/10.3389/fbioe.2022.869206.
42. Kang M, Choe D, Kim K, Cho BK, Cho S (2020). Synthetic biology approaches in the development of engineered therapeutic microbes, International Journal of Molecular Sciences 21(22); 8744, https://doi.org/10.3390/ijms21228744.
43. Zhang Y, Ding W, Wang Z, Zhao H, Shi S (2021). Development of host-orthogonal genetic systems for synthetic biology, Advanced Biology 5(3); e2000252, https://doi.org/10.1002/adbi.202000252.
44. Wang X, Zhang S, Liu Y, Du J, Huang H (2021). How pharmaceutical innovation evolves: The path from science to technological development to marketable drugs, Technological Forecasting and Social Change 167; 120698, ISSN 0040–1625, https://doi.org/10.1016/j.techfore.2021.120698.
45. Cresswell K, Cunningham-Burley S, Sheikh A (2018). Healthcare robotics: Qualitative exploration of key challenges and future directions, Journal of Medical Internet Research 4; 20(7), https://doi.org/10.2196/10410.
46. Deo N, Anjankar A (2023). Artificial intelligence with robotics in healthcare: A narrative review of its viability in India, Cureus 23; 15(5); e39416, https://doi.org/10.7759/cureus.39416.
47. Morgan AA, Abdi J, Syed MAQ, Kohen GE, Barlow P, Vizcaychipi MP (2022). Robots in healthcare: A scoping review, Current Robotics Reports 3(4); 271–280, https://doi.org/10.1007/s43154-022-00095-4.
48. Cumming D, Johan S, Oberst C, Uzuegbunam I (2020). The unintended consequences of biotechnology innovation adoption, Industry and Innovation 27(10); 1089–1109, https://doi.org/10.1080/13662716.2020.1731431.
49. Ranjha MMAN, Shafique B, Khalid W, Nadeem HR, Mueen-Ud-Din G, Khalid MZ (2022). Applications of biotechnology in food and agriculture: A mini-review, Proceedings of the National Academy of Sciences, India Section B: Biological Sciences 92(1); 11–15, https://doi.org/10.1007/s40011-021-01320-4.
50. Trump B, Cummings C, Klasa K, Galaitsi S, Linkov I (2023). Governing biotechnology to provide safety and security and address ethical, legal, and social implications, Frontiers in Genetics 13; 1052371, https://doi.org/10.3389/fgene.2022.1052371.
51. Lele U (2003). Biotechnology: Opportunities and challenges for developing countries, American Journal of Agricultural Economics 85(5); 1119–1125, https://doi.org/10.1111/j.0092–5853.2003.00518.x.
52. Kirillova A, Bushev S, Abubakirov A, Sukikh G (2020). Bioethical and legal issues in 3D bioprinting, International Journal of Bioprinting 28; 6(3); 272, https://doi.org/10.18063/ijb.v6i3.272.
53. Hefferon KL (2015). Nutritionally enhanced food crops; progress and perspectives, International Journal of Molecular Sciences 16(2); 3895–914, https://doi.org/10.3390/ijms16023895.
54. Tang G, Qin J, Dolnikowski GG, Russell RM, Grusak MA (2009). Golden rice is an effective source of vitamin A, The American Journal of Clinical Nutrition 89(6); 1776–1783, https://doi.org/10.3945/ajcn.2008.27119.
55. Aggarwal B, Vats S, Kaushal L, Singh A, Padalkar G, Yadav H, Kumar V, Sinha S, Umate SM (2023). Biofortification of maize (*Zea mays*). In: Deshmukh R, Nadaf A, Ansari WA, Singh K, Sonah H (eds) Biofortification in Cereals, Springer, Singapore, https://doi.org/10.1007/978-981-19-4308-9_8.
56. Zhong X, Wang J, Shi X, Bai M, Yuan C, Cai C, Wang N, Zhu X, Kuang H, Wang X, Su J (2024). Genetically optimizing soybean nodulation improves yield and protein content, Natural Plants, https://doi.org/10.1038/s41477-024-01696-x.
57. Duraiswamy A, Sneha ANM, Jebakani KS, Selvaraj S, Pramitha JL, Selvaraj R, Petchiammal KI, Kather Sheriff S, Thinakaran J, Rathinamoorthy S, Kumar PR (2023). Genetic manipulation of anti-nutritional factors in major crops for a sustainable diet in future, Frontiers in Plant Science 15(13); 1070398, https://doi.org/10.3389/fpls.2022.1070398.

Section I

Evolution and Revolutions in Healthcare Technologies

2 The Role of Traditional Medicine in Sustainable Development

Foziyah Zakir, Amrita Mishra, and Geeta Aggarwal*
School of Pharmaceutical Sciences, Delhi Pharmaceutical Sciences and Research University, New Delhi, India

2.1 INTRODUCTION

Sustainable development is a practice that ensures human development without compromising nature and environment; it can be categorized into human, social, economic and environmental. The sustainable development initiative was started in 1992 and the UN adopted Transforming Our World: 2030 Agenda for Sustainable Development. The 2030 Agenda called for achieving 17 Sustainable Development Goals (SDGs) from 2015 to 2030 (Figure 2.1) for the harmony of human/animal life and the planet. Seven of the SDGs can tie directly to traditional medicine: SDG1, no poverty; SDG2, zero hunger; SDG3, good health and wellbeing; SDG12, responsible consumption and production; SDG13, climate action; SDG15, life on land and SDG17, partnerships for the goals (Kumar et al., 2021).

Traditional medicine involves maintaining wellbeing through ancient knowledge and health practices that use plant-, animal- and mineral-based medicines for disease treatment and prevention. Ayurveda is one such traditional medicine system with roots dating more than 3000 years ago. Interest in the use and pursuit of traditional medicines has increased since the increasing costs of and rising resistance to contemporary medicines. More than half the countries around the world are either developing or underdeveloped and lack adequate healthcare facilities; in these regions, traditional medicines make medical care affordable and accessible.

2.2 AYURVEDA: A SUSTAINABLE SYSTEM OF MEDICINE

Ayurveda is an ancient science that involves a holistic approach toward life and health that balances the forces that surround us (Figure 2.2).

In Ayurveda, the human body is made up of five elements of nature: ether (*akash*), air (*vayu*), fire (*agni*), water (*jal/apas*) and earth (*prithvi*). Ayurveda in its each step follows sustainable practice starting from the diagnostics to the medicine preparation and administration. It follows the concepts of *tridosa* (*vata*, *pitta* and *kapha*) and pulse check (*nadi parikshan*) for primary diagnosis (Tiwari and Dhakad, 2022). The popular and common practice of detoxification (*panchkarma*) involves emesis (*vamana*), purgation (*virechana*), medicated enemas (*basti*), medicated nasal oils (*nasya*), toxic bloodletting (*raktamokshan*), oil massage (*snehana*) and steam bath (*svedhana*). All of these processes involve natural plant-based materials (Conboy et al., 2009) (Table 2.1).

In Ayurvedic medicines, cow urine and dung play significant roles. For instance, cow urine has proved to possess antimicrobial and wound-healing properties and in Ayurveda is an important constituent of *Panchagavya* (Bajaj et al., 2022). Cow urine is also used in the processing and purification (*shodhan*) of metal- and mineral-based ingredients in Ayurveda; Ayurveda employs various metal ores and minerals to prepare drugs that are purified with plant exudates and animal byproducts. Cow dung possesses antibacterial and antifungal activity and is also an ingredient in *Panchgavya* as well as being a fuel used to prepare *bhasma*. Ayurvedic formulations are prepared by processes including calcination, trituration, grinding, boiling, and fermentation using equipment and vessels mostly made of iron, clay and woods. Ayurvedic texts make no mention of the use of any organic or synthetic chemicals, making Ayurveda the most sustainable system of medicine.

2.3 HEALTH AND TRADITIONAL MEDICINE

Good health and wellbeing are among the foremost of the UN's Sustainable Development Goals; health and wellbeing directly contribute to a nation's economic development. The UN reports that approximately 400 million people across the globe do not have access to basic healthcare facilities (Ghebreyesus, 2017), posing significant impediments to achieving global health and wellbeing; major facilities tend to be in urban areas, making them inaccessible to poor, rural populations. Poor countries also lack the funds and expertise required to manufacture conventional drugs internally and must rely on imported medicines, which creates a financial burden on both the government and the public (Fokunang et al., 2011). Populations in underdeveloped countries or regions who cannot afford or even access expensive conventional treatments have long relied on indigenous medicines for their treatment. These have proven to be both affordable and easily available.

* Corresponding author: foziyahzakir@gmail.com

DOI: 10.1201/9781032660592-3

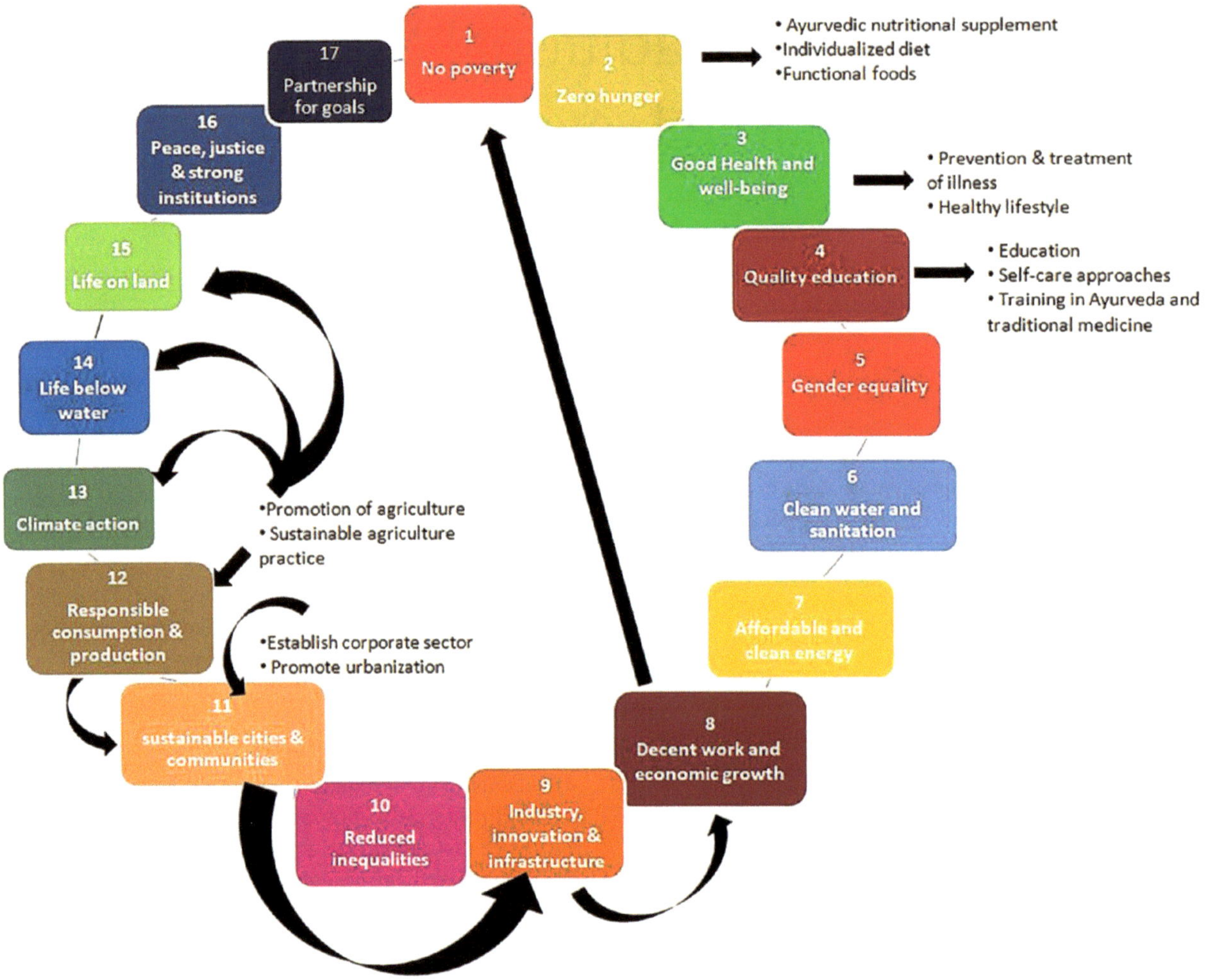

FIGURE 2.1 The UN 2030 Agenda for Sustainable Development Goals and the role of traditional medicine.

In fact, traditional knowledge has served as a backbone for modern medicine; many new drugs have been developed from traditional medicine. According to the WHO Traditional Medicine Centre, 98 of 122 modern chemical compounds originated from traditional practices (Yuan et al., 2016). In 1805, the first plant-based morphine was isolated from an opium poppy (Carlin et al., 2020). Later in 1967, artemisinin was identified and isolated from *Artemisia annua* for the treatment of malaria and many others in recent times.

Researchers have also identified synergistic effects of herbal and mineral-based drugs and synthetic medicines. Complex diseases like cancer and TB require multitarget treatment approaches, and allopathic and herbal drugs are usually combined to reduce the dose and improve the side effects of conventional drugs (Hurwitz et al., 2008). Additionally, components from different herbs can be combined to reduce resistance against microbes, target multiple sites for therapeutic action, diminish side effects, and improve bioavailability (Yuan et al., 2016). Drug discovery

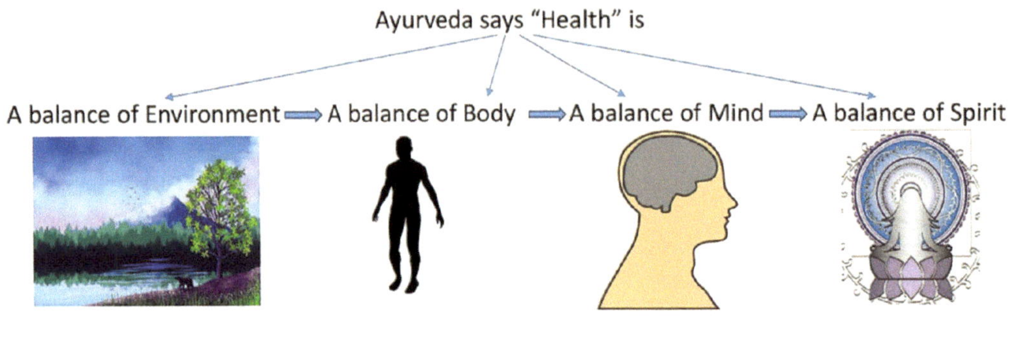

FIGURE 2.2 The Ayurvedic balance of the forces of health.

TABLE 2.1
Animal Products and Metals Used in Ayurvedic Medicine Preparations

Name	Uses	Formulations	Reference
Cow urine	Antimicrobial, wound healing, purification, constituent of *Panchgavya*	Swadeshi Gauvyam Ark, Snannadi Vilayan bathing liquid, Hetha Gir Cow Gaumutra or Gir Cow Urine	Sanganal et al. (2011), Suk et al. (2018)
Cow dung	Used in *bhasma*, as fuel, antibacterial and antifungal, constituent of *Panchgavya*	*Panchagavya ghritam*, *Panchgavya dhoop*	Dhama et al. (2014), Bajaj et al. (2022)
Cow milk and milk products (ghee, butter milk, curd)	For purification of minerals, metals and plants, immunomodulatory	*Panchgavya*, *Panchgavya ghrita*	Kadooka et al. (2010), Singh and Chauhan, 2004
Mica, iron, copper, zinc, gold, etc.	Hyperglycemia, anemia, immunomodulatory, rejuvenator, skin disorders	*Loha bhasma*, *Abhrak bhasma*, *Tamra bhasma*, *Yashad bhasma*	Galib et al. (2011)
Cassia fistula Linn., *Ficus rumphii* Bl., *Coccinia indica* W&A, *Bauhinia purpurea* Linn	As emetic in *panchkarma*	–	Kalburgi et al. (2022)
Picrorhiza kurroa, *Operculina turpethum*, Castor oil	As purgative in *panchkarma*	–	Rais and Bhatted (2013)
Sesame oil	As *nasya* in *panchkarma*	–	Gupta et al. (2012)
Leeches	Bloodletting toxins (*raktamokshan*)	–	Kumar et al. (2013)

is a cost-intensive process, and traditional medicine can contribute to considerably reducing costs, significantly contributing to the public's good health and a country's economic development.

2.4 TRADITIONAL MEDICINE AND ECONOMIC DEVELOPMENT

Most of the global population is dependent on primary sectors such as agriculture and forestry for their survival and livelihoods.

With the rising interest in traditional medicine, increasing plant cultivation will play a prominent role in human welfare by creating financial freedom, effectively contributing to achieving UN SDG 1, no poverty. Poor governments cannot on their own economically support this medicinal plant cultivation; however, most traditional medicine is indigenous and can be easily cultivated and commercialized. Integrating traditional medicine into the modern pharmaceutical industry will help establish corporate sectors and promote urbanization in developing nations and regions. The healthcare demands of the local community will be met in addition to fulfilling the needs of other communities and nations, which can form the backbone of trade sector. In line with this objective, the UN Conference on Trade and Development has developed initiatives to promote the production and commercialization of products obtained from biodiversity. Promoting biodiversity in this way sustainably supports worldwide economic development.

2.5 TRADITIONAL MEDICINE AND THE ENVIRONMENT

Sustainability is the key to survival, and the gravest threat faced by our environment is climate change. Activities such as deforestation and industrialization have destroyed or depleted natural resources, and sustainable agriculture practices are needed to combat climate change and ensure SDG 12, responsible consumption and production. Sustainable agriculture enhances soil quality and increases ground water levels, which increases resistance to natural disasters and supports carbon sequestration (Sud, 2020), which comports with SDG 15, diminish loss of biodiversity and sustain life on land. Indeed, bird populations declined in 2014 because of increased usage of insecticides and lack of crop fields (Hallman et al., 2014); sustainable agricultural practices encompass minimal use of chemical pesticides and fertilizers that leach into soil and water and accumulate in food chains.

2.6 FUTURE PROSPECTS

It is well established that contemporary healthcare depends nearly entirely on allopathic medicine; however, the escalating cost of these treatments and increasing drug resistance have forced the requirement for cheaper and more effective alternatives. Traditional medicine is a sustainable practice that promotes health and wellbeing while depending entirely on naturally available products. There have been concerns of overharvesting drawing large numbers of native species to near extinction, but with careful practices such as microfarms, practitioners can obtain high-quality medicine from natural resources without harming them. Traditional medicine can be an ecofriendly and biofriendly system of medicine and therefore should be encouraged.

REFERENCES

Bajaj, K. K., Chavhan, V., Raut, N. K., & Gurav, S. (2022). Panchgavya: A precious gift to humankind. *Journal of Ayurveda and Integrative Medicine*, *13*, 100525.

Carlin, M. G., Dean, J. R., & Ames, J. M. (2020). Opium alkaloids in harvested and thermally processed poppy seeds. *Frontiers in Chemistry*, *8*, 1–9.

Conboy, L. A., Edshteyn, I., & Garivaltis, H. (2009). Ayurveda and Panchakarma: Measuring the effects of a holistic health intervention. *Scientific World Journal*, *9*, 272–280.

Dhama, K., Khurana, R. K., Karthik, K., Tiwari, R., Malik, Y. P., & Chauhan, R. S. (2014). Panchgavya: Immune-enhancing and therapeutic perspectives. *Journal of Immunology and Immunopathology*, *16*, 1–11.

Fokunang, C. N., Ndikum, V., Tabi, O. Y., Jiofack, R. B., Ngameni, B., Guedje, N. M., Tembe-Fokunang, E. A., Tomkins, P., Barkwan, S., Kechia, F., & Asongalem, E. (2011). Traditional medicine: Past, present and future research and development prospects and integration in the National Health System of Cameroon. *African Journal of Traditional, Complimentary and Alternative Medicine*, *8*, 284–295.

Galib, B. M., Mashru, M., Jagtap, C., Patgiri, B. J., & Prajapati, P. K. (2011). Therapeutic potentials of metals in ancient India: A review through Charaka Samhita. *Journal of Ayurveda Integration Medicine*, *2*(2), 55–63. https://doi.org/10.4103/0975–9476.82523. PMID: 21760689; PMCID: PMC3131772.

Ghebreyesus, T. A. (2017, September). All roads lead to universal health coverage. *The Lancet Global Health*, pp. E839–E840.

Gupta, A. K., Shah, N., & Thakar, A. B. (2012). Effect of Majja basti (therapeutic enema) and Asthi shrinkhala (Cissus quadrangularis) in the management of osteoporosis (Asthi-Majjakshaya). *Ayu*, *33*(1), 110–113. https://doi.org/10.4103/0974–8520.100326. PMID: 23049194; PMCID: PMC3456847.

Hallman, C. A., Foppen, R. P., van Turnhout, C. A., de Kroon, H., & Jongejans, E. (2014). Declines in insectivorous birds are associated with high neonicotinoid concentrations. *Nature*, *511*, 341–343.

Hurwitz, J.L., Zhan, X., Brown, S.A., Bonsignori, M., Stambas, J., Lockey, T.D., Sealy, R., Surman, S., Freiden, P., Jones, B., & Martin, L. (2008). HIV-1 vaccine development: Tackling virus diversity with a multi-envelope cocktail. *Frontiers in Bioscience*, *13*, 609–620.

Kadooka, Y., Sato, M., Imaizumi, K., Ogawa, A., Ikuyama, K., Akai, Y., Okano, M., Kagoshima, M., & Tsuchida, T. (2010). Regulation of abdominal adiposity by probiotics (*Lactobacillus gasseri* SBT2055) in adults with obese tendencies in a randomized controlled trial. *European Journal of Clinical Nutrition*, *64*, 636e43. https://doi.org/10.1038/ejcn.2010.19.

Kalburgi, N., Kulkarni, D. H., Patil, D. V., Bhadargade, P., Kolhapure, M., Naik, S., & Rodd, M. (2022). Scientific review on Vamana Dravya (emetic drugs) of Charaka Samhita. *Journal of Ayurveda and Holistic Medicine (JAHM)*, *9*(6). http://www.jahm.co.in/index.php/jahm/article/view/481.

Kumar, A., Kumar, S., Komal, R. N., & Singh, P. (2021). Role of traditional ethnobotanical knowledge and indigenous communities in achieving sustainable development goals. *Sustainability*, *13*, 3062.

Kumar, S., Dobos, G. J., & Rampp, T. (2013). Clinical significance of leech therapy in Indian medicine. *Journal of Evidence Based Complementary Alternate Medicine*, *18*(2), 152–158. https://doi.org/10.1177/2156587212466675. PMID: 31216615.

Rais, A., & Bhatted, S. (2013). Clinical study to evaluate the effect of Virechanakarma on serum electrolytes. *Ayu*, *34*(4), 379–382. https://doi.org/10.4103/0974–8520.127719. PMID: 24696575; PMCID: PMC3968700.

Sanganal, J. S., Jayakumar, K., Jayaramu, G. M., Tikare, V. P., Paniraj, K. L., & Swetha, R. (2011). Effect of cow urine on wound healing property in Wister Albino rats. *Veterinary World*, *4*, 317–321.

Singh, B. P., & Chauhan, R. S. (2004). Cow Dahi (curd) or Matha (Butter Milk): As probiotic to control animal diseases. *Indian Cow Science Economics Journal*, *1*, 6e10.

Sud, M. (2020). *Managing the biodiversity impacts of fertiliser and pesticide use*. Paris: Organisation for Economic Cooperation and Development Environment Directorate.

Suk, K. H., Gopinath, S. C., Anbu, P., & Lakshmipriya, T. (2018). Cellulose nanoparticles encapsulated cow urine for effective inhibition of pathogens. *Powder Technology*, *328*, 140–147.

Tiwari, Y., & Dhakad, H. S. (2022). Nadi Pariksha: An ancient Ayurvedic method of diagnosis & prediction. *Journal of Ayurveda and Integrated Medical Science*, *7*, 59–63.

Yuan, H., Ma, Q., Ye, L., & Piao, G. (2016). The traditional medicine and modern medicine from natural products. *Molecules*, *21*, 559.

3 Application of Evolutionary Principles in Health Sciences

Jagdish Rai and RC Sobti*
Panjab University, Chandigarh, India

3.1 INTRODUCTION

"Nothing makes sense in biology except in the light of evolution," said the famous geneticist Theodosius Dobzhansky. The theory of evolution is becoming central to biology, and soon it is expected to be the unifying paradigm in the health sciences. Evolutionary biologist George C. Williams speculated on broad applications of evolutionary theory in geriatric and other diseases, first in 1957 in his article titled "Pleiotropy, Natural Selection, and the Evolution of Senescence" (1) and later in a book coauthored with Randolph Nesse in 1994 titled *Why We Get Sick: The New Science of Darwinian Medicine* (2). Paul Ewald further clearly stated the role that evolutionary perspective can play in health sciences in "Evolutionary Biology and the Treatment of Signs and Symptoms of Infectious Disease" (3). He looked at infections as the pathogen's adaptation to survive in the host and the host's response as an adaptation to prevent the infection with some side reactions that are not adaptive for either. Following on these lines of thinking, later researchers determined that fever is a host's adaptation to prevent infection, so that indiscriminate use of antipyretic drugs for sublethal fever delays recovery (4–6).

Anorexia is one of the observed symptoms during an infection and can be a defense strategy of the body; it is part of the body's acute phase response that also includes fever and iron sequestration. Anorexia can help fight infection by depriving the pathogen of the nutrients that are required for its growth; it also deprives the host cells of nutrients, but these cells can survive this stress. Anorexia also diverts the energy that would be required for digestion toward the body's immune response against pathogens (7).

Recent clinical trials have supported the benefits of anorexia in expedited recovery and reduced complications (8, 9). Similarly, iron sequestration is another evolved mechanism of defense against pathogens that reduces their growth by depriving them of iron. Indeed, administering iron supplements delayed recovery in one study (10). Vitamin C deficiency is observed in sepsis patients, and some clinical trials showed that vitamin supplementation produced anti-inflammatory and anti-oxidative properties in patients (11), but later, in a larger multicenter clinical trial, vitamin C was harmful for sepsis treatment (12). Many other immunotherapeutic interventions have been proposed, but they later failed in more rigorous clinical trials because such simple interventions are generally no match for our evolutionarily fine-tuned immune response (13).

Elevated levels of blood glucose in critically ill patients have also been shown to be good for the patient, contrary to earlier clinical trials showing the benefits of controlling hyperglycemia in patients in intensive care units (14). Stress-induced hyperglycemia is an essential evolved survival response for many stressful situations. In a recent study, VITamin D and OmegA-3 TriaL (VITAL), vitamin D supplementation showed no benefit in reducing the risk of frailty; the investigators concluded that healthy people should not be tested or treated for vitamin D deficiency, which is contrary to the practice so far. These observations endorse Samuel Shem's famous quote, "The delivery of good medical care is to do as much nothing as possible".

Postmenopausal decrease in estrogen and progestin has been thought to be the cause of many postmenopausal disorders like osteoporosis and heart diseases; therefore, menopause was once regularly treated with hormone replacement therapy. However, a recent clinical trial showed that estrogen plus progestin did not decrease the risk of coronary heart disease but rather increased the risk of both heart disease and breast cancer in the long run, although it was beneficial for a short initial duration (15). According to SN Austad, menopause is not a disease; rather, it evolved as an adaptation for women in the later years of life to care for children and grandchildren without wasting resources giving birth to more children. However, women are recently giving birth to fewer children than their natural fertility, which decreases their exposure to estrogen by around 10 times; therefore, young women close to menopause may benefit from short-duration hormone replacement therapy (16, 17).

In his Nobel Prize acceptance speech, Alexander Fleming described that overdose of antibiotics is not a concern but that underdose can drive bacteria's evolution of resistance to current antibiotics. The emergence of antibiotic resistance in bacteria can be reduced by not prescribing antibiotics to patients who do not need them, but may physicians will not take the risk of not prescribing at all. Another step is to reduce the duration of antibiotic treatment; duration is generally 7 or 14 days, but this is based on the arbitrary logic of the number of days in the week. New

* Corresponding author: jagdishrai@pu.ac.in

DOI: 10.1201/9781032660592-4

clinical trials based on evolutionary rationales have shown that shorter antibiotic therapy is better, for example, 3–5 days for pneumonia (18, 19).

Today, evolutionary medicine has come a long way. It has a dedicated scientific society (International Society for Evolution, Medicine, and Public Health) and a journal published by the Society (*Evolution, Medicine, and Public Health*). Stearns and Medzhitov's *Evolutionary Medicine* can be considered a standard textbook of the field (20). The advantage of the evolutionary perspective is that it gives a unifying paradigm in biomedical sciences that explains the emergence of diseases and helps frame better public health policies (21–25). The next sections detail the salient achievements of evolutionary medicine in specific diseases and health guidance.

3.2 EVOLUTIONARY INSIGHTS INTO DIABETES

Milind Watve gave an evolutionary perspective on diabetes in his *Doves, Diplomats, and Diabetes* (26). Watve explains the difference between the physiology of hawk and dove and then draws the analogy to explain the differences between a soldier and a diplomat in their requirements for diet and exercise. Specifically a hawk is an aggressive creature with low serotonin and higher dopamine activity in the brain and low plasma cholesterol, corticosteroids, and plasma insulin. In contrast, a dove has high serotonin, low dopamine, and higher plasma levels of corticosteroids and cholesterol. Comparatively, humans are physiologically closer to doves than to hawks, and the metabolic problem called diabetes arises when a diplomat consumes a soldier's diet. Watve suggests that diet and exercise are the two behavioral aspects that cause the insulin resistance observed in type 2 diabetes (27). Further, intensive physical activities can reverse insulin resistance independent of obesity; that is, obesity is a result of rather than the cause of insulin resistance (28–30). In a program called Behavioral Intervention for Lifestyle Disorders, these ideas are being clinically tested (23, 24).

The economic development in the last century has resulted in many changes in our food production, distribution, and consumption that have caused 'diabesity' (obesity and diabetes) in the developed world that in recent decades has been spreading fast in the developing world (31). Diabesity is also not limited to humans but has been observed in captive primates as well (32). Obesity is characterized by excess body fat and measured by basic metabolic index, whereas diabetes (type 2) is characterized by high levels of blood glucose due to insulin resistance of muscle tissue and reduced secretion of insulin by pancreatic beta cells (33). According to the thrifty phenotype hypothesis, undernutrition at the fetal and infant stages (34) reduces the growth of the pancreas and muscle mass as part of developmental adjustment, anticipating similar conditions later in life. However, adequate nutrition in adulthood causes obesity and diabetes; broadly, this can be understood as an imbalance between metabolic capacity and metabolic load (Figure 3.1). Metabolic capacity is the capacity of the pancreas to produce insulin and the capacity

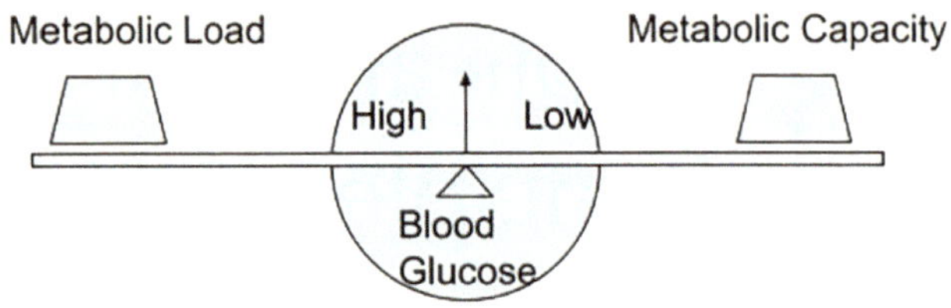

FIGURE 3.1 Diabetes as an imbalance between metabolic load and capacity.

of muscles to clear glucose from blood, and the metabolic load refers to dietary glycemic index, sedentary lifestyle, and abdominal adiposity.

Low birth weight is an indicator of low metabolic capacity, and adults who lead an unhealthy lifestyle that causes high metabolic load experience diabetes (35). Diabesity is more toxic to people who were undernourished in childhood than in people with better early-life nutrition (36), and indeed, shorter people have a higher diabetes risk (37). As we discussed earlier, under the thrifty phenotype, the body diverts metabolic resources to defense and reproduction at the cost of long-term health (38). Consequently, to control the diabesity pandemic, there is a need to control low birth weight and malnutrition in childhood as well as healthy lifestyles later (39). For instance, adolescent mothers generally give birth to underweight babies; therefore, the age of motherhood and consequently marriage need to be increased in developing countries. Further, to counter malnutrition in childhood, seasonal spikes in food insecurity and infection risk have to be controlled. Jonathan C Wells suggests the need to entirely redesign human food as consumerism is promoting unhealthy behavior for economic profit (40).

In comparison with other primates, humans have evolved low muscle mass and high adiposity, which makes them vulnerable to diabetes in recent obesogenic environments (38). This evolutionary trend on a long-term scale is similar to the life course adjustment of metabolic capacity according to metabolic load anticipated by the fetus or infant in later life. Within the human species also, there is divergence in body composition according to the ecological demands of different geographical locations. This diversity in the metabolic capacity of different ethnic groups and differences in metabolic load due to changes in economy or migration can explain the vulnerability to noncommunicable diseases.

3.3 EVOLUTIONARY INSIGHTS INTO CANCER

In multicellular organisms, the division of labor among cells encompasses many functions, including cell division to make new cells. Stearns and Medzinov classify cells into three types: the germs cells that make cells for the next generation, the stem cells that make cells for growth and repair of the body, and the somatic cells that do not divide and only perform the designated functions of the body (20). Cancer occurs when any cells break away from this arrangement, and the closest to cross this boundary are stem cells.

Stem cells are most often progenitors of cancer because they start dividing even when it's not required for growth or repair of the body, like in unicellular organisms. The frequency of metastatic cancer in different mammalian species is proportional to the degree of the invasive placenta (41). The stem cells from the fetus can cross the placenta to invade the mother's body and then manipulate the arteries to supply more nutrients to the fetus (42), and the many genes involved in this invasion and manipulation are the same as the genes activated in metastatic cancer. This tradeoff between more metastatic cancer and more nutrients to the fetus has resulted in many flip-flops in the invasiveness of the placenta in the evolutionary history of mammals: Very primitive mammals have very invasive placenta that is suppressed in some lineages like cows and horses. However, more invasive placentas evolved in dogs and cats and re-evolved in the ancestors of humans, chimpanzees, and gorillas about 15 million years ago.

An adaptation tradeoff between melanoma (skin cancer) and vitamin D deficiency is also observed. Sunlight exposure is the primary source of vitamin D, and low exposure to sunlight can lead to vitamin D deficiency that causes weakened bones, compromised immune function, and an increased risk of certain chronic diseases (43). However, more exposure to sunlight, specifically its ultraviolet component, can cause DNA damage and eventually skin cancer. Therefore, exposure to sunlight should be for short intermittent periods and preferably in the morning time or late afternoon. The melanin produced by melanocytes in the skin is responsible for our skin color; the differences in skin color among different ethnic groups are due to varying amounts of melanin, which varied as adaptations to differing amounts of sunlight exposure. Near the equator, people have dark skin to protect skin from too much sunlight, whereas away from the equator, skin color is increasingly light to maximize vitamin D synthesis. Therefore, people migrating between latitudes may face vitamin D deficiency or skin cancer if adequate precaution for vitamin D supplementation or sunlight protection is not taken (44).

Cancer is a collection of more than 300 types of diseases that share common features like uncontrolled cell division, genomic instability, and evasion from the immune system. The molecular and mechanistic explanations are far from comprehensive; therefore, an evolutionary ecological perspective is required that can explain our vulnerability to cancer as multicellular organisms and our enhanced vulnerability to cancer as humans specifically as well as to provide new strategies like adaptive therapy (45, 46). Cancer is an inevitable risk of multicellularity and has indeed been observed in all multicellular organisms including mice, whales, dinosaurs, and ancient human populations. Richard Peto observed a paradox, later known as Peto's paradox, that the incidence of cancer is not greater in larger animals than in smaller ones even though in theory, because larger animals have more cells, they have more chances of somatic mutations that fundamentally cause cancer (47).

A life history perspective can resolve Peto's paradox: Animals with more cells that live longer have evolved additional mechanisms to maintain genomic integrity like DNA repair and also better immune systems to clear cancer cells. For example, there are 20 copies of gene TP53 in elephants, which is associated with increased body size and enhanced DNA damage response (48). Data is showing that anthropogenic exposure of wildlife to toxins and pathogens is increasing the rate of cancer in wildlife (49). The new mutagens that have entered our environment have caught us unprepared as we have not yet evolved the DNA repair and immune system to tackle the increased rate of mutations, but other larger animals have already evolved additional mechanisms of DNA repair (50, 51) and immune system (52) response to increased susceptibility to mutation due to their longer life span and large number of cells. This environmental and evolutionary perspective can give insight into new preventive and therapeutic measures against cancer.

3.4 EVOLUTIONARY PERSPECTIVES ON ADDICTION AND DEPRESSION

Psychiatrist Randolph Nesse is the pioneer of the evolutionary perspective in psychiatry and also one of the pioneers in evolutionary medicine. He proposed that bad feelings also had good reasons, that is, adaptation to respective situations. He drew an evolutionary tree of feelings (Figure 3.2), in which arousal in simple organisms has two types (physical pain and pleasure) that ensure their survival; in complex organisms, these feelings of pain and pleasure have further evolved into jealousy, grief, love, and so on, to help in reproductive selection; further social selection has led to the evolution of more complex negative and positive feelings like shame, anger, friendship, and pride.

Our brain's reward system modulates our physiology and behavior through these diverse feelings. The positive as well as negative feelings make us more adapted to our respective situations. The liking for positive feelings and aversion to negative feelings is part of the mechanism of the brain's reward system for modulating our behavior, but

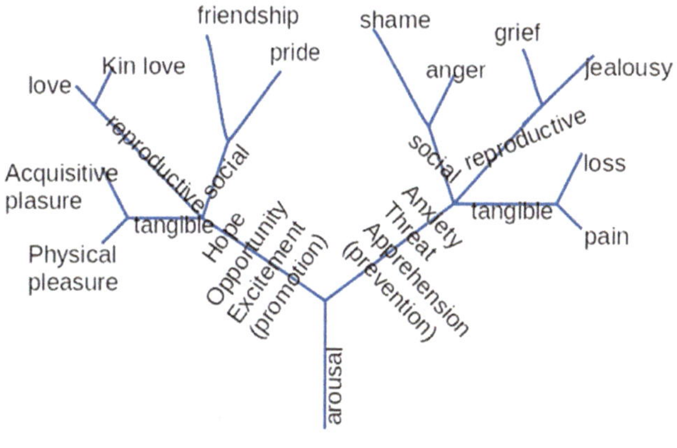

FIGURE 3.2 A phylogeny of feelings.

sometimes people are lured to psychotropic substances to artificially induce happiness or suppress sadness. This hacking of the brain hampers the adaptive function of the reward system. In Randolph Nesse's words, this is equal to placing a large magnet near the compass of the ship (53). In many surveys, people report that they began using psychotropic substances as self-medication to manage sleep or anxiety or to elevate their mood or just for socialization but then become dependent on them (54). According to Slavoj Žižek, the market economy drives consumption by "cultural injunction to happiness" through various channels of communication. This injunction to happiness may give a wrong self-evaluation to some people that they are enjoying less, which eventually can push people to drug use or seek psychiatric help (55). The overprescription of antidepressants can be offset by the realization that the brain's reward system lures us to happiness, but it's only one aspect; sad feelings can also be adaptive (56). Popularization of this evolutionary perspective on the brain's reward system can prevent drug use in youth and also help in logic-based de-addiction therapy (57).

Cognitive scientists have made extensive use of evolutionary principles to put various observations into a more meaningful theory, as well as get insights for further research directions. Evolutionary psychologist Robin Dunbar has given "Dunbar's number," the number of stable relationships people are cognitively able to maintain at once (58). This number, ~150, comes from the social group size in hunter–gatherer human societies. Among primates, the Dunbar number is correlated with brain size, specifically the size of the neocortex. Recently, urbanization has packed too many people together, which causes undue stress, and more so in the professions that require dealing with many unknown people. A picnic is psychologically relaxing because it recreates the situation when our ancestors in the Stone Age could relax. In the Stone Age, having a water body in sight, dense plantation, and members of the tribe around was a sign of, respectively, ample water, ample food, and security. Our brain is still receptive to these signals even though we have more water, food, and security in urban settings.

3.5 EVOLUTIONARY PERSPECTIVE ON GENETIC DISEASES

Many disease-causing alleles of genes are persistent in a population because they conferred or still confer some adaptive advantage in some situations. A popular example is the hemoglobin allele HbS, which causes sickle cell anemia.

3.5.1 Hemoglobin S

HbS is a mutant of the normal hemoglobin allele HbA with the glutamic amino acid at the sixth position in its polypeptide chain substituted with valine (59). This substitution makes the polypeptide chain more hydrophobic and causes it to polymerize into long fibrils. These protein fibrils distort the shapes of the red blood cells they form in, and the cells get stuck in narrow capillaries. However, this elongated shape of the red blood cells is also lethal for the malaria parasite *Plasmodium falciparum* (60), and therefore, people who are homozygous or heterozygous for HbS will survive malaria infection; heterozygous people have the best of both, as neither sickle cell anemia nor malaria is lethal for them.

This HbS is in higher frequency in populations in geographical areas where malaria is more prevalent; it protects the population from lethal malaria, but individuals who are homozygous for HbS will pay the price for this protection. Following Hardy–Weinberg's law of population genetics, p^2 will be the fraction of people afflicted with severe sickle anemia due to homozygosity, and 2pq will be the fraction of people protected by malaria as well as severe anemia. Here, p is the frequency of HbS alleles, and q is the normal HbA allele in the population. This population genetics and evolutionary perspective on the disease-causing allele can help in mitigating the social stigma that sickle cell anemic people face in some societies (59). There are many other examples of genetic diseases, and the adaptive advantage of the respective allele in special situations may be known or yet to be known, but the evolutionary reasoning indicates some adaptive advantage behind their long-time persistence in significant frequency.

3.5.2 Phenylketonuria

Phenylketonuria is a genetic defect that makes the phenylalanine hydroxylase enzyme dysfunctional. Consequently, the patient has a deficiency of the amino acid tyrosine but high levels of phenylalanine, a part of which also converts into phenyl ketones. This condition is very severe in the homozygous recessive genotype, leading to mental retardation, but heterozygous has no symptoms even with higher-than-normal levels of phenyl ketones in the blood. Louis Woolf found that high levels of phenylalanine in the blood caused by this defective allele protect people from the lethal toxic effects of ochratoxin A (61). Ochratoxin A is an N-acyl derivative of phenylalanine produced by fungi like *Aspergillus* and *Penicillium*. High levels of phenylalanine prevent ochratoxin from binding to phenylalanine t-RNA synthetase. Historically, there have been many outbreaks of ochratoxin toxicity in Eastern Europe due to fungal infections in food grains. People descending from these regions have a higher frequency of phenylketonuria because the ochratoxin toxicity worked as a positive selection pressure for defective phenylalanine hydroxylase enzymes in these populations.

3.5.3 Lactose Intolerance

Around 25% of the world's adult population, mainly in East Asia, is intolerant of the lactose sugar found in milk because they cannot make lactase, the enzyme required for digesting this disaccharide. Whereas all humans can make lactase at birth to digest the lactose in breast milk, in these populations, the lactase regulatory region switches off the expression of the gene in the adult stage (62). This could have been an adaptive feature for ancient humans living in the forest when milking animals were not yet domesticated; as milk was not available to adults, production of lactase was a waste of energy. After the domestication of milking animals, some humans may have had a mutation that rendered the lactase regulatory region nonfunctional. This mutation led to the continuous expression of lactase throughout life, allowing those humans to consume the milk of domesticated animals. This mutation was an adaptive advantage, and therefore, this allele spread except in lactose-intolerant populations (63).

3.5.4 Migraine

Migraine affects around 20% of the population, and around 60% of migraine cases are hereditary (64). Four genes are known to be associated with migraine, one a protein involved in exocytosis in neuron synapses and the other three ion transporters (65). The alleles causing migraine make the central nervous system hypersensitive to sensory stimuli. It has been observed that family members of migraineurs have enhanced visual sensitivity (66). Migraine could have been an evolutionary advantage in ancestral hunter–gatherer communities, but the modern environment has too many visual stimuli such as artificial light sources, screens, and visual advertisements that can confound the sensitive brain. Some migraineurs are also very sensitive to unpleasant smells, which helped with detecting toxic substances in wild settings but with increasing air pollution and synthetic scents in recent times has become a liability (67). The tendency to develop head pain in response to strong stimuli from various sensory neurons can be an adaptation to prevent dangerous dilation of cranial arteries, or it can be also to make an individual avoid the harmful situation, or both (68).

3.5.5 Schizophrenia

Heredity is a significant contributing factor in schizophrenia (69). In fact, around 1% of the population in almost every culture and ethnic group suffers from schizophrenia, indicating that it can have some adaptive advantages (70). People with the heterozygotic genotype for schizophrenia are carriers of the trait but do not suffer from the severe disease. Some of these carriers show advantageous traits such as fewer viral infections, higher IQ, and increased creativity and mathematical reasoning (71, 72). Geoffrey Miller proposed that the personality traits predictive of schizophrenia are also associated with artistic creativity functions to attract mates (73). Later, Daniel Nettle and Helen Clegg found this hypothesis plausible when they examined the association between schizotypal traits, creative activity, and mating success in a large sample of visual artists and poets in Britain. Poets and artists scored as high as schizophrenia patients in unusual experiences and impulsive nonconformity but not in introvertive anhedonia. Artists seem to make productive use of these creative traits, whereas schizophrenia patients succumb to their disorganizing effects due to other unfavorable conditions in life, lack of other complementary mental resources, and health issues (74).

3.6 EVOLUTIONARY INSIGHTS INTO A HEALTHY WAY OF LIFE

David A Raichlen and Daniel E Lieberman recently investigated the evolution of human walking behavior (75). They used a smartwatch to count the number of footsteps people walk today and compared it with humans in the Stone Age as well as our evolutionary cousins, the primates. The average steps per day are 1000 for orangutans, 1100 for gorillas, around 4000 for chimpanzees, 5000 for bonobos, 20,000 for humans in the Stone Age, and 5000 for humans in modern industrialized societies. This reduced number of steps from 20,000 in hunter–gatherer societies to 5000 in industrialized urban societies is an evolutionary mismatch because our genetics have not changed as fast as our economy and walking behavior. Our bipedal walk and long-distance travel gave us many adaptive advantages over our primate cousins, but our suddenly sedentary lifestyles have reduced our walking, which is at least partly contributing to the prevalence of diabetes and cardiovascular diseases.

The food we eat today is also very different from the average Stone Age diet, which our bodies are well tuned for because our ancestors spent a long time in Stone Age environments. In texture, today's food is very soft and easy to chew, which has led to a lack of exercise for our gums and teeth (76). Therefore, we see a rise in dental problems despite the extra care we take by brushing and antibacterial mouthwashes. Our food is sometimes also very sticky and high in easily available nutrients like sugars that give bacteria a chance to grow (77). A new field of evolutionary dentistry has emerged in which children do mouth exercises and wear devices that put stronger force on their growing jaws (78).

The nutrient composition of our diet has also changed drastically compared with the diets of our hunter–gatherer ancestors. Our ancestors craved salt and sugar because they were rare, and our primitive cravings have remained the same; however, now we have easy access to salt, sugar, and fat due to developments in transportation, agriculture, and food processing, and the excess consumption of salt, sugar, and fat has respectively increased incidences of hypertension, diabetes, and obesity. In ancient times, we could hardly afford to rest for long durations, but today, mechanization has made leisure possible; our evolutionary primitive tendency to seek rest to conserve calories has led us to take too much rest and eventually suffer many diseases caused by sedentary life.

Except for the mismatches we have described, our instincts have evolved to respond appropriately to most of the situations that are similar to what our hunter–gatherer ancestors faced. The nutritional value of a food item is generally proportional to how much we like it. The food that we do not like is generally not good for us, for example, unripe fruits or vegetables containing toxins taste bitter, and food infected with harmful bacteria smells bad. Even in a disease condition, our liking changes to expedite recovery. For example, children deficient in calcium like to eat clay or lime that is rich in calcium, canines are often seen eating grass to supplement fiber in case of indigestion, and people with fevers like lemons.

We can even use evolutionary arguments to crosscheck health recommendations. Ayurveda strongly recommends not drinking water after or during meals, but sometimes we feel thirsty during or after meals. The instinct of thirst has evolved over a very long time. Our thirst neurons in the hypothalamus make us feel thirsty based on water concentration in the blood and also monitor how hydrating a drink will be based on signals from the gut. Even in a complex situation, our thirst neurons make the right decision. For example, if we are thirsty and mistakenly take a cup of hot tea, then we do not take cold water, and we might not drink water for some time even if the water concentration in our blood has not changed significantly. Contrary to the recommendation of Ayurveda, later it was found that drinking water during a meal or after a meal has no adverse effect if one is indeed thirsty (79, 80). Rather, the recommendation of Ayurveda is based on a simplistic model of the stomach as a sack in which food and digestive secretions are mixed. In this model, water deletes the digestive secretions, leading to inefficient digestion, but the stomach is not a simple impervious sack. Rather, its thick walls can add or absorb extra water from the stomach.

3.7 ANCIENT DNA RESEARCH AND EVOLUTIONARY MEDICINE

The 2022 Nobel Prize in Physiology and Medicine was aptly given to Svante Pääbo for his work on ancient DNA. He sequenced the genome of the now-extinct hominins Neanderthal man and Denisova, which will help in advancing medicine. Today's *Homo sapiens* of European and Asian descent have around 1%–6% DNA, respectively, from Neanderthals and Denisova in a genetic admixture (Figure 3.3). Svante Pääbo was instrumental in establishing the Max Planck Institute for Evolutionary Anthropology to understand how archaic gene sequences from our extinct relatives influence the physiology of present-day humans. For example, it was discovered that a Denisovan version of the EPAS1 gene in the Tibetan population is crucial for adaptation to high altitude (81). The ethnic differences in the immune responses of Africans and Europeans is attributed to some Neanderthal DNA present in modern-day Europeans. When *Homo sapiens* migrated to Europe, DNA acquired from Neanderthals helped them fight new pathogens they encountered (82).

A comparison of ancient human genomes with the present-day population in Europe shows that increased susceptibility to autoimmune diseases is due to an evolutionary tradeoff between immunity to pathogens and autoimmune diseases that occurred around the Bronze Age (83). It was found that TYK2 P1104A increases the risk of tuberculosis infection and is a candidate variant for negative selection, but it protects against autoimmune/autoinflammatory disorders (84). Another allele, *TLR3* L412F, has been associated with mild protection against autoimmune thyroid disease, but it increases the risk of viral diseases. This also was a candidate for negative selection in ancient times. Another negatively selected allele in ancient times, *IL23R* R381Q, is associated with low risk of autoimmune inflammatory bowel disease and high risk of mycobacterial infection (85). These findings indicate an antagonistic pleiotropy between infectious and autoimmune disorders. The selection pressure on host defense genes in Europe in the last 10 millennia has probably contributed to present-day disparities in risk of infectious and inflammatory disease (83). Similarly, there are many practical implications of evolutionary insights in medicine and public health (86).

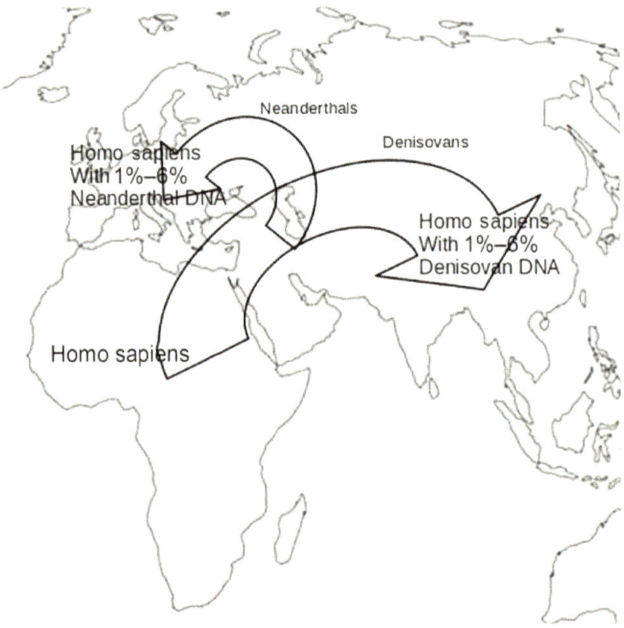

FIGURE 3.3 A small fraction of DNA inherited from extinct hominins has survived due to the adaptive advantage it confers on modern human populations.

3.8 CONCLUSION

From evolutionary perspectives, the cause of disease lies in either some tradeoff between adaptation to environmental parameters, a good gene having a bad effect, a mismatch between our present environment and our genes that evolved in different (ancient) environments, or pathogens evolved too fast. Biological systems have an evolutionary trajectory that necessitates that new adaptive features be compatible with the already evolved systems; they are often small incremental changes in existing components. This issue of compatibility of new features with existing systems is more pronounced in more complex creatures.

Biological systems are continuously adapting because the environment is changing continuously. Therefore, living beings are hardly a perfect fit for the environment, and this lag in adaptation can manifest as disease. Medical sciences aid in adapting to the environment and counter maladaptive biological traits.

REFERENCES

1. Williams GC. Pleiotropy, natural selection, and the evolution of senescence. Evolution. 1957;11(4):398–411.
2. Nesse RM, Williams GC. Why We Get Sick: The New Science of Darwinian Medicine [Internet]. Knopf Doubleday Publishing Group; 2012. (Vintage). Available from: https://books.google.co.in/books?id=2rhNZccHwDoC.
3. Ewald PW. Evolutionary biology and the treatment of signs and symptoms of infectious disease. J Theor Biol. 1980;86(1): 169–76.
4. Wrotek S, LeGrand EK, Dzialuk A, Alcock J. Let fever do its job: The meaning of fever in the pandemic era. Evol Med Public Health. 2021 Jan 1;9(1):26–35.
5. Young P, Saxena M, Bellomo R, Freebairn R, Hammond N, Van Haren F, et al. Acetaminophen for fever in critically ill patients with suspected infection. N Engl J Med. 2015 Dec 3;373(23):2215–24.
6. Drewry AM, Mohr NM, Ablordeppey EA, Dalton CM, Doctor RJ, Fuller BM, et al. Therapeutic hyperthermia is associated with improved survival in afebrile critically ill patients with sepsis: A pilot randomized trial. Crit Care Med. 2022 Jun;50(6):924–34.
7. Alcock J, LeGrand EK. Anorexia. Evol Med Public Health. 2014 Nov 27;2014(1):149.
8. Casaer MP, Mesotten D, Hermans G, Wouters PJ, Schetz M, Meyfroidt G, et al. Early versus late parenteral nutrition in critically ill adults. N Engl J Med. 2011 Aug 11;365(6):506–17.
9. Arabi YM, Tamim HM, Dhar GS, Al-Dawood A, Al-Sultan M, Sakkijha MH, et al. Permissive underfeeding and intensive insulin therapy in critically ill patients: A randomized controlled trial. Am J Clin Nutr. 2011 Mar;93(3):569–77.
10. Nairz M, Weiss G. Iron in infection and immunity. Mol Aspects Med. 2020 Oct;75:100864.
11. Muhammad M, Jahangir A, Kassem A, Sattar SBA, Jahangir A, Sahra S, et al. The role and efficacy of vitamin C in sepsis: A systematic review and meta-analysis. Adv Respir Med. 2022 Jul 28;90(4):281–99.
12. Lamontagne F, Masse MH, Menard J, Sprague S, Pinto R, Heyland DK, et al. Intravenous vitamin C in adults with sepsis in the intensive care unit. N Engl J Med. 2022 Jun 23;386(25): 2387–98.
13. Alcock J, Masters A. Cytokine storms, evolution and COVID-19. Evol Med Public Health. 2021;9(1):83–92.
14. NICE-SUGAR Study Investigators, Finfer S, Chittock DR, Su SYS, Blair D, Foster D, et al. Intensive versus conventional glucose control in critically ill patients. N Engl J Med. 2009 Mar 26;360(13):1283–97.
15. Manson JE, Hsia J, Johnson KC, Rossouw JE, Assaf AR, Lasser NL, et al. Estrogen plus progestin and the risk of coronary heart disease. N Engl J Med. 2003 Aug 7;349(6):523–34.
16. Rossouw JE, Prentice RL, Manson JE, Wu L, Barad D, Barnabei VM, et al. Postmenopausal hormone therapy and risk of cardiovascular disease by age and years since menopause. JAMA [Internet]. 2007 Apr 4 [cited 2024 Jan 18];297(13). Available from: http://jama.jamanetwork.com/article.aspx?doi=10.1001/jama.297.13.1465.
17. Eaton SB, Pike MC, Short RV, Lee NC, Trussell J, Hatcher RA, et al. Women's reproductive cancers in evolutionary context. Q Rev Biol. 1994 Sep;69(3):353–67.
18. Spellberg B, Rice LB. Duration of antibiotic therapy: Shorter is better. Ann Intern Med. 2019 Aug 6;171(3):210.
19. Wald-Dickler N, Spellberg B. Short-course antibiotic therapy: Replacing constantine units with "shorter is better." Clin Infect Dis. 2019 Oct 15;69(9):1476–9.
20. Stearns SC, Medzhitov R. Evolutionary Medicine. Sunderland, MA: Sinauer Associates, Inc., Publishers; 2016, p. 306.
21. Omenn GS. Evolution and public health. Proc Natl Acad Sci. 2010 Jan 26;107(suppl_1):1702–9.
22. Gluckman PD, Low FM, Buklijas T, Hanson MA, Beedle AS. How evolutionary principles improve the understanding of human health and disease. Evol Appl. 2011 Mar;4(2):249–63.
23. Benton ML, Abraham A, LaBella AL, Abbot P, Rokas A, Capra JA. The influence of evolutionary history on human health and disease. Nat Rev Genet. 2021 May;22(5):269–83.
24. Hood E, Jenkins KP. Evolutionary medicine: A powerful tool for improving human health. Evol Educ Outreach. 2008 Apr;1(2):114–20.
25. Nesse RM. Evolution: Medicine's most basic science. The Lancet. 2008 Dec;372:S21–7.
26. Watve M. Doves, Diplomats, and Diabetes: A Darwinian Interpretation of Type 2 Diabetes and Related Disorders. New York: Springer; 2013, p. 380.
27. Watve MG, Yajnik CS. Evolutionary origins of insulin resistance: A behavioral switch hypothesis. BMC Evol Biol. 2007;7(1):61.
28. Baig U, Lokhande L, Lalwani P, Chawla S, Watve M. Foraging theory and the propensity to be obese: An alternative to thrift. Homo Int Z Vgl Forsch Am Menschen. 2019 Nov 11;70(3):193–216.
29. Patil P, Lalwani P, Vidwans H, Kulkarni S, Bais D, Diwekar-Joshi M, et al. A multidimensional functional fitness score has a stronger association with type 2 diabetes than obesity parameters in cross sectional data. Palazón-Bru A, editor. PLoS One. 2021 Feb 5;16(2):e0245093.
30. Watve M, Diwekar-Joshi M. What to expect from an evolutionary hypothesis for a human disease: The case of type 2 diabetes. Homo Int Z Vgl Forsch Am Menschen. 2016 Oct;67(5):349–68.
31. Popkin BM, Adair LS, Ng SW. Global nutrition transition and the pandemic of obesity in developing countries. Nutr Rev. 2012 Jan;70(1):3–21.
32. Howard CF. Nonhuman primates as models for the study of human diabetes mellitus. Diabetes. 1982 Apr 1;31(Suppl_1):37–42.
33. Bergman RN, Ader M, Huecking K, Van Citters G. Accurate assessment of β-cell function. Diabetes. 2002 Feb 1;51(suppl_1): S212–20.
34. Hales CN, Barker DJP. Type 2 (non-insulin-dependent) diabetes mellitus: The thrifty phenotype hypothesis. Diabetologia. 1992 Jul;35(7):595–601.
35. Li Y, Ley SH, Tobias DK, Chiuve SE, VanderWeele TJ, Rich-Edwards JW, et al. Birth weight and later life adherence to unhealthy lifestyles in predicting type 2 diabetes: Prospective cohort study. BMJ. 2015 Jul 21;h3672.
36. Leon DA, Koupilova I, Lithell HO, Berglund L, Mohsen R, Vagero D, et al. Failure to realise growth potential in utero and adult obesity in relation to blood pressure in 50 year old Swedish men. BMJ. 1996 Feb 17;312(7028):401–6.
37. Wells JCK, Pomeroy E, Walimbe SR, Popkin BM, Yajnik CS. The elevated susceptibility to diabetes in India: An evolutionary perspective. Front Public Health [Internet]. 2016 Jul 7 [cited 2024 Jan 27];4. Available from: http://journal.frontiersin.org/Article/10.3389/fpubh.2016.00145/abstract.

38. Wells JCK. Body composition and susceptibility to type 2 diabetes: An evolutionary perspective. Eur J Clin Nutr. 2017 Jul;71(7):881–9.
39. Wells JCK. The diabesity epidemic in the light of evolution: Insights from the capacity–load model. Diabetologia. 2019 Oct;62(10):1740–50.
40. Wells JCK. The Metabolic Ghetto: An Evolutionary Perspective on Nutrition, Power Relations and Chronic Disease. Cambridge University Press; 2016, p. 625.
41. D'Souza AW, Wagner GP. Malignant cancer and invasive placentation: A case for positive pleiotropy between endometrial and malignancy phenotypes. Evol Med Public Health. 2014 Oct 15;2014(1):136–45.
42. Dawe GS, Tan XW, Xiao ZC. Cell migration from baby to mother. Cell Adhes Migr. 2007;1(1):19–27.
43. Holick MF. Sunlight, UV radiation, vitamin D, and skin cancer: How much sunlight do we need? In: Reichrath J, editor. Sunlight, Vitamin D and Skin Cancer [Internet]. Cham: Springer International Publishing; 2020 [cited 2024 Jan 28], pp. 19–36. (Advances in Experimental Medicine and Biology; Vol. 1268). Available from: http://link.springer.com/10.1007/978-3-030-46227-7_2.
44. Lucock MD. The evolution of human skin pigmentation: A changing medley of vitamins, genetic variability, and UV radiation during human expansion. Am J Biol Anthropol. 2023 Feb;180(2):252–71.
45. Dujon AM, Aktipis A, Alix-Panabières C, Amend SR, Boddy AM, Brown JS, et al. Identifying key questions in the ecology and evolution of cancer. Evol Appl. 2021 Apr;14(4):877–92.
46. Gatenby RA, Silva AS, Gillies RJ, Frieden BR. Adaptive therapy. Cancer Res. 2009 Jun 1;69(11):4894–903.
47. Peto R. Epidemiology, multistage models, and short-term mutagenicity tests. Int J Epidemiol. 2016 Jun;45(3):621–37.
48. Sulak M, Fong L, Mika K, Chigurupati S, Yon L, Mongan NP, et al. TP53 copy number expansion is associated with the evolution of increased body size and an enhanced DNA damage response in elephants. eLife. 2016 Sep 19;5:e11994.
49. Pesavento PA, Agnew D, Keel MK, Woolard KD. Cancer in wildlife: Patterns of emergence. Nat Rev Cancer. 2018 Oct;18(10):646–61.
50. Gomes NMV, Ryder OA, Houck ML, Charter SJ, Walker W, Forsyth NR, et al. Comparative biology of mammalian telomeres: Hypotheses on ancestral states and the roles of telomeres in longevity determination. Aging Cell. 2011 Oct;10(5):761–8.
51. Lorenzini A, Johnson FB, Oliver A, Tresini M, Smith JS, Hdeib M, et al. Significant correlation of species longevity with DNA double strand break recognition but not with telomere length. Mech Ageing Dev. 2009;130(11–12):784–92.
52. O'Connor EA, Cornwallis CK. Immunity and lifespan: Answering long-standing questions with comparative genomics. Trends Genet TIG. 2022 Jul;38(7):650–61.
53. Nesse RM. Natural selection and the elusiveness of happiness. Huppert FA, Baylis N, Keverne B, editors. Philos Trans R Soc Lond B Biol Sci. 2004 Sep 29;359(1449):1333–47.
54. Morgan CJ, Noronha LA, Muetzelfeldt M, Feilding A, Curran HV. Harms and benefits associated with psychoactive drugs: Findings of an international survey of active drug users. J Psychopharmacol (Oxf). 2013 Jun;27(6):497–506.
55. Žižek S. Enjoy your symptom! [Internet]. Routledge; 2013 [cited 2023 Nov 29]. Available from: https://www.taylorfrancis.com/books/9781135957551.
56. Nesse RM, Schulkin J. An evolutionary medicine perspective on pain and its disorders. Philos Trans R Soc B Biol Sci. 2019 Nov 11;374(1785):20190288.
57. Du Plessis G. An existential perspective on addiction treatment: A logic-based therapy case study. SSRN Electron J [Internet]. 2019 [cited 2023 Nov 29]. Available from: https://www.ssrn.com/abstract=3350606.
58. Dunbar RIM. Neocortex size as a constraint on group size in primates. J Hum Evol. 1992 Jun;22(6):469–93.
59. Inusa BPD, Hsu LL, Kohli N, Patel A, Ominu-Evbota K, Anie KA, et al. Sickle cell disease-genetics, pathophysiology, clinical presentation and treatment. Int J Neonatal Screen. 2019 Jun;5(2):20.
60. Luzzatto L. Sickle cell anaemia and malaria. Mediterr J Hematol Infect Dis. 2012;4(1):e2012065.
61. Woolf LI. The heterozygote advantage in phenylketonuria. Am J Hum Genet. 1986 May;38(5):773–5.
62. Ingram CJE, Mulcare CA, Itan Y, Thomas MG, Swallow DM. Lactose digestion and the evolutionary genetics of lactase persistence. Hum Genet. 2009 Jan;124(6):579–91.
63. Ségurel L, Bon C. On the evolution of lactase persistence in humans. Annu Rev Genomics Hum Genet. 2017 Aug 31; 18:297–319.
64. Bron C, Sutherland HG, Griffiths LR. Exploring the hereditary nature of migraine. Neuropsychiatr Dis Treat. 2021;17:1183–94.
65. Ducros A. [Genetics of migraine]. Rev Neurol (Paris). 2013 May;169(5):360–71.
66. Puca FM, de Tommaso M, Tota P, Sciruicchio V. Photic driving in migraine: Correlations with clinical features. Cephalalgia Int J Headache. 1996 Jun;16(4):246–50.
67. Snyder RD, Drummond PD. Olfaction in migraine. Cephalalgia Int J Headache. 1997 Nov;17(7):729–32.
68. Edvinsson L, Jansen Olesen I, Kingman TA, McCulloch J, Uddman R. Modification of vasoconstrictor responses in cerebral blood vessels by lesioning of the trigeminal nerve: Possible involvement of CGRP. Cephalalgia Int J Headache. 1995 Oct;15(5):373–83.
69. Picchioni MM, Murray RM. Schizophrenia. BMJ. 2007 Jul 14;335(7610):91–5.
70. Erlenmeyer-Kimling L, Paradowski W. Selection and schizophrenia. Am Nat. 1966 Nov;100(916):651–65.
71. Charlesworth D. Balancing selection and its effects on sequences in nearby genome regions. PLoS Genet. 2006 Apr;2(4):e64.
72. Polimeni J, Reiss JP. Evolutionary perspectives on schizophrenia. Can J Psychiatry Rev Can Psychiatr. 2003 Feb; 48(1):34–9.
73. Miller G, Boden MA. The mating mind: How sexual choice shaped the evolution of human nature. Nature. 2000;405 (6786):512.
74. Nettle D, Clegg H. Schizotypy, creativity and mating success in humans. Proc R Soc B Biol Sci. 2006 Mar 7;273(1586):611–5.
75. Raichlen DA, Lieberman DE. The evolution of human step counts and its association with the risk of chronic disease. Curr Biol. 2022 Nov;32(21):R1206–14.
76. Corruccini RS. An epidemiologic transition in dental occlusion in world populations. Am J Orthod. 1984 Nov;86(5):419–26.
77. Smith TM, Warinner C. Developmental, evolutionary and behavioural perspectives on oral health. In: Palaeopathology and Evolutionary Medicine [Internet]. Oxford University Press; 2022 [cited 2024 Jan 21], pp. 95–119. Available from: https://academic.oup.com/book/43111/chapter/361608086.
78. Gibbons A. Evolutionary biology: An evolutionary theory of dentistry. Science. 2012 May 25;336(6084):973–5.
79. Freuman TD, Kidd K. The Bloated Belly Whisperer: See Results Within a Week, and Tame Digestive Distress Once and For All. First Edition. New York: St. Martin's Press; 2018.

80. Valtin H, (With the Technical Assistance of Sheila A. Gorman). "Drink at least eight glasses of water a day." Really? Is there scientific evidence for "8 × 8"? Am J Physiol-Regul Integr Comp Physiol. 2002 Nov 1;283(5):R993–1004.
81. Huerta-Sánchez E, Jin X, Asan, Bianba Z, Peter BM, Vinckenbosch N, et al. Altitude adaptation in Tibetans caused by introgression of Denisovan-like DNA. Nature. 2014 Aug;512(7513):194–7.
82. Reardon S. Neanderthal DNA affects ethnic differences in immune response. Nature. 2016 Oct 20;nature.2016.20854.
83. Kerner G, Neehus AL, Philippot Q, Bohlen J, Rinchai D, Kerrouche N, et al. Genetic adaptation to pathogens and increased risk of inflammatory disorders in post-Neolithic Europe. Cell Genomics. 2023 Feb;3(2):100248.
84. Kerner G, Ramirez-Alejo N, Seeleuthner Y, Yang R, Ogishi M, Cobat A, et al. Homozygosity for *TYK2* P1104A underlies tuberculosis in about 1% of patients in a cohort of European ancestry. Proc Natl Acad Sci. 2019 May 21;116(21):10430–4.
85. De Lange KM, Moutsianas L, Lee JC, Lamb CA, Luo Y, Kennedy NA, et al. Genome-wide association study implicates immune activation of multiple integrin genes in inflammatory bowel disease. Nat Genet. 2017 Feb;49(2):256–61.
86. Hanson MA. Evolutionary medicine evolving. Front Sci. 2023 Feb 28;1:1147104.

4 Biotechnological Innovation for Sustainable Immunization

*Jaishree Vasudevan**
Department of Paediatrics, Chettinad Hospital and Research Institute, Chettinad Academy of Research and Education (CARE), Chennai, Tamil Nadu, India

4.1 INTRODUCTION

Biotechnology in its simplest form refers to the use of biological resources to develop new products or methods for improving human health and life. It encompasses techniques like plant hybridization or fermentation to manufacture curd.

Innovation refers to the development of a new idea or product; a vaccine is a product of biological origin that provides immune protection against a particular pathogen; and immunization is an extremely cost-effective method of preventing and controlling infectious diseases. From traditional vaccines containing inactivated or attenuated or killed organisms, biotechnological innovations have led to new and successful vaccines that are controlling or preventing infectious diseases and thus directly contributing to reaching UN SDG 3, health and wellbeing. Specifically, SDG 3.2 is to end the preventable deaths of newborns and children less than 60 months of age, but vaccination affects other SDGs as well.

4.2 IMMUNIZATION IN INDIA

In India, the Expanded Program of Immunization was launched in 1978 and relabeled the Universal Immunization Program (UPI) in 1985. The UPI is a vast public health program that at the time offered protection against six diseases: tuberculosis, diphtheria, pertussis, tetanus, polio, and measles. Over the years, vaccines against hepatitis B, *Haemophilus influenzae*, rotavirus and pneumococcal and human papilloma virus have been added to the immunization program.

4.3 BASICS OF IMMUNOLOGY

Immunization is the process by which a person develops protection against a disease through vaccination. Vaccination is the process of introducing an antigen into the body to stimulate immunity.[1] The term vaccine is derived from *variolae vaccinae* and is attributed to Jenner, who first described cowpox and demonstrated the protective role of cowpox against smallpox in 1798.

Most vaccines protect against diseases by inducing specific antibodies against an antigen of the pathogen or toxin; antibodies are secreted by plasma cells, which are activated by B lymphocytes. T lymphocytes mediate cell-mediated immunity and help B cells. Exposure to a natural infection or vaccine stimulates active immunity. Traditional vaccines use either a weakened or killed microorganism to produce an immune response. However, newer vaccines are produced through recombinant DNA technology, an innovation in biotechnology with far-reaching effects on human health.

4.4 THE ROLE OF BIOTECHNOLOGY IN TRANSFORMING HEPATITIS B VACCINATION

In 1976, Blumberg received the Nobel Prize for "discoveries concerning new mechanisms for the origin and dissemination of infectious diseases". He identified a previously unknown antigen in the blood of an Australian aborigine, hepatitis B surface antigen (HbsAg); until then, vaccines had been produced using killed or attenuated microbes, which was not possible with hepatitis B virus. Since HbsAg was a protein, Blumberg experimented with it using it to provoke an immune response.[2]

He invented a process for purifying HbsAg from the plasma of carriers of hepatitis B and rendering it safe for use as a vaccine; this was the first vaccine using only a part of the pathogen. Around this time period, a new biotechnology startup company Genentech was established. The scientists there had invented a novel method using recombinant microbes for protein production, and one of the applications of DNA recombinant technology was stated to be vaccine development. Scientists cloned pieces of DNA from hepatitis B virus, and not only was this partial genetic code expressed in *E. coli*, the whole viral DNA was also sequenced. Further work led to expression of strongly immunogenic HbsAg in yeast as well as in recombinant mammalian cells. The commercial success of hepatitis B vaccine helped to give a big boost to both biotechnology and vaccine industry and innovation.[2]

4.5 TYPES OF VACCINES

4.5.1 Live Attenuated Vaccines

Live attenuated vaccines contain weakened strains of the microbes. They do not cause the disease but stimulate the immune system to mount a defense as well as produce

* Corresponding author: vsdvn_jshr@yahoo.com

DOI: 10.1201/9781032660592-5

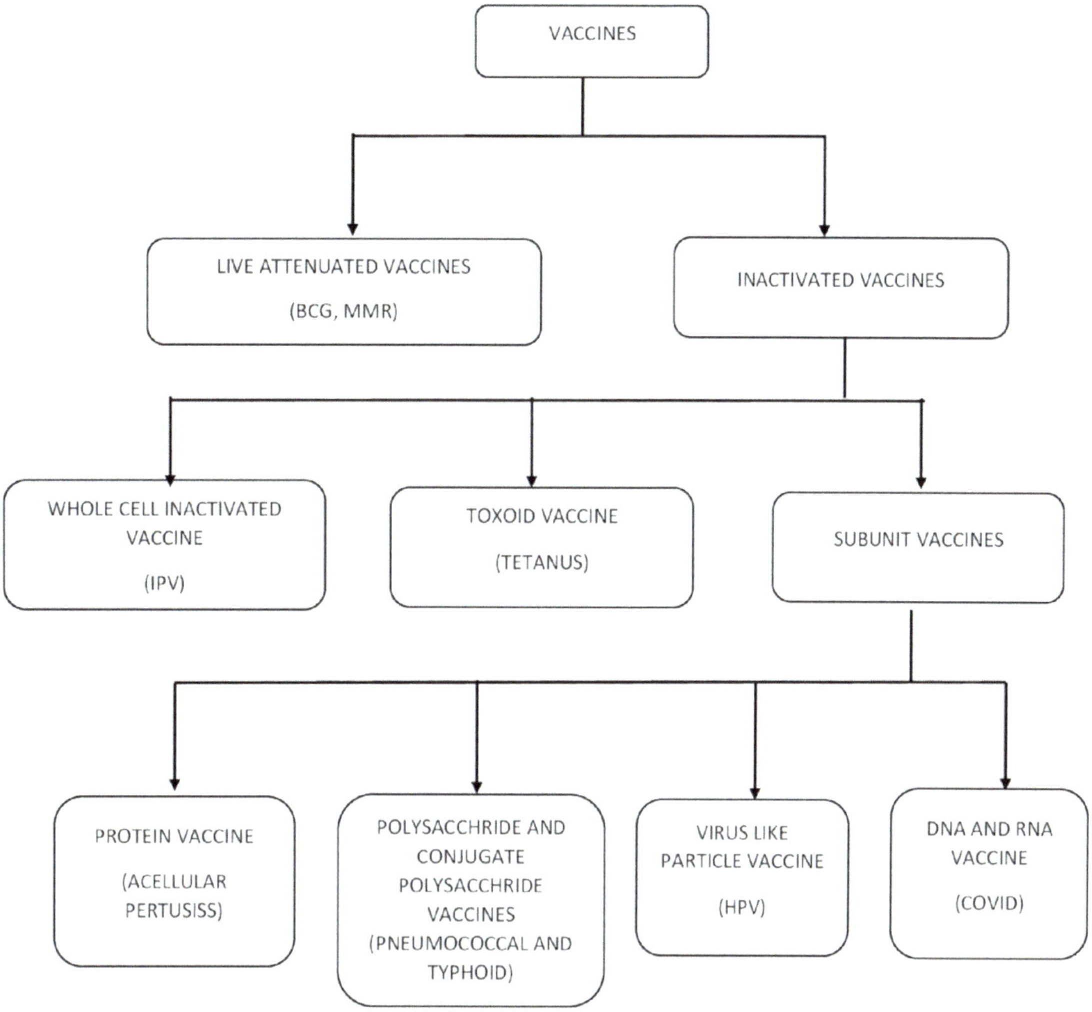

FIGURE 4.1 Types of vaccines.

memory cells for a long-lasting response; these include oral polio and chickenpox vaccines. However, these vaccines can cause disease if administered to immunosuppressed individuals. Figures 4.1 and 4.2 highlight some types of vaccines.

4.5.2 Inactivated Vaccines

Inactivated vaccines are mostly killed vaccines, toxoids, or subunit vaccines. The subunit vaccines contain only the antigens required for eliciting an immune response; for example, the hepatitis B vaccine is a protein subunit vaccine and was the first recombinant vaccine. It was developed by Maurice Hilleman, the father of modern vaccines, and his team, who used cloning. They inserted HbsAg into *Saccharomyces cerevisiae* (baker's yeast) which then expressed the antigen. This was then harvested and purified to be used as a vaccine.[2] Subunit vaccines can be polysaccharide vaccines or polysaccharide conjugate vaccines, which are conjugated to a protein to improve immune response. Subunit vaccines can be made either from the original pathogens or recombinantly; that is, the vaccine antigen is made using another microorganism.

4.5.3 Reverse Vaccinology

Meningococcus is a microbe with several subtypes that can cause meningitis and septicemia. The problem with group B meningococcus is that some target antigens are similar to some proteins found in the human tissue; use of these target antigens in a vaccine can lead to autoimmune damage to human tissue because of this molecular mimicry. Therefore, despite efforts lasting over nearly half a century, a vaccine against group B meningococcus has proved elusive.

However, through reverse vaccinology, the whole genome of the pathogen can be sequenced to explore genes, and bioinformatics and computer predictions can identify virulence factors and novel vaccine candidates. Scientists initially identified 600 candidate and finally included five in a vaccine against meningococcus type B. Using bioinformatics and functional genomics, the whole process took just 18 months. This technique has also been used for other vaccines, such as against antibiotic-resistant *Staphylococcus aureus* and can lead to the development of newer vaccines.[1]

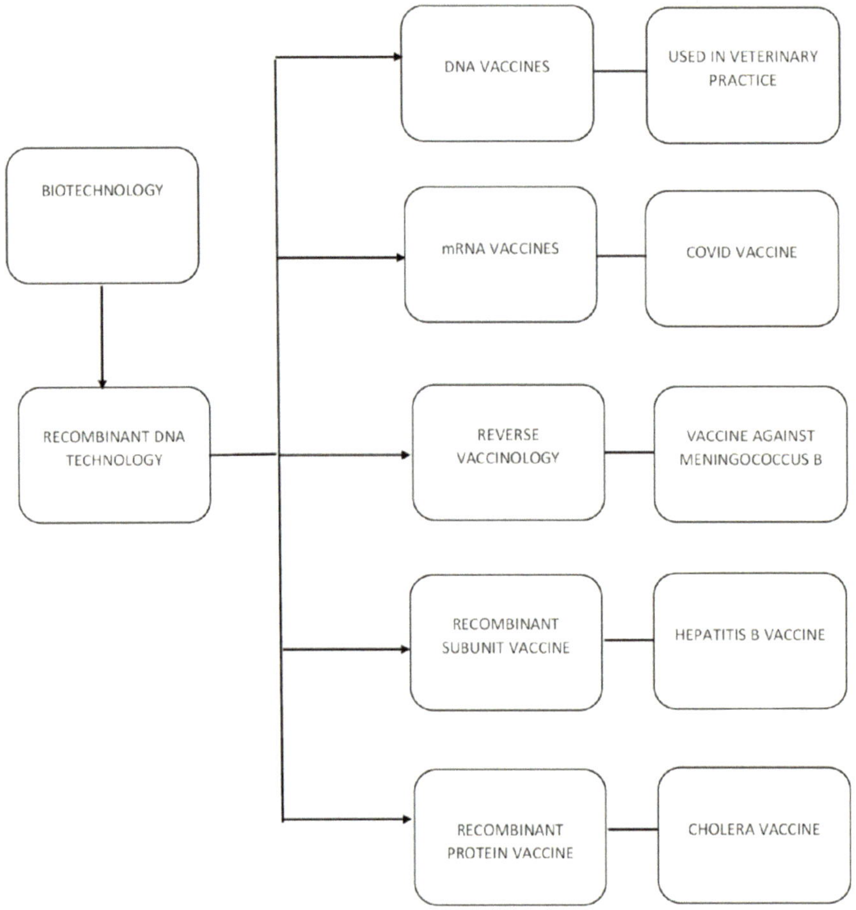

FIGURE 4.2 Types of vaccines using biotechnological innovations.

4.5.4 Viral-Like Particle and Nanoparticle Vaccines

While subunit vaccines are safer than traditional vaccines, they are generally less immunogenic, necessitating repeated doses. Viral-like particle (VLP) and nanoparticle (NP) vaccines have been devised to increase immunogenicity and stability. VLP vaccines do not have the viral genome and therefore are safer because they cannot multiply in the host; these vaccines produce particles that mimic antigens that can be taken up and displayed by dendritic cells, leading to a good immune response. The human papilloma virus (HPV) vaccine is an example; the major L protein of HPV is used to make virus-like particles that are expressed in the baculovirus. NP vaccines use carrier molecules crosslinked with antigen to improve the immune response, for instance Novavax's COVID-19 vaccine. NP vaccines against substance abuse elicit antibody response against the drug and destroy it before the drug can reach the brain to exert its effects. Both VLP and NP vaccines are used in toxoid vaccines.[3–4]

4.5.5 Recombinant Protein/DNA Vaccines

In recombinant protein vaccines, the gene encoding the protein or toxin responsible for producing an immune response is isolated from the pathogen and used to develop recombinant DNA in another organism such as yeast. When this is used to vaccinate a person, the host produces antibodies against the antigen without suffering from the disease. In addition to producing vaccines, recombinant DNA technology has been used to produce products like human insulin, growth hormone and interferons.

DNA vaccines were first tried in the 1990s by injecting plasmid DNA in hosts to elicit immune response to viral and nonviral antigens. DNA vaccines are at present available for veterinary practice for instance against the West Nile virus in horses and against canine melanoma. However, DNA vaccines tend to induce poor immune response in humans, probably because of the poor humoral responses. In addition, antigen expression occurs mainly at the site of administration.[4]

4.5.6 mRNA Vaccines

In mRNA vaccines, the mRNA encode for disease-specific antigens. The cell synthesize the antigens by translation, and these proteins induce the immune response. The Pfizer and Moderna COVID-19 vaccines are examples of mRNA vaccines. They can be of nonreplicating mRNA, *in vivo* self-replicating mRNA or *in vitro* dendritic cell nonreplicating types. These vaccines can be developed easily, which proved useful during the COVID-19 pandemic. However, disadvantages are the inherent instability of mRNA and the requirement of ultra cold storage for the vaccines. Unlike DNA vaccines, though, where the risk of modifying the parent genome (however theoretical) does exist, mRNA vaccines are safer because they cannot integrate into parent genomes.[4]

4.5.7 Viral Vector Vaccines

These vaccines such as the Oxford AstraZeneca COVID-19 and Ebola vaccines use a harmless, nonpathogenic virus to deliver the genetic code of an antigen against which the host mounts an immune response. The temperature requirement of these vaccines is less stringent than that for mRNA vaccines.

4.5.8 Futuristic Vaccines

Vaccines are a powerful tool in our weaponry to prevent infectious diseases. Recent approaches based on identifying disease-specific protein markers that normally escape immune response could lead to the development of therapeutic vaccines for treating disorders including even cancers. Table 4.1 presents some of the current and upcoming successes.

TABLE 4.1
Newer Vaccines Currently Available or in the Pipeline

Preventive Infection	Indication	Vaccine Type	Administration
Rotavirus	Gastroenteritis	Live attenuated	Oral
Human papilloma virus	Cervical cancer	Subunit	Im
Dengue virus	Dengue	Live attenuated	Im
Neisseria meningitidis Group B strain	Meningitis	Subunit	Im
HIV	AIDS	Subunit	Im
Respiratory syncytial virus	Pneumonia	Subunit	Im
Ebola virus	Ebola	Live attenuated	Im
Plasmodium falciparum	Malaria	Both subunit and live attenuated	Im (subunit) Iv (live)
Mycobacterium tuberculosis *M. vaccae*	Tuberculosis Shortens treatment for drug-susceptible TB	Live attenuated Killed	IdId
Shigella	Shigellosis	Live attenuated	Im
Enterotoxigenic *E. coli*	Traveller's diarrhoea	Subunit	Oral
Group B Streptococcus	Pneumonia	Polysaccharide	Im
Targeted against prostatic acid phosphatase	Prostate adenocarcinoma	Autologous	Im

Just as bacterial and viral antigens provoke an immune response, it might be expected that antigens expressed by cancer cells could also be detected and destroyed by our immune system. Why this does not occur could be because the cancer cells either hide from the defending mechanisms or overwhelm the immune system, enabling the cancer to spread unhindered. Today's research is concentrating on developing vaccines that will trigger and stimulate the immune system to fight cancer cells more effectively and rapidly.

Therapeutic cancer vaccines will treat cancers once they have developed as opposed to preventive cancer vaccines (such as hepatitis B and HPV vaccines), which prevent infections known to cause cancers. Therapeutic cancers are autologous or allogenic, and advances in biotechnology have led to personalized cancer treatment in the form of autologous vaccines. Here, a patient's own cancer cells are isolated and made more immunogenic; when these are reinjected into the patient's body, they provoke an immune response that not only destroys the cancer cells but also retains memory cells. These memory cells launch a defense in cases of cancer relapse. Another type of autologous cancer vaccine uses the patient's own immune system to manufacture the vaccine. Sipuleucel-T is an example of autologous prostate cancer vaccine that has been approved by the US FDA. Similarly, the University of Pennsylvania has developed an experimental breast cancer vaccine. Meanwhile, allogenic cancer vaccines are developed from laboratory-grown cells that are not from the patient's own body. Such vaccines have been developed against melanoma and pancreatic and prostate cancers, but none has been licensed yet.

Another area of research is HIV vaccines. As a virus, HIV has a high rate of mutability and variability, and there is lack of natural immunity to HIV; these are a few reasons developing a HIV vaccine is difficult. However, the RVIGG trial in 16,000 subjects in Thailand used a 'prime' boost strategy with two experimental HIV vaccines: priming with a canary pox virus vector with genes coding for HIV proteins and a booster vaccine using a recombinant surface protein from HIV, and vaccine candidates that protect against malaria and BCG are also in the pipeline. Another use of vaccines that is being pursued is for preventing some noninfectious diseases. In diabetes, the destruction of beta cells can be prevented by tolerance developed to diabetes-specific autoantigens, and immunization against amyloid can control Alzheimer's disease.[5]

4.6 CONCLUSION

At present, we have more than 20 vaccines protecting against life-threatening diseases. Immunization by reducing infant and childhood mortality has ensured child survival and is an excellent cost-effective health intervention. The WHO finds that vaccination prevents 3.5–5 million deaths every year. Indeed, COVID-19 highlights the specific relevance of vaccination:. *Lancet Infectious Diseases* reported that COVID-19 vaccines prevented over 42 lakh deaths in India in 2021, and globally, a mathematical model reported that COVID-19 vaccination prevented 14.4 million deaths in

2021. As our understanding of disease and microorganisms improves, biotechnological innovations like recombinant technology and viral vector vaccines will be of immense help in developing new vaccines not only against infectious diseases but also against cancers and lifestyle diseases. Technology is allowing for new vaccines to be developed more rapidly and efficiently, as evidenced by the COVID-19 vaccine manufacturing rush during the pandemic.

REFERENCES

1. Chen Y, Cheng H, Yang Y, Yeh MK. Biotechnologies Applied in Biomedical Vaccines [Internet]. Vaccines. InTech; 2017. Available from: http://dx.doi.org/10.5772/intechopen.69547.
2. Huzair F, Sturdy S. Biotechnology and the transformation of vaccine innovation: The case of the hepatitis B vaccines 1968–2000. Stud Hist Philos Biol Biomed Sci. 2017 Aug;64:11–21. Available from: http://dx.doi.org/10.1016/j.shpsc.2017.05.004. Epub 2017 May 13. PMID: 28511068; PMCID: PMC5541201.
3. Khan MY, Alsaadi M, Alghamdi K, Qadri I. Biotechnology and its applications in vaccine development. Biomed J Scient Tech Res. 2021;37(1):29114–6.
4. Brisse M, Vrba SM, Kirk N, Liang Y, Ly H. Emerging concepts and technologies in vaccine development. Front Immunol. 2020 Sep 30;11:583077.
5. Indra Shekhar Rao M, Srinivas G Kasi, IAP Guide Book on Immunization. 4th Edition. AVCIP Publications; 2022.

5 Biomedical Innovations in Telemedicine and Human Health

Mathangi Damal Chandrasekar[1]* *and Mathangi Krishnakumar*[2]
[1] Department of Mind Body Medicine and Lifestyle Sciences, Sri Ramachandra Institute of Higher Education and Research, Tamil Nadu, India
[2] Surgical and Neuro Intensive Care Unit, Department of Anaesthesia, St John's Medical College Hospital, Karnataka, India

5.1 INTRODUCTION

The one bright spot that saw us through the COVID-19 outbreak was telemedicine. Frequent hospital visits, travel expenses, and the shortage of specialists in rural and isolated places are all obviated by telemedicine, and indeed, it has been applied to many fields such as mental health counseling, chronic disease monitoring, medical diagnosis and treatment, professional consultations, and even remote patient monitoring.[1] The many advantages of telemedicine are driving its popularity, including better access to treatment – which is crucial for those living in remote locations or with limited mobility – lower costs, more convenience, and, most importantly, the ability to provide much-needed higher-quality care. Furthermore, telemedicine is used to manage chronic illnesses, provide palliative care, and alleviate the shortage of healthcare professionals.[2]

We are poised to embrace the fourth industrial revolution in healthcare, ushering in advanced technologies that enhance treatment and service delivery. This revolution, also known as "Medical 4.0," empowers healthcare professionals with real-time data access, enabling data-driven decision-making for optimal patient care.[3] The seamless connection between doctors, nurses, and patients facilitated by these technologies fosters continuous information exchange and ensures that all stakeholders remain informed. Equipped with the right tools and knowledge, empowered personnel can consistently excel in their roles.

However, while automation will undoubtedly continue to evolve, the human element will always remain at the core of healthcare, ensuring compassion and personalized care for patients.[4] Medical gadgets generate vast datasets that contribute to detailed patient profiles and enhance anatomical understanding. The growing focus on electronic health records simplifies predictive health analysis, allowing for informed decisions about medication, equipment, and overall care needs. Real-time technologies facilitate these assessments and enable remote communication between physicians and specialists regardless of location. Additionally, healthcare software solutions are being developed to improve patient satisfaction and streamline communication between hospitals and patients.[5,6] The emergence of medical technologies has ignited a transformative journey within healthcare, encompassing a spectrum of advancements from mobile medical applications to AI-powered software that empowers clinical decision-making for doctors.

Every effective healthcare innovation centers on three aspects: (a) The patient's needs are satisfied; (b) the patient is seen; and (c) the patient is heard. Effective management of quality, prices, safety, efficiency, and results should be made possible by these advancements.[7] Innovations arise when we seek solutions to embrace rather than problems to address. In this regard, telemedicine and biomedical innovation have been increasingly entwined, benefiting not just humanity but also each industry itself; each has fueled the other's creativity, which has improved healthcare through wearable technology and sensors that allow for remote monitoring as well as the increasing accessibility and availability of telemedicine.

Similarly, data analytics with AI and machine learning of medical data have assisted consultants with provisional diagnosis, treatment recommendations, and even predicting health risks, facilitating preventive and promotive healthcare. Big data can also be used to better understand different types of patients and figure out ways to prevent them from needing to come back. This data can also help hospitals predict how many patients they will have so they can accurately schedule staff, equipment, and hospital facilities.[8,9] This now marks a paradigm shift in healthcare, driving a future of individualized therapy, rapid symptom response, and minute-by-minute patient monitoring facilitated by cutting-edge technologies. This revolution hinges on breakthrough tracking technologies that offer continuous and consistent patient data, the cornerstone of effective treatment strategies (Figure 5.1).[10]

Advancements in robotics and remote control have enabled minimally invasive procedures to be performed remotely, increasing access to specialized care. By allowing remote patients to participate in clinical trials with diversity and inclusion and real-time, continuous data collecting that provides insightful information, this merging has also expedited the process of bringing new treatment protocols

* Corresponding author: dcmathangi@gmail.com

DOI: 10.1201/9781032660592-6

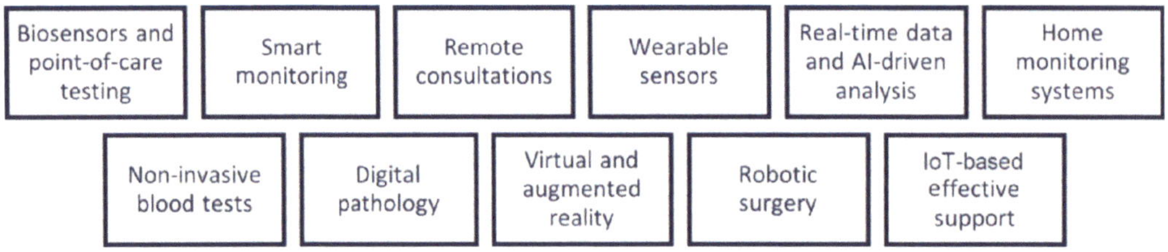

FIGURE 5.1 Technologies in healthcare.

to market. This mutually beneficial partnership has not only increased patient access to care but also improved patient satisfaction with affordable, precise, and customized medical services. The medical field is currently experiencing a significant transformation due to the combination of advanced biomedical advancements and the extensive availability of telecommunication technology.[11] Once considered a science fiction concept, telemedicine is quickly entering the mainstream and changing the way healthcare is provided. This chapter gives an overview of the intricate interplay between these innovations and their profound impacts on human health, paving the way for a future of personalized, ubiquitous, and preventive healthcare.

5.2 DIAGNOSTIC INNOVATIONS

Biosensors and point-of-care testing are examples of compact, portable medical equipment that has made it possible to diagnose patients quickly and accurately wherever they are.[12] This has made it possible to diagnose patients more quickly, which improves treatment results and lowers the risk of consequences. By enabling remote consultations and point-of-care tests at a reasonable cost, the combination of telemedicine and modern diagnostics with these small devices can result in less expensive and more accurate and informed diagnoses.

Our daily lives are becoming increasingly integrated with widely used wearable sensors such as fitness trackers and smartwatches, which continuously record vital indicators including heart rate, blood pressure, oxygen levels, sleep patterns, and activity levels in real time.[13] By supplying doctors with real-time data streams for early disease detection and preemptive intervention, these discreet monitors enable patients to take an active role in their treatment. The use of AI and machine learning algorithms to analyze medical data – including pictures, scans, and data from wearables – has been beneficial since it allows for identifying trends and suggesting diagnoses, increasing efficiency and accuracy.[14]

Consider a farmer in a far-off village who is having chest pains, for example. Their cardiac rhythm is transmitted to a cardiologist miles away via a straightforward, noninvasive ECG patch, saving the arduous trip to the closest clinic; equipped with real-time data and AI-driven analysis tools, the specialist can identify a possible arrhythmia quickly and start the right treatment, possibly saving the farmer's life.[15] Additionally, improvements in home monitoring systems are enabling patients to receive advanced medical testing right at their door. Imagine diabetics constantly monitoring their blood sugar levels with a finger prick or using sensors that are automatically uploaded to a cloud-based platform so their healthcare team can view the data in real time. With proactive condition management and customized insulin changes made possible by this ongoing monitoring, patients' outcomes improve, and hospital stays can be decreased or even avoided.

Diagnoses can be made early and with greater accuracy with picture analysis, which offers more thorough and accurate insights into possible issues. To determine a person's vulnerability to a disease and provide early, individualized therapy, picture analysis has also been used to analyze individual genetic profiles.[16] Noninvasive blood tests such as blood glucose monitoring have revolutionized the treatment of diabetes by helping patients understand the lifestyle choices and habits that cause spikes in blood parameters; this understanding fosters accountability among patients. In a similar vein, telepathology, or digital pathology, enables professionals to analyze tissue samples remotely, increasing access to professional diagnosis regardless of location.[17]

5.3 THERAPEUTIC INNOVATION

Biomedical innovation and telemedicine are coming together to create intriguing opportunities to transform therapeutic approaches. They provide novel paths for therapeutic innovation, overcoming physical barriers to specialized consultations and complicated medical procedures. For instance, robotic surgery provides underprivileged populations and patients in critical condition with access to top-notch care directed by proficient surgeons working remotely. With the help of robotic arms that can precisely replicate the movements of their subjects and high-definition cameras, a neurosurgeon in a large metropolis can now execute a delicate brain surgery on a patient hundreds of miles away.[18]

Telemedicine is bringing about a change in the field of mental health, which was previously stigmatized and had restricted access. Virtual and augmented reality are beginning to find their way into the therapeutic sphere. Anxiety and phobias can be treated with distant therapeutic interventions such as virtual reality therapy.[19] For example, a physical therapist can use VR simulations to walk a stroke patient through rehabilitation exercises, giving them motivation and sensory feedback no matter where they are in the world.[20] Personalized cognitive behavioral treatment (CBT) can also be remotely administered through digital therapy apps,[21] breaking down geographical and social boundaries in private, secure settings. Enhancing treatment compliance, telehealth systems can also be used to track medication adherence, refill prescriptions, and remotely resolve medication-related issues. Patients recovering from operations or injuries can benefit from individualized rehabilitation programs made possible using remote therapy consultations and monitoring.

5.4 BRINGING PUBLIC HEALTH TO ALL: THE SYNERGY OF TELEMEDICINE AND BIOMEDICAL INNOVATION

With the combined force of telemedicine and biomedical innovation, public healthcare that reaches even the most rural and underserved regions is more accessible than ever. This has made it possible for everyone to have fair access to healthcare, including preventative care and basic medical treatments, regardless of location or socioeconomic background. Better health outcomes have been achieved for populations through early detection, preventative measures, and prompt treatment.[22]

Patients are also becoming increasingly involved in their healthcare as connection improves and access to information becomes simple. These tools and technologies empower individuals to actively manage their health and wellbeing, increase their personal accountability, and engage more in their treatment decisions (Table 5.1). This promotes confidence, builds trust, and guarantees that patients receive immediate guidance throughout their disease. In addition, the overall cost of healthcare can be decreased through point-of-care diagnostics, individualized therapies, and remote consultations.[23] Technologies like Internet of Things, VR, AI, data sciences, edge computing, and others have blurred the borders between technologist and medical field for the benefit of mankind.[24] Real-time data sharing and virtual discussions between all nations and the World Health Organization during the recent COVID pandemic allowed predictive maps and quicker outbreak detection and response, particularly in remote areas, enabling us to handle the pandemic as one planet.[25–27]

TABLE 5.1
Impacts of Innovation

Innovation	Application	Impact
Wearable technology	Continuous health tracking	Improved patient monitoring
AI diagnostics	Predictive analytics in diagnostics	Enhanced diagnostic accuracy
Remote patient monitoring	Chronic condition management	Increased care accessibility
Teletherapy	Mental health services	Expanded mental health support

Benefits always have drawbacks that must be considered to fully realize the potential of telemedicine to transform healthcare delivery. Strong cybersecurity measures and patient education are necessary to address the critical concerns of data privacy and security. One major obstacle is the digital divide, which makes equal access to infrastructure and technology necessary.[28] Technology-related restrictions such as limited bandwidth and unstable internet access can reduce the efficacy of remote consultations and processes.

It is necessary to address the ethical issues of informed permission, accurate distant diagnosis, and the possibility of impersonalizing patient care. Clear rules and regulations must be established, and healthcare workers must get ongoing training to guarantee the ethical and responsible use of telemedicine technologies.[29] Providing long-term access to telemedicine and biomedical innovation in public health initiatives also will require sustainable funding methods combining the public and private sectors.

5.5 ETHICAL ASPECTS

The basic principle of bioethics is autonomy, emphasizing a person's right to make decisions about their own healthcare. In the context of telemedicine, additional consideration must be given to its digital nature to ensure patients' autonomy. Data privacy and security are of utmost concern, especially with the integration of AI, remote monitoring, and sensitive patient information. Maintaining privacy and implementing cybersecurity measures are of paramount importance when handling patient data.[30]

The biggest challenge in assessing patients and collecting information using telemedicine platforms is understanding how much is too much or too little to arrive at a medical decision; the regulations surrounding this are still evolving. Moving forward, the aim has to be standardized protocols that ensure uniformity, reduce interoperable variability, and ensure patient safety. Intensive training, quality management, and continuous compliance with changing healthcare policies and technology are imperative.[29] In the evolution of medicine, there is always the question of whether machine can replace man. The human touch in patient care has stood the test of time; the most significant factor in curing a disease is not the medicine but the doctor. Information about every aspect of medicine is available and accessible with the advent of technology, and AI in both diagnosis and treatment has made effective healthcare reachable to all. However, AI can only complement medical decisions; the patient–doctor relationship is critical to wellness. The responsibility of maintaining transparency and retaining human touch in healthcare is with us.[29]

5.6 CASE STUDIES

CASE STUDY 5.1: REMOTE MONITORING

India has emerged as the diabetes capital of the world owing to rapidly changing lifestyles amid the technology revolution. Chronic diseases such as hypertension, diabetes, kidney disease and heart disease require constant monitoring and continuous treatment. The main cause for worsening disease condition and eventual mortality is lack of followup, and telemedicine has now made this aspect of care possible in the comfort of home. This ensures compliance and better control of these diseases. Researchers who used remote monitoring to manage diabetes and found that their patients achieved better control of blood sugar levels than they achieved with the usual clinic care. The technology was a wearable blood sugar monitor that gave instant feedback on a mobile app to patient and doctor. This enabled timely decisions and personalized care plans.[31]

CASE STUDY 5.2: TELEPSYCHIATRY

Mental health has been an increasing concern throughout the 21st century, but the taboo surrounding the problem has been a major impediment to help seeking. One application of telemedicine that emerged during the COVID-19 pandemic was telepsychiatry; this played a critical role in keeping patients and their families together. Introducing this capacity in a rural community significantly increased patient satisfaction. Specifically, video conferencing to connect patients with healthcare providers significantly reduced wait times.[32]

5.7 FUTURE DIRECTIONS

The intersection of biomedical innovations and telemedicine represents a pivotal evolution in healthcare delivery, offering unprecedented opportunities to enhance access, quality, and cost-effectiveness of healthcare services. Through the integration of remote monitoring tools, AI-driven diagnostics, and secure data management systems, telemedicine has transcended geographical and logistical barriers, bringing specialized care to the doorsteps of patients worldwide. However, as we embrace the advancements in telemedicine, it is imperative to navigate the accompanying challenges and ethical considerations with vigilance and a commitment to patient-centered care. Data privacy, regulatory compliance, and the preservation of the human element in healthcare are paramount in ensuring that telemedicine continues to serve as a force for good in the healthcare ecosystem.

Looking forward, telemedicine is poised for further growth and innovation. Continued investment in research and development, coupled with collaborative efforts among stakeholders in the healthcare, technology, and regulatory sectors, will be crucial in harnessing the full potential of telemedicine. As we venture into this future, it is the responsibility of the global healthcare community to ensure that telemedicine evolves in a manner that is ethical, equitable, and focused on enhancing the wellbeing of patients around the world.

5.8 CONCLUSION

A new era in healthcare marked by proactive wellness management, accessibility, and customization is being ushered in by biomedical breakthroughs and telemedicine. The possibilities are boundless, ranging from online therapy and public health programs to robotic surgery and remote diagnostics. The future of healthcare seems promising, with everyone having the opportunity to live happier, more satisfying lives and distance no longer being a barrier to high-quality care as we embrace ethical issues and overcome obstacles.

REFERENCES

1. Smith AC, Thomas E, Snoswell CL, et al. Telehealth for global emergencies: Implications for coronavirus disease 2019 (COVID-19). J Telemed Telecare. 2020;26(5):309–313.
2. Dorsey ER, Topol EJ. State of telehealth. N Engl J Med. 2016;375(2):154–161.
3. Mhlanga D. The role of artificial intelligence and machine learning amid the COVID-19 pandemic: What lessons are we learning on 4IR and the sustainable development goals. Int J Environ Res Public Health. 2022 Feb 8;19(3):1879.
4. Haleem A, Javaid M. Medical 4.0 and its role in healthcare during COVID-19 pandemic: A review. J Ind Integr Manag. 2020;5:531–545.
5. Piwek L, Ellis DA, Andrews S, Joinson A. The rise of consumer health wearables: Promises and barriers. PLoS Med. 2016;13(2):e1001953.
6. Steinhubl SR, Muse ED, Topol EJ. The emerging field of mobile health. Sci Transl Med. 2015;7(283):283rv3.
7. Bhattacharyya O, Shapiro J, Schneider EC. Innovation centers in healthcare delivery systems: Structures for success. J Med Internet Res. 2022 Feb 10;24(2):e33961.
8. Esteva A, Kuprel B, Novoa RA, et al. Dermatologist-level classification of skin cancer with deep neural networks. Nature. 2017;542(7639):115–118.
9. Ting DSW, Cheung CY, Lim G, et al. Development and validation of a deep learning system for diabetic retinopathy and related eye diseases using retinal images from multiethnic populations with diabetes. JAMA. 2017;318(22):2211–2223.
10. Kvedar J, Coye MJ, Everett W. Connected health: A review of technologies and strategies to improve patient care with telemedicine and telehealth. Health Aff (Millwood). 2014; 33(2):194–199.
11. Morgan AA, Abdi J, Syed MAQ, Kohen GE, Barlow P, Vizcaychipi MP. Robots in healthcare: A scoping review. Curr Robot Rep. 2022;3(4):271–280.
12. St John A, Price CP. Existing and emerging technologies for point-of-care testing. Clin Biochem Rev. 2014 Aug;35(3): 155–167.

13. Masoumian Hosseini M, Masoumian Hosseini ST, Qayumi K, Hosseinzadeh S, SajadiTabar SS. Smartwatches in healthcare medicine: Assistance and monitoring; A scoping review. BMC Med Inform DecisMak. 2023 Nov 3;23(1):248.
14. Ahuja AS. The impact of artificial intelligence in medicine on the future role of the physician. PeerJ. 2019 Oct 4;7:e7702.
15. Sana F, Isselbacher EM, Singh JP, Heist EK, Pathik B, Armoundas AA. Wearable devices for ambulatory cardiac monitoring. J Am Coll Cardiol. 2020;75:1582–1592.
16. Kumar Y, Koul A, Singla R, Ijaz MF. Artificial intelligence in disease diagnosis: A systematic literature review, synthesizing framework and future research agenda. J Ambient Intell Humaniz Comput. 2023;14(7):8459–8486.
17. Chordia TD, Vikey A, Choudhary AB, Samdariya Y, Chordia DS. Current status and future trends in telepathology and digital pathology. J Oral Maxillofac Pathol. 2016 May–Aug;20(2):178–182.
18. Shah J, Vyas A, Vyas D. The history of robotics in surgical specialties. Am J Robot Surg. 2014 Jun 1;1(1):12–20. doi: 10.1166/ajrs.2014.1006.
19. Andersen NJ, Schwartzman D, Martinez C, Cormier G, Drapeau M. Virtual reality interventions for the treatment of anxiety disorders: A scoping review. J Behav Ther Exp Psychiatry. 2023 Dec;81:101851.
20. Khokale RS, Mathew G, Ahmed S, Maheen S, Fawad M, Bandaru P, Zerin A, Nazir Z, Khawaja I, Sharif I, Abdin ZU, Akbar A. Virtual and augmented reality in post-stroke rehabilitation: A narrative review. Cureus. 2023 Apr 14;15(4):e37559.
21. Canonico ME, Hsia J, Guthrie NL, Simmons M, Mehta P, Lupinacci P, Edwards K, Mosesso K, Gearhart M, Skuban A, Bonaca MP, Berman MA. Cognitive behavioral therapy delivered via digital mobile application for the treatment of type 2 diabetes: Rationale, design, and baseline characteristics of a randomized, controlled trial. Clin Cardiol. 2022 Aug;45(8):850–856.
22. Maroju RG, Choudhari SG, Shaikh MK, Borkar SK, Mendhe H. Role of telemedicine and digital technology in public health in India: A narrative review. Cureus. 2023 Mar 10;15(3):e35986.
23. Scott Kruse C, Karem P, Shifflett K, Vegi L, Ravi K, Brooks M. Evaluating barriers to adopting telemedicine worldwide: A systematic review. J Telemed Telecare. 2018;24(1):4–12.
24. Lakshminarayanan V, Ravikumar A, Sriraman H, Alla S, Chattu VK. Healthcare equity through intelligent edge computing and augmented reality/virtual reality: A systematic review. J Multidiscip Healthc. 2023 Sep 21;16:2839–2859.
25. Hollander JE, Carr BG. Virtually perfect? Telemedicine for covid-19. N Engl J Med. 2020;382(18):1679–1681.
26. Topol EJ. High-performance medicine: The convergence of human and artificial intelligence. Nat Med. 2019;25(1):44–56.
27. Fagherazzi G, Goetzinger C, Rashid MA, Aguayo GA, Huiart L. Digital health strategies to fight COVID-19 worldwide: Challenges, recommendations, and a call for papers. J Med Internet Res. 2020;22(6):e19284.
28. Luxton DD, Kayl RA, Mishkind MC. mHealth data security: The need for HIPAA-compliant standardization. Telemed J E Health. 2012;18(4):284–288.
29. Mittelstadt BD, Floridi L. The ethics of big data: Current and foreseeable issues in biomedical contexts. Sci Eng Ethics. 2016;22(2):303–341.
30. Luxton DD, Kayl RA, Mishkind MC. mHealth data security: The need for HIPAA-compliant standardization. Telemed J EHealth. 2012;18(4):284–288.
31. Or CK, Karsh BT. A systematic review of patient acceptance of consumer health information technology. J Am Med Inform Assoc. 2009;16(4):550–560.
32. Yellowlees P, Shore JH. Telepsychiatry and health technologies: A guide for mental health professionals. Am J Psychiatry. 2018;175(1):5–15.

6 Nanotechnology and Sustainable Health

A Pathway to Progress

Jyoti Rawat,[1] *Lokesh Kumar Tripathi,*[2] *and Veena Pande*[2*]

[1] Department of Biosciences, School of Liberal Arts and Sciences, Mody University of Science and Technology – Laxmangarh, Sikar, Rajasthan, India

[2] Department of Biotechnology, Sir JC Bose Technical Campus, Kumaun University Nainital (Bhimtal), Uttarakhand, India

6.1 INTRODUCTION

International efforts to promote sustainability can benefit greatly from the application of nanotechnology. Rather than merely enhancing current technologies, nanotechnology presents revolutionary, game-changing discoveries and breakthroughs that can offer prompt answers and solutions to benefit our planet, society, and environment. Through the creation of smart materials and linked devices, breakthroughs in nanotechnology are having an impact on resource management, energy, healthcare, and the environment. Additionally, the domains of nanotechnology and nanoscience have impacted many adjacent sectors by bringing scientific, technical, medical, and other communities together through communication (Zhao et al., 2019). Because it is now a general-purpose technology, the notion is not new. The speed at which nanoscience is developing indicates that nanoscale production will soon be integrated into nearly every field of study and technology.

It is impossible to overlook the origins of nanomedicine while discussing the vast domains of biology, biotechnology, and medicine. Successful uses of nanotechnology in medicine have improved human life quality over time, leading to the development of a whole new field called nanomedicine. This has enabled researchers to develop improved methods of disease prevention, screening, treatment, diagnosis, and proactive healthcare measures (Vogel et al., 2020). With progressions in nanobased genomics, gene therapy, and tissue engineering, these procedures will also include medication manufacture, design, conjugation, and efficient delivery procedures. Nanomedicine will likely become the top research area for future biologists to examine its benefits and hazards (Visakh, 2022).

Scientists are investigating the possible uses of nanotechnology in a variety of medical treatments. Medical instruments use nanoscale computer-derived robotic features such as diagnostic surfaces, sensor technologies, and sample purification kits (Park, 2014). The field of diagnostics is also undergoing changes with the advent of technologies that can operate, respond, and alter inside the human body for the express aim of early detection and therapy. Aside from cell therapy and tissue engineering, nanomanufacturing has also been influenced by regenerative medicine (Erkoc and Ulucan-Karnak, 2021). In addition, new technologies called "lab-on-a-chip" are bringing about significant changes in various industries, including nanomedicine, diagnostics, dentistry, and cosmetics (Dubey et al., 2022). Indeed, novel uses of nanotechnology in genomes and proteomics have led to molecular understandings of infectious disease pathogenesis, and new applications including surgical nanorobotics, nanobioelectrics, and drug delivery technologies are being developed through the merging of medicine, programming, nanoengineering, and biotechnology (Gibney et al., 2021). When taken as a whole, these aid medical researchers and practitioners in their quest to comprehend illness etiology and develop more effective treatments.

Nanodevices are quickly finding applications in diagnostics, where they facilitate the regulation of therapy options and the prediction of disease etiologic factors (Gupta et al., 2021) (Figure 6.1). *In vitro* diagnostics improve the efficacy of disease prevention efforts via devices that screen for disease states and react to toxins, carcinogens, and pathological abnormalities (Aslan et al., 2013). Diagnostics, disease treatment, regenerative medicine, gene therapy, dentistry, oncology, aesthetics, medication delivery, and medications themselves are benefiting from nanotechnology (Nikalje, 2015; Adeola et al., 2020). A more planned, distinct, and technically designed nanomedicine will emerge in the future with thorough collaborations between doctors, researchers, and technology. The pathophysiology of diseases is driving efforts to find solutions to the problems that have long plagued medical nanotechnology applications (Malik et al., 2023a). In this chapter, we describe nanotechnologies that scientists, engineers, and researchers can use for further studies.

* Corresponding author: veena.biotech@gmail.com

DOI: 10.1201/9781032660592-7

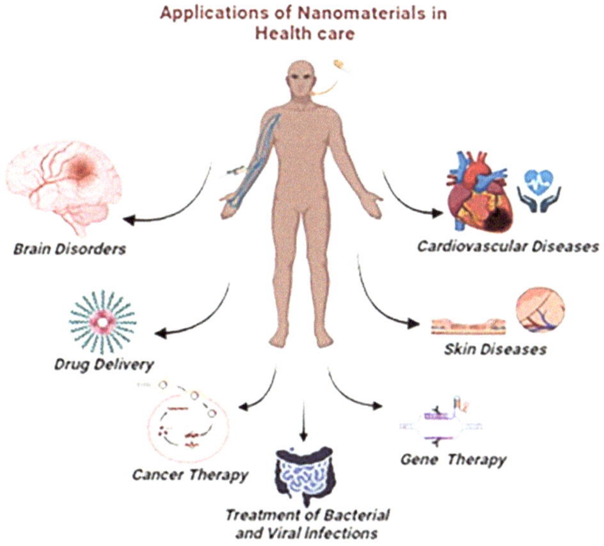

FIGURE 6.1 Therapeutic nanomaterials in targeted delivery applications.

6.2 NANOMEDICINES TRANSFORMING HEALTHCARE PRACTICES

A comprehensive branch of research and technology, nanomedicine integrates several medical application areas, including pain management, illness prevention, disease treatment and diagnostics, human health enhancement, and trauma damage prevention (Pramanik et al., 2020; Vaishampayan et al., 2023). The results of biomedical robotics, nanomaterials, biotechnology, and genetic engineering integrated under the umbrella of nanomedicine are being applied using an interdisciplinary approach (Mbunge et al., 2021). In general, most biological and chemical processes that produce medical materials benefit from the efficiency, quick response time, and functional effectiveness that nanoscale medical technologies offer. Therefore, there is always hope for the future and new uses of nanomedicine thanks to research. Certain nanoparticles are used in novel diagnostic equipment, imaging and methods, targeted drugs, pharmaceuticals, biomedical implants, and tissue engineering. Now, nanotechnology makes high-toxicity treatments like cancer medicines safer, and the technology can also detect infections, the states of cancer cells, and changes in vital signs. Predictive analytics and AI can be applied to biomedicine to treat patients.

Bone regeneration is one more way that nanotechnology is being used in the medical field. Researchers are developing bone graft technology to facilitate muscle reorganization and bone reformation (Khang et al., 2008; Nikalje, 2015). To transform the science of osteology and bone tissue engineering, fundamental research is being done on collagen-mimic coatings, biomineralization, collagen fibers, and artificial joints and muscles (Walmsley et al., 2015). Similarly, to enhance drug delivery stability as well as pharmacodynamic and pharmacokinetic profiles on a broad scale, drug delivery technologies are focusing on nano-scaling alternatives (Jahangirian et al., 2017). One crucial tool in enabling medication to traverse the circulatory system and deliver drug molecules to precise target zones is nanorobots (Park, 2014). Researchers are even developing wireless intracellular and intranuclear surgery using nanorobots to treat a variety of cancers that would otherwise be incurable (Gibney et al., 2021). Such nanorobotics being at a small scale can even slice a sole neuronic dendrite deprived of damaging intricate neural networks (Park, 2013).

Additionally, nanotechnology is being used in oncology. Researchers have created nanosized agents, fluorescent materials, molecular diagnostic kits, and specifically targeted medications that aid in cancer detection and treatment (Aslan et al., 2013). To improve therapeutic specificity and organ targeting, scientists are experimenting with attaching already-available medicines with nanoparticulate conjugation (Aslan et al., 2013; Mazayen et al., 2022; Misra et al., 2010). Hundreds of distinct antitumor compounds can be delivered to tumor sites by nanomedicine. Additionally, tumor imaging and immunotherapy techniques associated with nanomedicine present promising avenues for managing cancer treatment (Misra et al., 2010; Buabeid et al., 2020). Efforts are also being made to improve the pharmacokinetic and pharmacodynamic qualities of chemotherapeutic medications and increase their effectiveness in targeting tumors to decrease their side effects (Nikalje, 2015). Similarly, gene therapy procedures combined with heat-induced ablation treatment for cancer cells are also being combined with nanorobotics (Yoon and Kim, 2020). Successful anticancer medications take advantage of enhanced permeation and retention through the application of nanoassemblies, such as liposomes, albumin nanospheres, micelles, and gold nanoparticles (Wu, 2021).

Most of the processes used to create nanomedicines involve the physical and chemical degradation of particles to the micro- and nanoscales. However, investigators are now exploring green chemistry and green engineering to produce nanomedicines due to worries about the harmful effects on the environment and health (Jahangirian et al., 2017). This green technology creates nanoassemblies with fewer environmental and health risks. The subsequent step to advancing the science of green nanomedicine will be to combine green nanoassemblies with medications, vaccinations, or diagnostic markers. Green engineering and nanotechnology have been used in the production of numerous inorganic nanoassemblies that have been released onto the market (Kanwar et al., 2019). Quantum dots, organic polymeric nanoparticles, mesoporous silica nanoparticles, dendrimers, nanostructured lipid carriers, solid lipid nanoparticles, and gold and silver nanoparticles are a few examples (Kanwar et al., 2019; Nasrollahzadeh et al., 2019).

6.3 NANOMATERIALS FOR IMPROVED DRUG DELIVERY

Because nanoscience makes it possible to produce medications with increased therapeutic efficacy and decreased toxicity, the pharmaceutical sector has undergone a revolution. Nanoparticles enhance drug pharmacokinetics by increasing drugs' stability, solubility, and bioavailability (Mazayen et al., 2022). Additionally, they can target particular cells and tissues, which lessens adverse effects and increases their effectiveness (Malik et al., 2023c). Due to their distinct physicochemical characteristics and nanoscale size, drugs must be administered according to exact parameters (Adir et al., 2020; Buya et al., 2020). The size, shape, surface characteristics, and mode of administration of nanoparticles all affect their dosage (Buya et al., 2020). For example, due to variations in absorption and biodistribution, oral delivery might need a greater dose to have the same effect as intravenous administration (Hu et al., 2018; Buya et al., 2020).

Medication distribution has undergone a revolutionary change because of nanotechnology, which offers more efficient and tailored drug administration while reducing adverse effects and boosting therapeutic efficacy. Using nanoparticles engineered to transport medications to the intended site of action is one way that nanotechnology is being applied to the delivery of pharmaceuticals (Sahu et al., 2021). There are various benefits to using nanotechnology for medication delivery. First, it makes it possible to deliver medications to particular parts of the body, like tumors, inflammatory tissues, and contaminated regions, in a focused and regulated manner; this minimizes adverse effects and reduces the number of medications needed (Sahu et al., 2021). Second, medications can be stabilised and made more soluble by using nanoparticles, which increases treatment efficacy (Oroojalian et al., 2015). Third, by improving the absorption and distribution of medications within the body, nanotechnology increases bioavailability. This makes it possible to employ lower medicine dosages, which reduces toxicity (Enrico, 2019).

Medical nanorobots are now being employed for medication delivery, sending antitumor medications through veins to precise locations. Researchers are even developing wireless intracellular and intranuclear nanoscale operations to treat a variety of illnesses including cancers (Sahu et al., 2021; Malik et al., 2023a), and the production of respirocytes, a mechanical red blood cell, is an amazing scientific feat as well. Compared with natural red blood cells, nanorobots can provide bodily tissues with oxygen 200 times more quickly (Enrico, 2019; Suhail et al., 2022). This raises the possibility of using nanotechnology in the future to identify, treat, and cure a range of blood-related diseases (Zhang et al., 2019). In conclusion, the delivery of drugs via nanotechnology has completely changed the medical industry. It has reduced adverse effects, improved drug therapeutic efficacy, and allowed for more efficient and focused drug administration. The ongoing development of medication delivery systems based on nanotechnology holds the key to the future of drug delivery.

6.4 NANOTECHNOLOGY IN DISEASE DIAGNOSIS

Nanodevices are being used to identify diseases early and quickly, leading to suggestions for additional medical procedures (Dessale et al., 2022). By increasing the precision, sensitivity, and speed of medical testing, nanotechnology has the potential to completely transform diagnostics in the healthcare sector (Fox et al., 2019). Nanoparticle-based diagnostic imaging improves positron emission tomography (PET), magnetic resonance imaging (MRI) and computed tomography (CT) scans by attaching nanoparticles to biomarkers and making them more sensitive, accurate, and specific (Singh and Amiji, 2022). Similar to this, point-of-care diagnostic tests enabled by nanotechnology can swiftly and accurately diagnose infectious infections, malignancies, and other ailments for treatment and prevention (Singh and Amiji, 2022).

Another area where nanotechnology is being applied is in generating highly sensitive biosensors that can identify even minute amounts of biomolecules in physiological fluids like urine and blood. This technology makes disease management and early detection easier (Xu et al., 2022; Sadeghi et al., 2023). Similar uses include the isolation and analysis of particular cells, proteins, and genetic material using microfluidic devices that include nanomaterials, enabling quick and precise illness detection (Rajput et al., 2023). Nanopore sequencing, a cutting-edge system that employs nanopores to identify the sequence of DNA and RNA molecules, also enables the quick and precise identification of hereditary diseases like cancer (Su et al., 2025).

Recent developments indicate that nomedicine in *in vitro* diagnostics can improve the accuracy and dependability of illness detection (Dang and Guan, 2020) using samples derived from human tissue, cell culture, bodily fluids, and other sources (Malik et al., 2023b; Welch et al., 2021). Nanomedicine is being applied to *in vivo* diagnostics to make devices that can function, react, and change inside the human body for the sole aim of early detection of anomalies that can cause toxicity or lead to tumor development (Rajput et al., 2023). Quantum dots, nanocrystals, nanoshells, paramagnetic nanoparticles, and nanosomes are a few of the nanoparticles that are now being used for diagnostic reasons (Zhou et al., 2017; Bardhan, 2022). In general, nanotechnology has enormous potential in healthcare diagnostics and is expected to play a significant role in the development of personalized medicine.

For the upcoming generation of medical scientists, nanotechnology offers potential applications in the design of medical instruments and procedures. For instance, the high surface area-to-volume ratios of nanoscale materials and structures enable more effective interactions with biological molecules; even in the earliest stages of diseases, this enhanced sensitivity allows for finding very low amounts of biomarkers or chemicals linked to disease. The major applications of nanomaterials in several fields are briefed in Table 6.1. Nanotechnology also allows for early disease detection; diagnostic nanosized instruments can identify diseases even before symptoms

TABLE 6.1
Medical Applications of Nanomaterials

S. No.	Nanomaterial	Applications	Size (nm)	References
1.	Polybutylcyanoacrylate/insulin	Diabetes	500	Radwan (2001)
2.	AuNPs	Chemotherapy	14	Song et al. (2012)
3.	AuNPs	Antibacterial agent	17–100	Tao et al. (2015)
4.	Au(I)thiolate oligomers	Optical imaging	12	Cantelli et al. (2016)
5.	Superparamagnetic iron oxide nanoparticle covered with dextran	Anemia in chronic kidney disease	30	Bullivant et al. (2013)
6.	PEGylated factor VIII	Hemophilia		Chelle et al. (2020)
7.	Radioactive gold	Nanotherapeutic agent in oncology	80	Shukla et al. (2012)
8.	Poly(3,4-ethylenedioxythiophene):poly(4-styrenesulfonate)	NIR-mediated hyperthermic therapy	80	Cheng et al. (2012)

appear, and better patient outcomes and successful treatment depend on early identification. Nanotechnology makes it easier to create quick and portable diagnostic tools that can be utilized in situations with limited resources, including at home or in a doctor's office, to support point-of-care testing. This makes healthcare more accessible, particularly in rural or underdeveloped areas.

6.5 CHALLENGES AND OPPORTUNITIES IN THE CONVERGENCE OF NANOTECHNOLOGY AND HEALTH

Even though nanotechnology has many potential applications in healthcare, there are still issues and moral dilemmas that need to be resolved. These include worries over how long-term exposure to nanoparticles will affect human health, how they will affect the environment, and how fair access to treatment based on nanotechnology will be. To guarantee the responsible development and application of nanotechnology in medicine, researchers and policymakers need to collaborate.

Regarding the effectiveness of delivery and clinical translation, nanoparticles face several challenges. The accumulation of nanocarriers in reticuloendothelial system organs, especially the liver, hinders clinical translation by capturing most of the injected dose, preventing adequate dose delivery to the targeted disease site and potentially causing toxicity (Bao et al., 2023). To tackle this problem, scientists have devised strategies such as preconditioning macrophages with chloroquine, flooding the organs of the reticuloendothelial system with drug-free nanocarriers, and temporarily stealthily coating scavenger cells to increase the effectiveness of drug-loaded nanoparticles that reach the diseased tissue (Wolfram et al., 2017; Saunders et al., 2020; Dirisala et al., 2020). Furthermore, adding specific cells to the surfaces of nanocarriers like those using the "do not eat us" tactic helps them avoid being caught by the reticuloendothelial system, which enhances the accumulation of nanodrugs at the intended location (Tang et al., 2019; Saunders et al., 2020).

However, even though surface shielding of nonionic hydrophilic polymers like polyethylene glycol (PEG) on nanocarriers improves colloidal stability and stealth in a biological context, it also decreases cellular uptake and endosomal escape, leading to low delivery efficacy (Dirisala et al., 2020). Targeting ligands are positioned carefully at the distal end of PEG segments to enable particular ligand receptor-mediated absorption to get around this stealth issue (Dirisala et al., 2014). Another technique to facilitate endosomal escape is to wrap negatively charged polymers around positively charged polymers in response to the acidic endo/lysosomal pH to facilitate endosomal escape (Dirisala et al., 2022). Even when injected intratumorally and intramuscularly, the hepatic protein expression of mRNA-loaded lipid nanoparticles limits their usage (Pardi et al., 2015). It might be beneficial for cancer immunotherapies and protein replacement therapies to minimize off-target liver expression. To specifically stop therapeutic mRNAs from being expressed in the liver, one method is to include microRNA target sites in them (Pardi et al., 2015). In general, nanotechnology has great potential in healthcare diagnostics and will help build tailored medication.

6.6 CONCLUDING REMARKS

Nanotechnology holds immense promise as a pathway to sustainable health. The technology has immense potential to revolutionize the approach to medical treatment concerning targeted drug delivery and advanced diagnostics. To explore the possibilities and address ethical considerations, nanotechnology remains a key player in the ongoing quest for sustainable and effective healthcare solutions. By increasing the bioavailability and solubility of drugs, nanotechnology can reduce the raw materials required for manufacturing, leading to more efficient and ecofriendly production. Additionally, the use of nanomaterials can facilitate the development of personalized medicine, tailoring treatments to individual patient profiles and minimizing unnecessary medical interventions.

REFERENCES

Adeola, H. A., Sabiu, S., Adekiya, T. A., Aruleba, R. T., Aruwa, C. E., & Oyinloye, B. E. Prospects of nanodentistry for the diagnosis and treatment of maxillofacial pathologies and cancers. Heliyon. 2020;6(9).

Adir, O., Poley, M., Chen, G., Froim, S., Krinsky, N., Shklover, J., . . . & Schroeder, A. Integrating artificial intelligence and nanotechnology for precision cancer medicine. Adv Mater. 2020;32:1901989. doi: 10.1002/adma.201901989.

Aslan, B., Ozpolat, B., Sood, A. K., & Lopez-Berestein, G. Nanotechnology in cancer therapy. J Drug Target. 2013; 21(10):904–913.

Bao, L., Cui, X., & Chen, C. Toxicology for nanotechnology. In: *Nanomedicine*. Singapore: Springer Nature; 2023:157–177.

Bardhan, N. Nanomaterials in diagnostics, imaging and delivery: Applications from COVID-19 to cancer. MRS Commun. 2022;12(6):1119–1139.

Buabeid, M. A., Arafa, E. S. A., & Murtaza, G. Emerging prospects for nanoparticle-enabled cancer immunotherapy. J Immunol Res. 2020;(1):9624532.

Bullivant, J. P., Zhao, S., Willenberg, B. J., Kozissnik, B., Batich, C. D., & Dobson, J. Materials characterization of Feraheme/ferumoxytol and preliminary evaluation of its potential for magnetic fluid hyperthermia. Int J Mol Sci. 2013;14(9):17501–17510.

Buya, A. B., Beloqui, A., Memvanga, P. B., & Préat, V. Self-nano-emulsifying drug-delivery systems: From the development to the current applications and challenges in oral drug delivery. Pharmaceutics. 2020;12:1194. doi: 10.3390/pharmaceutics12121194.

Cantelli, A., Battistelli, G., Guidetti, G., Manzi, J., Di Giosia, M., & Montalti, M. Luminescent gold nanoclusters as biocompatible probes for optical imaging and theranostics. Dyes Pigm. 2016;135:64–79.

Chelle, P., Yeung, C. H., Croteau, S. E., Lissick, J., Balasa, V., Ashburner, C., . . . & Iorio, A. Development and validation of a population-pharmacokinetic model for rurioctacog alfa pegol (Adynovate®): A report on behalf of the WAPPS-Hemo Investigators Ad Hoc Subgroup. Clin Pharmacokinet. 2020;59:245–256.

Cheng, L., Yang, K., Chen, Q., & Liu, Z. Organic stealth nanoparticles for highly effective in vivo near-infrared photothermal therapy of cancer. ACS Nano. 2012;6(6):5605–5613.

Dang, Y., & Guan, J. Nanoparticle-based drug delivery systems for cancer therapy. Smart Mater Med. 2020;1:10–19. doi: 10.1016/j.smaim.2020.04.001.

Dessale, M., Mengistu, G., & Mengist, H. M. Nanotechnology: A promising approach for cancer diagnosis, therapeutics and theragnosis. Int J Nanomed. 2022;17:3735–3749. doi: 10.2147/IJN.S378074.

Dirisala, A., Osada, K., Chen, Q., Tockary, T. A., Machitani, K., Osawa, S., . . . & Kataoka, K. Optimized rod length of polyplex micelles for maximizing transfection efficiency and their performance in systemic gene therapy against stroma-rich pancreatic tumors. Biomaterials. 2014;35:5359–5368. doi: 10.1016/j.biomaterials.2014.03.037.

Dirisala, A., Uchida, S., Li, J., Van Guyse, J. F., Hayashi K., Vummaleti S. V., . . . & Kataoka K. Effective mRNA protection by poly (l-ornithine) synergizes with endosomal escape functionality of a charge-conversion polymer toward maximizing mRNA introduction efficiency. Macromol Rapid Commun. 2022;43:2100754. doi: 10.1002/marc.202100754.

Dirisala A., Uchida S., Toh K., Li J., Osawa S., Tockary T. A., . . . & Kataoka K. Transient stealth coating of liver sinusoidal wall by anchoring two-armed PEG for retargeting nanomedicines. Sci Adv. 2020;6:eabb8133. doi: 10.1126/sciadv.abb8133.

Dubey, S. K., Dey, A., Singhvi, G., Pandey, M. M., Singh, V., & Kesharwani, P. Emerging trends of nanotechnology in advanced cosmetics. Colloids Surf B Biointerfaces. 2022;214:112440.

Enrico, C. Nanotechnology-based drug delivery of natural compounds and phytochemicals for the treatment of cancer and other diseases. Stud Nat Prod Chem. 2019;62:91–123.

Erkoc, P., & Ulucan-Karnak, F. Nanotechnology-based antimicrobial and antiviral surface coating strategies. Prosthesis. 2021;3:25–52.

Fox, K. E., Tran, N. L., Nguyen, T. A., Nguyen, T. T., & Tran, P. A. Surface modification of medical devices at nanoscale: Recent development and translational perspectives. In: *Biomaterials in Translational Medicine*. Cambridge, MA: Academic Press; 2019:163–189.

Gibney, S., Hicks, J. M., Robinson, A., Jain, A., Sanjuan-Alberte, P., & Rawson, F. J. Toward nanobioelectronic medicine: Unlocking new applications using nanotechnology. WIREs Nanomed Nanobiotechnol. 2021;13(3):e1693.

Gupta, N., Bahl, S., Bagha, A. K., Vaid, S., Javaid, M., & Haleem, A. Nanomedicine technology and COVID-19 outbreak: Applications and challenges. J Ind Integr Manag. 2021;6(2):161–174.

Hu, Q., Li, H., Wang, L., Gu, H., & Fan, C. DNA nanotechnology-enabled drug delivery systems. Chem Rev. 2018;119:6459–6506. doi: 10.1021/acs.chemrev.7b00663.

Jahangirian, H., Lemraski, E. G., Webster, T. J., Rafiee-Moghaddam, R., & Abdollahi, Y. A review of drug delivery systems based on nanotechnology and green chemistry: Green nanomedicine. Int J Nanomed. 2017;12:2957–2978. doi: 10.2147/IJN.S127683.

Kanwar, R., Rathee, J., Salunke, D. B., & Mehta, S. K. Green nanotechnology-driven drug delivery assemblies. ACS Omega. 2019;4:8804–8815. doi: 10.1021/acsomega.9b00304.

Khang, D., Carpenter, J., Chun, Y. W., Pareta, R., & Webster, T. J. Nanotechnology for regenerative medicine. Biomed Microdevices. 2008;12:575–587.

Malik, S., Muhammad, K., & Waheed, Y. Emerging applications of nanotechnology in healthcare and medicine. Molecules. 2023a;28(18):6624.

Malik, S., Muhammad, K., & Waheed, Y. Nanotechnology: A revolution in modern industry. Molecules. 2023b;28:661. doi: 10.3390/molecules28020661.

Malik, S., Niazi, M., Khan, M., Rauff, B., Anwar, S., Amin, F., & Hanif, R. Cytotoxicity study of gold nanoparticle synthesis using aloe vera, honey, and Gymnema Sylvestre leaf extract. ACS Omega. 2023c;8:6325–6336. doi: 10.1021/acsomega.2c06491.

Mazayen, Z. M., Ghoneim, A. M., Elbatanony, R. S., Basalious, E. B., & Bendas, E. R. Pharmaceutical nanotechnology: From the bench to the market. Future J Pharm Sci. 2022;8:12. doi: 10.1186/s43094-022-00400-0.

Mbunge, E., Muchemwa, B., & Batani J. Sensors and healthcare 5.0: Transformative shift in virtual care through emerging digital health technologies. Glob Health J. 2021;5:169–177. doi: 10.1016/j.glohj.2021.11.008.

Misra, R., Acharya, S., & Sahoo, S. K. Cancer nanotechnology: Application of nanotechnology in cancer therapy. Drug Discov Today. 2010;15:842–850. doi: 10.1016/j.drudis.2010.08.006.

Nasrollahzadeh, M., Sajjadi, M., Sajadi, S. M., & Issaabadi, Z. Green nanotechnology. In: *Interface Science and Technology*, Vol. 28. Amsterdam, The Netherlands: Elsevier; 2019:145–198.

Nikalje, A. P. Nanotechnology and its applications in medicine. Med Chem. 2015;5:81–89.

Oroojalian, F., Charbgoo, F., Hashemi, M., Amani, A., Yazdian-Robati, R., Mokhtarzadeh, A., . . . & Weissman, D. Expression kinetics of nucleoside-modified MRNA delivered in lipid nanoparticles to mice by various routes. J Control Release. 2015;217:345–351. doi: 10.1016/j.jconrel.2015.08.007.

Pardi, N., Tuyishime, S., Muramatsu, H., Kariko, K., Mui, B. L., Tam, Y. K., . . . &; Weissman, D. Expression kinetics of nucleoside-modified mRNA delivered in lipid nanoparticles to mice by various routes. J Control Release. 2015;217: 345–351.

Park, K. Facing the truth about nanotechnology in drug delivery. ACS Nano. 2013;7:7442–7447.

Park, K. Controlled drug delivery systems: Past forward and future back. J Control Release. 2014;190:3–8.

Pramanik, P. K. D., Solanki, A., Debnath, A., Nayyar, A., El-Sappagh, S., & Kwak, K. S. Advancing modern healthcare with nanotechnology, nanobiosensors, and internet of nano things: Taxonomies, applications, architecture, and challenges. IEEE Access. 2020;8:65230–65266. doi: 10.1109/ACCESS.2020.2984269.

Radwan, M. A. Enhancement of absorption of insulin-loaded polyisobutylcyanoacrylate nanospheres by sodium cholate after oral and subcutaneous administration in diabetic rats. Drug Dev Ind Pharm. 2001;27(9):981–989.

Rajput, A., Sevalkar, G., Pardeshi, K., & Pingale, P. Computational nanoscience and technology. OpenNano. 2023;12:100147. doi: 10.1016/j.onano.2023.100147.

Sadeghi, M., Sadeghi, S., Naghib, S. M., & Garshasbi, H. R. A comprehensive review on electrochemical nano biosensors for precise detection of blood-based oncomarkers in breast cancer. Biosensors. 2023;13:481. doi: 10.3390/bios13040481.

Sahu, T., Ratre, Y. K., Chauhan, S., Bhaskar, L. V. K. S., Nair, M. P., & Verma, H. K. Nanotechnology based drug delivery system: Current strategies and emerging therapeutic potential for medical science. J Drug Deliv Sci Technol. 2021;63:102487. doi: 10.1016/j.jddst.2021.102487.

Saunders, N. R., Paolini, M. S., Fenton, O. S., Poul, L., Devalliere, J., Mpambani, F., . . . & Langer, R. A nanoprimer to improve the systemic delivery of siRNA and mRNA. Nano Lett. 2020;20:4264–4269. doi: 10.1021/acs.nanolett.0c00752.

Shen, L., Wang, P., & Ke, Y. DNA nanotechnology-based biosensors and therapeutics. Adv Healthcare Mater. 2021;10:2002205. doi: 10.1002/adhm.202002205.

Shukla, R., Chanda, N., & Zambre, A. Laminin receptor specific therapeutic gold nanoparticles (198AuNPEGCg) show efficacy in treating prostate cancer. Proc Natl Acad Sci USA. 2012;109:12426–12431.

Singh, A., & Amiji, M. M. Application of nanotechnology in medical diagnosis and imaging. Curr Opin Biotechnol. 2022;74:241–246. doi: 10.1016/j.copbio.2021.12.011.

Song, J., Zhou, J., & Duan, H. Self-assembled plasmonic vesicles of SERS-encoded amphiphilic gold nanoparticles for cancer cell targeting and traceable intracellular drug delivery. J Am Chem Soc. 2012;134:13458–13469.

Su, X., Lin, Q., Liu, B., Zhou, C., Lu, L., Lin, Z., . . . & Duan, S. The promising role of nanopore sequencing in cancer diagnostics and treatment. Cell Insight. 2025;100229.

Suhail, M., Khan, A., Rahim, M. A., Naeem, A., Fahad, M., Badshah, S. F., . . . & Janakiraman, A. K. Micro and nanorobot-based drug delivery: An overview. J Drug Target. 2022;30:349–358. doi: 10.1080/1061186X.2021.1999962.

Tang, Y., Wang, X., Li, J., Nie, Y., Liao, G., Yu, Y., & Li, C. Overcoming the reticuloendothelial system barrier to drug delivery with a "don't-eat-us" strategy. ACS Nano. 2019;13:13015–13026. doi: 10.1021/acsnano.9b05679.

Tao, Y., Ju, E., Ren, J., & Qu, X. Bifunctionalized mesoporous silica-supported gold nanoparticles: Intrinsic oxidase and peroxidase catalytic activities for antibacterial applications. Adv Mater. 2015;27:1097–1104.

Vaishampayan, V., Kapoor, A., & Gumfekar, S. P. Enhancement in the limit of detection of lab-on-chip microfluidic devices using functional nanomaterials. Can J Chem Eng. 2023;101: 5208. doi: 10.1002/cjce.24915.

Visakh, P. M. *Nanomaterials and Nanotechnology in Medicine*. Hoboken, NJ: Wiley; 2022.

Vogel, H. G., Maas, J., & Gebauer, A. Nanotechnology in medicine. In: *Drug Discovery and Evaluation: Methods in Clinical Pharmacology*. Berlin/Heidelberg, Germany: Springer; 2020:533–546.

Walmsley, G. G., McArdle, A., Tevlin, R., Momeni, A., Atashroo, D., Hu, M. S., . . . & Wan, D. C. Nanotechnology in bone tissue engineering. Nanomed Nanotechnol Biol Med. 2015;11:1253–1263.

Welch, E. C., Powell, J. M., Clevinger, T. B., Fairman, A. E., & Shukla, A. Advances in biosensors and diagnostic technologies using nanostructures and nanomaterials. Adv Funct Mater. 2021;31:2104126. doi: 10.1002/adfm.202104126.

Wolfram, J., Nizzero, S., Liu, H., Li, F., Zhang, G., Li, Z., . . . & Ferrari, M. A chloroquine-induced macrophage-preconditioning strategy for improved nanodelivery. Sci Rep. 2017;7:13738. doi: 10.1038/s41598-017-14221-2.

Wu, J. The enhanced permeability and retention (EPR) effect: The significance of the concept and methods to enhance its application. J Pers Med. 2021;11(8):771.

Xu, Q. Y., Tan, Z., Liao, X. W., & Wang, C. Recent advances in nanoscale metal-organic frameworks biosensors for detection of biomarkers. Chin Chem Lett. 2022;33:22 32. doi: 10.1016/j.cclet.2021.06.015.

Yoon, H.-J., & Kim, S.-W. Nanogenerators to power implantable medical systems. Joule. 2020;4:1398–1407.

Zhang, N., Wei, M. Y., & Ma, Q. Nanomedicines: A potential treatment for blood disorder diseases. Fron Bioeng Biotechnol. 2019;7:369.

Zhao, Y., Bai, C., Brinker, C. J., Chi, L., Dawson, K. A., Gogotsi, Y., . . . & Weiss, P. S. NanoasaRosettaStone: The Global Roles and Opportunities for Nanoscience and Nanotechnology. ACS Nano. 2019;13(10):10853–10855.

Zhou, Y., Du, J., Wang, L., & Wang, Y. Nanocrystals technology for improving bioavailability of poorly soluble drugs: A mini-review. J Nanosci Nanotechnol. 2017;17:18–28. doi: 10.1166/jnn.2017.13108.

7 Theranostics as an Emerging Technology for Targeted Cancer Drug Delivery

Anjali Kumari,[1] *Amanpreet Kaur,*[1,2] *and Geeta Aggarwal*[1,2]*
[1] School of Pharmaceutical Sciences, Delhi Pharmaceutical Sciences and Research University, New Delhi, India
[2] Centre for Advanced Formulation Technology, Delhi Pharmaceutical Sciences and Research University, New Delhi, India

7.1 INTRODUCTION

Cancer is a significant global health challenge due to its widespread occurrence and substantial impact on mortality rates. According to The Pan American Health Organization, a regional office of the WHO, an estimated 20 million new cancer cases are diagnosed each year, and it accounts for 10 million deaths worldwide. In the United States alone, cancer was responsible for an estimated 609,360 deaths in 2023 (Sathishkumar et al., 2023), with men experiencing more favorable incidence trends than women (Siegel et al., 2023).

Given these statistics, it is imperative to develop new treatments and medications that provide targeted and effective cancer therapy to effectively tackle the growing cancer death rates (Ferlay et al., 2022). Cancer treatment options are also on the rise along with the increasing cases of cancer. The challenges in cancer treatment include multidrug resistance in cancer cells, poor solubility of hydrophobic anticancer drugs, biocompatibility issues, and harmful effects of radiation. Nanomedicines have emerged as promising solutions to the various limitations encountered in cancer treatment.

Nanomedicines are based on coupling nanotechnology with advanced formulations, carefully selecting biocompatible excipients, and functionalizing nanocarriers, all of which helps in overcoming the challenges in cancer treatment. The field of nanomedicine has advanced significantly to the cutting-edge approach called theranostics: the integration of diagnostics and treatments into a unified system for personalized cancer treatment. This has opened numerous opportunities for advancing cancer diagnosis and treatment in the future. The aim of theranostics is to select the right molecular probe (diagnostic/therapeutic) for the particular patient to maximize treatment efficacy while minimizing side effects and toxicity. Traditionally, the approach to medicine is a sequential process, where patients undergo diagnostic tests to identify the presence of a disease, followed by the administration of treatment based on the diagnosis. However, theranostics aims to break this sequential approach by integrating diagnosis and therapy into a single cohesive process.

The key concept behind theranostics is the use of diagnostic tools that can provide real-time information about a patient's disease's status; this information is then used to tailor the treatment strategy to the specific needs of the individual patient. By utilizing molecular diagnostics, imaging techniques, and biomarkers, theranostics allows physicians to identify the molecular profile of a disease and predict the patient's response to various treatment options. By utilizing nanoparticles (NPs), this method addresses the challenges faced by conventional therapies in diagnosing and treating late-stage cancers. Ultimately, it aims to enhance treatment outcomes and improve the overall survival rate. Nanocarriers (NCs) play a pivotal role in this approach, serving as both diagnostic tools and therapeutic agents; this nanotheranostics is a promising approach in various fields including cancer treatment due to its numerous benefits. Rapid progress in nanotheranostics holds promise for improved diagnostic accuracy, timely intervention, and enhanced therapeutic outcomes in the battle against cancer. Moreover, nanotheranostics allows real-time drug response monitoring, resulting in fewer adverse effects and less drug toxicity in patients, leading to more effective and safer cancer treatments.

In this chapter, we aimed to investigate the recent progresses made in the field of cancer theranostics and nanotheranostics in current clinical trials. The primary focus is on highlighting advancements in theranostics as contrast agents in various imaging modalities and their potential as a versatile platform for cancer therapy. Studies of significant importance regarding cancer theranostics have been examined to provide insights into the promising role of theranostics and nanotheranostics in improving cancer treatment outcomes and enabling precise diagnostic imaging.

7.2 CHALLENGES IN CANCER TREATMENT

Cancer is a complex and multifaceted disease that has many challenges for healthcare providers and researchers. Significant advancements have been made in cancer

* Corresponding author: geetaaggarwal17@gmail.com

DOI: 10.1201/9781032660592-8

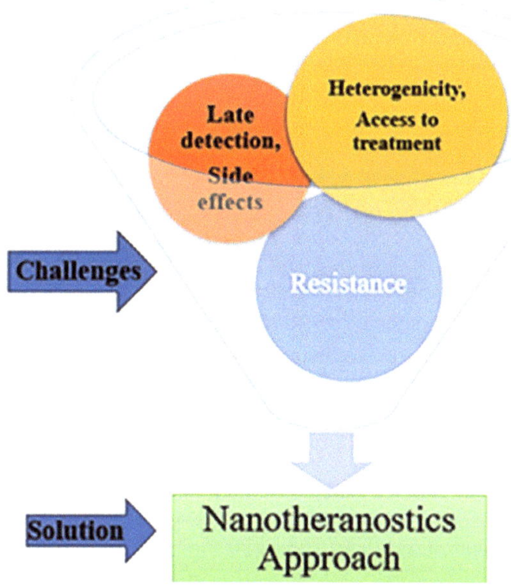

FIGURE 7.1 Challenges in cancer treatment.

research and treatment, but challenges remain, which we discuss here as well as graphically displaying Figure 7.1.

The primary challenge to cancer treatment is its heterogeneity. Cancer is not a singular disease but a collection of diseases, each with its own unique characteristics. Tumors differ in their genetic makeup, cellular behavior, and response to treatment, and this heterogeneity makes it challenging to develop standardized treatment approaches that work for all types of cancer (Chavda, 2023). Personalized approaches that take into account individual genetic profiles and tumor characteristics have gained attention, but implementing them on a larger scale remains a challenge.

Early detection is crucial for the successful treatment of cancer, but many cancers are diagnosed only at advanced stages when the disease has already manifested. Developing reliable and easily accessible screening methods that can detect cancer at its earliest stages is an ongoing challenge (Aminolroayaei et al., 2021). Improving awareness and access to screening programs, as well as investing in research for innovative detection technologies, are vital to addressing this challenge. However, what truly makes cancer so difficult to treat is its ability to develop resistance to various treatment modalities, such as chemotherapy and radiation therapy. This resistance can arise due to genetic mutations, tumor heterogeneity, or the presence of cancer stem cells, and it is a significant challenge that requires a deeper understanding of the mechanisms involved and the development of novel therapeutic strategies.

Cancer treatment when it is available brings about disabling side effects. Although cancer treatments have become more targeted and sophisticated, they often have adverse side effects that can significantly impact patients' quality of life. Chemotherapy, for example, can cause nausea, fatigue, hair loss, and a compromised immune function (Aminolroayaei et al., 2021). Finding ways to minimize treatment-related side effects without jeopardizing efficacy is an ongoing challenge that requires continued research and innovation. And there is also the matter of access to cancer treatment. Disparities exist in the access to cancer treatment both globally and within countries; expensive specialized medications and treatments are not evenly accessible, and global efforts are needed to make cancer treatment accessible and affordable to everyone worldwide.

While the challenges in the treatment of cancer are formidable, literature has shown that most of these challenges can be overcome with the help of theranostics or nanotheranostics. Through continued research, collaboration, and innovation, theranostics and nanotheranostics are contributing to significant strides in improving cancer prevention, detection, and treatment.

7.3 CURRENT TECHNOLOGIES IN CANCER TREATMENT (NANOMEDICINE AND DRUG TARGETING)

In recent years, significant advancements have been made in the field of cancer treatment. These technologies have revolutionized the way we approach cancer care, enabling more precise, targeted, and personalized treatment strategies.

Nanomedicine is offering more precise and targeted therapies that combine cancer immunotherapy with advanced functionalized nanotechnology and have led to significant advancements in the field of immunology. Nanomedicines can overcome obstacles such as cancer cell drug resistance, hydrophobicity/poor solubility, hazardous radiation, and biocompatibility. NPs can be utilized as medicines, diagnostics, or both combined and play a crucial role in the treatment of cancer. The rapid development of cancer nanomedicine has elevated immunotherapy outcomes to a new level in modern medical research. Some effective nanomedicine-based cancer immunotherapies include immune checkpoint inhibitors, CRISPR/Cas9 NPs, core–shell NPs for combination cancer immunotherapy, biomimetic NPs, and CAR-T cells in cancer nano-immunotherapy. The findings demonstrate the potential of nanomedicine as a versatile platform for innovative therapeutic applications and intelligent healthcare management (Lakshmanan et al., 2021). Some cancer nanomedicines, such as DOXil, Thermo DOX, and Myocet, have received FDA approval for specific cancer therapies (Rodríguez et al., 2022).

Researchers are also utilizing the promising ligand-target NCs in cancer theranostics. These sophisticated agents have the potential to target specific cancer sites and aid in the detection of diverse metastatic areas inside tumor masses. This novel approach not only provides targeted cancer treatment but also allows for simultaneous diagnosis and therapeutic action, making it an advanced strategy in the fight against cancer (Patel et al., 2023a). The future of cancer nanomedicine lies in the development of multimodal nanoplatforms that integrate therapeutic components with imaging, diagnostic, and targeting ligands in a single functionalized NP. To ensure that nanotheranostics continue to progress in the right direction, it is crucial to improve the platforms in a way that optimizes their performance for achieving the best clinical outcomes.

Moreover, nanotheranostics enable targeted therapy by incorporating targeting ligands on NP surfaces. These ligands specifically bind to tumor cells or molecular targets, increasing the accumulation of the drug at the tumor site while minimizing off-target effects; this targeted approach enhances treatment effectiveness and reduces side effects. Additionally, nanotheranostics facilitates combination therapies by integrating multiple therapeutic modalities into a single NP. This approach allows for synergistic effects and helps overcome drug resistance mechanisms, leading to improved therapeutic outcomes.

7.4 THERANOSTICS AS AN EMERGING TECHNOLOGY FOR DRUG-TARGETED CANCER TREATMENT

Theranostics is an emerging field in medicine that combines the principles of therapies and diagnostics to provide personalized and targeted treatments for patients. The word "theranostics" combines the two concepts to emphasize the field's dual function in both the treatment and diagnosis of illnesses. Note that the term theranostics is also written as theragnostics (Chen and Wong, 2014).

The term "theranostic" was likely coined by John Funkhouser in 1998 when he introduced a test to monitor the effectiveness of a novel anticoagulant drug (Kelkar and Reineke, 2011). Theranostics involves the use of NC systems that can simultaneously deliver therapeutic agents and imaging agents to cancer cells. By incorporating imaging agents, such as radioactive isotopes or fluorescent dyes, into the carrier system, physicians can noninvasively visualize the tumor and track the distribution of the therapeutic agent in real time. This imaging capability enables the selection of patients who are most likely to benefit from the treatment and allows for monitoring the response to therapy. The therapeutic component of theranostics involves the delivery of targeted drugs to cancer cells while minimizing damage to healthy tissues. This targeted approach enhances the efficacy of the treatment and reduces the risk of systemic toxicity.

Theranostics also offers the potential for personalized medicine by tailoring treatment strategies to individual patients. By analyzing the genetic and molecular characteristics of a tumor, clinicians can identify specific targets for therapy and then use theranostics to deliver drugs that are specifically designed to act on those targets. This personalized approach maximizes the likelihood of treatment success and minimizes unnecessary side effects. While theranostics is still in the early stages of development, it holds immense potential for revolutionizing cancer treatment. By integrating diagnostics and therapeutics into a single platform, theranostics offers a new paradigm for targeted drug delivery and precision medicine. Ongoing research and clinical trials are focused on optimizing the design and delivery of theranostic agents, expanding the range of therapeutic options, and exploring their application across various cancer types. With further advancements, theranostics has the potential to significantly improve patient outcomes and transform the landscape of cancer treatment.

7.5 MECHANISMS OF DRUG TARGETING BY THERANOSTICS

The mechanisms of drug targeting by theranostics involve the use of NC systems to deliver therapeutic agents specifically to cancer cells. These carrier systems are designed to have specific properties that enable them to recognize and bind to cancer cells, allowing for targeted drug delivery. We next describe the steps in the process, which are also depicted in Figure 7.2.

First is the design and synthesis of theranostic agents. These agents and NC systems in general are carefully engineered to possess specific properties that include size, shape, surface charge, and functional groups that facilitate targeted drug delivery. Additionally, imaging agents such as radioactive isotopes or fluorescent dyes are incorporated into the carrier system for diagnostic purposes.

Next is target recognition and binding. Theranostic agents are designed to recognize and bind to specific molecular markers or receptors that are overexpressed on the surface of cancer cells. These markers can be identified through various diagnostic techniques including genetic analysis, proteomic profiling, or imaging studies. The binding can be achieved through specific interactions such as between ligand and receptor or between antibody and antigen.

Agents then require internalization and drug release. Once theranostic agents bind to cancer cells, they are internalized through processes like receptor-mediated endocytosis. The therapeutic agents release carrier systems within the cancer cells that include chemotherapy drugs, immunotherapeutic agents, gene therapy vectors, and other targeted therapies. The release of the therapeutic agents can be triggered by various stimuli such as changes in pH, temperature, or enzymatic activity within the cancer cells.

After administration, the released therapeutic agents exert their intended effects on the cancer cells, leading to apoptosis, inhibition of tumor growth, or immune activation against the tumor. Simultaneously, the imaging agents within the theranostic agents help in noninvasive tumor visualization and monitoring. This imaging component enables medical practitioners to assess the efficacy of the treatment, monitor how therapeutic chemicals are disseminated, and make well-informed treatment decisions.

FIGURE 7.2 The mechanisms of theranostics.

The combination of drug targeting and imaging capabilities in theranostics offers a powerful approach for personalized and precise cancer treatment. Theranostics has the potential to improve therapy efficacy while decreasing adverse effects and improving patient outcomes by directly delivering therapeutic agents to cancer cells while causing minimal damage to healthy tissues. Ongoing research is aimed at improving the design and functionality of theranostic agents in order to maximize drug targeting and therapeutic advantages.

7.6 DIAGNOSTIC AND THERAPEUTIC APPLICATIONS OF NANOTHERANOSTICS PLATFORMS

Figure 7.3 illustrates the potential of cancer nanotheranostics, depicting the leveraging of light (photodynamic therapy [PDT], photothermal therapy [PTT]), magnetism (MRI), and sound (ultrasound) for effective cancer diagnosis and treatment (Chen and Wong, 2014).

7.6.1 Diagnostic Applications of Nanotheranostics

Scientists are actively exploring different imaging agents for the purpose of visualizing internal organs and studying drug distribution within the body. Some diagnostic techniques are MRI, sonography, CT, X-ray scatter imaging, optical imaging, PET, SPECT, and multimodal imaging, which we describe in more detail.

7.6.1.1 MRI

MRI is a noninvasive imaging method that uses radio waves and magnetic fields, but it suffers from low sensitivity and lengthy imaging times. To address these limitations, contrast agents are employed to enhance the MRI signal. Ferromagnetic and paramagnetic materials such as superparamagnetic iron oxide nanoparticles (SPIONs) possess a core with a radius of 5–15 nm and a total radius, including a shell and water coat, ranging from 20 to 150 nm (Frantellizzi et al., 2020).

Both SPIONs and gadolinium (Gd) are commonly employed as contrast agents in MRI. Gd works as a T1 contrast agent, amplifying signals in fatty tissue, and SPIONs work as a T2 contrast agent, enhancing signals in water-based tissues. Different contrast agents based on Gd and iron oxide NPs have been created for T1/T2-weighted imaging to increase the sensitivity of MRI. Due to its short blood circulation duration and possible toxicity, which might result in nephrogenic systemic fibrosis, the usage of Gd has been restricted. However, because of their susceptibility to and tendency for dark signals, artefacts in MRI and SPIONs, particularly iron oxide NPs, have proved to be unusable in therapeutic settings. Because of these constraints, it is difficult to discern between small, early-stage tumors and hypointense regions. Because they produce clear signals and have strong biocompatibility, manganese-based agents are viewed as attractive options (Cai et al., 2019).

In terms of targeting specific molecular markers, NCs have been explored for their ability to enhance imaging sensitivity in MRI. For example, SPIONs conjugated with antibodies targeting proteins like Enolase 1 and epidermal growth factor receptor (EGFR) (Salehnia et al., 2019) have been investigated for early diagnosis and evaluation of pancreatic ductal adenocarcinomas and abnormal EGFR expression levels, respectively (Chen et al., 2019).

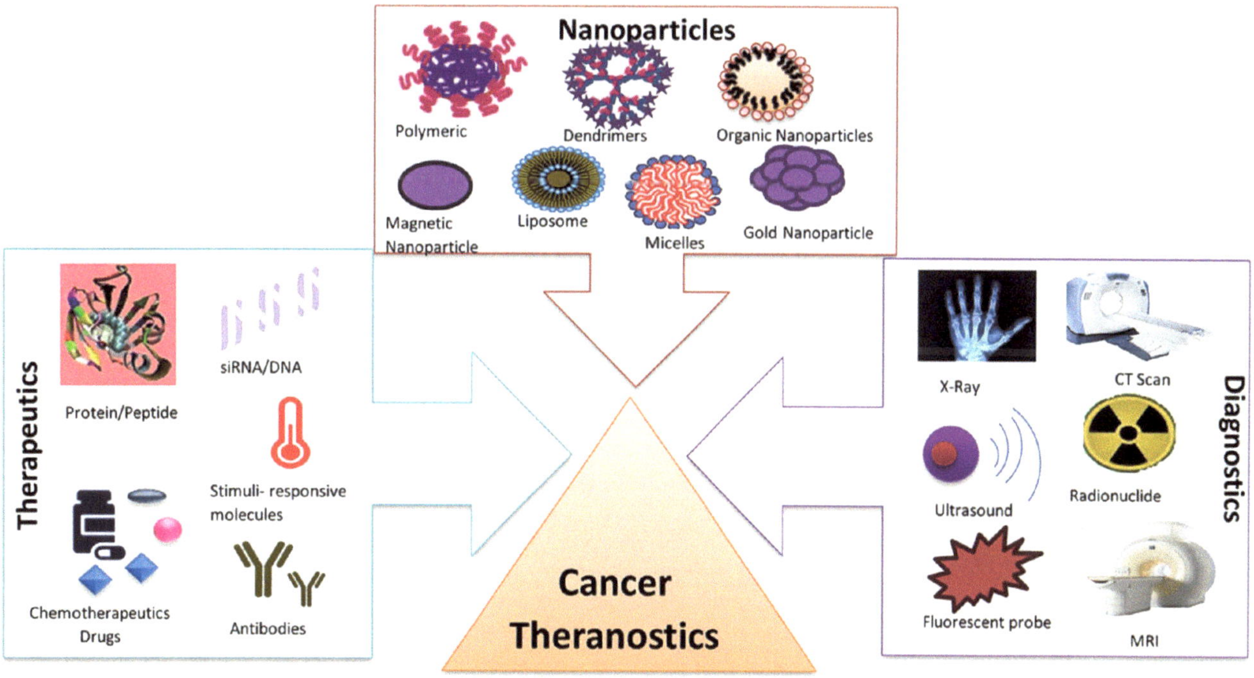

FIGURE 7.3 An overview of cancer nanotheranostics.

7.6.1.2 Sonography

Sonography is a real-time imaging technique that uses ultrasound to provide excellent contrast for soft tissues. Sonography uses a variety of contrast agents, including solid NPs, perfluorocarbon nanodroplets, and non-microbubble agents like echogenic liposomes as well as microbubbles filled with gas. Microbubbles consist of a gaseous core surrounded by a shell made of albumin, lipids, or synthetic polymers, offering high sensitivity and specificity. The highly sensitive of B-mode ultrasound imaging for diagnosing malignant tumors can be limited despite its great sensitivity. To address this, the combination of targeted microbubbles with ultrasound imaging can enhance specificity (Berg et al., 2009). Cavitation, which involves the formation, oscillation, and collapse of gas bubbles, can be leveraged to improve the accumulation of NPs and the distribution of drugs within tumors. Ultrasound-guided delivery of NPs, such as PEG-PDLA (polyethylene glycol-poly(D-lactide)) NPs, has been utilized to overcome solubility barriers and improve the targeted delivery of drugs such as paclitaxel to tumor sites (Rapoport et al., 2013).

7.6.1.3 Computed Tomography

CT imaging is highly regarded for its ability to swiftly collect high-resolution and penetrate deep into tissues, allowing for the reconstruction of 3D images. Nonetheless, it encounters drawbacks such as a low signal-to-noise ratio and difficulty in distinguishing soft tissues with similar densities. To overcome these limitations, CT contrast media are employed to improve the contrast of soft tissues. Typically, elements such as tungsten (W), barium (Ba), and iodine (I) are employed as contrast agents in CT (Carlsson, 1999).

In recent years, nanoscale metal organic frameworks and gold nanoparticles (AuNPs) have gained popularity as CT contrast media. Ultrasmall AuNPs showed much greater sensitivity for hepatocellular carcinoma (HCC) detection (Rand et al., 2011; Chavda et al., 2023). The use of NPs allows for using lower doses of CT contrast agents than molecular contrast agents (Keshtkar et al., 2020). Consequently, NPs exhibit lower number density, viscosity, and osmolality at the same concentration, resulting in shorter imaging times and reduced renal toxicity when administered as CT contrast agents.

7.6.1.4 X-Ray Scatter Imaging

X-ray scatter imaging relies on detecting variations in X-ray penetration intensity, primarily caused by differences in tissue density and thickness. This imaging modality has garnered significant attention due to its ability to exploit these differences. In the context of contrast media, AuNPs have emerged as promising candidates for X-ray scatter imaging. AuNPs are a great option for this imaging approach because they have a higher atomic number than iodine NPs and are less toxic (Kojima et al., 2010).

Researchers utilized polyelectrolyte-coated AuNPs as contrast agents in X-ray scatter imaging of HCC cell pellets. They compared the performance of 10 nm and 50 nm AuNPs and demonstrated that ultrasmall AuNPs offered higher sensitivity as contrast agents in X-ray scatter imaging. Additionally, they showed that ultrasmall AuNPs outperformed traditional X-ray diffraction in the diagnosis of small clusters of HCC (Rand et al., 2011).

7.6.1.5 Optical Imaging

Optical imaging is a noninvasive method that utilizes different colors of light to visualize and measure various characteristics of organs. However, it has limitations in terms of penetration depth. Near-infrared (NIR) light, ranging from 650 to 900 nm, has been introduced as a solution to overcome this limitation. Several molecular imaging modalities have emerged in optical imaging, including Raman spectroscopy, endoscopy, photoacoustic imaging (PAI), optical coherence tomography, super-resolution microscopy, and diffuse optical tomography. NPs have been increasingly explored in these modalities.

Zhang et al. (2019) investigated the use of UiO-66 (Universiteteti Oslo), a metal organic framework composed of [Zr6O4(OH)4] with 1,4-benzodicarboxylic acid, within a 1,2-dioleoyl-sn-glycero-3-phosphate (DOPA) lipid bilayer (UiO-66-DOPA-LB), as a contrast agent. The results showed greater accumulation in tumors, faster blood flow, and improved stability. The researchers were able to image breast cancer tumors and detect tumors early by labeling UiO66-DOPA-LB with a NIR dye, IR dye 800 nm CW.

7.6.1.6 Positron Imaging Tomography

PET imaging exhibits exceptional sensitivity and is not limited by depth penetration, although it lacks precise anatomical information required for the accurate detection of molecular events. To overcome this limitation, PET is commonly integrated with other imaging modalities to enhance its diagnostic capabilities such as CT and MRI. By utilizing multimodality imaging techniques like PET/CT and PET/MRI, it becomes possible to detect lesions or tumors from both functional and anatomical perspectives (Xiao et al., 2012).

In a study conducted in 2019, SPIONs were coated with N,N,N-trimethyl chitosan and targeted them with bombesin ligands. The NPs were further labeled with 68Ga using S-2–(4-isothiocyanato benzyl)-1,4,7,10-tetraazacyclododecane tetraacetic acid (DOTA) as chelators. The results of the study demonstrated that these NPs were suitable for PET/MRI imaging for the diagnosis of lung, breast, and prostate cancers (Hajiramezanali et al., 2019). Combining PET and MRI produced both functional and anatomical information for improved cancer diagnosis.

7.6.1.7 Single-Photon Emission Computed Tomography

SPECT imaging is a method that depends on detecting gamma (γ) rays and typically offers lower sensitivity and spatial resolution than PET; however, in preclinical examinations, micro-SPECT achieved higher spatial resolution than PET. In SPECT imaging, 99mTc is commonly used as a valuable radionuclide for this purpose (Franc et al., 2020).

Zhao et al. (2016) investigated AuNPs doped with 199Au using SPECT and demonstrated that when AuNPs were doped with 199Au and D-Ala1-peptide T-amide was attached to them, these NPs could function as targeted nanoprobes for C-C chemokine receptor 5. This nanoprobe exhibited high sensitivity and specificity in detecting both triple-negative breast cancer and its metastasis. The study highlighted the potential of using doped AuNPs in SPECT imaging for the detection of specific cancer types and their spread.

7.6.1.8 Multimodal Imaging

Multimodal imaging approaches have gained considerable attention as they address the limitations of individual imaging modalities. Researchers have explored the use of core–shell NPs to enhance disease diagnosis and imaging.

In one study, SPIONs were coated in PEG and labeled with cyanine 7.5 NHS ester to improve circulation and minimize opsonization. The labeled SPIONs were injected into mice with implanted U87 MG cells, and subsequent MRI and fluorescence imaging were performed. The results showed a strong MRI signal in the brain tumor region, and fluorescence imaging revealed higher fluorescence intensity in the brains of mice that received the labeled SPIONs than in the control group (Liu et al., 2019). This study demonstrated the effectiveness of labeled SPIONs for imaging brain tumors.

7.6.2 Therapeutic Applications of Nanotheranostics

Theranostics includes traditional cancer therapies such as chemotherapy, radiation, and surgery while attempting to reduce their side effects, risks trauma, and difficulties. Nanotechnology enables theranostics to diversify therapeutic techniques such as chemotherapy, PDT, PTT, imaging-guided therapy, and radiation therapy.

7.6.2.1 Chemotherapy

The use of anticancer drugs has significantly improved the survival rates of cancer patients. Theranostics extensively utilizes various clinical anticancer drugs that can be classified into six types based on their structure and source: alkylating agents, antimetabolites, antitumor antibiotics, phytogenic anticarcinogens, hormones, and miscellaneous anticancer drugs. Some cancer nanomedicines, such as DOXil, Thermo DOX, and Myocet, have received FDA approval for specific cancer therapies (Rodríguez et al., 2022). Prodrugs are also employed in theranostic systems to minimize drug toxicity and selectively target tumor sites. For example, platinum (IV) prodrugs can be entered into cells by endocytosis and activated to highly toxic platinum (II) complexes using NIR light converted from up-conversion NPs (Dai et al., 2013). To enhance therapeutic efficacy, theranostics relies on drug delivery systems with high drug payloads. Multidrug resistance is another challenge addressed by theranostics. Strategies include combining anticancer drugs with P-glycoprotein -reversing agents or using synergistic combinations of multiple anticancer drugs loaded on NCs, resulting in enhanced anticancer effects.

7.6.2.2 PDT

PDT involves utilizing light to remotely activate drug delivery and has gained significant attention in cancer theranostics. It offers advantages in terms of precise spatial and temporal control over the release of drug compounds. Cancer theranostic platforms encompass various approaches, involving PDT, PTT, and phototriggered chemotherapy. PDT involves the use of a photosensitizer (PS) and light of specific wavelengths. When activated by light in the presence of oxygen, the PS generates cytotoxic reactive oxygen species leading to tumor cell and tissue destruction (Jain et al., 2021). PDT provides targeted treatment with minimal damage to surrounding healthy tissue, and NPs are incorporated into PDT to address the problem of photobleaching, where the photoactive agents lose their effectiveness over time (Jaque et al., 2014).

The most commonly used agents for PDT are members of the tetrapyrrole family such as porphyrins and phthalocyanines as well as other dyes including phenothiazines, squaraines, indocyanine green, and phenalenones (Abrahamse and Hamblin, 2016). Inorganic nanotheranostics like graphene quantum dots can also be employed for PDT (Zhang et al., 2016). Dendrimers, which are delivery systems for various theranostic agents, have been utilized in PDT research; researchers created a new dendrimer-based delivery system to encapsulate a phthalocyanine derivative that enabled the effective delivery of multimodal imaging agents and enhanced PDT efficacy. This approach led to higher cytotoxicity, sustained PDT effect for 24 hours, and efficient imaging agent uptake in tumor tissues. Thus, PDT offers a promising approach to selectively target and destroy tumor cells using tumor-specific agents and external light stimulation, and integrating NPs and dendrimers enhances its therapeutic potential by improving delivery and efficacy.

7.6.2.3 PTT

PTT involves the use of contrast agents or nanotheranostic agents capable of transforming light energy into thermal energy to result in tumor cellular necrosis through hyperthermia induction. PTT utilizes electromagnetic radiation, typically in the NIR range, to activate these photosensitive agents and initiate the release of heat. When these agents are exposed to external irradiation, their electrons absorb photons and undergo excitation, transitioning to an unstable state. Upon returning to the ground state, they release energy in the form of heat, and this heat induces thermal damage to cancerous cells (Nomura et al., 2020). This thermal effect can also cause biological changes in the tissue, ranging from protein structural alterations to tissue carbonization.

PTT selectively targets cancer tissues by utilizing agents that accumulate in cancer cells and protecting healthy cells from damage. When these photosensitive agents are irradiated, the temperature in the tumor rises to around 40 to 60 °C, causing coagulation necrosis (West et al., 2019). PTT can employ organic, inorganic, or hybrid substances. Both organic agents including porphyrin-based agents and cyanine-based agents are used as theranostics for PTT, as are other organic agents such as Prussian blue, gallic acid, and polyaniline, which also serve as light-triggered nanotheranostics agents for PTT (Zhang et al., 2016).

Inorganic substances are frequently used as well, including Au, iron oxide (Fe_2O_3), graphene, silver (Ag), and carbon nanotubes due to their optical and thermal properties. However, hybrid nanoparticulate agents are preferable to inorganic substances because the latter can have problems with bioavailability and toxicity (Chitgupi et al., 2017). Hybrid nanotheranostics are preferred due to their ability to surpass limitations, good distribution and elimination profiles, and low toxicity. Additionally, hybrid agents facilitate the synergetic effects of PTT and PDT, resulting in increased tumor cell death and improved sensitivity of tumors to other treatment modalities (Gong et al., 2014). When placed on the surfaces of biopolymers, magnetic NPs like iron oxide or certain carbohydrates enhance the hyperthermia effect caused by PTT. Ovarian tumor cells have been demonstrated to be destroyed by polypropyleneimine dendrimers due to the synergistic interaction between PDT and PTT (Taratula et al., 2015).

7.6.2.4 Imaging-Guided Therapy

Theranostic NPs have gained significant attention in recent years for imaging-guided therapy. These NPs have the ability to monitor the pharmacokinetics process, guide treatments, and analyze therapeutics outcomes. They offer a range of applications, such as visualizing and quantifying drug delivery systems for purposes like studying the pharmacokinetics and biodistribution of NCs, drug release processes, and metabolic responses. In this section, we describe common techniques for using NPs to enhance detectability in various imaging modalities (Dennahy et al., 2022).

SPIONs primarily composed of iron oxide, known as magnetite, are commonly used either alone or integrated into NP matrices. Notably, iron oxide NPs, like Feraheme, are clinically approved and extensively researched. They are available in a variety of sizes and have variable intrinsic relaxivities. Si et al. (2019) employed iron oxide NPs to create a novel contrast agent for MRI. Moreover, due to their small size, SPIONs have been selected as cargo in nanodrug delivery systems; until IONs serve as therapeutic agent carriers, this approach holds great promise for developing improved and versatile contrast agents.

7.6.2.5 Radiation Therapy

The traditional treatment used 2D fluoroscopic views, employing two or three radiation beams from different angles to irradiate the target tissue widely. In modern treatment, multiple radiation beams approach the tumor from various angles and planes, precisely converging on the target to administer the planned dose while minimizing radiation exposure to surrounding healthy tissues (Alfouzan, 2021).

Researchers investigated a nanodrug system that comprised poly(allylamine)-citraconic anhydride/doxorubicin NPs. They used fluorescence lifetime imaging microscopy (FLIM) to observe dynamic changes in DOX fluorescence lifetime within intracellular environments. Results showed that DOX released from the NPs upon acidic pH stimulation in endosomes and lysosomes displayed varying fluorescence lifetimes in different cellular compartments. Phasor-FLIM analysis successfully separated subcellular compartments based on the difference in DOX fluorescence lifetime, indicating the release and distribution of DOX within cells (Zhou et al., 2018). An injectable nanoparticle generator was loaded with polymeric doxorubicin in 2016.

Researchers employed a hollow-structured CuS-Cu2S-Au nanohybrid to enhance the efficacy of PTT and enable photo-switchable targeting for theranostic applications. They found improved treatment efficacy with DOX loaded onto the nanohybrid. Combining PDT with chemotherapy increased the effectiveness of tumor treatment, and this combination approach showed promise for enhancing the therapeutic outcomes in cancer treatment. By incorporating a plasmonic metal core/semiconductor shell, they observed improved photothermal conversion with 808 nm laser radiation. This NP offered significant advantages such as large cavities and a mesoporous shell. These features were found to be beneficial for loading and releasing drugs, thus providing potential applications in drug delivery systems (Deng et al., 2017). That is, NPs provide a platform for targeted drug delivery in cancer therapy, allowing for site-specific drug release, improved therapeutic outcomes, decreased side effects, and improved drug stability. These improvements hold great potential for the development of more efficient and personalized treatments of cancer.

7.7 NANOTHERANOSTICS AS THE FUTURE OF CANCER TREATMENT

The integration of diagnostic and therapeutic functions within a single agent using NCs is gaining significance in modern medicine. NCs possess the ability to penetrate deep into tumors, enabling targeted delivery of therapeutic or diagnostic agents. This has led to high expectations for nanoparticulate agents to serve as optimal theranostic agents. Nanoparticulate agents have potential therapeutic applications in gene silencing, targeted drug delivery, tumor-specific targeting, and phototherapy, among others (Chavda, 2023).

The versatile theranostic concept spans various scientific domains, and its categorization includes nanotheranostics, optotheranostics, magnetotheranostics, and radiotheranostics approaches employing nanoparticles, optical probes, magnetic particles, and radionuclides, respectively (Lee and Li, 2011). In the context of treating cancer, molecular imaging has gained recognition because of its diagnostic contrast capabilities; various imaging modalities have their strengths and limitations. For example, CT imaging provides high resolution and facilitates 3D reconstruction but lacks sufficient contrast for effectively distinguishing soft tissues. On the other hand, magnetic NCs are used as contrast agents in MRI, but the inherent dark signals of contrast agents can sometimes be mistaken for other areas with low signal intensity. To address these challenges, multimodal imaging has emerged as a solution, allowing for the complementary use of different imaging techniques to overcome these limitations.

Through active preclinical research, several new molecular targets have also emerged with strong potential for translation into clinical theranostics, for instance, fibroblast activation protein inhibitors (FAPI): Cancer-associated fibroblasts constitute 90% of the tumor mass and exhibit abundant expression of membrane-bound glycoprotein FAP in epithelial carcinomas (Scanlan et al., 1994). A first-in-human study using 90Y-FAPI-04 for a patient with metastasized breast cancer demonstrated significant reduction in pain medication despite the rapid clearance observed despite high tumor uptake. Neurotensin is a neuropeptide found in the gastrointestinal and central nervous systems as well as the myocardium that shows overexpression of neurotensin receptor 1 in various cancers (breast, pancreatic, etc.). A first-in-human study of 177Lu-3BP-227 treatment in six pancreatic adenocarcinoma patients demonstrated feasibility, tolerability, and positive outcomes including partial response and symptom improvement, with one patient surviving 11 months post-treatment (Baum et al., 2018). Immuno-agents like CXCR-4 were significantly overexpressed in cancers such as leukemia, lymphoma, and multiple myeloma, demonstrating promising results with 68Ga-labeled pentixafor for imaging (Mousavi, 2020). A first-in-human study using 177Lu- and 90Y-labeled pentixather showed favorable responses in treating relapsed multiple myeloma through CXCR4-directed therapy (Herrmann et al., 2016). The current focus of theranostics is on disease stabilization and enhanced quality of life. The upcoming goals of theranostics encompass not only discovering novel molecular targets but also enhancing existing treatments to provide better therapeutic options.

7.8 APPLICATIONS OF THERANOSTICS AND NANOTHERANOSTICS IN CANCER TREATMENT IN THE LITERATURE

Researchers by now have used a variety of different nanotheranostics in treating various cancers such as breast, lung, brain, and prostate. Table 7.1 depicts applications of theranostics and nanotheranostics in the treatment of cancer.

TABLE 7.1
Theranostics and Nanotheranostics in Cancer Treatment

Diagnostic	Targeting Strategy	Consequences (Application)	Ref.
Breast Cancer			
A NIR dye (IR780 iodide) (pre/intraoperative image-guided)	Liposome loaded with perfluorooctyl bromide	Liposome NPs were developed with dye and targeting substance. NP-IR780 showed remarkable tumor-targeting capabilities and improved the symptoms of breast cancer.	(Song et al., 2019)
GQDs	Biosurfactant conjugates with FA using carbodiimide	Conjugation of GQDs with biosurfactant reduced the possibility of MCF-7 breast cancer cells.	(Bansal et al., 2019)
GQDs NIR (CT, MRI dual-modal imaging)	Iron oxide-bismuth oxide GQDs (PTT)	NIR laser irradiation induced programmed cell death in MCF-7 breast cancer cells.	(Badrigilan et al., 2019)
Fluorescence imaging	Targeted theranostics human RBC membrane-coated polymeric NPs (TT-RBC-NPs) delivered DOX.	TT-RBC-NPs demonstrated sustained drug release (mostly within 5 days), increased cytotoxicity against EpCAM+ MCF-7 breast cancer, improved cellular uptake, reduced breast cancer symptoms.	(Marshall et al., 2022)
FDOT and NIR irradiation	ICG & CPT co-loaded perfluorocarbon double-layer nanocomposite (ICPNC)	In MDA-MB-231 cancer-bearing mice, ICPNC was combined with 1-min NIR irradiation. After a 21-day treatment, the tumor size decreased by 80% without causing any harm to organs or systemic toxicity.	(Lee et al., 2021)
Caspase 3-responsive imaging probe with quencher and fluorophore	SPNs (supramolecular polysaccharide nanotheranostics) formed by polysaccharide-constructs and kinase inhibitor-β-cyclodextrin conjugates	SPNs demonstrated improved cytotoxicity and inhibited kinase signaling and tumor growth in mice on fluorescence imaging. This adaptable design had the ability to deliver small molecule inhibitors to tumors while also monitoring their activity.	(Deshpande et al., 2020)
MnO NPs (fluorescent agent)	lyotropic LCNs	LCNs inhibited 96.5% tumor growth, highlighting the potential of novel, smart, endogenic stimuli-sensitive nanosystems for advanced cancer nanotheranostics. These nanosystems eliminate the need for external stimuli to modulate the tumor microenvironment.	(Urandur et al., 2020)

(Continued)

TABLE 7.1
Theranostics and Nanotheranostics in Cancer Treatment (Continued)

Diagnostic	Targeting Strategy	Consequences (Application)	Ref.
Fluorescence and SERS dual modalities	Plasmonic Au-graphene hybrids with FA ligated chitosan modified photosensitizer and DOX.	The nanoprobe's strong light–matter interactions allowed for improved therapeutic and imaging effects, targeted chemotherapy, and tumor regression without causing harm to other organs.	(Jibin et al., 2020)
Dual-mode imaging (NIR and MRI)	Targeted multi-responsive (magnetic hyperthermia and pH) nano-micelles incorporated with NIR fluorescent SPIONs	A self-assembled nanotheranostics technology for dual-mode imaging and dual toxicity that successfully targeted and killed CTCs in the bloodstream. *In vitro* and *in vivo* studies revealed that it had the potential to reduce CTCs and distant organ metastasis, minimizing secondary tumor growth at distant sites.	(Dhandapani et al., 2023)
MRI	Manganese-based and hypericin-loaded polyester dendrimer NPs	NPs administering hypericin provided MRI-guided PDT suppression of breast tumors and improved ROS generation by alleviating the Mn2+-mediated hypoxic microenvironment.	(Zhou et al., 2023a)
Highly fluorescent QDs	Gelatin/chondroitin layer-by-layer assembled nanocapsules co-loaded with CXB & RAP	The NCs fluoresced brightly, allowing them to be easily internalized by cancer cells. They showed increased *in vitro* cytotoxicity against breast cancer cells and superior antitumor effectiveness *in vivo* as well as a nonimmunogenic response.	(AbdElhamid et al., 2018)
PAI/MRI dual-model imaging	Her2-DOX-SPIOs- (PLGA)-Au NPs (Her2-DSG NPs) emulsion	Targeted drug delivery with photothermal-responsive drug release demonstrated good biosafety *in vivo*. Her2-targeted NPs displayed a strong antitumor effect in the BT474 nude mouse model.	(Zheng et al., 2020)
GQDs, NIR	BNN6-loaded AuS QD nanodevice for combination photothermal/NO therapy	PTT evidenced significant tissue thermal ablation and inhibited tumor growth. The PAI-guided NIR-II triggered thermal/NO combination therapy effectively suppressed tumor growth.	(Vayalakkara et al., 2022)
MRI	Loaded with DOX & Fe_3O_4 caramelized nanospheres (Fe_3O_4/DOX-CNSs)	Fe_3O_4/DOX-CNSs with NIR had the greatest tumor suppression *in vitro* and *in vivo* due to pH/heat-induced DOX release and showed effective tumor inhibition. They also exhibited enhanced T2 MRI contrast for real-time imaging during treatment.	(Wang et al., 2023)
CdSe/ZnS quantum dots	DOX loaded pH-responsive stealth niosomes	A drug-loaded nanoplatform was created using thin-film hydration, and the candidate for future cancer theranostic due to bioactivity assessments showed nearly 55% cytotoxicity via induced cell apoptosis.	(Saharkhiz et al., 2023)
CRISPR	CRISPR/Cas9 and amp system, CAR-T cell	Conjunction with nanomaterials to target AKT improved cancer diagnosis. Incorporating modified CAR-T cells into multifunctional nanodelivery systems inhibited aggressive TNBC & CSCs in a tailored manner.	(Nayak et al., 2023)
Lung Cancer			
DNA biodots (BDs)	Loaded cetuximab conjugated and etoposide in liposomes	BDs showed improved delivery of theranostic liposomes to NSCLC cells over non-targeted and control formulations and suppressed the growth of cancer cells. They showed promise as carriers for lung theranostics due to their nanosize.	(Jha et al., 2020)
Superparamagnetic nanotheranostic agent (MRI)	Loaded with erlotinib against EGFR-wt	The nanotheranostic agent co-delivered erlotinib and bevacizumab to EGFR-wt tumors and markedly inhibited tumor growth.	(Wang et al., 2022)
NIRF (biological window II, PAI, MRI)	CuO/Cu_2O nanocube-based theranostic nanomedicine	The nanomaterial sensitized the formation of singlet oxygen for PDT and absorbed long NIR light for PTT. This effectively destroyed multidrug-resistant lung tumors.	(Shanmugam et al., 2021)

(Continued)

TABLE 7.1
(Continued)

Diagnostic	Targeting Strategy	Consequences (Application)	Ref.
AuNP imaging	AuNPs	MSC-mediated delivery of AuNPs showed substantially increased tumor retention and penetration, a 2.4- to 9.3-fold increase over plain AuNPs via the EPR effect, a promising result.	(Xu et al., 2023)
H-dots	Loaded with gefitinib and genistein	H-dot complexes offer enhanced therapeutic efficacy with fewer adverse effects and low drug resistance risk; they showed promising suppression of lung cancer.	(Yin et al., 2022)
NIR	PEGylated ^{131}I-labeled CuS nanotheranostics	CuS-^{131}I-PEG nanotheranostics were injected into tumors, followed by NIR laser irradiation to induce mild hyperthermia. The multiple mild hyperthermia strategy combined with photothermal brachytherapy completely inhibited tumor growth.	(Cui et al., 2022)
MRI	SDP-GEM/PEI-PEG-anti EGFR	The multifunctional nanotheranostic was properly administered to tumor locations within 3 hours *in vivo* and resulted in extended lifetime, reduced tumor volume, and suppressed tumor growth.	(Tang et al., 2020)
BPsQDs, fluorescence imaging	Co-loaded onto PLGA, conjugation with ginsenoside Rg3 (G-Rg3)	BPs/G-Rg3-PLGA exhibited excellent controllable PTT with near-light irradiation, leading to enhanced photothermal tumor ablation. Rapid G-Rg3 release upon heating induced apoptosis, resulting in reduced lung cancer metastasis.	(Xiong et al., 2021)
Colorectal Cancer			
Fluorescence imaging	Lipid micellar NPs with co-loaded PTX QDs and EGFR antibodies/ EGFR aptamers conjugation	The anti-EGFR immuno-PTX-QDMs and anti-EGFR aptamo-PTX-QDMs hold great potential as targeted theranostic delivery systems for cancer treatment in clinical settings.	(Kang et al., 2018)
MRI-guided Gd neutron capture therapy	Hybridized PEGpolyanion and gadolinium-diethylenetriaminepentaacetic acid with calcium phosphate micelles.	After subjecting colon carcinoma cells to thermal neutron irradiation, around 60% of these cells were selectively eliminated, while the viability of neighboring normal cells remained unaffected. The localization of micelles was visualized using MRI.	(Mi et al., 2015)
NIR imaging	CatCry-AgNP-DOX	In murine models, CatCry-AgNP-DOX displayed effective PTT and improved chemotherapy in colon cancer cells, eliminating tumors with few adverse effects.	(Zhou et al., 2022)
MRI	Iron oxide-hydroxide nanospindles (FeOOH NSs)	FeOOH NSs scavenged H_2S, hindering CT26 colon cancer growth. Overexpressed H_2S created FeS for PTT and Fe2+-mediated ferroptosis. FeOOH NSs were successfully employed as a theranostic for specifically treating colon cancer.	(Li et al., 2020)
GoQDs	GoQDs-CH-PEG nanoconjugate	The MUC-1 receptor-selective binding of the aptamer triggered strand separation in acidic conditions, leading to sustained drug release and photoluminescence recovery. Suppression of MUC-1-overexpressed tumor cells over negative controls.	(Esmaeili et al., 2022)
Brain Cancer			
Photoluminescent nanoprobes, fluorescent inorganic semiconductor (Ag-In-S, AIS)	Nanostructures combining CMC, pro-apoptotic peptide (KLA), and cell-penetrating moiety (cysteine)	Cysteine-bearing CMC conjugates with pro-apoptotic KLA peptides showed significant killing activity against GBM cells *in vitro*. Importantly, these novel nanohybrids demonstrated higher lethality to U-^{87} MG cancer cells than DOX.	(Mansur et al., 2022)
AuNPs	Liposomes based on gold, encapsulating DCX and glutathione-reduced AuNOs	*In vivo* brain distribution study revealed 70% sustained drug release over 72 hrs. The delivery of DCX into the brain by targeted liposomes was higher than that of the marketed formulation.	(Sonkar et al., 2021)

(Continued)

TABLE 7.1
Theranostics and Nanotheranostics in Cancer Treatment (Continued)

Diagnostic	Targeting Strategy	Consequences (Application)	Ref.
MRI	Ti-FeAu core-shell NPs conjugated with angiopep-2	Superparamagnetic Ti-FeAu core–shell NPs enhanced negative contrast in glioma and achieved a 12 °C temperature increase with magnetic stimulation, suggesting applications in MRI and hyperthermia-based cancer therapy. Angiopep-2 decoration led to selective glioma targeting, resulting in 85% improved cytotoxicity via hyperthermia.	(Thirumurugan et al., 2022)
Photosensitizer, IR-780 (NIR light), fluorescence imaging	Liposomal IR-780 by U87MG (PTT & PDT)	Loaded liposomes and NIR laser treatment induced HSP70 overexpression and intracellular ROS generation in U87MG cells, subsequently increasing cytotoxicity and apoptosis. Convection-enhanced delivery delivered stable liposomal IR-780 specifically to brain tumors, enhancing glioma treatment efficacy through fluorescence imaging and diagnostic tools.	(Lu et al., 2021)
Neodymium magnets	MDT of magnetized temozolomide	In contrast with the other treatment groups, this intervention notably decreased tumor growth and significantly extended mouse survival to 48 days.	(Patel et al., 2023b)
CT/MRI	Encapsulated AuNPs and SPION with polyethylene glycol and polycaprolactone (PEG-PCL) micelles	Under radiation exposure, the AuNPs demonstrated anticancer effects on glioblastoma multiforme by causing DNA impairment. MRI was used to visualize micelle localization.	(Sun et al., 2016)
Prostate Cancer			
Fluorescence imaging and T_2-weighted MRI (T_2-W MRI)	PSMA-targeted magnetic NPs (MNPs)	After PDT, T2-W MRI showed an 8-fold increase in PSMA-targeted MNPs delivery in pretreated tumors over untreated ones, along with enhanced peripheral intratumor delivery.	(Ngen et al., 2021)
Radiolabeled with 99mTc radionuclide	PSMA ligand (PSMA), PEG2000, and palmitate encapsulated liposomes, co-loaded with DOX	99mTc-labeled P3-liposomes showed cytotoxic activity against LNCaP prostate cancer cells through PSMA targeted and thereby suppressed the tumor.	(Yari et al., 2019)
TME	ZnS-TA, utilizing TME-responsive Zn2+-interference and H_2S gas synergistic therapy	This combinatorial action increased cell apoptosis, decreasing tumor growth and limiting prostate cancer cell migration and invasion. Both *in vitro* and *in vivo* studies demonstrated this strategy's extraordinary antitumor performance.	(Zhou et al., 2023b)
Laser scanning microscopy	AuNPs coated with Dendrigraft Poly-L-Lysine and siRNA	The nanocomplexes functionalized with FA to facilitate targeted cellular uptake by prostate cancer and suppressed the growth of cancer.	(Minassian et al., 2023)
Photoluminescent polymer	TFP conjugated it on the surface of iron oxide MNPs	In cell studies, TFP-MNPs demonstrated compatibility with human dermal fibroblasts and prostate epithelial cells. They were effectively internalized by cancer cells, resulting in enhanced tumor cell destruction.	(Pandey et al., 2020)
MRI	Targeted delivery of nanobubbles to tumors using PMSA as a transport mechanism	Active molecular targeting of nanobubbles to PSMA biomarkers extended tumor uptake and retention throughout the tumor volume, though variability was expected due to tumor heterogeneity, which was helpful for tumor suppression.	(Perera et al., 2022)
Ovarian Cancer			
Image-guided therapy	PTXeNMTSLs are liposomes that contain Paclitaxel, NIR QDs, and magnetic nanoparticle	By utilizing an alternating magnetic field to induce heat generation, PTX-NMTSLs demonstrated cytotoxic effects against both MCF-7 cells and SKOV-3 ovarian cancer cells.	(Deng et al., 2020)

(*Continued*)

TABLE 7.1
(Continued)

Diagnostic	Targeting Strategy	Consequences (Application)	Ref.
Gd_2O_3, multimode imaging (FL/MRI/IR)	Hollow mesoporous organosilica-based nanoplatform (HMON)- CuS	The HMON-CuS/Gd_2O_3 nanoplatform was particularly taken up and trapped in endo/lysosomes by SKOV-3 cells. *In vivo* investigations revealed improved FL/MRI/IRT imaging capabilities, allowing for selective and safe phototherapy to limit ovarian cancer growth without systemic toxicity.	(Li et al., 2022)
NIR (PAI) signal	PEG-loaded reduced-GO AuNPs	The rGO-AuNP-PEG nanocomposite exhibited robust SERS signal, NIR-II, PA signal, and excellent photothermal efficacy when subjected to 1061 nm laser irradiation. This exceptional performance resulted from the plasmonic coupling of AuNPs and the enhanced response of rGO and AuNPs and suppression of tumor cells.	(Wang et al., 2020)
NIR (PAI) signal	Gold nanorods (GNRs) with polydopamine (PDA)	GNR-PDA hybrids by coating GNRs with PDA, forming enhanced photo-mediated theranostic agents. *In vitro* experiments showed that the PDA coating strategy improved the PA imaging and PTT through nanoparticles and inhibited tumor cell growth.	(Yim et al., 2022)

Other Cancers	Diagnosis	Therapeutics	Application	Ref.
Cervical adenocarcinoma	Multicolor cell imaging	([1E,6E]-3,5-dioxohepta-1,6-diene-1,7-diyl) bis (2- methoxy-4,1-phenylene) diacetate acetyl curcumin-loaded PLGA shell liposome NPs.	Demonstrated dose-dependent cytotoxic and growth inhibitory effects against HeLa cervical adenocarcinoma cells, without affecting normal fibroblasts.	(Reddy et al., 2019)
Fibrosarcomas	Fluorescence imaging	PTX and CXB encapsulated within a polymeric micelle and with peptide linker (PLGLAG).	The PLGLAG peptide linker was digested by MMP-2 derived from HT-1080 fibrosarcoma cells. This enzymatic digestion led to the liberation of CXB, which subsequently facilitated the targeted delivery of PTX to the tumor site and suppressed the growth of cancer cells.	(Huang et al., 2019)

Abbreviations: NIR: Near-infrared, GQDs: Graphene quantum dots, MCF-7: Michigan Cancer Foundation, RBC: Red blood cells, DOX: Doxorubicin, Go: Graphene oxide, FDOT: Fluorescence with optical tomography, ICG: Indocyanine green, CPT: Camptothecin, AuNPs: Gold nanoparticles, MUC-1: Mucin-1, Apt: Aptamer, PAI: Photoacoustic imaging, CT: Computed tomography, SERS: Surface-enhanced Raman spectroscopy, FA: Folic acid, HA: Hyaluronic acid, CTCs: Circulating tumor cells, MRI: Magnetic resonance Imaging, Mn: Manganese, MnO: Manganese oxide, LCNs: Liquid crystalline nanoparticles, TME: Tumor microenvironment, PDT: Photodynamic therapy, ROS: Reactive oxygen species, QDs: Quantum dots, CXB: Celecoxib RAP: Rapamycin, SPN: Supramolecular polysaccharide nanotheranostics, SPIONs: superparamagnetic iron oxide nanoparticles, PLGA: Poly (D, L-lactide-co-glycolide) acid, siRNA: small interfering ribonucleic acid, FL: Fluorescence, IRT: Infrared, CuS: copper sulfide, Gd_2O_3: Gadolinium(III) oxide, CdSe: Cadmium Selenide, ZnS: Zinc sulfide, CRISPR: Clustered Regularly Interspaced Short Palindromic Repeats, TNBC: Triple-negative breast cancer, CSCs: cancer stem cells, CAR-T cells: Chimeric antigen receptor T, CatCry: Catalase nanocrystals, EGFR: Epidermal growth factor receptor, CH: Chitosan, CD: Carbon Dots, PEG: Polyethylene glycol, CMC: carboxymethylcellulose, GBM: glioblastoma cancer, Ti: Titanium, DCX: Docetaxel MDT: Magnetic drug targeting, PSMA: Prostate-specific membrane antigen, TA: Tannic acid, MMP-2: Matrix metalloproteinase-2, COX-2: cyclooxygenase-2, TFP: thermo-responsive fluorescent polymer, MNPs: Magnetic nanoparticles, PMSA: prostate-specific membrane antigen, BPs: Black phosphorus, NIRF: Triple modal near-infrared fluorescence.

7.9 POTENTIAL OF NANOTHERANOSTICS FOR PRECISION CANCER TREATMENT

In cancer, molecular changes lead to the expression or upregulation of diverse targets suitable for imaging and treatment with nanotheranostics. This nanotheranostic method has gained recognition in personalized medicine because it allows for the effective treatment of advanced tumors with a very few side effects (Langbein et al., 2019).

Precision medicine, also known as personalized medicine, is an up-and-coming approach in modern oncology facilitated by advanced tools like next-generation and RNA sequencing, and it has revolutionized disease management. The identification of molecular modifications that are both prognostic and predictive, such as gene mutations, amplifications, and fusions, has led to significant improvements in localized and metastatic conditions, although challenges arise due to the complex structure of tumor cells, with various molecular modifications and tumor heterogeneity affecting the efficacy of targeted agents (Gambardella et al., 2020). The goal of precision medicine is to eliminate the one-size-fits-all approach, recognizing that individuals respond differently to treatments based on their specific genetic variations and other factors. This information allows physicians to develop targeted therapies and interventions that are tailored to an individual's specific genetic profile.

Precision medicine has the potential to revolutionize healthcare by improving patient outcomes, minimizing adverse effects, and optimizing resource utilization. It can help individuals who face an elevated risk of developing specific diseases, allowing for early interventions and even preventive measures. Additionally, precision medicine can guide the selection of medications and dosages that are most likely to be effective and safe for each patient, reducing trial-and-error approaches and improving treatment response rates. Recent advancements in personalized treatment for solid tumor patients have shown promising results. Offering a molecular-based personalized approach in oncology holds great potential. However, to achieve significant improvements in clinical outcomes, a deeper understanding of molecular biology is essential. To comprehensively understand diseases, integrating multi-omic methods like DNA and RNA alterations, proteomics, and metabolomics will be essential. Utilizing translational studies with liquid biopsy, organoids, or xenografts can help identify resistance mechanisms and propose innovative combination therapies.

However, the widespread implementation of precision medicine faces challenges, including the high cost of genetic testing, the interpretation and integration of complex data, the need for standardized guidelines, and ethical considerations related to patient privacy and data sharing. Urgent further research is required to enhance precision and optimize personalized cancer treatments. Despite these challenges, precision medicine holds great promise for advancing healthcare by providing personalized and targeted approaches to cancer diagnosis, treatment, and prevention.

7.10 CLINICAL TRIALS OF THERANOSTICS AND NANOTHERANOSTICS IN CANCER TREATMENT

Nanotheranostics platforms have diverse applications beyond their diagnostic and therapeutic functions. Some clinical uses of nanotheranostics are patient risk identification and subpopulation stratification, monitor drug distribution within tumors, evaluating CT imaging responses, and imaging-guided therapies. These applications are particularly valuable in addressing tumor heterogeneity and facilitating personalized therapy, where individualized treatment approaches are essential (Chavda, 2023). The multifaceted nature of nanotheranostics platforms allows for their utilization in various aspects of patient care and treatment optimization. Recent clinical trials that have had significant effects on cancer treatment are listed in Table 7.2 (Clinical trial.gov).

7.11 CONCLUSION AND FUTURE PROSPECTS

Rapid advancements in theranostics have significantly impacted cancer therapy. The future of cancer nanotheranostics lies in the development of multimodal nanoplatforms that integrate diagnostic tools, imaging agents, targeting ligands, and therapeutic components into a single functionalized nanoparticle. Despite ongoing endeavors, conventional cancer treatment approaches encounter challenges such as cancer cells' multidrug resistance, hydrophobicity/limited solubility, biocompatibility issues, and adverse effects of radiation. In response, nanotheranostics offer a promising avenue for the future of cancer treatment.

Theranostics in medicine not only allows for early diagnosis but also offers a paradigm shif from trial and error to precise, targeted, and personalized therapies. Extensive *in vitro* and *in vivo* studies in cancer theranostics demonstrate hopeful outcomes that will significantly enhance cancer treatment and patient survival and pave the way for a new era of personalized medicine. The key advantage of nanotheranostics over traditional methods is their multifunctionality, allowing them to perform multiple tasks simultaneously. Evidence supports their efficient and noninvasive *in vivo* action without causing any systemic toxicity. Although theranostics holds tremendous promise, there are still challenges to overcome. First, applying theranostics requires integrating various disciplines, including genomics, proteomics, imaging, and bioinformatics, which can be complex and costly. Standardizing the diagnostic tests and ensuring their accessibility to a wide range of healthcare settings is another important consideration. Additionally, regulatory and reimbursement frameworks need to be adaptive to accommodate the unique nature of theranostics. To optimize nanotheranostics for the best clinical outcome, ongoing development should focus on improving the delivery with effective targeting and minimal side effects. This includes enhancing their performance and efficacy to address the challenges.

TABLE 7.2
Clinical Trials of Theranostics for Cancer Treatment

Cancer Type	Drug Used	Clinical Trial ID	Brief Summary	Interventions	Sponsor	Status/Start Date
Neuroendocrine tumors, prostate cancer	Somatostatin antagonists	NCT05680675	The objective of this clinical trial is to acquire imaging data using PET combined with x-ray CT for each pair of tracers. This data will serve as a foundation for the development of novel techniques and processing algorithms that enable simultaneous dual-tracer imaging using these specific tracer pairs.	*Device*: PET/CT Imaging *Radiation*: ^{68}Ga-DOTA-JR11 and ^{177}Lu-DOTA-JR11	University of Utah, United states	Active/2022
Theranostics & metastatic castration-resistant prostate cancer	Abiraterone, enzalutamide, & apalutamide	NCT04769817	In this clinical trial is a descriptive, observational, prospective, open-ended registry that utilizes electronic data capture to gather information on the outcomes of men undergoing treatment with PSMA theranostics.	*Device*: PET/CT *Radiation*: ^{177}Lu-PSMA	Peter MacCallum Cancer Centre, Australia	Active/2021
Neuroendocrine tumors	Somatostatin antagonists	NCT02609737	Radiolabeled somatostatin antagonists, ^{68}Ga-DOTA-JR11, and ^{177}Lu-DOTA-JR11 were evaluated.	*Device*: PET/CT Imaging *Radiation*: ^{68}Ga-DOTA-JR11 and ^{177}Lu-DOTA-JR11	Memorial Sloan Kettering Cancer Center New York	Completed/2015
Malignant tumors	–	NCT05798273	This clinical trial validates PSMA as a diagnostic and potential theranostic target in intra-axial brain tumors using ^{68}Ga-PSMA-11 PET imaging. Additionally, its authors investigate whether intra-arterial injection enhances PSMA uptake in tumors that exhibit uptake after intravenous injection of ^{68}Ga-PSMA–11.	IA injection *Device*: PET/MRI *Other*: Radiopharmacon contrast agent	Erasmus Medical Center	Active/2020
GIT	–	NCT05547919	This clinical trial explores the applicability of PSMA-directed *in vivo* imaging in GIT, specifically HCC. The objectives include evaluating PSMA expression in biopsy-derived tissue, assessing PSMA-PET/CT for *in vivo* PSMA expression, comparing molecular imaging with conventional imaging for disease detection, and determining if PSMA PET signal predicts clinical course and outcomes in guideline-based treatment.	*Radiation*: ^{18}F-PSMA *Device*: PET/CT	Wuerzburg University Hospital	Active/2022
Gliomas	–	NCT05644080	This clinical pilot study aims to assess the effectiveness of ^{68}Ga/177Lu-PSMA theranostics as a treatment option for patients. It offers a potentially well-tolerated radionuclide treatment to enhance both overall survival and quality of life.	*Radiation*: ^{177}Lu-PSMA I&T	St. Olavs Hospital	Active/2023
Metastatic cancer		NCT04849247	In this Phase I trial, patients with various forms of cancer, including locally advanced and metastatic instances, will be evaluated for safety and efficacy of the ^{68}Ga-DOTA-FAPI and ^{177}Lu-DOTA-FAPI theranostic combination. The goal of this clinical trial examines the potential of this cutting-edge strategy for both diagnosis and treatment of various cancer types.	*Drug*: ^{68}Ga-DOTA-FAPI, ^{177}Lu-DOTA-FAPI	The First Affiliated Hospital of Xiamen University	Active/2021

(*Continued*)

TABLE 7.2
Clinical Trials of Theranostics for Cancer Treatment (Continued)

Cancer Type	Drug Used	Clinical Trial ID	Brief Summary	Interventions	Sponsor	Status/Start Date
Prostate cancer, metastatic cancer	Cabazitaxel	NCT03392428	This clinical trial evaluates the activity and safety of LuPSMA radionuclide therapy administered to metastatic prostate cancer patients. The trial compares LuPSMA therapy with cabazitaxel chemotherapy, assessing endpoints such as PSA response rate, pain response, progression-free survival, quality of life, and adverse events.	*Radiation*: ^{177}Lu-PSMA617	Prostate Cancer Trials Group and Australian and New Zealand Urogenital	Completed/2018
Metastatic medullary and thyroid cancer	–	NCT02088645	This clinical trial evaluates the use of ^{177}Lu-PP-F11N, a gastrin analog for imaging and therapy in advanced MTC. In this Phase I study, the maximum tolerated dose, treatment response, and organ radiation doses will be determined, allowing for personalized therapy planning.	*Drug*: ^{177}Lu-PP-F11N	University Hospital, Basel, Switzerland	Active/2015
Medullary thyroid carcinoma	Co-administration of gelofusine/ gelaspan	NCT03246659	This phase I clinical trial aims to assess the safety and efficacy of ^{111}In-CP04, a radiolabeled peptide, for imaging and therapy in patients with MTC. The study involves establishing a clinically useful formulation, evaluating tracer biodistribution and dosimetry, and assessing the potential of CCK_2 receptor scintigraphy. The goal is to improve diagnostic accuracy and treatment efficacy and minimize nephron and myelotoxicity.	*Drug*: ^{111}In-CP04	Paola Anna Erba Universitaria Pisana	Completed/2016
Prostate cancer	–	NCT05146973	The trial aims to evaluate the therapeutic benefits when ^{177}Lu-TLX591 is administered with EBRT to target PSMA expression.	*Biological*: ^{177}Lu-DOTA-TLX591	Telix International Pty Ltd	Active/2022
Soft tissue sarcoma	–	NCT05420727	The goal of this clinical trial is to assess the feasibility of theranostics PSMA receptor targeting in STS patients through Ga-68-PSMA-11 PET/CT and Lu-177-ITG-PSMA-1 therapy, with a focus on vascular disruption utilizing translation.	*Radiation*: Ga-68-PSMA-11, Lu-177-ITG-PSMA-1 Device: PET/CT	John O. Prior	Active/2023

Abbreviations: NCT: National Clinical Trial, ADKP: Pancreatic adenocarcinoma, TNEP: Pancreatic neuroendocrine tumors, FDG: Fludeoxyglucose F1, GIT: Gastrointestinal tumors, FAPi: Fibroblast activation protein inhibitor, DOPA: 1,4,7,10-tetraazacyclododecane-1,4,7,10-tetraacetic acid, IA: Intra-arterial, LuPSMA: lutetium-177 PSMA, MTC: Medullary thyroid carcinoma, GEP-NET: Gastroenteropancreatic neuro-endocrine tumors, CCK_2: cholecystokinin B receptor, EBRT: External beam radiation therapy, ALK: Anaplastic lymphoma kinase, STS: Soft tissue sarcomas.

Furthermore, the integration of targeting ligands enables specific drug delivery to cancer cells, increasing treatment efficacy and reducing off-target effects. The imaging agents aid in accurate disease detection and monitoring treatment progress; the diagnostic agents provide valuable information for treatment decision-making and assessing treatment response. Thus, NPs have the potential to significantly reduce cancer-related mortality, including existing cases, relapses, and subsequent malignancies. This field of research is important and promising for improving cancer treatment outcomes.

REFERENCES

AbdElhamid AS, Helmy MW, Ebrahim SM, Zayed DG, Dein EAZ, Gizawy SAE, et al. (2018). Layer-by-layer gelatin/chondroitin quantum dots-based nanotheranostics: combined rapamycin/celecoxib delivery and cancer imaging. Nanomedicine, 13(14), 1707–1730. https://doi.org/10.2217/nnm-2018–0028.

Abrahamse H, Hamblin MR (2016). New photosensitizers for photodynamic therapy. Biochem J, 473(4), 347–364. https://doi.org/10.1042/BJ20150942.

Alfouzan AF (2021). Radiation therapy in head and neck cancer. Saudi Med J, 42(3), 247–254. https://doi.org/10.15537/smj.2021.42.3.20210660.

Aminolroayaei F, Shahbazi-Gahrouei D, Shahbazi-Gahrouei S, Rasouli N (2021). Recent nanotheranostics applications for cancer therapy and diagnosis: a review. IET Nanobiotechnol, 15(3), 247–256. https://doi.org/10.1049/nbt2.12021.

Badrigilan S, Shaabani B, Gharehaghaji N, Mesbahi A (2019). Iron oxide/bismuth oxide nanocomposites coated by graphene quantum dots: "three-in-one" theranostic agents for simultaneous CT/MR imaging-guided in vitro photothermal therapy. Photodiagn Photodyn Ther, 25, 504–514. https://doi.org/10.1016/j.pdpdt.2018.10.021.

Bansal S, Singh J, Kumari U, Kaur IP, Barnwal RP, Kumar R, et al. (2019). Development of biosurfactant-based graphene quantum dot conjugate as a novel and fluorescent theranostic tool for cancer. Int J Nanomed, 14, 809–818. https://doi.org/10.2147/IJN.S188552.

Baum RP, Singh A, Schuchardt C, Kulkarni HR, Klette I, Wiessalla S, et al. (2018). 177 Lu-3BP-227 for neurotensin receptor 1–targeted therapy of metastatic pancreatic adenocarcinoma: first clinical results. J Nucl Med, 59(5), 809–814. https://doi.org/10.2967/jnumed.117.193847.

Berg WA, Blume JD, Cormack JB, Mendelson EB, Lehrer D, Pisano ED, et al. (2009). Combined screening with ultrasound and mammography vs mammography alone in women at elevated risk of breast cancer: result of the first-year screen in ACRIN 6666. JAMA, 299(18), 2151–2163. http://doi.org/10.1001/jama.299.18.2151.

Cai X, Zhu Q, Zeng Y, Zeng Q, Chen X, Zhan Y (2019). Manganese oxide nanoparticles as MRI contrast agents in tumor multimodal imaging and therapy. Int J Nanomed, 14, 8321–8344. https://doi.org/10.2147/IJN.S218085.

Carlsson CA (1999). Imaging modalities in X-ray computerized tomography and in selected volume tomography. Phys Med Biol, 44(3), 23–56. http://iopscience.iop.org/0031–9155/44/3/011.

Chavda VP (2023). Current status of cancer nanotheranostics: emerging strategies for cancer management. Nanotheranostics, 7(4), 368–379. https://doi.org/10.7150/ntno.82263.

Chavda VP, Balar PC, Patel SB (2023). Interventional nanotheranostics in hepatocellular carcinoma. Nanotheranostics, 7(2), 128–141. https://doi.org/10.7150/ntno.80120.

Chen W, Shen B, Sun X (2019). Analysis of progress and challenges of EGFR-targeted molecular imaging in cancer with a focus on affibody molecules. Mol Imaging, 18. https://doi.org/10.1177/1536012118823473.

Chen X, Wong STC (2014). Cancer theranostics: an introduction. In: *Cancer Theranostics*. Academic Press, 3–8. https://doi.org/10.1016/B978-0-12-407722-5.00001-3.

Chitgupi U, Qin Y, Lovell JF (2017). Targeted nanomaterials for phototherapy. Nanotheranostics, 1(1), 38–58. https://doi.org/10.7150/ntno.17694.

Cui Y, Yan H, Wang H, Zhang Y, Li M, Cui K, et al. (2022). CuS-131I-PEG nanotheranostics-induced "multiple mild-hyperthermia" strategy to overcome radio-resistance in lung cancer brachytherapy. Pharmaceutics, 14(12), 2669. https://doi.org/10.3390/pharmaceutics14122669.

Dai Y, Xiao H, Liu J, Yuan Q, Ma P, Yang D, et al. (2013). In vivo multi-modality imaging and cancer therapy by near-infrared light triggered trans-platinum pro-drug conjugated upconverison nanoparticles. J Am Chem Soc, 135(50), 18920–18929. http://doi.org/10.1021/ja410028q.

Deng X, Li K, Cai X, Liu B, Wei Y, Deng K, et al. (2017). A hollow-structured CuS@Cu_2S@Au nanohybrid: synergistically enhanced photothermal efficiency and photoswitchable targeting effect for cancer theranostics. Adv Mater, 29(36). https://doi.org/10.1002/adma.201701266.

Deng Y, Huang H, Chen M, Chen G, Zou W, Zhao Y, et al. (2020). Comprehensive effects of near-infrared multifunctional liposomes on cancer cells. Molecules, 25(5). https://doi.org/10.3390/molecules25051098.

Dennahy IS, Han Z, MacCuaig WM, Chalfant HM, Condacse A, Hagood JM, et al. (2022). Nanotheranostics for image-guided cancer treatment. Pharmaceutics, 14(5), 917. https://doi.org/10.3390/pharmaceutics14050917.

Deshpande N, Ramesh A, Nandi D, Nguyen A, Brouillard A, Kulkarni A (2020). Supramolecular polysaccharide nanotheranostics that inhibit cancer cells growth and monitor targeted therapy response. Nanotheranostics, 4(3), 156–172. https://doi.org/10.7150/ntno.44703.

Dhandapani R, Sethuraman S, Krishnan UM, Subramanian A (2023). Self-assembled multifunctional nanotheranostics against circulating tumor clusters in metastatic breast cancer. Acta Pharm Sin B, 13(4), 1711–1725. https://doi.org/10.1016/j.apsb.2022.12.003.

Esmaeili Y, Seyedhosseini GH, Ghasemi F, Shariatim R, Bidram E, Zarrabi A, et al. (2022). Graphene oxide quantum dot-chitosan nanotheranostic platform as a pH-responsive carrier for improving curcumin uptake internalization: in vitro & in silico study. Biomater Adv, 139, 213017. https://doi.org/10.1016/j.bioadv.2022.213017.

Ferlay J, Ervik M, Lam F, Colombet M, Mery L, Piñeros M (2022). *Cancer Global Cancer Observatory: Cancer Today*. Lyon: International Agency for Research on Cancer, World Health Organization.

Franc BL, Seo Y, Flavell R, Aparici CM (2020). Preclinical SPECT and SPECT-CT in oncology. Recent Res Cancer Res, 216, 359–404. https://doi.org/10.1007/978-3-030-42618-7_11.

Frantellizzi V, Conte M, Pontico M, Pani A, Pani R, De VG (2020). New frontiers in molecular imaging with superparamagnetic iron oxide nanoparticles (SPIONs): efficacy, toxicity, and future applications. Nucl Med Mol Imaging, 54, 65–80. https://doi.org/10.1007/s13139-020-00635-w.

Gambardella V, Tarazona N, Cejalvo JM, Lombardi P, Huerta M, Roselló S, et al. (2020). Personalized medicine: recent progress in cancer therapy. Cancers, 12(4), 1009. https://doi.org/10.3390/cancers12041009.

Gong H, Dong Z, Liu Y, Yin S, Cheng Li, Xi W, et al. (2014). Engineering of multifunctional nano-micelles for combined photothermal and photodynamic therapy under the guidance of multimodal imaging. Adv Funct Mater, 24(41), 6492–6502. https://doi.org/10.1002/adfm.201401451.

Hajiramezanali M, Atyabi F, Mosayebnia M, Akhlaghi M, Geramifar P, Jalilian AR, et al. (2019). 68 Ga-radiolabeled bombesin-conjugated to trimethyl chitosan-coated superparamagnetic nanoparticles for molecular imaging: preparation, characterization and biological evaluation. Int J Nanomed, 14, 2591–2605. https://doi.org/10.2147/IJN.S195223.

Herrmann K, Schottelius M, Lapa C, Osl T, Poschenrieder A, Hanscheid H, et al. (2016). First-in-human experience of CXCR4-directed endoradiotherapy with 177 Lu- and 90 Y-labeled pentixather in advanced-stage multiple myeloma with extensive intra- and extramedullary disease. J Nucl Med, 57(2), 248–251. https://doi.org/10.2967/jnumed.115.167361.

Huang J, Xu Y, Xiao H, Xiao Z, Guo Y, Cheng D, et al. (2019). Core shell distinct nanodrug showing on-demand sequential drug release to act on multiple cell types for synergistic anticancer therapy. ACS Nano, 13(6), 7036–7049. https://doi.org/10.1021/acsnano.9b02149.

Jain NK, Chathoth BM, Bhaskar VS, Meena H, Prasad R, Srivastava R (2021). Nanoengineered photoactive theranostic agents for cancer. Nanophotonics, 10, 2973–2997. https://doi.org/10.1515/nanoph-2021–0205.

Jaque D, Martínez ML, Del RB, Haro-Gonzalez P, Benayas A, Plaza JL (2014). Nanoparticles for photothermal therapies. Nanoscale, 6, 9494–530. https://doi.org/10.1039/C4NR00708E.

Jha A, Viswanadh MK, Burande AS, Mehata AK, Poddar S, Yadav K, et al. (2020). DNA biodots based targeted theranostic nanomedicine for the imaging and treatment of non-small cell lung cancer. Int J Biol Macromol, 150, 413–425. https://doi.org/10.1016/j.ijbiomac.2020.02.075.

Jibin K, Prasad JS, Saranya G, Shenoy SJ, Maiti KK, Jayasree RS (2020). Optically controlled hybrid metamaterial of plasmonic spiky gold inbuilt graphene sheets for bimodal imaging guided multimodal therapy. Biomater Sci, 8(12), 3381–3391. https://doi.org/10.1039/d0bm00312c.

Kang SJ, Jeong HW, Kim MW, Jeong IH, Choi MJ, You YM, et al. (2018). Anti-EGFR lipid micellar nanoparticles co-encapsulating quantum dots and paclitaxel for tumor-targeted theranosis. Nanoscale, 10(41), 19338–19350. https://doi.org/10.1039/c8nr05099f.

Kelkar SS, Reineke TM (2011). Theranostics: combining imaging and therapy. Bioconjug Chem, 22(10), 1879–1903. https://doi.org/10.1021/bc200151q.

Keshtkar M, Shahbazi-Gahrouei D, Mahmoudabadi A (2020). Synthesis and application of Fe_3O_4@Au composite nanoparticles as magnetic resonance/computed tomography dual-modality contrast agent. JMSS, 10(3), 201–207. https://doi.org/10.4103/jmss.JMSS_55_19.

Kojima C, Umeda Y, Ogawa M, Harada A, Magata Y, Kono K (2010). X-ray computed tomography contrast agents prepared by seeded growth of gold nanoparticles in PEGylated dendrimer. Nanotechnology, 21(24), 245104. https://doi.org/10.1088/0957–4484/21/24/245104.

Lakshmanan VK, Jindal S, Packirisamy G, Ojha S, Lian S, Kaushik A, et al. (2021). Nanomedicine-based cancer immunotherapy: recent trends and future perspectives. Cancer Gene Ther, 28, 911–923. https://doi.org/10.1038/s41417-021-00299-4.

Langbein T, Weber WA, Eiber M (2019). Future of theranostics: an outlook on precision oncology in nuclear medicine. J Nucl Med, 60(2), 13S–19S. https://doi.org/10.2967/jnumed.118.220566.

Lee DY, Li KCP (2011). Molecular theranostics: a primer for the imaging professional. AJR Am J Roentgenol, 197(2), 318–324. https://doi.org/10.2214/AJR.11.6797.

Lee Y-H, Kuo PW, Chen CJ, Sue CJ, Hsu YF, Pan MC (2021). Indocyanine green-camptothecin co-loaded perfluorocarbon double-layer nanocomposite: a versatile nanotheranostics for photochemotherapy and FDOT diagnosis of breast cancer. Pharmaceutics, 13(9), 1499. https://doi.org/10.3390/pharmaceutics13091499.

Li P, Lin B, Chen Z, Liu P, Liu J, Li W, et al. (2022). Biodegradable hollow mesoporous organosilica nanotheranostics (HMONs) as a versatile platform for multimodal imaging and phototherapeutic-triggered endolysosomal disruption in ovarian cancer. Drug Deliv, 29(1), 161–173. https://doi.org/10.1080/10717544.2021.2021322.

Li Y, Chen W, Qi Y, Wang S, Li L, Li W, et al. (2020). S-scavenged and activated iron oxide-hydroxide nanospindles for MRI-guided photothermal therapy and ferroptosis in colon cancer. Small, 16(37), 2001356. https://doi.org/10.1002/smll.202001356.

Liu X, Du C, Li H, Jiang T, Luo Z, Pang Z, et al. (2019). Engineered superparamagnetic iron oxide nanoparticles (SPIONs) for dual-modality imaging of intracranial glioblastoma via EGFRvIII targeting. Beilstein J Nanotechnol, 10, 1860–1872. https://doi.org/10.3762/bjnano.10.181.

Lu Y-J, Anilkumar T, Chuang CC, Chen JP (2021). Liposomal IR-780 as a highly stable nanotheranostic agent for improved photothermal/photodynamic therapy of brain tumors by convection-enhanced delivery. Cancers, 13(15), 3690. https://doi.org/10.3390/cancers13153690.

Mansur AAP, Paiva MRB, Cotta OAL, Silva LM, Carvalllho IC, Capanema NSV, et al. (2022). Carboxymethylcellulose biofunctionalized ternary quantum dots for subcellular-targeted brain cancer nanotheranostics. Int J Biol Macromol, 210, 530–544. https://doi.org/10.1016/j.ijbiomac.2022.04.207.

Marshall SK, Angsantikul P, Pang Z, Nasongkla N, Hussen RSD, Thamphiwatana SD (2022). Biomimetic targeted theranostic nanoparticles for breast cancer treatment. Molecules, 27(19), 6473. https://doi.org/10.3390/molecules27196473.

Mi P, Dewi N, Yanagie H, Kokuryo D, Suzuki M, Sakurai Y, et al. (2015). Hybrid calcium phosphate-polymeric micelles incorporating gadolinium chelates for imaging-guided gadolinium neutron capture tumor therapy. ACS Nano, 9(6), 5913–5921. http://pubs.acs.org; https://doi.org/10.1021/acsnano.5b00532.

Minassian G, Ghanem E, Hage RE, Rahme K (2023). Gold nanoparticles conjugated with dendrigraft poly-L-lysine and folate-targeted poly(ethylene glycol) for siRNA delivery to prostate cancer. Nanotheranostics, 7(2), 152–166. https://doi.org/10.7150/ntno.79050.

Mousavi A (2020). CXCL12/CXCR4 signal transduction in diseases and its molecular approaches in targeted-therapy. Immunol Lett, 217, 91–115. https://doi.org/10.1016/j.imlet.2019.11.007.

Nayak V, Patra S, Singh KR, Ganguly B, Kumar DN, Panda D, et al. (2023). Advancement in precision diagnosis and therapeutic for triple-negative breast cancer: harnessing diagnostic potential of CRISPR-cas & engineered CAR T-cells mediated therapeutics. Environ Res, 235, 116573. https://doi.org/10.1016/j.envres.2023.116573.

Ngen EJ, Chen Y, Azad BB, Boinapally S, Jacob D, Lisok A, et al. (2021). Prostate-specific membrane antigen (PSMA)-targeted photodynamic therapy enhances the delivery of PSMA-targeted magnetic nanoparticles to PSMA-expressing prostate tumors. Nanotheranostics, 5(2), 182–196. https://doi.org/10.7150/ntno.52361.

Nomura S, Morimoto Y, Tsujimoto H, Arake M, Harada M, Saitoh D, et al. (2020). Highly reliable, targeted photothermal cancer therapy combined with thermal dosimetry using a near-infrared absorbent. Sci Rep, 10(1), 9765. https://doi.org/10.1038/s41598-020-66646-x.

Pandey N, Menon JU, Takahashi M, Hsieh JT, Yang J, Nguyen KT, et al. (2020). Thermo-responsive fluorescent nanoparticles for multimodal imaging and treatment of cancers. Nanotheranostics, 4(1), 1–13. https://doi.org/10.7150/ntno.39810.

Patel P, Alghamdi A, Shaw G, Legge C, Glover M, Freeman D, et al. (2023b). Development of a personalised device for systemic magnetic drug targeting to brain tumours. Nanotheranostics, 7(1), 102–116. https://doi.org/10.7150/ntno.76559.

Patel P, Kumar K, Jain VK, Popli H, Yadav AK, Jain K (2023a). Nanotheranostics for diagnosis and treatment of breast cancer. Curr Pharm Des, 29(10), 732–747. https://doi.org/10.2174/1381612829666230329122911.

Perera RH, Abenojar E, Nittayacharn P, Wang X, Ramamurthy G, Peiris P, et al. (2022). Intracellular vesicle entrapment of nanobubble ultrasound contrast agents targeted to PSMA promotes prolonged enhancement and stability *in vivo* and *in vitro*. Nanotheranostics, 6(3), 270–285. https://doi.org/10.7150/ntno.64735.

Rand D, Ortiz V, Liu Y, Derdak Z, Wands JR, Tatíček M, et al. (2011). Nanomaterials for X-ray imaging: gold nanoparticle enhancement of X-ray scatter imaging of hepatocellular carcinoma. Nano Lett, 11(7), 2678–2683. https://doi.org/10.1021/nl200858y.

Rapoport N, Payne A, Dillon C, Shea J, Scaife C, Gupta R (2013). Focused ultrasound-mediated drug delivery to pancreatic cancer in a mouse model. J Ther Ultrasound, 1(1), 11. https://doi.org/10.1186/2050-5736-1-11.

Reddy AS, Lakshmi BA, Kim S, Kim J (2019). Synthesis and characterization of acetyl curcumin-loaded core/shell liposome nanoparticles via an electrospray process for drug delivery, and theranostic applications. Eur J Pharm Biopharm, 142, 518–530. https://doi.org/10.1016/j.ejpb.2019.07.024.

Rodríguez F, Caruana P, De la Fuente N, Español P, Gámez M, Balart J, et al. (2022). Nano-based approved pharmaceuticals for cancer treatment: present and future challenges. Biomolecules, 12(6), 784. https://doi.org/10.3390/biom12060784.

Saharkhiz S, Zarepour A, Zarrabi A (2023). A new theranostic pH-responsive niosome formulation for doxorubicin delivery and bio-imaging against breast cancer. Int J Pharm, 122845. https://doi.org/10.1016/j.ijpharm.2023.122845.

Salehnia Z, Shahbazi-Gahrouei D, Akbarzadeh A, Baradaran B, Farajnia S, Naghibi M (2019). Synthesis and characterisation of iron oxide nanoparticles conjugated with epidermal growth factor receptor (EGFR) monoclonal antibody as MRI contrast agent for cancer detection. IET Nanobiotechnol, 13(4), 400–406. https://doi.org/10.1049/iet-nbt.2018.5285.

Sathishkumar K, Chaturvedi M, Das P, Stephen S, Mathur P (2023). Cancer incidence estimates for 2022 & projection for 2025: result from National Cancer Registry Programme, India. Indian J Med Res, 156(4–5), 598–607. https://doi.org/10.4103/ijmr.ijmr_1821_22.

Scanlan MJ, Raj BK, Calvo B, Garin-Chesa P, Sanz-Moncasi MP, Healey JH, et al. (1994). Molecular cloning of fibroblast activation protein alpha, a member of the serine protease family selectively expressed in stromal fibroblasts of epithelial cancers. Proc Natl Acad Sci USA, 91(12), 5657–5661. https://doi.org/10.1073/pnas.91.12.5657.

Shanmugam M, Kuthala N, Vankayala R, Chiang CS, Kong X, Hwang KC (2021). Multifunctional CuO/Cu_2O truncated nanocubes as trimodal image-guided near-infrared-III photothermal agents to combat multi-drug-resistant lung carcinoma. ACS Nano, 15(9), 14404–14418. https://doi.org/10.1021/acsnano.1c03784.

Si Y, Zhang G, Wang D, Zhang C, Yang C, Bai G, et al. (2019). Nanostructure-enhanced water interaction to increase the dual-mode MR contrast performance of gadolinium-doped iron oxide nanoclusters. Chem Eng J, 360, 289–298. https://doi.org/10.1016/j.cej.2018.11.219.

Siegel RL, Miller KD, Wagle NS, Jemal A (2023). Cancer statistics, 2023. CA: Cancer J Clin, 73(1), 17–48. https://doi.org/10.3322/caac.21763.

Song J, Zhang L, Yi H, Huang J, Zhang N, Zhong Y, et al. (2019). NIR-responsive nanoplatform for pre/intraoperative image-guided carcinoma surgery and photothermal ablation of residual tumor tissue. Nanomedicine, 20. https://doi.org/10.1016/j.nano.2019.102020.

Sonkar R, Sonali, Jha A, Viswanadh MK, Burande AS, Pawde DM, et al. (2021). Gold liposomes for brain-targeted drug delivery: formulation and brain distribution kinetics. Mater Sci Eng C Mater Biol Appl, 120, 111652. https://doi.org/10.1016/j.msec.2020.111652.

Sun L, Joh DY, Al-Zaki A, Stangl M, Murty S, Davis JJ, et al. (2016). Theranostic application of mixed gold and superparamagnetic iron oxide nanoparticle micelles in glioblastoma multiforme. J Biomed Nanotechnol, 12(2), 347–356. https://doi.org/10.1166/jbn.2016.2173.

Tang J, Zheng F, Zhao J, Zhao J (2020). Self-assembled multifunctional nanotheranostics loading GEM for targeted lung cancer therapy. Mater Sci Eng C Mater Biol Appl C, 112, 110786. https://doi.org/10.1016/j.msec.2020.110786.

Taratula O, Patel M, Schumann C, Naleway MA, Pang AJ, He H, et al. (2015). Phthalocyanine-loaded graphene nanoplatform for imaging-guided combinatorial phototherapy. Int J Nanomed, 10, 2347–2362. https://doi.org/10.2147/IJN.S81097.

Thirumurugan S, Dash P, Liu X, Tseng YY, Huang WJ, Li Y, et al. (2022). Angiopep-2-decorated titanium–alloy core–shell magnetic nanoparticles for nanotheranostics and medical imaging. Nanoscale, 14(39), 14789–14800. https://doi.org/10.1039/D2NR03683E.

Urandur S, Banala VT, Shukla RP, Gautam S, Marwaha D, Rai N, et al. (2020). Theranostic lyotropic liquid crystalline nanostructures for selective breast cancer imaging and therapy. Acta Biomater, 113, 522–540. https://doi.org/10.1016/j.actbio.2020.06.023.

Vayalakkara RK, Lo CL, Chen HH, Shen MY, Yang YC, Sabu A, et al. (2022). Photothermal/NO combination therapy from plasmonic hybrid nanotherapeutics against breast cancer. J Control Release, 345, 417–432. https://doi.org/10.1016/j.jconrel.2022.03.030.

Wang D, Zhou J, Fang W, Huang C, Chen Z, Fan M, et al. (2022). A multifunctional nanotheranostic agent potentiates erlotinib to EGFR wild-type non-small cell lung cancer. Bioact Mater, 13, 312–323. https://doi.org/10.1016/j.bioactmat.2021.10.046.

Wang F, Li N, Wang W, Ma L, Sun Y, Wang H, et al. (2023). A multifunctional, highly biocompatible, and double-triggering caramelized nanotheranostic system loaded with Fe_3O_4 and DOX for combined chemo-photothermal therapy and real-time magnetic resonance imaging monitoring of triple negative breast cancer. Int J Nanomed, 18, 881–897. https://doi.org/10.2147/IJN.S393507.

Wang Z, Sun X, Huang T, Song J, Wang YA (2020). A sandwich nanostructure of gold nanoparticle coated reduced graphene oxide for photoacoustic imaging-guided photothermal therapy in the second NIR window. Front Bioeng and Biotechnol, 8, 655. https://doi.org/10.3389/fbioe.2020.00655.

West CL, Doughty ACV, Liu K, Chen WR (2019). Monitoring tissue temperature during photothermal therapy for cancer. J BioX Res, 2(4), 159–168. https://doi.org/10.1097/JBR.0000000000000050.

Xiao Y, Hong H, Javadi A, Engle JW, Xu W, Yang Y, et al. (2012). Multifunctional unimolecular micelles for cancer-targeted drug delivery and positron emission tomography imaging. Biomaterials, 33(11), 3071–3082. https://doi.org/10.1016/j.biomaterials.2011.12.030.

Xiong J, Yuan H, Wu H, Cheng J, Yang S, Hu T (2021). Black phosphorus conjugation of chemotherapeutic ginsenoside Rg3: enhancing targeted multimodal nanotheranostics against lung cancer metastasis. Drug Deliv, 28(1), 1748–1758. https://doi.org/10.1080/10717544.2021.1966129.

Xu L, Xu M, Sun X, Feliu N, Feng L, Parak WJ, et al. (2023). Quantitative comparison of gold nanoparticle delivery *via* the enhanced permeation and retention (EPR) effect and mesenchymal stem cell (MSC)-based targeting. ACS Nano, 17(3), 2039–2052. https://doi.org/10.1021/acsnano.2c07295.

Yari H, Nkepang G, Awasthi V (2019). Surface modification of liposomes by a lipopolymer targeting prostate specific membrane antigen for theranostic delivery in prostate cancer. Materials, 12(5), 756. https://doi.org/10.3390/ma12050756.

Yim W, Borum RM, Zhou J, Mantri Y, Wu Z, Zhou J, et al. (2022). Ultrasmall gold nanorod-polydopamine hybrids for enhanced photoacoustic imaging and photothermal therapy in second near-infrared window. Nanotheranostics, 6(1), 79–90. https://doi.org/10.7150/ntno.63634.

Yin X, Cui Y, Kim RS, Stiles WR, Park SH, Wang H, et al. (2022). Image-guided drug delivery of nanotheranostics for targeted lung cancer therapy. Theranostics, 12(9), 4147–4162. https://doi.org/10.7150/thno.72803.

Zhang P, Hu C, Ran W, Meng J, Yin Q, Li Y. (2016). Recent progress in light-triggered nanotheranostics for cancer treatment. Theranostics, 6(7), 948–968. https://doi.org/10.7150/thno.15217.

Zhang R, Qiao C, Jia Q, Wang Y, Huang H, Chang W, et al. (2019). Highly stable and long-circulating metal-organic frameworks nanoprobes for sensitive tumor detection in vivo. Adv Healthc Mater, 8(19). https://doi.org/10.1002/adhm.201900761.

Zhao Y, Pang B, Luehmann H, Detering L, Yang X, Sultan D, et al. (2016). Gold nanoparticles doped with 199Au atoms and their use for targeted cancer imaging by SPECT. Adv Healthc Mater, 5(8), 928–935. https://doi.org/10.1002/adhm.201500992.

Zheng D, Wan C, Yang H, Xu L, Dong Q, Du C, et al. (2020). Her2-targeted multifunctional nano-theranostic platform mediates tumor microenvironment remodeling and immune activation for breast cancer treatment. Int J Nanomed, 15, 10007–10028. https://doi.org/10.2147/IJN.S271213.

Zhou B, Jia BX, Zhang MJ, Tan YJ, Liang WY, Gan X, et al. (2023b). Zn2+-interference and H_2S-mediated gas therapy based on ZnS-tannic acid nanoparticles synergistic enhancement of cell apoptosis for specific treatment of prostate cancer. Colloids Surf B Biointerfaces, 226, 113313. https://doi.org/10.1016/j.colsurfb.2023.113313.

Zhou R, Ohulchanskyy TY, Xu Y, Ziniuk R, Xu H, Liu L, et al. (2022). Tumor-microenvironment-activated NIR-II nanotheranostic platform for precise diagnosis and treatment of colon cancer. ACS Appl Mater Interfaces, 14(20), 23206–23218. https://doi.org/10.1021/acsami.2c04242.

Zhou T, Luo T, Song J, Qu J (2018). Phasor-fluorescence lifetime imaging microscopy analysis to monitor intercellular drug release from a pH-sensitive polymeric nanocarrier. Anal Chem, 90(3), 2170–2177. https://doi.org/10.1021/acs.analchem.7b04511.

Zhou X, Xu X, Hu Q, Wu Y, Yu F, He C, et al. (2023a). Novel manganese and polyester dendrimer-based theranostic nanoparticles for MRI and breast cancer therapy. J Mater Chem B, 11(3), 648–656. https://doi.org/10.1039/D2TB01855A.

Section II

Biotechnological Applications in Regenerative Medicine and Tissue Engineering

8 Stem Cells, Cancer Stem Cells, and Regenerative Medicine

Kafayat Yusuf,[1] *Shahid Umar,*[1] *and Ishfaq Ahmed*[2]*
[1] Department of Surgery, University of Kansas Medical Center, Kansas City, KS, USA
[2] Department of Biology, Kansas City Community College, Kansas City, KS, USA

LIST OF ABBREVIATIONS

CSCs	Cancer stem cells
ESCs	Embryonic stem cells
iPSCs	Induced pluripotent stem cells,
Lgr5	Leucine-rich repeat-containing G-protein-coupled receptor 5
Oct4	Octamer binding transcription factor
SCD	Sickle cell disease
Sox2	Sex determining region Y-box 2
VEGF	Vascular endothelial growth factor

8.1 INTRODUCTION

Stem cells (SCs) are undifferentiated cells that are present in our bodies from embryo stage to adulthood that enable building every tissue in the body. These cells are present in very low quantities and can grow and differentiate into more than 200 different types of body cells [1, 2]. Stem cells (a) usually arise from single cells and (b) have unlimited potential to self-renew and divide under the right conditions to (c) give rise to cells with specialized functions [3, 4]. These properties distinguish SCs from progenitor cells, which can differentiate but not self-renew, or somatic cells, which can proliferate but not differentiate. Self-renewal and differentiation make stem cells produce an infinite quantity of cell types with a clinical interest [5].

Following fertilization of an oocyte by a spermatozoon, a one-celled zygote is formed that on successive divisions gives rise to 32–64 totipotent cells called a morula, and on further division, the morula develops into blastocysts. The inner walls of blastocysts are lined with embryonic stem cells that give rise to all types of body cells and tissues and eventually the adult organism [6]. Scientists have isolated various types of stem cells and grown them in nutrient media to then induce them to differentiate into multiple cell types [7, 8] (Figure 8.1). Stem cells provide an opportunity for regenerative medicine to treat many incurable diseases [5]

The field of regenerative medicine deals with developing new treatments to functionally repair, regenerate, and restore specific body tissues or organs damaged due to disease, injuries, aging or defects where body's own regenerative response is inadequate [9]. Stem cell therapy uses the self-renewal potential and biological properties of human stem cells to regenerate and potentially heal the body. The clinical potential of SCs in regenerative medicine has sparked scientific and commercial interest, raised public expectations, and offered hope for treating diseases for which current treatments only relieve symptoms or slow disease progression [10].

8.2 CLASSIFICATION

SCs are classified into two broad categories according to their potency and tissue of origin.

8.2.1 Potency

8.2.1.1 Totipotent Cells

Totipotent or omnipotent stem cells are undifferentiated cells found during early development. After the fertilization of oocyte by spermatozoon, the zygote forms, a totipotent cell with the highest differentiation potential. It develops into three primary germ layers of both embryonic and extraembryonic tissues, forming cells of the embryo and the placenta. Totipotent stem cells have the developmental capacity to develop into individual organisms [11, 12].

8.2.1.2 Pluripotent Cells

Pluripotent cells can give rise to cells from all three germ layers but not extraembryonic structures such as the placenta [13]. Pluripotent stem cells (PSCs) have been used for clinical research to study early development of fetuses and therapeutic applications [14]. In 2006, somatic cells were induced to develop into pluripotent stem cells (iPSCs) [15]. A close resemblance between PSCs and iPSCs has been observed that is promising for regenerative medicine.

8.2.1.3 Multipotent Stem Cells

Multipotent stem cells have the ability to differentiate into discrete cells of one specific lineage [16]. Hematopoietic stem cells present in blood and bone marrow are an example of MSCs that differentiate into both myeloid and lymphoid lineages to give rise to all the blood cells [17]. Mesenchymal and neural stem cells are other examples of multipotent stem cells.

* Corresponding author: iahmed@kckcc.edu

DOI: 10.1201/9781032660592-10

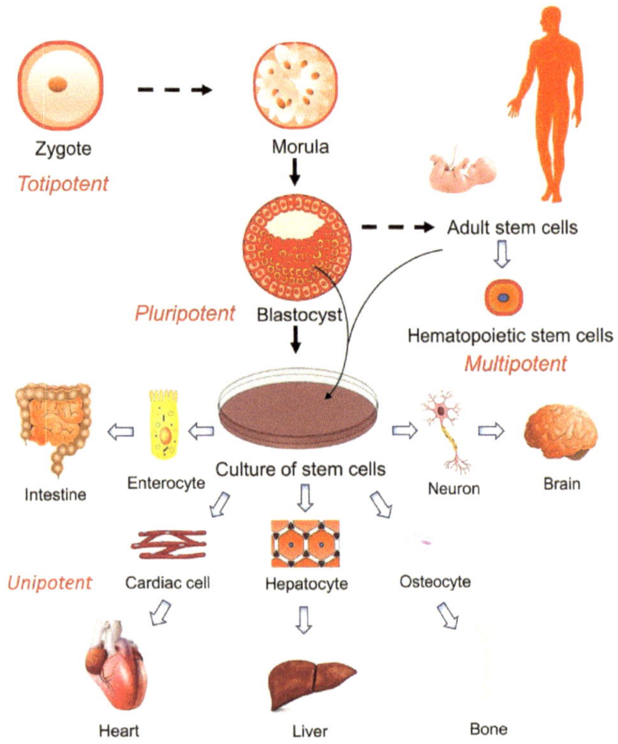

FIGURE 8.1 Cells isolated from the inner cell mass of human blastocysts or adult tissues can be grown and expanded in culture. Under the proper growth conditions, these stem cells can be stimulated to differentiate into virtually all cell types in the body and used for transplant.

8.2.1.4 Oligopotent Stem Cells

Oligopotent stem cells can differentiate into a limited number of other cell types. For instance, myeloid SCs can differentiate into white but not red blood cells [18].

8.2.1.5 Unipotent Stem Cells

These cells have the narrowest differentiation capabilities and can differentiate into only one cell type. Muscle stem cells, for example, give rise to muscle cells only [19].

8.2.2 Origin

8.2.2.1 Embryonic Stem Cells

Embryonic stem cells are self-renewing PSCs derived from cells in early embryos that are five to six days old. These cells can differentiate into cells from all three germ layers, ectoderm, mesoderm, and endoderm [20] into any body cell [21]. Immortality, unlimited proliferation, and the ability to differentiate into any somatic cell type makes ESCs an attractive source for regenerative medicine; they have been cultured from mice [22, 23], monkeys [24], and humans [8, 25]. Despite some notable differences, there are many similarities between human and mouse ESCs in terms of culture conditions [26]. The largest source of ESCs are donated preimplantation embryos from *in vitro* fertilization clinics.

8.2.2.2 Adult Stem Cells

Adult stem cells (ASCs), also called somatic stem cells, are undifferentiated cells that are found in many different tissues throughout the bodies of nearly all organisms including humans. Their main role is tissue repair and regeneration. ASCs have shown promising results in terms of treating various diseases [27]. Mesenchymal SCs are examples of ASCs used in regenerative medicine.

8.2.2.3 Tissue-Resident Stem Cells

Tissue-resident SCs are tissue-specific cells that are involved in the maintenance and repair of adult tissues. They are generally believed to remain in non-dividing state in the stem cell niche [28] and to proliferate and differentiate only after specific stimuli from the microenvironment to replace naturally dying cells or repair wounds in response to injury [29–32]. The epithelial lining of the intestine is the best example of stem cell-driven tissue renewal. Intestinal stem cells like Lgr5, located at the base of the crypts, produce progenitor cells that proliferate vigorously giving rise to all the cell types of the villus epithelium required for its normal function [33, 34].

8.2.2.4 Induced Pluripotent Cells

Induced pluripotent stem cells (iPSCs) are cells generated from adult somatic cells that have been genetically altered [35]; they are engineered to have the same physiologic qualities as embryonic stem cells, such as morphology, proliferation, gene expression, self-renewal, and differentiation into numerous cell types [36, 37]. Because iPSCs can imitate the roles and features of embryonic stem cells while preserving a high level of plasticity, they are suitable for use in regenerative medicine [35, 37] without the ethical concerns that come with employing embryonic stem cells in clinical research [36]. The ectopic injection of pluripotency-inducible transcription factors such as c-Myc/Oct4/Sox2 and Klf4 or Oct4/Nanog/LIN28 and Sox2 into somatic cells to stimulate their differentiation into an embryonic stem cell-like phenotype is widely used to create iPSCs [35–38].

8.2.3 Cancer Stem Cells

Despite the current various anticancer therapies, cancer recurrence, relapse, and metastases remain challenges. The current understanding is that therapeutic drugs can kill proliferative cells but not the quiescent cells called cancer stem cells (CSCs) [39]. CSCs are pluripotent cells seen in both solid and liquid cancers that have the potential to self-renew and produce heterogeneous cell populations. According to the stem cell theory of cancer, a tumor comprises a subpopulation of CSCs that act as stem cells and drive tumor growth by reproducing and differentiating, much like normal stem cells [40] (Figure 8.2).

CSCs cause cellular diversity that results in vexing resistance to therapies and gives rise to metastases. CSCs act as a reservoir of cancer cells by giving rise to additional CSCs and other cell types that form the tumor mass and promote

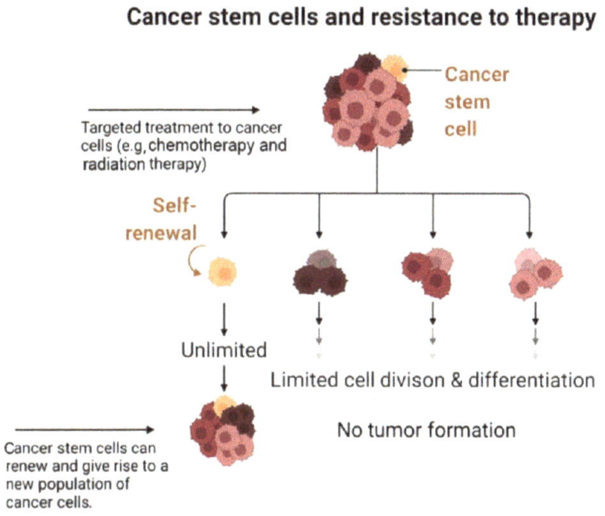

FIGURE 8.2 The role of cancer stem cells in development of treatment resistance.

(Adapted from "Cancer Stem Cell Templates," by BioRender.com (2022). Retrieved from https://app.biorender.com/biorender-templates.)

stemness in non-CSCs [41]. Tumors resemble their cells of origin morphologically or by specific gene expression. Non-CSCs can switch to CSCs due to cellular plasticity, for example during epithelial-to-mesenchymal transition (EMT), rendering them resistant to therapies [42]. EMT has been shown to promote migration, invasiveness, and metastasis [39]. During colorectal cancer treatment with anticancer drugs, $LGR5^+$ stem cells convert into $Lgr5^-$ cells while absence of drug drives $Lgr5^-$ cells back to $Lgr5^+$, suggesting CSC plasticity. In another study, hippo signaling inhibition reprogrammed $Lgr5^+$ cells and suppressed metastatic colorectal cancer [43, 44]. These studies suggest the role of stem cells in sustaining tumors.

Normal stem cells are in a quiescent state until they receive a proper stimulus and start to divide in a controlled manner; cancer-initiating mutations occur in stem cells that make them independent of stimulating signals, allowing them to divide continuously and thus converting them into cancer-initiating cells [45, 46]. Lapidot et al. [47] provided direct evidence that CSCs initiate cancer. They discovered that leukemia-initiating stem cells derived from acute myelogenous leukemia (AML) patients induced AML when transplanted into immune-deficient mice. The presence of CSCs in various solid cancers including of the breast, brain, prostate, colon, pancreas, ovaries, and lung have been reported, providing the evidence that normal SCs or progenitor cells can give rise to CSCs [48]. Normal stem cells and CSCs share some important signaling pathways including Wnt, NF-κB (nuclear factor-κB), and Notch [49].

The observation that many cancers relapse after seeming eradication offers significant support for the CSC hypothesis. CSCs are resistant to therapies and can be generated from non-CSCs or from therapy-induced senescent SCs [50]. Thus, therapies that target CSCs are presently an active area of research that promises to cure various cancers. In fact, various molecular targets in CSCs have been identified and modulated including signaling proteins, cell surface molecules, tumor microenvironments, apoptotic pathways, microRNA, stem cell differentiation, and drug resistance markers [51]. New therapies specifically targeting CSCs have been developed with some already in clinical trials [49].

8.2.3.1 Cancer and Precision Medicine

Recent progress in lineage tracing and cell ablation have provided insights into CSC plasticity, quiescence, renewal, and therapeutic response. CSCs are promising therapeutic targets for cancer therapy and have been a focus of research for several years. Several regulators of CSCs have been identified, and drugs, vaccines, antibodies and CAR-T cells targeting these pathways have been developed. However, it is confronted with many hurdles and challenges that are being studied [49].

8.2.4 Stem Cells in Regenerative Medicine

8.2.4.1 Hematopoietic Stem Cells

Hematopoietic SCs are quite peculiar in their ability to regenerate and produce all blood lineages (red blood cells, platelets, lymphocytes, and myeloid cells) of an organism [52]. For decades, hematopoietic SC transplants have remained in the forefront for their role in the treatment of several diseases (Figure 8.3). They have been utilized as the gold standard for the treatment of hematologic malignancies and hematopoietic system abnormalities and as a potential therapy for some solid tumors [53]. In a retrospective study conducted by Gratwohl and colleagues, nearly one million HSC transplants were performed as of 2014 by around 1500 transplant centers from 75 countries for the treatment of various diseases [54].

8.2.4.1.1 Hematopoietic Stem Cells in Cancer Treatment

In aggressive blood cancers such as leukemias, lymphomas, and multiple myeloma, the first line of treatment is usually chemotherapy or radiotherapy to eliminate rapidly dividing cancer cells, followed by infusion of autologous stem cells (from the patient's own bone marrow) or allogeneic hematopoietic stem cells (from another healthy donor) [55–57]. The goal of this treatment regimen is to replenish the various blood cell types damaged during the initial stages of treatment, which is usually accomplished through differentiation of engrafted stem cells [55].

The choice of autologous or allogeneic transplantation (Figure 8.3) is determined by a number of factors including cancer type, stage, and grade; patient age; donor availability; ability to collect a tumor-free autograft; cancer sensitivity to either radiation or chemotherapy; and cancer cell susceptibility to graft vs host immune response [58].

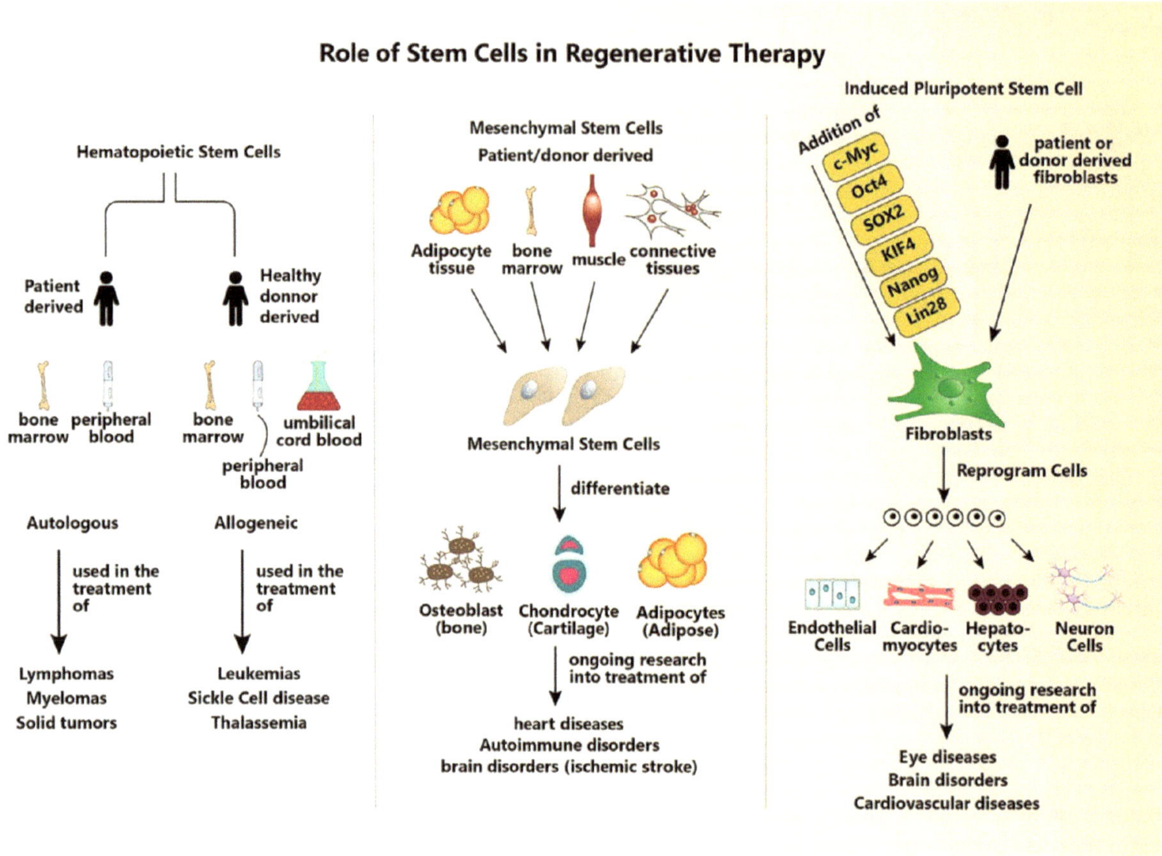

FIGURE 8.3 The role of stem cells in regenerative therapy.

In autologous stem cell transplantation, the SCs are removed from the blood and set aside while the remaining blood is returned to the patient. The patient then receives high doses of chemotherapy or radiation to kill cancer cells followed by implantation of the patient's own bone marrow stem cells in blood, which proliferate and produce all normal blood cells [59]. As no external donor is required, this approach is convenient and readily available, and there is less graft rejection and treatment-related mortality with autologous transplants [58]. However, one limitation of using autologous SCs is the possibility of clonogenic tumor cell contamination of the graft, which can result in disease relapse over time [58]. Autologous transplantation is commonly used to treat lymphomas, myelomas, and, more recently, solid tumors [55, 58, 60].

An allogeneic transplant usually involves a donor and a recipient who are genetically unrelated [59]. In this case, hematopoietic SCs can be obtained from peripheral blood containing mobilized stem cells or bone marrow or umbilical cord blood of a fit donor who is human leukocyte antigen (HLA) type matched to the recipient [61, 62]. Several studies have suggested that allogeneic transplants can not only help patients repair their blood cells but also stimulate host immune responses, which in turn reduces cancer cell proliferation [57, 59, 62, 63], in a process known as the immune graft-versus-malignancy effect [58]. Allogeneic transplants have the advantage of employing a healthy donor with a graft free of cancer cells, lowering the likelihood of disease recurrence. There are also donor-derived immunocompetent cells in the graft that can result in graft versus malignancy [58, 59, 62]. Allogeneic HSC transplants have traditionally been used to treat leukemias and myelodysplastic diseases, but they are generally limited by graft failure, graft-vs-host disease, and the probable hazard of opportunistic infections [55, 56, 58].

8.2.4.1.2 HSCs in Noncancerous Blood Diseases

Sickle cell disease, also called sickle cell anemia, is a homozygous recessive familial anomaly that is one of the world's most common hemoglobinopathies [64–66]. Sickle cell disease (SCD) is caused by a single nucleotide mutation that occurs when glutamine is substituted with valine at the sixth position in hemoglobin A's β-globin chain, resulting in the formation of a pathological hemoglobin known as hemoglobin S (HbS) [64–67]. This mutation causes HbS to polymerize, resulting in the creation of sickle-shaped red blood cells, with homozygous Hb SS individuals having high morbidity, anemia, vaso-occlusion of arteries, and severe organ and tissue damage [64–68]. Sickle cell anemia affects around 20 million people globally [65], with approximately 300,000 newborns diagnosed annually [67]. According to recent statistics, patients in the United States with the homozygous SCD have a median life expectancy of less than 40 years [68].

For patients with sickle cell, allogeneic stem cell transplantation is the only therapeutic alternative that completely cures the disease [66–69]. Treatment is entirely dependent on a human leukocyte antigen-paired donor, with overall survival after transplant reaching up to 90% [67, 69]. According to Özdoğu and Boğa, patients have a better central nervous system, less organ damage, and less discomfort following transplantation. They may also have up to 80% disease-free survival [67].

The first successful HSC transplant in SCD was described in 1984, but more than three decades later, only about 1200 patients have received a successful transplant [65, 68, 69]. The difficulties in obtaining a matched donor, the high risk, and concerns linked to the surgery such as toxicities, graft vs. host disease, and infertility, as well as cost, have all contributed to the minimal use of HSCs in SCD [64, 66–68]. To increase the use of HSCs in SCD treatment, researchers must focus their efforts on expanding donor sources, redefining HSC transplant eligibility in SCD, and increasing public understanding of the treatment's benefits and risks [64, 67–69].

Meanwhile, thalassemia is a category of hereditary blood disorders that cause inefficient erythropoiesis by disrupting the formation of hemoglobin's alpha and beta chains. Beta thalassemia is caused by defective or missing synthesis of the β-globin chains that make up adult hemoglobin molecules, whereas alpha thalassemia is caused by deletions of α-globin genes [66, 70]. According to a report by De Sanctis and colleagues, there are approximately 270 million people with abnormal hemoglobin and thalassemia [71]. Hyperactive bone marrow, inadequate erythropoiesis, and iron burden owing to repeated blood transfusions are common problems for beta thalassemia patients. Retarded growth, bone abnormalities, and jaundice are also common in affected infants. Alpha thalassemia, on the other hand, might result in the lysis of red blood cells and anemia [72].

The only curative treatment for thalassemia currently available is allogeneic HSC transplantation from a donor who has been paired with the patient [66, 73]. In thalassemia patients, HSC transplant replaces the complete hematopoietic system, allowing for efficient erythropoiesis [74]. With over three decades of use and over 3000 patients successfully treated [74], HSC transplants have provided patients suffering from the condition hope for a better life. Patients should, however, be appropriately educated about potential risks associated with the transplant including graft vs. host disease, prolonged immunosuppressive medications, graft failure, and transplantation-related mortality [70]. HSCs also have applications with other blood disorders. Patients with aplastic anemia and diseases induced by DNA repair abnormalities (such as Fanconi anemia, Shwachman–Diamond syndrome, dyskeratosis congenita, and Bloom syndrome) have all benefited from hematopoietic stem cell transplants [75, 76].

8.2.4.1.3 HSCs in Autoimmune Diseases

Autoimmune diseases are a set of disorders that occur when the immune system produces antibodies to assault itself. It is often triggered by a series of genetic, pathogenic, and environmental factors. Celiac disease, Graves' disease, rheumatoid arthritis, systemic lupus, arteriosclerosis, inflammatory bowel disease, schizophrenia, and multiple sclerosis are examples of autoimmune disorders that might be organ specific or systemic [77]. Autoimmune disorders affect roughly 6% of the global population, causing much pain for affected individuals besides being a massive global burden [78].

Autologous HSC transplantation has built its reputation as an efficient treatment strategy for inducing long-term, symptom-free remission in a variety of autoimmune diseases [79]. This treatment strategy has been used successfully for numerous years, with a five-year survival rate of over 80% and treatment-related fatalities of only 5% [80]. The goal of the transplant is to reduce auto-reactive immune cells and regenerate an ingenious, self-tolerant immune system. This process is complemented by the renewal of both the CD4+ and regulatory T cell populations, which then restore the role and heterogeneity of T cell receptors. Similarly, the B cell compartment becomes reinstated, and the number of autoreactive antibodies declines, leading to clinical remission of the disease [79].

8.2.4.2 Mesenchymal Stem Cells

MSCs are multipotent cells derived from human adult bone marrow, skeletal muscle, adipose tissue, umbilical cord, and other connective tissues (Figure 8.3). These cells are non-hematopoietic, have potential for self-renewal and multidirectional differentiation, have anti-inflammatory properties, and can aid in the regeneration of mesoderm-derived bone, cartilage, and muscle tissues as well as liver and pancreatic islets [38]. There has been significant development in the study of MSCs for use in regenerative medicine, immune disorders, and anticancer therapies because of the ease with which they can be harvested, cultured, and expanded *in vitro* [81, 82].

MSCs are believed to mediate tissue repair and regeneration. Once injected intravenously, they proliferate in bone marrow and migrate (through chemoattractant signals) to wounded tissues or inflamed areas and secrete broad spectrum of bioactive agents such as growth factors, cytokines, or chemokines that repress the local immune system, promote angiogenesis, reduce free radical production, and inhibit fibrosis while still allowing tissue-resident stem cells to proliferate and differentiate [83, 84]. The role of MSCs in potential cancer treatment is controversial as some studies show proliferative and others inhibitory effects on tumor progression [85, 86]. However, MSCs could be manipulated and used against cancer cells [87].

8.2.4.2.1 MSCs in Cancer

Transplanted MSCs have been demonstrated to have antitumor effects, the feature currently being exploited for cancer therapeutics. These cells have the potential to migrate toward tumors via chemoattractant cues and once there can release a variety of therapeutic chemicals including growth factors, extracellular vesicles, microRNAs, and cytokines that can reduce cancer cell proliferation, angiogenesis, and cell survival [84, 88]. Naïve MSCs inhibited the growth of solid tumor and leukemia cell lines [89], and the inhibition of the Wnt signaling pathway by secreted Dickkopf-related protein 1 and the activation of the apoptotic pathway in tumor cells are thought to suppress cell growth [90, 91]. Akimoto and colleagues discovered that MSCs procured from umbilical cord blood lowered glioblastoma proliferation by secreting tumor necrosis factor (TNF)-related apoptosis-inducing ligand, which triggers cell death [92]. MSCs have been shown to have anti-angiogenic effects and tumor development inhibition in a mouse model of melanoma [93, 94]. There are currently quite a number of proposed MSC therapies for cancer treatment in clinical trials [84].

Formulated MSCs have been shown to express various therapeutic factors including paclitaxel or cisplatin [95], proapoptotic and suicide genes [96, 97], anti-angiogenesis factors [98], and immunomodulatory cytokines like interleukin-12, TNF-α, and interferons β and γ [99–101], and MSC-derived exosomes have played roles in targeted antitumor therapy. Paracrine cell regulation and the homing to tumors are the properties of MSCs that have been exploited to use them as vehicles for targeted therapy. Exosomes derived from MSCs, loaded with bioactive proteins, mRNAs, or miRNAs can target tumor cells due to their tumor-tropic behavior [102, 103]. Several studies have shown that exosomes derived from MSCs suppressed angiogenesis and tumor progression in breast cancer cells by modulating VEGF signaling [104, 105]. In hepatocellular carcinoma, ovarian cancer and Kaposi's sarcoma, extracellular vesicles (EVs) from human bone marrow MSCs cause apoptosis and growth suppression by activating cell cycle negative regulators [106]. In similar manner, human umbilical cord MSC-derived EVs inhibited bladder carcinoma [107], while human adipose MSC-derived exosomes inhibited ovarian cancer progression [108]. Yudintceva et al. [109] were recently able to target brain tumors with MSCs in the rat model of the orthotopic glioblastoma.

MSCs have also been used to deliver oncolytic viruses (OVs) into tumor sites to cause oncolysis [110, 111]. Strong tropism into tumor microenvironment and support for viral replication make MSCs attractive candidates for delivering OVs [112]. This therapy has already been tested in glioma; colon, prostate, ovary, breast, liver, and lung cancer; and lymphoblastic leukemia [113–117]. Recently, MSCs loaded with oncolytic adenoviruses (OAd-MSCs) or the combination of OAd-MSCs with G-CSF significantly reduced tumor growth in immunocompetent mice with osteosarcoma [113].

Virus-loaded MSCs are showing promising results in treating metastases. In a xenograft model of breast cancer metastases to lungs, intravenous injection of conditionally replicative adenovirus-based agent loaded-hMSCs reduced tumor burden, indicating the efficacy of virus-loaded MSCs in killing tumor cells and thereby using them to treat human cancers [118, 119]. In another study, MSCs loaded with oncolytic herpes simplex virus (MSC-oHSV) significantly tracked metastatic lesions and prolonged the survival of mice with melanoma metastatic brain tumors [120]. This study not only demonstrates the utility of MSCs as OV carriers to disseminated brain lesions but also provides therapeutic guidance to target melanoma brain metastasis.

8.2.4.2.2 MSCs in Cardiovascular Diseases

MSCs can differentiate into osteoblasts, chondrocytes, and adipocytes [121], and they have found numerous applications in the treatment of cardiovascular disorders via processes like somatic reorganization, trans-differentiation, and paracrine signaling [122]. In animal models of myocardial infarction, transplanted MSCs establish themselves and differentiate into cardiomyocytes and vasculature cells, coordinate with intrinsic cardiac stem cells, and secrete a wide range of factors that can be used to remodel an injured ischemic ventricle. As a result, MSC treatment for ischemic heart disease and other cardiac diseases is now being explored in clinical studies [121].

8.2.4.2.3 MSCs in Autoimmune Diseases

MSCs are often regarded as being immunosuppressive, with the capacity of evading graft rejection in animal models; as such, their role in autoimmune illnesses is currently being investigated [123]. A patient with severe treatment-resistant graft-vs-host disease (GvHD) of some internal organs was treated using *ex vivo* grown haplo-identical human mesenchymal cells in 2004 and continued to show excellent outcomes after a year, making the trial a major accomplishment [123]. This finding served as a basis for other examinations of MSCs in the treatment of GvHD, to a large extent after allogeneic stem cell transplantation [124].

Our understanding of MSCs and their therapeutic potential has advanced significantly in the last few decades. MSCs are increasingly recognized as important cells that maintain tissue homeostasis through modulation of the microenvironment. This property, coupled with the ability to differentiate into several multipotent forms, has greatly benefited the development of these cells as a potential treatment for a variety of diseases [121].

8.2.4.3 Induced Pluripotent Stem Cells

These cells can be created in humans in a patient-specific manner [125] by collecting somatic cells (e.g., fibroblasts) from the patient (to be utilized as an autologous transplant) and reconfiguring them to achieve the required plasticity and differentiation state (Figure 8.3) [37]. Human iPSCs offer a promising option for stem cell therapy in a wide range of diseases that lack effective treatments, including neurological diseases, heart conditions, alongside lung, liver and kidney diseases [125]. Another significant benefit of iPSC technology is the capacity to undergo directed differentiation into specific cell types such as hepatocytes, vascular endothelial cells, brain cells, and cardiomyocytes, with research utilizing the diverse cell types in disease treatment [35, 126].

The use of iPSCs in the clinic has been approved for selected diseases in Japan. In 2014, surgeons implanted retinal tissues created after remodeling the patient's own skin cells to pluripotent state. The Japanese woman in her 70s suffered from age-related macular degeneration and was the first person to receive iPSCs. The procedure took place at the Foundation for Biomedical Research and Innovation (FBRI) in Kobe, Japan, where ophthalmologist Masayo Takahashi developed and tested the epithelial sheets [127]. Similarly, a Japanese man in his 60s underwent allogeneic stem cell transplantation using iPSCs in 2017. Skin cells from an anonymous donor were reprogrammed into iPSCs, transformed into a type of retinal cell, and then transplanted into the patient's retina, who suffered from age-related macular degeneration. Physicians believe that the cells will halt the disease's progression and potentially rescue the patient from blindness [128]. In addition, iPSCs have been endorsed for use in the treatment of Parkinson's [129] and cardiovascular diseases [130].

8.3 CONCLUSION

Stem cells can self-renew and differentiate into various cell types. Because of increasing understanding of stem cell biology, it has been possible to isolate various stem cells, culture them *in vitro*, modify them, and transplant them into patients to regenerate tissues to treat various defects, disorders and diseases. Cancer stem cells are also being intensively explored for therapeutic targets. The results are promising despite various challenges and obstacles. It is hoped that in the near future, stem cell-based therapies will be able to treat a range of heretofore incurable diseases.

REFERENCES

1. Blanpain, C. and E. Fuchs, *Stem cell plasticity: plasticity of epithelial stem cells in tissue regeneration*. Science, 2014. **344**(6189): p. 1242281.
2. Amit, M., et al., *Feeder layer- and serum-free culture of human embryonic stem cells*. Biol Reprod, 2004. **70**(3): p. 837–45.
3. Weissman, I.L., *Stem cells: units of development, units of regeneration, and units in evolution*. Cell, 2000. **100**(1): p. 157–68.
4. Kolios, G. and Y. Moodley, *Introduction to stem cells and regenerative medicine*. Respiration, 2013. **85**(1): p. 3–10.
5. Teo, A.K. and L. Vallier, *Emerging use of stem cells in regenerative medicine*. Biochem J, 2010. **428**(1): p. 11–23.
6. Larijani, B., et al., *Stem cell therapy in treatment of different diseases*. Acta Med Iran, 2012. **50**(2): p. 79–96.
7. Rubin, L.L., *Stem cells and drug discovery: the beginning of a new era?* Cell, 2008. **132**(4): p. 549–52.
8. Thomson, J.A., et al., *Embryonic stem cell lines derived from human blastocysts*. Science, 1998. **282**(5391): p. 1145–7.
9. Mason, C. and P. Dunnill, *A brief definition of regenerative medicine*. Regen Med, 2008. **3**(1): p. 1–5.
10. Aly, R.M., *Current state of stem cell-based therapies: an overview*. Stem Cell Investig, 2020. **7**: p. 8.
11. Rossant, J., *Stem cells from the Mammalian blastocyst*. Stem Cells, 2001. **19**(6): p. 477–82.
12. Zakrzewski, W., et al., *Stem cells: past, present, and future*. Stem Cell Res Ther, 2019. **10**(1): p. 68.
13. DeMiguel, M.P., S. Fuentes-Julián, and Y. Alcaina, *Pluripotent stem cells: origin, maintenance and induction*. Stem Cell Rev Rep, 2010. **6**(4): p. 633–49.
14. Smith, A.G., *Embryo-derived stem cells: of mice and men*. Annu Rev Cell Dev Biol, 2001. **17**: p. 435–62.
15. Takahashi, K. and S. Yamanaka, *Induction of pluripotent stem cells from mouse embryonic and adult fibroblast cultures by defined factors*. Cell, 2006. **126**(4): p. 663–76.
16. Khanlarkhani, N., et al., *Multipotent stem cell and reproduction*. J Stem Cells, 2016. **11**(4): p. 219–229.
17. Marone, M., et al., *Cell cycle regulation in human hematopoietic stem cells: from isolation to activation*. Leuk Lymphoma, 2002. **43**(3): p. 493–501.
18. Sarkar, A., et al., *Understanding stem cells and its pivotal role in regenerative medicine*. Life Sci, 2021. **273**: p. 119270.
19. Bentzinger, C.F., et al., *The emerging biology of muscle stem cells: implications for cell-based therapies*. Bioessays, 2013. **35**(3): p. 231–41.
20. Kalra, K., et al., *Advances in the generation of functional β-cells from induced pluripotent stem cells as a cure for diabetes mellitus*. Curr Drug Targets, 2018. **19**(13): p. 1463–1477.
21. Biswas, A. and R. Hutchins, *Embryonic stem cells*. Stem Cells Dev, 2007. **16**(2): p. 213–22.
22. Evans, M.J. and M.H. Kaufman, *Establishment in culture of pluripotential cells from mouse embryos*. Nature, 1981. **292**(5819): p. 154–6.
23. Martin, G.R., *Isolation of a pluripotent cell line from early mouse embryos cultured in medium conditioned by teratocarcinoma stem cells*. Proc Natl Acad Sci USA, 1981. **78**(12): p. 7634–8.
24. Thomson, J.A., et al., *Isolation of a primate embryonic stem cell line*. Proc Natl Acad Sci U S A, 1995. **92**(17): p. 7844–8.
25. Shamblott, M.J., et al., *Derivation of pluripotent stem cells from cultured human primordial germ cells*. Proc Natl Acad Sci USA, 1998. **95**(23): p. 13726–31.
26. Tesar, P.J., et al., *New cell lines from mouse epiblast share defining features with human embryonic stem cells*. Nature, 2007. **448**(7150): p. 196–9.
27. Prentice, D.A., *Adult stem cells*. Circ Res, 2019. **124**(6): p. 837–839.
28. Kiefer, J.C., *Primer and interviews: the dynamic stem cell niche*. Dev Dyn, 2011. **240**(3): p. 737–43.
29. Voog, J. and D.L. Jones, *Stem cells and the niche: a dynamic duo*. Cell Stem Cell, 2010. **6**(2): p. 103–15.
30. He, X.C., et al., *BMP signaling inhibits intestinal stem cell self-renewal through suppression of Wnt-beta-catenin signaling*. Nat Genet, 2004. **36**(10): p. 1117–21.
31. Kulkarni, V., et al., *Asrij maintains the stem cell niche and controls differentiation during Drosophila lymph gland hematopoiesis*. PLoS One, 2011. **6**(11): p. e27667.
32. Ahmed, I., et al., *Dietary interventions ameliorate infectious colitis by restoring the microbiome and promoting stem cell proliferation in mice*. Int J Mol Sci, 2021. **23**(1).
33. Barker, N., S. Bartfeld, and H. Clevers, *Tissue-resident adult stem cell populations of rapidly self-renewing organs*. Cell Stem Cell, 2010. **7**(6): p. 656–70.
34. Umar, S., *Intestinal stem cells*. Curr Gastroenterol Rep, 2010. **12**(5): p. 340–8.
35. Ye, L., C. Swingen, and J. Zhang, *Induced pluripotent stem cells and their potential for basic and clinical sciences*. Curr Cardiol Rev, 2013. **9**(1): p. 63–72.

36. Singh, V.K., et al., *Induced pluripotent stem cells: applications in regenerative medicine, disease modeling, and drug discovery*. Front Cell Dev Biol, 2015. **3**(2).
37. GoisBeghini, D., et al., *Induced pluripotent stem cells: hope in the treatment of diseases, including muscular dystrophies*. Int J Mol Sci, 2020. **21**(15): p. 5467.
38. Nelson, T.J., et al., *Stem cell platforms for regenerative medicine*. Clin Transl Sci, 2009. **2**(3): p. 222–7.
39. Chang, J.C., *Cancer stem cells: role in tumor growth, recurrence, metastasis, and treatment resistance*. Medicine (Baltimore), 2016. **95**(1 Suppl 1): p. S20–S25.
40. Ayob, A.Z. and T.S. Ramasamy, *Cancer stem cells as key drivers of tumour progression*. J Biomed Sci, 2018. **25**(1): p. 20.
41. Vermeulen, L., et al., *The developing cancer stem-cell model: clinical challenges and opportunities*. Lancet Oncol, 2012. **13**(2): p. e83–9.
42. Guha, M., et al., *Mitochondrial retrograde signaling induces epithelial-mesenchymal transition and generates breast cancer stem cells*. Oncogene, 2014. **33**(45): p. 5238–50.
43. de Sousae Melo, F., et al., *A distinct role for Lgr5(+) stem cells in primary and metastatic colon cancer*. Nature, 2017. **543**(7647): p. 676–680.
44. Cheung, P., et al., *Regenerative reprogramming of the intestinal stem cell state via hippo signaling suppresses metastatic colorectal cancer*. Cell Stem Cell, 2020. **27**(4): p. 590–604.e9.
45. Tan, B.T., et al., *The cancer stem cell hypothesis: a work in progress*. Lab Invest, 2006. **86**(12): p. 1203–7.
46. Sell, S., *On the stem cell origin of cancer*. Am J Pathol, 2010. **176**(6): p. 2584–594.
47. Lapidot, T., et al., *A cell initiating human acute myeloid leukaemia after transplantation into SCID mice*. Nature, 1994. **367**(6464): p. 645–8.
48. Tang, D.G., *Understanding cancer stem cell heterogeneity and plasticity*. Cell Res, 2012. **22**(3): p. 457–72.
49. Yang, L., et al., *Targeting cancer stem cell pathways for cancer therapy*. Signal Transduct Target Ther, 2020. **5**(1): p. 8.
50. Milanovic, M., et al., *Senescence-associated reprogramming promotes cancer stemness*. Nature, 2018. **553**(7686): p. 96–100.
51. Prasad, S., et al., *Cancer cells stemness: a doorstep to targeted therapy*. Biochim Biophys Acta Mol Basis Dis, 2020. **1866**(4): p. 165424.
52. Adair, J.E., S.P. Kubek, and H.P. Kiem, *Hematopoietic stem cell approaches to cancer*. Hematol Oncol Clin North Am, 2017. **31**(5): p. 897–912.
53. Henig, I. and T. Zuckerman, *Hematopoietic stem cell transplantation-50 years of evolution and future perspectives*. Rambam Maimonides Med J, 2014. **5**(4): p. e0028.
54. Gratwohl, A., et al., *One million haemopoietic stem-cell transplants: a retrospective observational study*. Lancet Haematol, 2015. **2**(3): p. e91–100.
55. Chu, D.T., et al., *Recent progress of stem cell therapy in cancer treatment: molecular mechanisms and potential applications*. Cells, 2020. **9**(3).
56. Zhang, C.L., et al., *Stem cells in cancer therapy: opportunities and challenges*. Oncotarget, 2017. **8**(43): p. 75756–66.
57. Copelan, E.A., *Hematopoietic stem-cell transplantation*. N Engl J Med, 2006. **354**(17): p. 1813–26.
58. Champlin, R., *Selection of autologous or allogeneic transplantation*, in Kufe DW, Pollock RE, Weichselbaum RR, et al., editors. *Holland-Frei Cancer Medicine*. 6th edition. Hamilton (ON): BC Decker, 2003.
59. Thomas, E., et al., *Bone-marrow transplantation (first of two parts)*. N Engl J Med, 1975. **292**(16): p. 832–43.
60. Ljungman, P., et al., *Allogeneic and autologous transplantation for haematological diseases, solid tumours and immune disorders: current practice in Europe 2009*. Bone Marrow Transplant, 2010. **45**(2): p. 219–34.
61. Lennard, A.L. and G.H. Jackson, *Stem cell transplantation*. West J Med, 2001. **175**(1): p. 42–46.
62. Barriga, F., N. Rojas, and A. Wietstruck, *Alternative donor sources for hematopoietic stem cell transplantation*, in *Innovations in Stem Cell Transplantation*. InTech, 2013.
63. Singh, A.K. and J.P.McGuirk, *Allogeneic stem cell transplantation: a historical and scientific overview*. Cancer Res, 2016. **76**(22): p. 6445–51.
64. Shenoy, S., *Hematopoietic stem-cell transplantation for sickle cell disease: current evidence and opinions*. Therap Adv Hematol, 2013. **4**(5): p. 335–44.
65. Ashorobi, D. and R. Bhatt, *Bone Marrow Transplantation In Sickle Cell Disease*. Treasure Island (FL): StatPearls Publishing, 2021 January.
66. Lucarelli, G., et al., *Hematopoietic stem cell transplantation in thalassemia and sickle cell anemia*. Cold Spring Harbor Perspect Med, 2012. **2**(5): p. a011825.
67. Özdoğu, H. and C. Boğa, *Hematopoietic stem cell transplantation in adult sickle cell disease: problems and solutions*. Turk J Hematol, 2015. **32**(3): p. 195–205.
68. Bhatia, M. and S. Sheth, *Hematopoietic stem cell transplantation in sickle cell disease: patient selection and special considerations*. J Blood Med, 2015: p. 229.
69. Walters, M.C., *Update of hematopoietic cell transplantation for sickle cell disease*. Curr Opin Hematol, 2015. **22**(3): p. 227–33.
70. Bajwa, H. and H. Basit, *Thalassemia*. StatPearls [Internet]. Treasure Island (FL): StatPearls Publishing, 2021 January.
71. De Sanctis, V., et al., *β-thalassemia distribution in the old world: an ancient disease seen from a historical standpoint*. Mediterr J Hematol Infect Dis, 2017. **9**(1): p. e2017018.
72. Muncie, H.L. and J. Campbell, *Alpha and beta thalassemia*. Am Fam Physician, 2009. **80**(4): p. 339–44.
73. Angelucci, E., et al., *Hematopoietic stem cell transplantation in thalassemia major and sickle cell disease: indications and management recommendations from an international expert panel*. Haematologica, 2014. **99**(5): p. 811–20.
74. Angelucci, E., *Hematopoietic stem cell transplantation in thalassemia*. Hematology Am Soc Hematol Educ Program, 2010. **2010**: p. 456–62.
75. Khaddour, K., C.K. Hana, and P. Mewawalla, *Hematopoietic Stem Cell Transplantation*. StatPearls [Internet]. Treasure Island (FL): StatPearls Publishing, 2021 January.
76. Tolar, J., P.A. Mehta, and M.C. Walters, *Hematopoietic cell transplantation for nonmalignant disorders*. Biol Blood Marrow Transplant, 2012. **18**(1 Suppl): p. S166–71.
77. Smith, D.A. and D.R. Germolec, *Introduction to immunology and autoimmunity*. Environ Health Perspect, 1999. **107**(Suppl 5): p. 661–5.
78. Fugger, L., L.T. Jensen, and J. Rossjohn, *Challenges, progress, and prospects of developing therapies to treat autoimmune diseases*. Cell, 2020. **181**(1): p. 63–80.
79. Swart, J.F., et al., *Haematopoietic stem cell transplantation for autoimmune diseases*. Nat Rev Rheumatol, 2017. **13**(4): p. 244–56.
80. Hügle, T. and T. Daikeler, *Stem cell transplantation for autoimmune diseases*. Haematologica, 2010. **95**(2): p. 185–8.
81. Sherman, L.S., et al., *Mesenchymal stromal/stem cells in drug therapy: new perspective*. Cytotherapy, 2017. **19**(1): p. 19–27.
82. Rajabzadeh, N., E. Fathi, and R. Farahzadi, *Stem cell-based regenerative medicine*. Stem Cell Investig, 2019. **6**: p. 19.

83. Joyce, N., et al., *Mesenchymal stem cells for the treatment of neurodegenerative disease*. Regen Med, 2010. **5**(6): p. 933–46.
84. Hmadcha, A., et al., *Therapeutic potential of mesenchymal stem cells for cancer therapy*. Front Bioeng Biotechnol, 2020. **8**: p. 43.
85. Wong, R.S., *Mesenchymal stem cells: angels or demons?* J Biomed Biotechnol, 2011. **2011**: p. 459510.
86. Gomes, J.P.A., et al., *Deepening a simple question: can MSCs be used to treat cancer?* Anticancer Res, 2017. **37**(9): p. 4747–58.
87. Javan, M.R., A. Khosrojerdi, and S.M. Moazzeni, *New insights into implementation of mesenchymal stem cells in cancer therapy: prospects for anti-angiogenesis treatment*. Front Oncol, 2019. **9**: p. 840.
88. Prockop, D.J., *The exciting prospects of new therapies with mesenchymal stromal cells*. Cytotherapy, 2017. **19**(1): p. 1–8.
89. Ramasamy, R., et al., *Mesenchymal stem cells inhibit proliferation and apoptosis of tumor cells: impact on in vivo tumor growth*. Leukemia, 2007. **21**(2): p. 304–10.
90. Qiao, L., et al., *Dkk-1 secreted by mesenchymal stem cells inhibits growth of breast cancer cells via depression of Wnt signalling*. Cancer Lett, 2008. **269**(1): p. 67–77.
91. Sun, B., et al., *Therapeutic potential of mesenchymal stromal cells in a mouse breast cancer metastasis model*. Cytotherapy, 2009. **11**(3): p. 289–98, 1 p following 298.
92. Akimoto, K., et al., *Umbilical cord blood-derived mesenchymal stem cells inhibit, but adipose tissue-derived mesenchymal stem cells promote, glioblastoma multiforme proliferation*. Stem Cells Dev, 2013. **22**(9): p. 1370–86.
93. Khakoo, A.Y., et al., *Human mesenchymal stem cells exert potent antitumorigenic effects in a model of Kaposi's sarcoma*. J Exp Med, 2006. **203**(5): p. 1235–47.
94. Okada, H. and I.F. Pollack, *Cytokine gene therapy for malignant glioma*. Expert Opin Biol Ther, 2004. **4**(10): p. 1609–20.
95. Lin, W., et al., *Mesenchymal stem cells and cancer: clinical challenges and opportunities*. Biomed Res Int, 2019. **2019**: p. 2820853.
96. Mueller, L.P., et al., *TRAIL-transduced multipotent mesenchymal stromal cells (TRAIL-MSC) overcome TRAIL resistance in selected CRC cell lines in vitro and in vivo*. Cancer Gene Ther, 2011. **18**(4): p. 229–39.
97. Altaner, C., et al., *Complete regression of glioblastoma by mesenchymal stem cells mediated prodrug gene therapy simulating clinical therapeutic scenario*. Int J Cancer, 2014. **134**(6): p. 1458–65.
98. Chu, Y., et al., *Human placenta mesenchymal stem cells expressing exogenous kringle1–5 protein by fiber-modified adenovirus suppress angiogenesis*. Cancer Gene Ther, 2014. **21**(5): p. 200–8.
99. Ryu, C.H., et al., *Gene therapy of intracranial glioma using interleukin 12-secreting human umbilical cord blood-derived mesenchymal stem cells*. Hum Gene Ther, 2011. **22**(6): p. 733–43.
100. Shahrokhi, S., S. Daneshmandi, and F. Menaa, *Tumor necrosis factor-α/CD40 ligand-engineered mesenchymal stem cells greatly enhanced the antitumor immune response and lifespan in mice*. Hum Gene Ther, 2014. **25**(3): p. 240–53.
101. Zhang, L., et al., *Bone marrow mesenchymal stem cells suppress metastatic tumor development in mouse by modulating immune system*. Stem Cell Res Ther, 2015. **6**(1): p. 45.
102. Yuan, Q.L., Y.G. Zhang, and Q. Chen, *Mesenchymal stem cell (MSC)-derived extracellular vesicles: potential therapeutics as MSC trophic mediators in regenerative medicine*. Anat Rec (Hoboken), 2020. **303**(6): p. 1735–42.
103. Weng, Z., et al., *Therapeutic roles of mesenchymal stem cell-derived extracellular vesicles in cancer*. J Hematol Oncol, 2021. **14**(1): p. 136.
104. Lee, J.K., et al., *Exosomes derived from mesenchymal stem cells suppress angiogenesis by down-regulating VEGF expression in breast cancer cells*. PLoS One, 2013. **8**(12): p. e84256.
105. Pakravan, K., et al., *MicroRNA-100 shuttled by mesenchymal stem cell-derived exosomes suppresses in vitro angiogenesis through modulating the mTOR/HIF-1α/VEGF signaling axis in breast cancer cells*. Cell Oncol (Dordr), 2017. **40**(5): p. 457–70.
106. Bruno, S., et al., *Microvesicles derived from human bone marrow mesenchymal stem cells inhibit tumor growth*. Stem Cells Dev, 2013. **22**(5): p. 758–71.
107. Wu, S., et al., *Microvesicles derived from human umbilical cord Wharton's jelly mesenchymal stem cells attenuate bladder tumor cell growth in vitro and in vivo*. PLoS One, 2013. **8**(4): p. e61366.
108. Reza, A., et al., *Human adipose mesenchymal stem cell-derived exosomal-miRNAs are critical factors for inducing anti-proliferation signalling to A2780 and SKOV-3 ovarian cancer cells*. Sci Rep, 2016. **6**: p. 38498.
109. Yudintceva, N., et al., *Targeting brain tumors with mesenchymal stem cells in the experimental model of the orthotopic glioblastoma in rats*. Biomedicines, 2021. **9**(11).
110. Shah, K., *Mesenchymal stem cells engineered for cancer therapy*. Adv Drug Deliv Rev, 2012. **64**(8): p. 739–48.
111. Zong, C., et al., *The distinct roles of mesenchymal stem cells in the initial and progressive stage of hepatocarcinoma*. Cell Death Dis, 2018. **9**(3): p. 345.
112. Ong, H.T., et al., *Systemically delivered measles virus-infected mesenchymal stem cells can evade host immunity to inhibit liver cancer growth*. J Hepatol, 2013. **59**(5): p. 999–1006.
113. Morales-Molina, A., et al., *Combination immunotherapy using G-CSF and oncolytic virotherapy reduces tumor growth in osteosarcoma*. J Immunother Cancer, 2021. **9**(3).
114. Stuckey, D.W. and K. Shah, *Stem cell-based therapies for cancer treatment: separating hope from hype*. Nat Rev Cancer, 2014. **14**(10): p. 683–91.
115. Ramírez, M., et al., *Patient-derived mesenchymal stem cells as delivery vehicles for oncolytic virotherapy: novel state-of-the-art technology*. Oncolytic Virother, 2015. **4**: p. 149–55.
116. Nowakowski, A., et al., *Engineered mesenchymal stem cells as an anticancer trojan horse*. Stem Cells Dev, 2016. **25**(20): p. 1513–31.
117. Russell, L., et al., *Oncolytic viruses: priming time for cancer immunotherapy*. BioDrugs, 2019. **33**(5): p. 485–501.
118. Stoff-Khalili, M.A., et al., *Mesenchymal stem cells as a vehicle for targeted delivery of CRAds to lung metastases of breast carcinoma*. Breast Cancer Res Treat, 2007. **105**(2): p. 157–67.
119. Hakkarainen, T., et al., *Human mesenchymal stem cells lack tumor tropism but enhance the antitumor activity of oncolytic adenoviruses in orthotopic lung and breast tumors*. Hum Gene Ther, 2007. **18**(7): p. 627–41.
120. Du, W., et al., *Stem cell-released oncolytic herpes simplex virus has therapeutic efficacy in brain metastatic melanomas*. Proc Natl Acad Sci USA, 2017. **114**(30): p. e6157–65.
121. Williams, A.R. and J.M. Hare, *Mesenchymal stem cells: biology, pathophysiology, translational findings, and therapeutic implications for cardiac disease*. Circ Res, 2011. **109**(8): p. 923–40.
122. Patel, D.M., J. Shah, and A.S. Srivastava, *Therapeutic potential of mesenchymal stem cells in regenerative medicine*. Stem Cells Int, 2013. **2013**: p. 496218.
123. Le Blanc, K., et al., *Treatment of severe acute graft-versus-host disease with third party haploidentical mesenchymal stem cells*. Lancet, 2004. **363**(9419): p. 1439–41.

124. Dazzi, F. and M. Krampera, *Mesenchymal stem cells and autoimmune diseases*. Best Pract Res Clin Haematol, 2011. **24**(1): p. 49–57.
125. Moradi, S., et al., *Research and therapy with induced pluripotent stem cells (iPSCs): social, legal, and ethical considerations*. Stem Cell Res Ther, 2019. **10**(1): p. 341.
126. Hirschi, K.K., S. Li, and K. Roy, *Induced pluripotent stem cells for regenerative medicine*. Annu Rev Biomed Eng, 2014. **16**(1): p. 277–94.
127. Cyranoski, D., *Japanese woman is first recipient of next-generation stem cells*. Nature, 2014. **12**.
128. Cyranoski, D., *Japanese man is first to receive 'reprogrammed' stem cells from another person*. Nature, 2017.
129. Cyranoski, D., *Reprogrammed' stem cells implanted into patient with Parkinson's disease*. Nature, 2018.
130. Cyranoski, D., *Reprogrammed' stem cells approved to mend human hearts for the first time*. Nature, 2018.

9 Mesenchymal Stem Cells and Their Exosomes
Emerging Therapeutics in Cancer Treatment

Keerthi N, Amit Dey, Ameena Farhana, Antara Banerjee, and Surajit Pathak*
Medical Biotechnology lab, Faculty of Allied Health Sciences, Chettinad Academy of Research and Education, Chettinad Hospital and Research Institute, Tamil Nadu, India

9.1 INTRODUCTION

Adult stem cells with the ability to self-renew and differentiate into many lineages are known as mesenchymal stem cells (MSCs) (Jiang et al. 2002). Paracrine factors secreted from the MSCs are essential for mediating the intercellular communication that preserves physiological homeostasis and impacts disease (Hmadcha et al. 2020).

Emerging evidence indicates that most biological activities of stem cells are mediated by their secreted extracellular vesicles (EVs), which exert paracrine actions to influence the injured microenvironment of the damaged tissue, leading to differentiation of cells and organogenesis. Exosomes have an intrinsic role in protein machinery that allows them to move their loads of biomaterials between target cells (Li et al. 2022). They transport bioactive cargo across biological barriers and are biodegradable and low in toxicity and immunogenicity. Additional agents that influence MSCs' migration toward cancerous tissue include diffusible cytokines like IL-8, growth factors like TGF-β1 or platelet derived growth factor, and extracellular matrix (ECM) molecules like matrix metalloproteinase-2 (MMP-2) (Bhoopathi et al. 2011). MSCs interact directly and indirectly with cancer cells once they have entered the tumor niche, which influences the growth of tumors. Exosomes contain a variety of materials, including lipids, proteins, enzymes, nucleic acids, and miRNAs (Bian et al. 2022) and are increasingly being used as drug transporters (Mukherjee et al. 2022). They also function as biomarkers for the early identification and advancement of cancer. As a result, they have been proposed for use in cancer treatments as either drug delivery systems or diagnostic tools (Stefańska et al. 2023). The most suitable exosome-producing sources need to be identified with the goal of preventing any adverse effects that promote tumor growth, depending on the purpose of medication administration (Kibria et al. 2018).

9.2 THE ROLE OF EXOSOMES IN CANCER

Exosome heterogeneity in source, size, contents, and roles results from variations in their physiological microenvironments (Keerthikumar et al. 2016). Researchers have demonstrated that exosomes are an effective therapeutic agent and delivery method based on their inherent benefits including accumulation in acidic tumor microenvironments (TMEs), specificity to certain cancer types, immunological effects, and high stability (Wang et al. 2023).

Tumor exosomes play a role in the communication between tumor and normal cells. Additionally, they trigger changes in the cellular landscape that promote tumor proliferation and invasion by activating MAPK/ERK signaling and other essential cellular pathways. Exosomes released by tumor cells also guide adipocytes by fostering an environment conducive to tumor growth and facilitating the propagation of any tumor-promoting downstream effects (Jella et al. 2018). Exosomes might impose an effect in encouraging the interaction between immune cells – such as dendritic cells, neutrophils, macrophages, natural killer (NK) cells, and T cells – and cancer cells. On the contrary, exosomes generated from tumors additionally govern macrophage polarization (Tai et al. 2018). Exosomes have alluring qualities as a drug delivery method for cancer therapies, including their nano-size, biocompatibility, capacity to target specific tumors, and ease of customization. Additionally, exosomes can be quickly transferred via the lymphatic system from the periphery to the lymph nodes, where they deliver drugs for cancer immunotherapy (Srinivasan et al. 2016).

9.3 THE BIOGENESIS OF MSC EXOSOMES

MSCs are the only identified cells that generate exosomes on a large scale. The discharge of exosomes has been linked to intraluminal vesicle internal budding and plasma membrane fusion. Molecular motors including dynein and kinesin, membrane fusion proteins, and microtubule and microfilament components of the cytoskeleton are among the main agents required for this procedure to take place. Through endocytosis, binding to the host receptor, and direct attachment to the plasma membrane, exosomes can be picked up by the recipient cells (Weng et al. 2016).

MSC exosomes must be cultivated under good manufacturing practices (GMP) settings for use in clinical trials. GMP guidelines must be closely adhered to while

* Corresponding author: antarabanerjee@care.edu.in

DOI: 10.1201/9781032660592-11

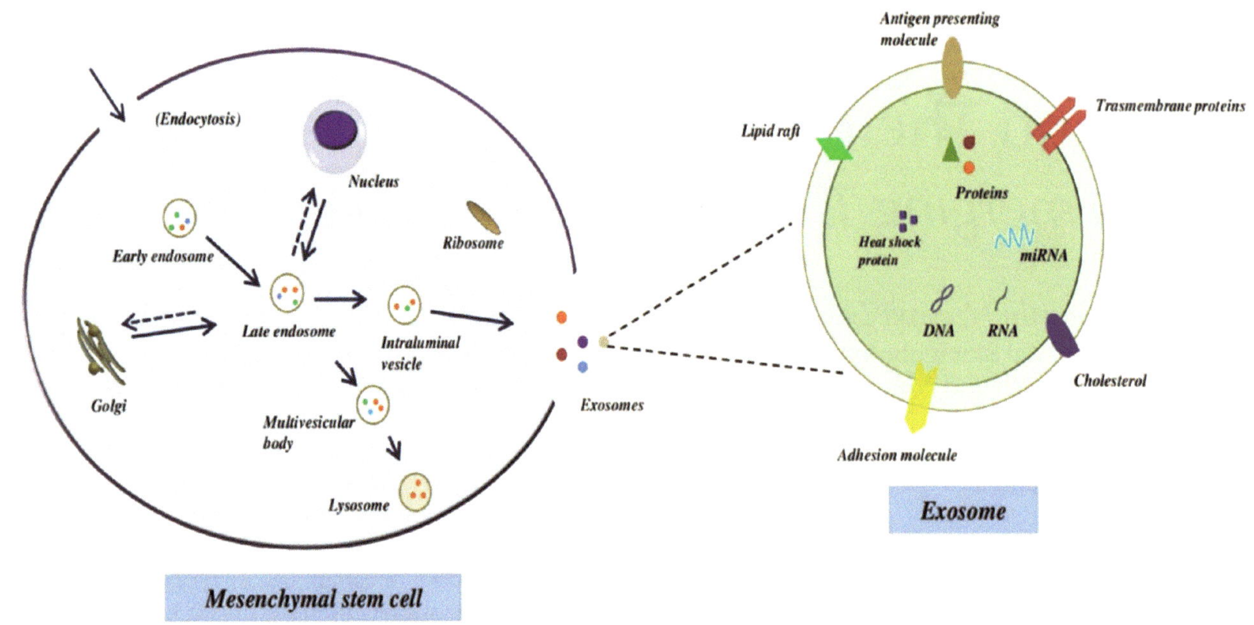

FIGURE 9.1 The biogenesis of exosomes.

monitoring the media, dissipation enzyme, cultivation method, and cell culture environment. The exosomes must be refined, examined for physical structure and bioactivity, and given authorization for product release before they can be used. Ultracentrifugation is the most often used method in clinical research of separating MSC exosomes (Busatto et al. 2018). A sterile hollow-fiber polyether-sulfone membrane with a certain size (measured in μm) is filtered through a cell culture medium to eliminate cell debris and protect the biomolecules. The filter is then repeatedly cleaned using sterile phosphate-buffered saline (PBS).

9.3.1 Isolation of MSC Exosomes

Over the years, several techniques have been employed for isolating exosomes such as ultracentrifugation (UC), capture, precipitation, and microfluidics. UC is the traditional method for isolating exosomes based on its pelleting characteristics (Helwa et al. 2017). The technique entails repeated centrifugations that isolate the exosomes from body fluids. The UC exosomes are isolated as pellets and then resuspended in PBS buffer and stored at −80°C for future use (Greening et al. 2015). High-purity exosomes are frequently created via capture-based methods, which are intimately connected to immunoaffinity (Oksvold et al. 2015). Antibody-coated magnetic beads are used to enrich the surfaces of exosomes with a range of membrane proteins, including CD9, CD63, ALIX, and Ep-CAM (Greening et al. 2015). Polymers are used to precipitate the exosomes, and the most frequently used polymer for exosome isolation is PEG, which produces high exosome yields (Weng et al. 2016; Helwa et al. 2017). In contrast, microfluidics isolation uses devices for precise, fast, and cost-effective separation, which is a perfect tool for isolating exosomes from other nanosized particles (Salafi et al. 2016).

Scaling up production is one of the main challenges in creating MSC exosome-mediated drug delivery systems. One of the most important things to take into account when designing exosome-based medication delivery systems is the MSC source. Exosomal release was enhanced by prolonged culture and maintaining cells at a low pH (Ban et al. 2015). Finally, large-scale exosome production can be achieved by using tissue engineering-style manufacturing bioreactors (Kink et al. 2024).

9.3.2 Purification of MSC Exosomes

Tangential flow filtration (TFF) has been recommended as one of the most effective processes for large-scale production of extracellular organic matter such as MSC exosomes. More importantly, a new study showed that exosomes separated by TFF had higher activity and efficacy than those separated by UC. Using diafiltration by TFF with appropriate pore diameters and parameters, such as transmembrane pressure, flow rate, and diafiltration factor, can achieve high-purity separation of MSC exosomes (Naseri et al. 2018).

9.3.3 Storage of MSC Exosomes

To preserve exosomes' biological activity for use in transportation and therapeutics, further investigation must focus on exosome storage techniques (Liu et al. 2021). Currently, freezing, freeze drying, and spray drying are the most common methods of protection (Yi Zhang et al. 2020). MSC exosomes are often suspended in phosphate buffer upon extraction (Kusuma et al. 2017).

Cryopreservation is the most widely utilized technique of storage; however, it can cause numerous layers of vesicles to develop and aggregate, as well as alterations in the structure and physical characteristics of the exosomes (Zhang et al. 2020). Standard laboratory storage temperatures for MSC exosomes are 4 °C, −20 °C, and −80 °C. Researchers studied exosomes held at those three temperatures for up to 28 days and found that in the same storage environment, exosomes from diverse sources exhibit significantly varied levels of stability because of their heterogeneity. Although storage at 4 °C easily destroyed proteins and nucleic acids, it also prevented the freeze–thaw process from destroying the vesicles (Liu et al. 2021).

9.4 THE IMPACT OF MSC EXOSOMES ON TUMOR MICROENVIRONMENTS

The complex and diverse network called the TME comprises endothelial cells, tumor-associated macrophages (TAMs), myeloid-derived suppressor cells (MDSCs), tumor-related stromal cells, and cancer-associated fibroblasts. The two primary cell types that comprise a tumor are "non-transformed" cells and general tumor cells. Cytokines, chemokines, growth factors, inflammation-related factors, and other cell-to-cell communication mechanisms are involved. MSC exosomes are responsible for mediating the interactions that are established between the two primary cell types (Maman et al. 2018; Xiong and Xu 2016), and the interactions among these cell groups characterize the TMEs. Additionally, MSC exosomes modify TME activity by delivering regulatory cargos to tumor cells, TAMs, MDSCs, and endothelial cells, among other target cells (Whiteside 2018).

In TMEs, distinct soluble signals generated by both cell types and paracrine signaling mediated by exosomes create the contact between MSC exosomes and tumor cells. The exosomes enhance the systemic entry and progression of cancer cells along the metastatic cascade. Exosomes are released by the mass of tumor cells and the associated tumor stroma, and they promote various biological processes including proliferation, resistance to apoptosis, and angiogenesis (Vallabhaneni et al. 2016). This phenomenon emphasizes how crucial it is to comprehend the biology of exosomes and how they contribute to the spread of metastatic illness.

During cancer cell metastasis, exosomes indirectly affect the endothelial–mesenchymal transition, preconditioning BM-derived cells, like MSCs, to migrate to and prepare the target tissue for the cancer cells (Kidd et al. 2009). By transferring components in a variety of settings and cancer stages, MSC exosomes can both stimulate and inhibit tumor formation in recipient cells (Li et al. 2021). Remarkably, exosomes derived from various tissues have varying impacts on malignancies; for instance, by transmitting distinct microRNAs, bone marrow MSC exosomes promote cancer cell growth (Chang et al. 2021). Tumor exosomes work in concert with the inflammatory cytokines generated by nearby proinflammatory microenvironments to attract and facilitate the formation of suppressive immune cells in the pre-metastatic niche. Such cells include TAMs, tumor-associated neutrophils, and regulatory T cells (Liu et al. 2021). Exosomes derived from MSCs have the capacity to sustain tumor growth *in vivo*, enhance migration, aid in the development of apoptosis resistance, and stimulate angiogenesis, all of which contribute to the development of a vascular microenvironment prior to metastasis (Vallabhaneni et al. 2015). Tumor cells release various factors and exosomes that enable MSCs to take on a chemo resistance phenotype with an increase in pluripotency markers, thus allowing them to endure harsh environments and continuing the release of chemicals to guarantee the survival of cancer cells (El-Badawy et al. 2017).

9.5 THERAPEUTIC ROLES OF MSC-DERIVED EXOSOMES IN CANCER

Exosomes generated from MSCs assist in identifying the environmental modifications that encourage the neoplastic transformations of various tissues (De Luca et al. 2019). Many strategies have aimed to use MSC exosomes to deliver therapeutic compounds because of their strong affinity for malignancies (Bolontrade et al. 2012). Because MSC exosomes are essential in forming the premetastatic niche, it is important to carefully evaluate their impacts on these conditions and thoroughly understand their contents and effects on metastatic cells as the tumor progresses. With this insight, potential therapeutic approaches that enable the modification of MSC exosomes can potentially be better understood, as might up-regulated molecules with promise as prognostic biomarkers (Bouchlaka et al. 2017). Studies have also revealed that variations in the MSC passage number, culture medium components, and expansion times have significant influence on outcomes of using exosomes for tumor treatment (Lyu et al. 2021).

9.5.1 Angiogenesis

MSCs emit exosomes that transport mRNAs and miRNAs to target cells, promoting the proliferation of endothelial cells to generate angiogenic effects, improve circulation reconstruction, and build capillary networks (Zhu et al. 2012; Subbaraj et al. 2021). Exosomes and paracrine factors secreted from MSCs prevent certain transcription factors that regulated angiogenesis, thereby leading to remodeling of the antivascular system.

The distinctive functions of miRNAs in tumor angiogenesis are believed to be associated with MSC exosomes (Ferguson et al. 2018). MSC-derived exosomes stimulate angiogenesis by enhancing the synthesis of vascular endothelial growth factor (VEGF) and by triggering the ERK1/2 and p38 mitogen-activated protein kinase pathways (MAPK) in tumor cells. The primary target of exosomes in the induction of angiogenesis could be a component of the angiopoietin network given that genes that regulate angiogenesis and vessel formation have been demonstrated to be targeted by the miRNA content of exosomes produced from MSCs (Huang et al. 2015).

Breast cancer cells were discovered to have the opposite effects, where human bone marrow MSC-derived exosomes inhibited tumor growth and angiogenesis both *in vivo* and *in vitro* by introducing miR-16 into tumor cells, making miR-16 target VEGF and lower its expression in breast cancer cells. The significant role of VEGF and its receptors, VEGFR-1 and VEGFR-2, in vasculogenesis and tumor angiogenesis is widely recognized. MSC exosomes derived from human bone marrow promoted tumor angiogenesis *in vivo* (Zhu et al. 2012; Vallabhaneni et al. 2015).

9.5.2 Metastasis and Invasion

MSCs encapsulate miRNA in exosomes that hinder the migration and invasion of tumors. Researchers used exosomes to effectively deliver artificial miR-145 and miR-124 mimics to co-cultured glioma cells using both gap junction-dependent and independent methods. The movement of gliomas was drastically inhibited by the miR-145 and miR-124 mimics that were administered (Lee et al. 2013). When MSC-derived exosomes were studied, Wnt signaling had shown a role in regulating the proliferation of breast tumor cells and thus in invasion as well (Maffey et al. 2017). Additionally, MSC-derived EVs contributed to the invasiveness of breast cancer cells produced by MSCs (Lin et al. 2013).

9.5.3 Proliferation and Apoptosis

MSC-derived exosomes regulate tumor growth both *in vitro* and *in vivo* via transferring their contents to neighboring cancer cells present in the TME or by altering the phenotype of recipient cells (Camussi et al. 2010; Girigoswami and Girigoswami 2023). The presence of MMP-2 in exosomes released from MSCs alters how cells function while encouraging the reorganization of the TME, presenting a unique strategy for accelerating tumor progression (Yang et al. 2015). Additionally, MSC exosomes function as transporters of miRNA, lipids, metabolites, and tumor-supporting proteins, all of which are crucial to breast cancer progression. Human MSC exosomes inhibited Akt protein kinase phosphorylation, upregulated p53/p21, cleaved caspase 3, and showed antiproliferative and apoptosis-inducing actions on human bladder cancer T24 cells *in vitro* and bladder cancer experimental models (Wu et al. 2013).

9.5.4 Immune Modulation

MSC-derived exosomes have been found to be critical regulators of immune responses in the TME. These nanoscale vesicles convey a wide variety of bioactive molecules, such as miRNAs, proteins, and lipids, which have the ability to regulate the behavior of a wide range of immune cells. A well-documented mechanism is enhancing the expansion of regulatory T cells (Treg) and polarization of macrophages to an anti-inflammatory M2 phenotype and hence creating an immunosuppressive environment suitable for tumor growth (Zhang et al. 2022). For example, exosomal miR-146a has been shown to suppress pro-inflammatory responses, whereas miR-155 can increase immune activation (Kwon et al. 2020). On the other hand, under some circumstances, MSC-derived exosomes can increase antitumor immunity. They are capable of enhancing the maturation and activation of dendritic cells and activating NK

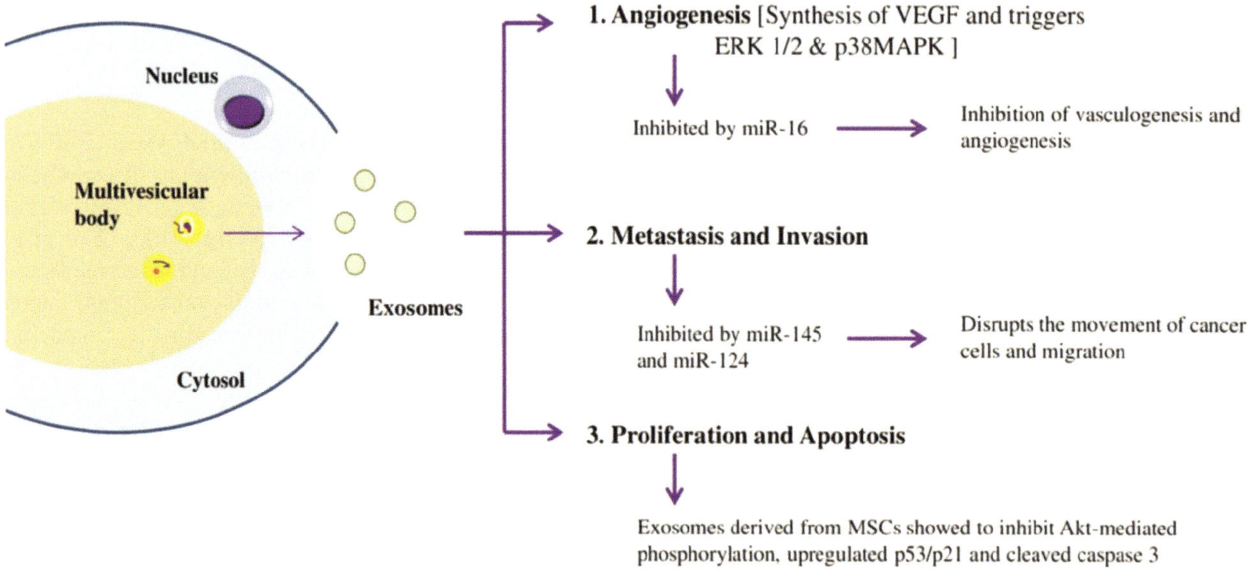

FIGURE 9.2 Therapeutic roles of exosomes in cancer.

cell function, thus enhancing the innate immune response of the body against tumor cells (Abbasi et al. 2022). These twin roles reflect the intricacy of MSC-derived exosomes in immune modulation, projecting them as potential therapeutic and target agents in cancer immunotherapy.

9.5.5 Tumor Microenvironment (TME) Remodeling

The TME is a dynamic and heterogeneous network of cancer cells, stromal cells, immune cells, and extracellular matrix elements. Exosomes derived from MSCs play an important role in remodeling this niche and affect the growth of tumors and metastasis. These exosomes have the ability to modify the phenotype of stromal cells within the TME. For instance, they can make fibroblasts take on a cancer-related phenotype, which is marked by elevated secretion of extracellular matrix proteins and pro-tumorigenic factors (Ringuette Goulet et al. 2018). Furthermore, MSC-derived exosomes may induce angiogenesis by transferring pro-angiogenic miRNAs and proteins into endothelial cells, allowing for the development of new blood vessels that provide nutrients and oxygen to the tumor (Gong et al. 2017). In addition, MSC-derived exosomes are able to modify the metabolic terrain of the TME. They have been found to deliver metabolic enzymes and miRNAs that reprogram the recipient cells' metabolism, promoting increased glycolysis and lactate production, which can sustain tumor growth and inhibit immune responses (Vallabhaneni et al. 2015). These complex interactions highlight the role of MSC-derived exosomes in remodeling the TME to promote tumor development. Targeting such exosomal mechanisms could provide new therapeutic approaches for interfering with supportive tumor stroma and preventing cancer progression.

9.6 MSC EXOSOMES AS A CANCER DRUG DELIVERY SYSTEM

The field of regenerative medicine has been focusing on exosomes formed from MSCs in recent years due to their ability to hold a diverse range of cargo molecules and factors (Angulski et al. 2017). The primary underlying mechanisms through which MSC-derived exosomes readily pass across physical barriers like the blood–brain barrier is transcytosis, which makes it possible for drugs to be administered at therapeutic concentrations (Lin et al. 2022).

MSC exosome-based cancer treatments have gained attention because they block exosomal release from cancer, carry genetic material, act as anticancer medicines, and utilize naturally occurring exosomes from immune cells to suppress cancer cells. MSC exosomes have been promising drug delivery vehicles because of their excellent biocompatibility, low toxicity, low immunogenicity, and capacity to overcome natural barriers (Didiot et al. 2016). Exosomal drug delivery also improves cancer treatment by lowering medication toxicity. The efficacy of targeted drug delivery lies in its ability to focus on specific tissue sites, thereby reducing the side effects of the drugs; most anticancer medications currently in use are harmful to healthy tissues and cause considerable damage.

Another effective treatment approach is injecting small-molecule medications into MSC-derived exosomes to treat inflammation and tissue regeneration. Recent studies show that small-molecule anticancer medications can be effectively loaded into MSC-derived exosomes for effective transport to target sites. The distinct ligand interaction and homing abilities of MSC-derived exosomes enhance their targeting potential (Sun et al. 2021). Exosomes, as naturally produced nanocarriers, exhibit high biocompatibility and can applied in therapeutic settings, such as for transporting medications, mRNAs, miRNAs, lipids, and proteins. MSC exosomes are ideal for delivering therapeutic agents such therapeutic miRNA and anticancer drugs because of their membrane compositions and adhesive proteins (Syn et al. 2017). Genetically modified exosomes experimentally carried suicide mRNAs and proteins that stopped tumor growth. When delivered *in vivo*, the exosomes showed an intrinsic propensity to target tumor and inflammatory tissues (Erkan et al. 2017). Because exosomes are anti-phagocytic, they can successfully evade phagocytosis or breakdown by macrophages for prolonged periods of time. Additionally, they avoid being broken down by lysosomes or the endosomal route when antitumor medicines are delivered to the target cells (Kalluri and LeBleu 2020).

9.7 CHALLENGES IN EXOSOME-BASED DRUG DELIVERY

Exosome-based delivery of anticancer drugs is currently undergoing clinic studies, but relatively few exosome-based anticancer drug delivery systems have entered clinical trials; the field still has many obstacles to overcome (Ingato et al. 2016).

The ability to effectively integrate MSC-derived exosomes into clinical practice is impeded by the inefficiency of large-scale exosome production. The individual protocols for isolating and purifying yield EVs with differential content, components, and physiological activities (Chen et al. 2022). Environmental factors such as the pH and hypoxic state of the culture medium influence the amount of miRNA cargo generated by exosomes. Therefore, for future applications and trials, it is crucial to enhance the stability and resistance of therapeutic cargo in MSC-derived exosomes to withstand the challenges posed by the human body (Luo et al. 2021). Additionally, careful attention to detail is required with cell culture settings; careless handling might lead to cell contamination, which can result in cell subtypes and variation. It is challenging to strictly regulate and preserve the genetic integrity of donor cells (Barok et al. 2018).

TABLE 9.1
Characteristics of MSC Exosome Sources in Cancer Therapy

Source	Cancer Type	Mechanism	Reference
Human umbilical cord Wharton's jelly	Bladder	Suppressed tumor growth by upregulating cleaved caspase-3 and downregulating Akt protein kinase phosphorylation	Wu et al. 2013
Adipose tissues	Prostate	Encouraged apoptosis via the caspase-3/7 pathway and inhibited tumor growth	Takahara et al. 2016
Human bone marrow	Hepatoma, Kaposi's sarcoma, Ovarian	Stopped tumor growth	Bruno et al. 2013
Human bone marrow	Pancreas	Facilitated potent antitumorigenic effects	Pascucci et al. 2014
Human bone marrow	Osteosarcoma	Reduced osteosarcoma migration	Shimbo et al. 2014
Human bone marrow	Bladder	Effective vehicle for delivering PLK-1 siRNA to bladder cancer cells	Greco et al. 2016
Human bone marrow	Glioma	Decreased migration of tumor cells and decreased glioma characteristics	Lee et al. 2013
Human bone marrow	Glioma	Decreased glioma invasion, migration, viability, and EGFR expression; blocked NF-κB and its upstream regulator TRAF6	Katakowski et al. 2013
Human bone marrow	Breast	Downregulated VEGF expression and suppressed angiogenesis	Lee et al. 2013
Human bone marrow	Oral/mouth	Suppressed tumor growth, invasion, migration, and proliferation; downregulated the type X collagen gene	Xie et al. 2019
Adipose tissues	Hepatocytes	Reduced expression of cyclin G1 and suppressed the PI3K/Akt pathway; increased tumor cell susceptibility to chemotherapeutic drugs; activated apoptosis	Lou et al. 2015

It is imperative to provide a cutting-edge method for separating MSC exosomes for drug delivery that is low cost, high quality, and efficient. One approach involves modifying the exosome surfaces using chemical alteration or genetic engineering (Jiang et al. 2017). Although chemical alteration frequently affects exosome structure and function, special equipment is not needed, and the exosomes' structures are rarely harmed (Barok et al. 2018). Nevertheless, this loading strategy demonstrated limited drug-loading efficiency. While electroporation and ultrasound can increase the effectiveness of drug loading, they also damage the exosome membrane and cause the exosomes to aggregate. Even though cell engineering has been well established, it remains difficult (Qin et al. 2021). Table 9.1 describes some of the MSC exosomes that are being incorporated into anticancer drug delivery.

9.8 BENEFITS OF MSC EXOSOMES IN CANCER THERAPY

MSC exosomes have garnered growing interest as potential markers for diagnosis and treatment; in particular, they are promising candidates for anticancer treatment with the advantages of low immunogenicity, stability, pharmacokinetics, biodistribution, and cellular absorption. These advantages can reduce unanticipated adverse effects while preferentially targeting tumor cells, increasing the therapeutic index of exosome-based cancer treatments (Nam et al. 2020). Preconditioned MSC exosomes enhance paracrine potency and therapeutic potential. Pretreatment with growth factors, heat shock protein (Hsp), pharmacological agents, chemical agents, serum deprivation, or inflammatory stimulation increases the potency and survival of MSC exosomes by promoting the secretion of reparative factors and increasing the expression levels of cytoprotective genes (Jung et al. 2022). In an experimental study employing bladder cancer cell line T24, Wharton's jelly MSC exosomes from human cord blood promoted cell cycle arrest and death via upregulating caspase-3 cleavage and downregulating Akt phosphorylation pathways (Wu et al. 2013).

MSC exosomes impede the growth of cancerous human bone marrow MSCs; the direct interaction between the tumor cells and the bone marrow MSCs inhibited tumor growth by preventing Akt protein kinase activation (Khakoo et al. 2006). They also prevented angiogenesis, thereby reducing tumor cell growth. Their angiogenic properties make them valuable in regenerative medicine (Marfy-Smith and Clarkin 2017). MSC-derived exosomes also prevent the migration of malignant cells. PTEN, a gene that suppresses tumor growth and functions as a negative regulator of the PI3K/Akt signaling cascade, is overexpressed in gliomas cocultured with hUC-MSCs. This downregulates Akt, which inhibits the migration of tumor cells (Shojaei et al. 2019).

9.9 LIMITATIONS OF MSC EXOSOMES IN CANCER THERAPY

Despite current findings, research on cancer has demonstrated that MSC exosomes in the TME can exhibit both pro- and antitumor functions (Fischer et al. 2009). Through their impacts on tissue remodeling activity in inflammatory sites, these cells decrease immune responses, which in turn promotes tumor progression (Kidd et al. 2009).

Subcutaneous co-injection of MSC exosomes with human gastric cancer cell SW480 and colon cancer cell SGC-7901 increased tumor growth and burden, and only the mRNA and protein expressions of CXCR4 and VEGF were elevated *in vivo* and *in vitro*. There were no appreciable alterations in cell cycle or proliferation (Cheshomi and Matin 2019). Therefore, even while MSC exosomes promote angiogenesis, which impacts tumor seeding and progression, the increased tumor burden seen *in vivo* is likely an unintended consequence of exosome activity (Zhu et al. 2012).

Lastly, improving the therapeutic efficacy of MSC exosomes can be achieved by selectively overexpressing specific molecules in the parent MSCs that are known to produce unique therapeutic impacts. Exosomes and other EVs produced by MSCs should therefore be researched and used in accordance with particular guidelines (Moore et al. 2017). The most challenging issue is the heterogeneity of MSC exosomes; existing preparations of extra vesicles, particularly MSC exosomes, are highly variable, and it is still challenging to isolate a particular exosome population. Because of this, till the biogenesis process is fully understood, the International Society for EVs advises using the collective term "EVs" (Miller and Grunewald 2015). The diversity and preparation of MSCs, different techniques for exosome synthesis and separation, a lack of standardized quality assurance assays, and the low repeatability of *in vitro* and *in vivo* functional assays are some of the major obstacles. In particular, measurable metrics should be used to characterize MSC exosomes in order to determine their biological origin, the existence of lipid–membrane vesicles, and their physical and biochemical integrity (Cheshomi and Matin 2019).

9.10 FUTURE PERSPECTIVES

MSC exosome therapy has emerged for the management of numerous illnesses in recent years (Mendt et al. 2018). They are an excellent tiny particle delivery method due to their natural ability to transmit genetic materials, shield them from destruction by diverse causes, and deliver these materials to recipient cells with great selectivity. As a result, they are useful in regenerative medicine and in supporting gene therapies for cancer and other crippling conditions (Keshtkar et al. 2018). Furthermore, a number of preclinical investigations have demonstrated that MSC-exosome therapy offers advantages over cell therapy in terms of safety, effectiveness, and adaptability (Wu et al. 2013).

To increase the practical use of MSC-exosome treatments, it is necessary to investigate and manage their side effects (Gimona et al. 2017). Therefore, additional preclinical research is required to determine the best MSC sources, dosages, and delivery methods. Producing more homogenous populations of MSCs and their exosomes can be facilitated by using established culture conditions and by producing tested immortalized MSC lines. The necessary long-term preservation of MSC exosomes is another difficulty. Scientists have studied lyophilizing the exosomes for extended storage or adding stabilizing compounds such sugars or PEG (Guo et al. 2022; Driscoll et al. 2022). Additionally, recent research focuses on encapsulating them to increase their persistence in the body (Khayambashi et al. 2021). These research efforts will build on the potential of MSC exosomes as novel drug delivery systems for anticancer medicines and cancer immunotherapy. Even though the function of exosomes has been extensively studied, more genetic models are required to fully understand their function in the TME. Furthermore, parameters like age and immune-senescence that are determined by the host during the administration of exosomes must be revealed. The physiological and biochemical changes that occur when exosomes are in the delivery system need to be greatly improved (Mendt et al. 2019).

9.11 CONCLUSION

In several ongoing clinical trials for different diseases, exosomes from MSCs are emerging as a unique and promising cell-free treatment approach. A small number of published clinical studies have shown MSC exosomes to be both safe and effective. However, there are drawbacks to their clinical use including product heterogeneity, fast evacuation from the body, and instability during long-term preservation. Standardizing the therapeutic dosages, mode of administration, and parent MSC sources for the exosomes is also crucial. Therefore, advancing MSC exosome therapy into therapeutic applications requires much more preclinical and clinical research. For instance, to increase the production of exosomes, comparable bioreactors should be built and combined with improved cell culture methods. It will be critical to develop a reliable, efficient, and low-cost method for producing, isolating, and purifying exosomes for use in drug delivery systems.

REFERENCES

Abbasi B, Shamsasenjan K, Ahmadi M, Beheshti SA, Saleh M. Mesenchymal stem cells and natural killer cells interaction mechanisms and potential clinical applications. Stem Cell Res Ther. 2022;13(1):97.

Angulski AB, Batista M, Marcon BH, Senegaglia AC, Stimamiglio MA, Correa A. The protein content of extracellular vesicles derived from expanded human umbilical cord blood-derived CD133+ and human bone marrow-derived mesenchymal stem cells partially explains why both sources are advantageous for regenerative medicine. Stem Cells Rev. 2017;13:244–257.

Ban J-J, Lee M, Im W, Kim M. Low pH increases the yield of exosome isolation. Biochem Biophys Res Commun. 2015;461(1): 76–79. https://doi.org/10.1016/j.bbrc.2015.03.172.

Barok M, Puhka M, Vereb G, Szollosi J, Isola J, Joensuu H. Cancer-derived exosomes from HER2-positive cancer cells carry trastuzumab-emtansine into cancer cells leading to growth inhibition and caspase activation. BMC Cancer. 2018;18(1):504. https://doi.org/10.1186/s12885-018-4418-2.

Bhoopathi P, Chetty C, Gogineni VR, Gujrati M, Dinh DH, Rao JS, Lakka SS. MMP-2 mediates mesenchymal stem cell tropism towards medulloblastoma tumors. Gene Ther. 2011;18:692–701. https://doi.org/10.1038/gt.2011.14.

Bian D, Wu Y, Song G, Azizi R, Zamani A. The application of mesenchymal stromal cells (MSCs) and their derivative exosome in skin wound healing: A comprehensive review. Stem Cell Res Ther. 2022;13:24. https://doi.org/10.1186/s13287-021-02697-9.

Bolontrade MF, Sganga L, Piaggio E, Viale DL, Sorrentino MA, Robinson A, Sevlever G, Garcia MG, Mazzolini G, Podhajcer O. A specific subpopulation of mesenchymal stromal cell carriers overrides melanoma resistance to an oncolytic adenovirus. Stem Cells Dev. 2012;21:2689–2702.

Bouchlaka MN, Hematti P, Capitini CM. Therapeutic purposes and risks of ex vivo expanded mesenchymal stem/stromal cells. In Bolontrade MF, Garcia MG, Eds. *Mesenchymal Stromal Cells as Tumor Stromal Modulators*. Amsterdam, The Netherlands: Elsevier, 2017, pp. 551–576.

Bruno S, Collino F, Deregibus MC, Grange C, Tetta C, Camussi, G. Microvesicles derived from human bone marrow mesenchymal stem cells inhibit tumor growth. Stem Cells Dev. 2013;22:758–771. https://doi.org/10.1089/scd.2012.0304.

Busatto S, Vilanilam G, Ticer T, Lin W-L, Dickson DW, Shapiro S, Bergese P, Wolfram J. Tangential flow filtration for highly efficient concentration of extracellular vesicles from large volumes of fluid. Cells. 2018;7:273.

Camussi G, Deregibus M, Bruno S, Cantaluppi V, Biancone L. Exosomes/microvesicles as a mechanism of cell-to-cell communication. Kidney Int. 2010;78:838–848.

Chang L, Gao H, Wang L, Wang N, Zhang S, Zhou X, Yang H. Exosomes derived from miR-1228 overexpressing bone marrow-mesenchymal stem cells promote growth of gastric cancer cells. Aging (Albany NY). 2021;13(8): 11808–11821.

Chen J, Li P, Zhang T, Xu Z, Huang X, Wang R, Du L. Review on strategies and technologies for exosome isolation and purification. Front Bioeng Biotechnol. 2022;9:811971.

Cheshomi H, Matin MM. Exosomes and their importance in metastasis, diagnosis, and therapy of colorectal cancer. J Cell Biochem. 2019 Feb;120(2):2671–2686.

De Luca LLI, Trino S, Lamorte D, Caivano A, Musto P. An update on extracellular vesicles in multiple myeloma: A focus on their role in cell-to-cell cross-talk and as potential liquid biopsy biomarkers. Expert Rev Mol Diagn. 2019;19:249–258.

Didiot MC, Hall LM, Coles AH, Haraszti RA, Godinho BM, Chase K, Sapp E, Ly S, Alterman JF, Hassler MR, Echeverria D. Exosome-mediated delivery of hydrophobically modified siRNA for huntingtin mRNA silencing. Mol Ther. 2016;24:1836–1847. https://doi.org/10.1038/mt.2016.126.

Driscoll J, Yan IK, Patel T. Development of a lyophilized off-the-shelf mesenchymal stem cell-derived acellular therapeutic. Pharmaceutics. 2022;14(4).

El-Badawy AGMA, Gabr MM, Ayman Salah R, Mohamed IK, Amer M, El-Badri N. Cancer cell-soluble factors reprogram mesenchymal stromal cells to slow cycling, chemoresistant cells with a more stem-like state. Stem Cell Res Ther. 2017;8:254.

Erkan EP, Senfter D, Madlener S, Jungwirth G, Ströbel T, Saydam N, Saydam O. Extracellular vesicle-mediated suicide mRNA/protein delivery inhibits glioblastoma tumor growth in vivo. Cancer Gene Ther. 2017;24(1):38–44.

Ferguson SW, Wang J, Lee CJ, Liu M, Neelamegham S, Canty JM, Nguyen J. The microRNA regulatory landscape of MSC-derived exosomes: A systems view. Scient Rep. 2018 Jan 23;8(1):1419.

Fischer UM, Harting MT, Jimenez F, Monzon-Posadas WO, Xue H, Savitz SI, Laine GA, Cox CS Jr. Pulmonary passage is a major obstacle for intravenous stem cell delivery: The pulmonary first-pass effect. Stem Cells Dev. 2009;18(5): 683–692. https://doi.org/10.1089/scd.2008.0253.

Gimona M, Pachler K, Laner-Plamberger S, Schallmoser K, Rohde E. Manufacturing of human extracellular vesicle-based therapeutics for clinical use. Int J Mol Sci. 2017;18(6):1190.

Girigoswami A, Girigoswami K. Cell death mechanisms elicited by multifactorial stress inducers-A minireview. Coronaviruses. 2023 Sep 1;4(3):70–82.

Gong M, Yu B, Wang J, Wang Y, Liu M, Paul C, Millard RW, Xiao DS, Ashraf M, Xu M. Mesenchymal stem cells release exosomes that transfer miRNAs to endothelial cells and promote angiogenesis. Oncotarget. 2017 Apr 1;8(28): 45200.

Greco KA, Franzen CA, Foreman KE, Flanigan RC, Kuo PC, Gupta GN. PLK-1 silencing in bladder cancer by siRNA delivered with exosomes. Urology. 2016 May 1;91:241-e1.

Greening DW, Xu R, Ji H, Tauro BJ, Simpson RJ. *A Protocol for Exosome Isolation and Characterization: Evaluation of Ultracentrifugation, Density-Gradient Separation, and Immunoaffinity Capture Methods*. Proteomic Profiling: Methods and Protocols, 2015, pp. 179–209.

Guo G, Tan Z, Liu Y, Shi F, She J. The therapeutic potential of stem cell-derived exosomes in the ulcerative colitis and colorectal cancer. Stem Cell Res Ther. 2022 Apr 1;13(1):138.

Helwa I, Cai J, Drewry MD, Zimmerman A, Dinkins MB, Khaled ML, Seremwe M, Dismuke WM, Bieberich E, Stamer WD, Hamrick MW. A comparative study of serum exosome isolation using differential ultracentrifugation and three commercial reagents. PloS One. 2017 Jan 23;12(1): e0170628.

Hmadcha A, Martin-Montalvo A, Gauthier BR, Soria B, Capilla-Gonzalez V. Therapeutic potential of mesenchymal stem cells for cancer therapy. Front Bioeng Biotechnol. 2020 Feb 5;8:43.

Huang L, Ma W, Ma Y, Feng D, Chen H, Cai B. Exosomes in mesenchymal stem cells, a new therapeutic strategy for cardiovascular diseases? Int J Biol Sci. 2015;11(2):238.

Ingato D, Lee JU, Sim SJ, Kwon YJ. Good things come in small packages: Overcoming challenges to harness extracellular vesicles for therapeutic delivery. J Control Release Off J Control Release Soc. 2016;241:174–185. https://doi.org/10.1016/j.jconrel.2016.09.016.

Jella KK, Nasti TH, Li Z, Malla SR, Buchwald ZS, Khan MK. Exosomes, their biogenesis and role in inter-cellular communication, tumor microenvironment and cancer immunotherapy. Vaccines. 2018 Sep 26;6(4):69.

Jiang XC, Gao JQ. Exosomes as novel bio-carriers for gene and drug delivery. Int J Pharmaceut. 2017;521(1–2):167–175. https://doi.org/10.1016/j.ijpharm.2017.02.038.

Jiang Y, Jahagirdar BN, Reinhardt RL, Schwartz RE, Keene CD, Ortiz-Gonzalez XR, Reyes M, Lenvik T, Lund T, Blackstad M, Du J. Pluripotency of mesenchymal stem cells derived from adult marrow. Nature. 2002;418:41–49.

Jung SY, Kim YE, Park WS, Ahn SY, Sung DK, Sung SI, Joo KM, Kim SG, Chang YS. Thrombin preconditioning improves the therapeutic efficacy of mesenchymal stem cells in severe intraventricular hemorrhage induced neonatal rats. Int J Mol Sci. 2022 Apr 18;23(8):4447.

Kalluri R, LeBleu VS. The biology, function, and biomedical applications of exosomes. Science. 2020 Feb 7;367(6478): eaau6977.

Katakowski M, Buller B, Zheng X, Lu Y, Rogers T, Osobamiro O, Shu W, Jiang F, Chopp M. Exosomes from marrow stromal cells expressing miR-146b inhibit glioma growth. Cancer Lett. 2013 Jul 10;335(1):201–204.

Keerthikumar S, Chisanga D, Ariyaratne D, Al Saffar H, Anand S, Zhao K, Samuel M, Pathan M, Jois M, Chilamkurti N, Gangoda L. ExoCarta: A web-based compendium of exosomal cargo. J Mol Biol. 2016 Feb 22;428(4):688–692.

Keshtkar S, Azarpira N, Ghahremani MH. Mesenchymal stem cell-derived extracellular vesicles: Novel frontiers in regenerative medicine. Stem Cell Res Ther. 2018 Dec;9:1–9.

Khakoo AY, Pati S, Anderson SA, Reid W, Elshal MF, Rovira II, Nguyen AT, Malide D, Combs CA, Hall G, Zhang J. Human mesenchymal stem cells exert potent antitumorigenic effects in a model of Kaposi's sarcoma. J Exp Med. 2006 May 15;203(5):1235–1247.

Khayambashi P, Iyer J, Pillai S, Upadhyay A, Zhang Y, Tran SD. Hydrogel encapsulation of mesenchymal stem cells and their derived exosomes for tissue engineering. Int J Mol Sci. 2021;22(2).

Kibria G, Ramos EK, Wan Y, Gius DR, Liu H. Exosomes as a drug delivery system in cancer therapy: Potential and challenges. Mol Pharmaceut. 2018 May 17;15(9):3625–3633.

Kidd S, Spaeth E, Dembinski JL, Dietrich M, Watson K, Klopp A, Battula VL, Weil M, Andreeff M, Marini FC. Direct evidence of mesenchymal stem cell tropism for tumor and wounding microenvironments using in vivo bioluminescent imaging. Stem Cells. 2009 Oct 1;27(10):2614–2623.

Kink JA, Bellio MA, Forsberg MH, Lobo A, Thickens AS, Lewis BM, Ong IM, Khan A, Capitini CM, Hematti P. Large-scale bioreactor production of extracellular vesicles from mesenchymal stromal cells for treatment of acute radiation syndrome. Stem Cell Res Ther. 2024 Mar 13;15(1):72.

Kusuma GD, Carthew J, Lim R, Frith JE. Effect of the microenvironment on mesenchymal stem cell paracrine signaling: Opportunities to engineer the therapeutic effect. Stem Cell Dev. 2017;26. https://doi.org/10.1089/scd.2016.0349.

Kwon Y, Kim M, Kim Y, Jung HS, Jeoung D. Exosomal microRNAs as mediators of cellular interactions between cancer cells and macrophages. Front Immunol. 2020 Jun 11;11:1167.

Lee HK, Finniss S, Cazacu S, Bucris E, Ziv-Av A, Xiang C, Bobbitt K, Rempel SA, Hasselbach L, Mikkelsen T, Slavin S. Mesenchymal stem cells deliver synthetic microRNA mimics to glioma cells and glioma stem cells and inhibit their cell migration and self-renewal. Oncotarget. 2013a;4:346–361.

Lee JK, Park SR, Jung BK, Jeon YK, Lee YS, Kim MK, Kim YG, Jang JY, Kim CW. Exosomes derived from mesenchymal stem cells suppress angiogenesis by down-regulating VEGF expression in breast cancer cells. PloS One. 2013 Dec 31;8(12):e84256.

Li M, Fang F, Sun M, Zhang Y, Hu M, Zhang J. Extracellular vesicles as bioactive nanotherapeutics: An emerging paradigm for regenerative medicine. Theranostics. 2022 Jun 21;12(11):4879–4903. https://doi.org/10.7150/thno.72812. PMID: 35836815; PMCID: PMC9274746.

Li T, Wan Y, Su Z, Li J, Han M, Zhou C. Mesenchymal stem cell-derived exosomal microRNA-3940–5p inhibits colorectal cancer metastasis by targeting integrin α6. Dig Dis Sci. 2021;66(6):1916–1927.

Lin R, Wang S, Zhao RC. Exosomes from human adipose-derived mesenchymal stem cells promote migration through Wnt signaling pathway in a breast cancer cell model. Mol Cell Biochem. 2013 Nov;383:13–20.

Lin Z, Wu Y, Xu Y, Li G, Li Z, Liu T. Mesenchymal stem cell-derived exosomes in cancer therapy resistance: Recent advances and therapeutic potential. Mol Cancer. 2022 Sep 13;21(1):179.

Liu J, Wu X, Lu J, Huang G, Dang L, Zhang H, Zhong C, Zhang Z, Li D, Li F, Liang C. Exosomal transfer of osteoclast-derived miRNAs to chondrocytes contributes to osteoarthritis progression. Nat Aging. 2021 Apr;1(4):368–384.

Lou G, Song X, Yang F, Wu S, Wang J, Chen Z, Liu Y. Exosomes derived from miR-122-modified adipose tissue-derived MSCs increase chemosensitivity of hepatocellular carcinoma. J Hematol Oncol. 2015;8:122.

Luo T, von der Ohe J, Hass R. MSC-derived extracellular vesicles in tumors and therapy. Cancers (Basel). 2021;13:5212.

Lyu, T., Zhang, B., Li, M., Jiao, X., & Song, Y. Research progress on exosomes derived from mesenchymal stem cells in hematological malignancies. Hematological oncology. 2021;39(2):162–169. https://doi.org/10.1002/hon.2793.

Maffey A, Storini C, Diceglie C, Martelli C, Sironi L, Calzarossa C, Tonna N, Lovchik R, Delamarche E, Ottobrini L, Bianco F. Mesenchymal stem cells from tumor microenvironment favour breast cancer stem cell proliferation, cancerogenic and metastatic potential, via ionotropic purinergic signalling. Sci Rep. 2017 Oct 13;7(1):13162.

Maman S, Witz IP. A history of exploring cancer in context. Nat Rev Cancer. 2018;18:359–376.

Marfy-Smith SJ, Clarkin CE. Are mesenchymal stem cells so bloody great after all? Stem Cells Transl Med. 2017;6(1):3–6.

Mendt M, Kamerkar S, Sugimoto H, McAndrews KM, Wu CC, Gagea M, Yang S, Blanko EVR, Peng Q, Ma X, Marszalek JR. Generation and testing of clinical-grade exosomes for pancreatic cancer. JCI Insight. 2018;3(8).

Mendt M, Rezvani K, Shpall E. Mesenchymal stem cell-derived exosomes for clinical use. Bone Marrow Transplant. 2019;54 (Suppl 2):789–792.

Miller IV, Grunewald TGP. Tumour-derived exosomes: Tiny envelopes for big stories. Biol Cell. 2015;107(9):287–305.

Moore C, Kosgodage U, Lange S, Inal JM. The emerging role of exosome and microvesicle- (EMV-) based cancer therapeutics and immunotherapy. Int J Cancer. 2017;141(3): 428–436.

Mukherjee A, Bisht B, Dutta S, Paul MK. Current advances in the use of exosomes, liposomes, and bioengineered hybrid nanovesicles in cancer detection and therapy. Acta Pharmacol Sin. 2022;43:2759–2776. https://doi.org/10.1038/s41401-022-00902-w.

Nam GH, Choi Y, Kim GB, Kim S, Kim SA, Kim IS. Emerging prospects of exosomes for cancer treatment: From conventional therapy to immunotherapy. Adv Mater. 2020 Dec;32(51):2002440.

Naseri Z, Oskuee RK, Jaafari MR, Moghadam MF. Exosome-mediated delivery of functionally active miRNA-142–3p inhibitor reduces tumorigenicity of breast cancer in vitro and in vivo. Int J Nanomed. 2018;13:7727–7747. https://doi.org/10.2147/IJN.S182384.

Oksvold MP, Neurauter A, Pedersen KW. Magnetic bead-based isolation of exosomes. In *RNA Interference: Challenges and Therapeutic Opportunities*, 2015, pp. 465–481.

Pascucci L, Coccè V, Bonomi A, Ami D, Ceccarelli P, Ciusani E, Viganò L, Locatelli A, Sisto F, Doglia SM, Parati E. Paclitaxel is incorporated by mesenchymal stromal cells and released in exosomes that inhibit in vitro tumor growth: A new approach for drug delivery. J Control Rel. 2014 Oct 28;192:262–270.

Qin Z, Duan L, Xu L, Xu X, Zhou X, Xiao Y, Liang Y, Xia J. Exosome-mediated delivery of gene vectors for gene therapy. Nanoscale. 2021;13(3):1387–1397. https://doi.org/10.1039/D0NR07622H.

Ringuette Goulet C, Bernard G, Tremblay S, Chabaud S, Bolduc S, Pouliot F. Exosomes induce fibroblast differentiation into cancer-associated fibroblasts through TGFβ signaling. Mol Cancer Res. 2018 Jul 1;16(7):1196–204.

Salafi T, Zeming KK, Zhang Y. Advancements in microfluidics for nanoparticle separation. Lab on a Chip. 2017;17(1):11–33. https://doi.org/10.1039/C6LC01045H.

Shimbo K, Miyaki S, Ishitobi H, Kato Y, Kubo T, Shimose S, Ochi M. Exosome-formed synthetic microRNA-143 is transferred to osteosarcoma cells and inhibits their migration. Biochem Biophys Res Commun. 2014 Mar 7;445(2):381–387.

Shojaei S, Hashemi SM, Ghanbarian H, Salehi M, Mohammadi-Yeganeh S. Effect of mesenchymal stem cells-derived exosomes on tumor microenvironment: Tumor progression versus tumor suppression. J Cell Physiol. 2019 Apr;234(4):3394–3409. https://doi.org/10.1002/jcp.27326.

Srinivasan S, Vannberg FO, Dixon JB. Lymphatic transport of exosomes as a rapid route of information dissemination to the lymph node. Sci Rep. 2016 Apr 18;6(1):24436.

Stefańska K, Józkowiak M, Angelova Volponi A, Shibli JA, Golkar-Narenji A, Antosik P, Bukowska D, Piotrowska-Kempisty H, Mozdziak P, Dzięgiel P, Podhorska-Okołów M. The role of exosomes in human carcinogenesis and cancer therapy: Recent findings from molecular and clinical research. Cells. 2023;12:356.

Subbaraj GK, Kumar YS, Kulanthaivel L. Antiangiogenic role of natural flavonoids and their molecular mechanism: An update. Egypt J Intern Med. 2021 Dec;33:1–10.

Sun Y, Liu G, Zhang K, Cao Q, Liu T, Li J. Mesenchymal stem cells-derived exosomes for drug delivery. Stem Cell Res Ther. 2021 Dec;12:1–5.

Syn N, Wang L, Chow E, Lim C, Goh B. Exosomes in cancer nanomedicine and immunotherapy: Prospects and challenges. Trends Biotechnol. 2017;35:665–676.

Tai YL, Chen KC, Hsieh JT, Shen TL. Exosomes in cancer development and clinical applications. Cancer Sci. 2018 Aug;109(8):2364–2374.

Takahara K, Inamoto T, Ibuki N, Uchimoto T, Saito K, Takai T, Tanda N, Hirano H, Nomi H, Kiyama S, Azuma H. 245 MicroRNA-145 mediates the inhibitory effect of adipose-derived stem cells on androgen-independent prostate cancer. Eur Urol Suppl. 2016;3(15):e245.

Vallabhaneni KC, Hassler MY, Abraham A, Whitt J, Mo YY, Atfi A, Pochampally R. Mesenchymal stem/stromal cells under stress increase osteosarcoma migration and apoptosis resistance via extracellular vesicle mediated communication. PLoS One. 2016;11(11):e0166027. https://doi.org/10.1371/journal.pone.0166027.

Vallabhaneni KC, Penfornis P, Dhule S, Guillonneau F, Adams KV, Mo YY, Xu R, Liu Y, Watabe K, Vemuri MC, Pochampally R. Extracellular vesicles from bone marrow mesenchymal stem/stromal cells transport tumor regulatory microRNA, proteins, and metabolites. Oncotarget. 2015;6:4953–4967.

Wang T, Fu Y, Sun S, Huang C, Yi Y, Wang J, Deng Y, Wu M. Exosome-based drug delivery systems in cancer therapy. Chin Chem Lett. 2023;34(2):107508.

Weng Y, Sui Z, Shan Y, Hu Y, Chen Y, Zhang L, Zhang Y. Effective isolation of exosomes with polyethylene glycol from cell culture supernatant for in-depth proteome profiling. Analyst. 2016;141(15):4640–4646.

Whiteside TL. Exosome and mesenchymal stem cell cross-talk in the tumor microenvironment. Semin Immunol. 2018;35:69–79.

Wu S, Ju G-Q, Du T, Zhu Y-J, Liu G-H. Microvesicles derived from human umbilical cord Wharton's jelly mesenchymal stem cells attenuate bladder tumor cell growth in vitro and in vivo. PLoS One. 2013;8(4):e61366.

Xie C, Du LY, Guo F, Li X, Cheng B. Exosomes derived from microRNA-101–3p-overexpressing human bone marrow mesenchymal stem cells suppress oral cancer cell proliferation, invasion, and migration. Mol Cell Biochem. 2019;458:11–26.

Xiong GF, Xu R. Function of cancer cell-derived extracellular matrix in tumor progression. J Cancer Metastasis Treat. 2016;2:357–364. https://doi.org/10.20517/2394–4722.2016.08.

Yang Y, Bucan V, Baehre H, von der Ohe J, Otte A, Hass R. Acquisition of new tumor cell properties by MSC-derived exosomes. Int J Oncol. 2015;47:244–252.

Zhang F, Guo J, Zhang Z, Qian Y, Wang G, Duan M, Zhao H, Yang Z, Jiang X. Mesenchymal stem cell-derived exosome: A tumor regulator and carrier for targeted tumor therapy. Cancer Lett. 2022 Feb 1;526:29–40.

Zhang Y, Bi J, Huang J, Tang Y, Du S, Li P. Exosome: A review of its classification, isolation techniques, storage, diagnostic and targeted therapy applications. Int J Nanomed. 2020;15:6917–6934.

Zhu W, Huang L, Li Y, Zhang X, Gu J, Yan Y, Xu X, Wang M, Qian H, Xu W. Exosomes derived from human bone marrow mesenchymal stem cells promote tumor growth in vivo. Cancer Lett. 2012 Feb 1;315(1):28–37.

10 Bioprinting in Human Health

Nidhi Mahajan,[1] *Heena Kansal,*[1] *Siddharth Sharma,*[1]* *and RC Sobti*[2]

[1] Department of Biotechnology, Thapar Institute of Engineering and Technology, Patiala, India

[2] Department of Biotechnology, Panjab University, Chandigarh, India

10.1 INTRODUCTION

Bioprinting is a new technology that creates functional tissues and organs to replace injured tissues. It offers excellent reproducibility and exact control over manufactured structures in an automated setting, possibly enabling high-throughput production. To create the structures, the process uses bioinks, solutions of various biomaterials in hydrogel form that usually encapsulate the desired cell types. This bioink can be crosslinked or stabilized after bioprinting to develop the desired construct (Groll et al., 2016).

Bioinks can be created entirely of natural or synthetic biomaterials, or as a blend of the two, and some cell aggregates without any extra biomaterials can be used as bioinks. An ideal bioink should have the mechanical, rheological, and biological qualities of the target tissues, which are required for the bioprinted tissues and organs to operate properly (Gungor-Ozkerim et al., 2018). Although bioprinting is still in its early stages, its versatility has accelerated its uses in tissue engineering to generate regenerative scaffolds tailored to the needs of the patient.

Bioprinting is a new area of regenerative medicine that creates cell-laden, 3D structures that mimic physiological tissues, and it has significant implications not just in tissue engineering but also in drug delivery and cancer research. To create tissue engineering scaffolds, bioprinting can provide patient-specific spatial geometry, controlled microstructures, and the positioning of diverse cell types. The ultimate goal of bioprinting is to replace animal testing in disease research and treatment development as well as to give an alternative to autologous and allogenic tissue implants (Li et al., 2016). In this chapter, we are going to discuss the different techniques, biomaterials, and bioreactors used for fabricating 3D models and their applications in the different fields of medicine and drug development. We include discussions of the ethical issues, the future, and the limitations of bioprinting.

10.2 STEPS FOR BIOPRINTING

The bioprinting process begins with using imaging modalities such as CT, MRI, and ultrasound to construct a computerized 3D biomodel of the targeted organ or tissue using medical data, then this data is transferred for CAD (computer-aided design) modeling. After determining the CAD model, bioinks are prepared, and finally models are constructed *in situ* using printers under sterile conditions and incubated to mature. After maturation, these constructs can be used as a disease model or transplanted into the patient (Norotte et al., 2009). These processes of bioprinting are divided into three steps: pre-printing, printing, and post-printing (Datta et al., 2018; Loai et al., 2019), which we describe in more detail in this section.

10.2.1 Pre-Bioprinting

Pre-bioprinting involves planning for the bioprinting modeling, which includes selecting the model using MRI or CT scan data specific to the patient's tissue morphology, selecting the material, and cultivating the cells. The 3D models are converted to standard tessellation language files and saved in g-code file format (van Lenthe et al., 2007). The printer easily follows the file format and constructs the model using biological material.

10.2.2 Bioprinting

This process involves selecting the bioink to mixed with the cells as well as the bioprinter and the correct bioprinting process. The bioinks are selected according to their rheological and biological properties, which directly affect printing outcomes. The bioprinter is selected based on the printer resolution and nozzle size, the model dimensions, price, cell viability, and the bioink to be used (Blaeser et al., 2016). Once the printer and materials are selected, the bioprinting takes place in a highly controlled environment. The deposition of bioink onto the scaffold to create and 3D biological structure is a critical step in bioprinting. This process requires precise resolution, repeatability, and biocompatibility to ensure the viability and function of the resulting 3D biological model (Ji and Guvendiren et al., 2017) (see Figure 10.1).

10.2.3 Post-Bioprinting

This is the last step of the bioprinting process; it aims to maintain the biological tissue or organ until it matures. Post-bioprinting is performed under controlled conditions specifically in incubators or bioreactors to mimic an *in vivo* environment (Yu et al., 2016). The simulations provided at this step facilitate the differentiation of cells in the printed model and help with tissue growth and maintenance.

* Corresponding author: Siddharthsharma.phd@thapar.edu

DOI: 10.1201/9781032660592-12

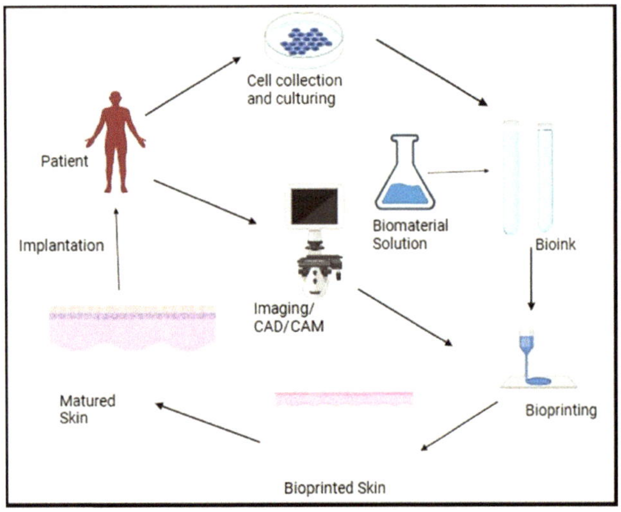

FIGURE 10.1 An overview of the bioprinting process.

10.3 BIOPRINTING TECHNIQUES

The 3D bioprinting techniques used for tissue fabrication are based on extrusion, inkjet, or laser (Kačarević et al., 2018) (Table 10.1); the techniques have different approaches and working principles, but they all follow the same three printing steps. Each technique works layer by layer to obtain 2D stacks to build the 3D images.

10.3.1 Extrusion Printing

Extrusion is the most common method of 3D bioprinting because it is user-friendly and affordable. In this technique, the bioink is extruded through a nozzle tip by applying mechanical or pneumatic pressure to print a layer-by-layer construct (Derakhshanfar et al., 2018). In the mechanical method, the printing is through rotational and vertical forces, and via piston extrusion, the linear motion squeezes out the bioink through the nozzle with air pressure. In screw-driven extrusion, the pressure is in a continuous flow. and the bioink is distributed in a microliter range through a nozzle. In the pneumatical method, the bioink is squeezed out in a forceful way through air pressure, which increases the shear stress and increases the chances of cell deformation (Zeng et al., 2018).

For extrusion bioprinting, the biomaterial commonly utilized is hydrogels (collagen, alginate, gelatine, and hyaluronic acid) as they are naturally bioactive and have high water content, which helps to preserve the cell shape (Gillispie et al., 2020). The drawback of extrusion bioprinting is the use of a nozzle tip; it exposes cells to mechanical forces and shear stress for a longer duration, which reduces the viability of the cell during the printing process (Chia and Wu, 2015). Extrusion bioprinting is widely used in tissue engineering such as for cartilage and bone tissues and skeletal muscles.

10.3.2 Inkjet Printing

The inkjet printer was the first bioprinter. Inkjet bioprinting works on the drop-on-demand principle, in which the bioink is dropped on the substrate precisely in a picolitre range under controlled computer-instructed conditions. Inkjet bioprinting exhibits high resolution and excellent cell viability for developing neural networks in printed tissues, although it brings the drawbacks of shear stress to the cells and cell encapsulation that leads to cell death. Using cell-loaded hydrogels for cell encapsulation and controlling the pattern of droplets to be ejected from the nozzle are two methods for overcoming these drawbacks of this technique (Singh et al., 2010).

Biomaterials are also printed using thermal, piezoelectric, electrostatic, and electrohydrodynamic inkjet methods. In thermal inkjet bioprinting, a heated element nucleates a bubble to build pressure within the printer to expel droplets, whereas in piezoelectric printing, acoustic waves are

TABLE 10.1
Principles of 3D Bioprinting

Bioprinting Types	Extrusion Bioprinting	Inkjet Bioprinting	Laser-Assisted Bioprinting	References
Printing principle	Biomaterial deposition by extruder	Droplets of bioinks	Laser is used for bioink deposition	Mandrycky et al. (2016)
Printing speed	Slow	Fast	Average	Billiet et al. (2012)
Viability of cells	40%–90%	>85%	>95%	Mandrycky et al. (2016)
Resolution	Moderate	High	High	Billiet et al. (2012)
Bioink used	Agarose, gelatin, alginate, and PEG	Fibrinogen, gelatin, alginate, and agarose	Collagen and alginate	Ashammakhi et al. (2019)
Advantages	High cell density and viscous printing	Good control over drop size and high resolution	Accurate and microscale resolution and speed	Xing et al. (2020)
Disadvantages	Low speed and moderate resolution	Low mechanical properties and limited options for biomaterials	Costly and time-consuming	Xing et al. (2020)
Applications	Mimics soft tissue, cartilage, and bone transplants	Cartilage tissues and skin regeneration	Pulmonary cell, breast cancer cell, and dermal fibroblast bioprinting	Saini et al. (2021)

generated to expel the bioink (Murphy et al., 2020). Inkjet bioprinters are commonly used to regenerate functional cartilage tissue and skin as well as in drug development and for drug coatings. Inkjet bioprinting can also be used to fabricate and print DNA (Li et al., 2020).

10.3.3 Laser-Assisted Printing

Laser-assisted bioprinting uses a laser to deposit the bioink on a substrate, relying on three components: the source of the laser, the biomaterial for deposition on substrate, and the receiving substrate. UV wavelengths are used for nanosecond lasers to print cells, hydrogels, and proteins (Zhang et al., 2021). Biomaterials are preferred in liquid or gel state for easy absorbance and spread on the substrate. The receiving substrate is made up of biopolymer material that supports cell adhesion and sustainable growth. Laser-assisted bioprinting works by inducing a laser to transfer biomaterial on the substrate. The laser irradiates the ribbon, which evaporates the liquid biomaterial in a droplet form to reach the receiving substrate (Koch et al., 2013). Stereolithography is another form of laser-assisted bioprinting that uses visible or UV light based on photosensitive polymers; this technique helps to eliminate shear stress (Wang et al., 2015). Laser-assisted bioprinting is mainly used to construct cellularized skin, pulmonary artery endothelial cells, dermal fibroblasts, and breast cancer cells.

10.4 BIOINKS FOR 3D BIOPRINTING

Bioinks are the hydrogel biomaterial that is used to create 3D constructs in bioprinting, specifically to encapsulate desired cell types during the printing. They can be prepared using natural or synthetic materials or combinations of the two (Decante et al., 2021). A bioink should possess flawless mechanical, rheological, physiochemical, and biological properties; should facilitate cell growth, attachment, and proliferation; and should be convenient enough to deliver biochemical signals. Other important aspects of bioink include industrial scalability, high resolution during printing, immunological compatibility, low cost, biomimicking, and elasticity (Gu et al., 2022). An ideal bioink requires properties such as shear thinning, oxygen and metabolic waste permeability, biodegradation, biocompatibility, viscoelasticity, and tissue regeneration for successful 3D printing (Ashammakhi et al., 2019). We discuss bioinks in more detail in the next section. Hydrogel-based bioinks are widely used as they are obtained from natural resources and can mimic structures; they are also biodegradable and biocompatible (Benwood et al., 2021). Synthetic bioinks offer better control over mechanical stability, pH, and temperature and have photo-crosslinking ability but are less preferred (Pekkanen et al., 2017).

10.4.1 Natural Bioinks

Natural biomaterials for tissue scaffolds include collagen, agarose, gelatin, fibrinogen, alginate, carboxymethyl cellulose, silk, hyaluronic acid, chitosan, and decellularized extracellular matrix (ECM) (Rijal, 2023).

10.4.1.1 Protein-Based Bioinks

Collagen, gelatin, fibrinogen, and decellularized ECM are protein-based bioink materials that provide a suitable microenvironment for cell growth. Collagen is a naturally occurring protein in the human body that is highly expressed in the ECM; it contains self-aggregating polypeptide chains and is used widely in scaffold bioprinting as it has excellent biocompatible, rheological, and viscoelastic properties. Collagen biomaterial can be printed by controlling its pH and temperature, which govern its utilization in printing blood vessels, heart, nerve, skin, bone, bladder, and adipose cells (Mori et al., 2013). Although collagen is thermally unstable, it is mixed with biomaterials such as alginate, gelatin, and hyaluronic acid to improve its mechanical properties.

Gelatin has the same composition as collagen; it is a product of denatured collagen but is more thermosensitive and has low mechanical properties. Extrusion-based bioprinting is commonly used to achieve stable gelatin printing, with other biomaterials mixed with gelatin during the bioprinting process. Ease of processing, low cost, biodegradability, no harmful byproducts, and biocompatibility are the advantages of using gelatin for printing (Rajabi et al., 2021).

Fibrin or fibrinogen is a soluble glycoprotein present in ECM and that forms naturally during blood clotting. Fibrin contains a large number of binding sites for cells and growth factors, which encourage platelet spreading, angiogenesis, cell infiltration, and fibroblast proliferation. Fibrinogen and fibrin are commonly employed in tissue engineering to accelerate wound healing and create functional tissue constructions because they are biodegradable, biocompatible, and nonimmunogenic, and ultimately result in cell adhesion, proliferation, and ECM (de Melo et al., 2020).

10.4.1.2 Agarose

Agarose is a naturally occurring, water-soluble polymer that can be crosslinked and de-crosslinked with temperature control; it is extracted from red algae. Agarose gels to printability between 26 and 30 °C. It possess admirable mechanical properties and biocompatibility, but it has limited capacity to promote cell development. To overcome this drawback, researchers use functional biomaterial blends in addition to agarose gel (Dravid et al., 2022).

10.4.1.3 Alginate

Alginate is a naturally occurring anionic polysaccharide that is extracted from brown seaweed and resembles the glycosaminoglycans that are present in the ECM. It is an inexpensive, moderate-gelation material with low cytotoxicity and biocompatibility, widely used in biomedicine. Specifically, its quick gelation under physiological settings without producing toxic byproducts has made it a popular bioink. Divalent cations such as $Ca2^+$ and $Ba2^+$ can readily cause alginate to gel by building bridges between polymer chains that allow for physical crosslinking and solidification (Axpe and Oyen, 2016). The majority of bioprinting techniques use alginate or modified alginate alone, or alginate combined with other biomaterials, as the quick gelation property of alginate results in excellent bioprinting.

10.4.2 Synthetic Bioinks

To create more stable structures with adjustable qualities for 3D bioprinting, natural biomaterials are mixed with synthetic bioinks. Synthetic bioinks are promising options for enhancing mechanical properties, printability, and crosslinking, even though they might not encourage cellular adhesion as natural polymers. Polyvinylpyrrolidone, PEG, polylactic acid (PLA), and polycaprolactone (PCL) are synthetic bioinks (Khoeini et al., 2021). Synthetic bioinks are frequently employed to create a robust mechanical scaffold framework, and then cell-incorporating hydrogels are printed or impregnated into the frameworks to create hybrid scaffolds.

10.5 BIOREACTORS FOR POST-BIOPRINTING

Static cultivation is the primary method used for scaffold maturation after bioprinting. In the incubators, 3D-bioprinted scaffolds are grown while being coated with manually swappable media. The different types of bioreactors for scaffold maturation allow dynamic cultivation of single cells or tissues. To mimic the tissue environment, dynamic bioreactors are a valuable source that allows for control and monitoring of pH, O_2 saturation, flow rate, temperature, and mechanical stimulation (Raggatt and Partridge, 2010). Selecting a bioreactor that can grow 3D cell-filled scaffolds following bioprinting is contingent upon the tissue to be created and its intended biomechanical environment. There are three categories of bioreactors used in tissue engineering: rotating wall vessels, perfusion systems, and spinner flasks (Massai et al., 2013).

10.5.1 Rotating Bioreactor

The rotating bioreactor is designed for applications in simulated microgravity conditions; they are horizontal tissue culture vessels composed of rotating concentric cylinders filled with culture medium, provided via a coaxial tubular silicon membrane. The rotating-wall vessel bioreactor allows cells to grow in 3D in a physiological low-fluid shear environment (Takebe et al., 2012). More than 50 rotating-wall vessel-derived tissue models have been created to date, including bone, cartilage, liver, neural tissue, cardiovascular muscle, adipose, and epithelial tissues.

10.5.2 Perfusion Bioreactor

Perfusion bioreactors include one or more perfusion chambers where cell–scaffold constructs are placed, a medium reservoir, a tubing circuit, and a pump that allows mass transport of nutrients and oxygen throughout the perfusion chamber. Perfusion bioreactors are described as either direct or indirect based on whether the culture medium is perfused around or throughout the cell–scaffold constructions (Gaspar et al., 2012). Perfusion bioreactors are advantageous in large tissue mass designs because they provide more exact control of the culture environment and are broadly used for bone tissue engineering.

10.5.3 Spinner-Flask Bioreactors

Spinner flasks are the simplest bioreactor systems, consisting of a glass or plastic tank in which cell–scaffold structures are linked to vertical needles dangling from the vessel's top and immersed in culture media. The vessel's top is typically employed for gas exchange and medium oxygenation. A stir bar at the bottom of the vessel or other mixing mechanisms keeps the medium mixed (Zhang et al., 2021). Spinner-flask bioreactors have been widely used in large-scale production of *in vitro* chondrogenic differentiation and subsequent *in vivo* cartilage production of human adipose-derived stem cells (Rosser and Thomas, 2018). Spinner flasks hold large amounts of mass and are the most used bioreactors in 3D constructs.

10.6 3D BIOPRINTING IN HUMAN HEALTH

Using proteins, nucleic acids, growth factors, medications, and living cells, 3D bioprinting has proven to be a potent tool for tissue synthesis and development and organ patterning. There are multiple healthcare applications of 3D bioprinting (Ozbolat et al., 2016) that we discuss in this section.

10.6.1 Tissue Transplantation

10.6.1.1 Blood Vessels

The major causes of death and morbidity worldwide are coronary heart disease and peripheral artery disease, with a projected 23.3 million more deaths annually by 2030 (Mohammad et al., 2021). There has therefore been high demand for bioengineered blood vessels, and regenerative medicine and 3D printing have emerged as methods for creating blood vessels and vascularized tissue constructs of varied topologies and sizes for transplantation and regeneration (Leonov et al., 2021). Researchers have studied a wide variety of organic and synthetic materials alone and in combination to create bioinks that closely approach the physicochemical characteristics of ECM. Gelatin, collagen, elastin, fibrin, hyaluronic acid, agarose, alginate, and Matrigel are only a few of the natural biomaterials that have been extensively employed for *in vitro* vascular tissue engineering. In addition to offering a favorable milieu for cell adhesion, development, and proliferation, natural biomaterial-based hydrogel often does not result in persistent inflammation or toxicity to the host (Cao et al., 2021).

10.6.1.2 Bone

The different 3D printing technologies produce products with diverse properties depending on the material. To produce bone, bioprinters use gypsum powder, plastics, resins, aluminide, ceramics, sand, metal, polyether ether ketone, and graphene supported by other materials (Brachet et al., 2023). Bone tissue engineering using 3D inkjet printing technology makes the necessary cell dispersion within the scaffold easier; the process uses hydrogels and viscous fluids as the bioinks (Midha et al., 2019) and is essential for producing mineralized bone tissue as well for cell adhesion, differentiation, proliferation, and migration. Additionally, bone tissue engineering offers alternatives for autograft and allograft, which can enhance and restore function (Haleem et al., 2020).

However, despite their promise, the current biomaterials frequently fall short of clinical standards in terms of controlled biodegradability, bone induction, and structural support. Orthopedic professionals are paying close attention to 3D porous scaffolds as a potential treatment for bone deformities (Chen et al., 2021). Three-dimensional printed complex bone scaffolds resemble human bone in terms of microstructure and appearance. Based on this, it is advantageous to combine active ingredients, such as cells and growth factors, to repair bone tissue, which is crucial for the individualized treatment of bone abnormalities.

10.6.1.3 Skin

The intricate physiological regulation of wound healing requires interaction between numerous cell types and mediators at extremely specific time intervals (Bhatt et al., 2022). For instance, burns and extensive skin injuries frequently do not fully heal and instead leave scars due to altered microenvironments that cause aberrant cell development and matrix formation and decreased perspiration and hair follicles, but novel 3D bioprinting technologies achieve high cell viability and printability (Kang et al., 2022). Because shrinkage is a common problem for this kind of material, most printed skin uses collagen type I as the supporting matrix because collagen combined with polymeric mesh scaffolds (like PCL) stops the shrinkage. The simplest skin model, made up entirely of keratinocytes, has been extensively used to examine the penetration and absorption of medicines and other substances (Yan et al., 2018).

The demand for skin biofabrication is increasing due to cosmetic surgeries and the rise in obesity, diabetes, and aging populations. Skin bioprints are used in clinical applications like regenerative medicine, for modeling physiology and pathology, and in the pharmaceutical and cosmetic industries. The ultimate goal of 3D bioprinted skin is to recapitulate biomimetic ECM, encouraging differentiation into mature skin tissues that closely resemble native skin including stratified epidermis, dermis, hypodermis, blood vessels, nerves, muscles, and skin (Olejnik et al., 2021).

10.6.1.4 Dentistry

The field of dentistry is being revolutionized by 3D printing; worldwide, the most cutting-edge dental practices and laboratories are equipped with 3D printers. They enable creating for patients biting splints, surgical guides, detachable complete prostheses (with and without implants), orthodontic devices, temporary fixed prosthetic restorations (crowns and bridges, whole arches), print studies, and working models in resin (Ramasamy et al., 2013). Additionally, in metal printing, methods such as laser sintering and melting enable the production of customized implants (such as root-analog, subperiosteal, and craniofacial implants), surgical guides, and meshes for bone regeneration. In the near future, 3D printers are going to make it possible to print fixed prosthetic reconstructions (crowns and bridges) made of monolithic zirconia, veneers made of lithium disilicate, and specially designed scaffolds made of hydroxyapatite/beta-tricalcium phosphate for regenerating small vertical and horizontal bone defects (Mangano et al., 2023). Investigators have conducted restricted clinical dentistry trials in the pediatric, restorative, and endodontic domains. To define and create standards and guarantee that the new technology is quickly implemented into a dental practice, long-term clinical research is required (Balhaddad et al., 2023).

10.6.2 Drug Development and Screening

In addition to being used for transplantation, 3D-bioprinted tissue constructions are also being created for studying chemical, biological, and toxicological agents and for fundamental research. An efficient and secure medicine normally takes 10 to 15 years to develop using the conventional method (Reddy et al., 2023). As a result, the idea of retooling the drug development pipeline using cutting-edge biofabrication technology has become popular, in particular 3D bioprinting, which stands apart from previous biofabrication technologies due to its premise of layer-by-layer building and on-demand spatial patterning of many materials. Therapeutic compounds have been deposited as oral tablets or implantable forms in 3D bioprinted medications. In particular, 3D bioprinting excels in customizing medications by fusing diverse therapeutic elements in various amounts in accordance with individual conditions (Yi et al., 2021).

Moreover, bioprinted vehicles can deliver antibiotics, antiinflammatory medications, and medications that encourage cell differentiation and proliferation to wounds; deliver anticancer medications directly to tumors; and promote or inhibit angiogenesis and vascularization to treat cancer, arterial diseases, heart diseases, and arthritis. Furthermore, customized implants that carry medications directly to the intended tissues can be printed in various shapes and configurations (Ramadan and Zourob, 2021). These 3D bioprinters can also produce personalized medication that is more effective than conventional pharmaceutical products while decreasing material waste, providing customized dosage based on patient needs, and using fewer water-soluble proteins and peptides, and they can be used with various active components (Gara et al., 2024).

Another application for 3D bioprinting is testing a drug's safety and efficacy in intricate human organs like the liver and heart. The objective is to avoid the time-consuming process of animal testing by creating microtissues that will generate identical physiologic responses on a tiny scale, as seen in a full organ. A deeper understanding of the mechanisms underlying illness progression and medication screening is made possible by 3D models. This approach is ideal for making 3D organ-on-a-chip devices due to the extremely fine control over the microfeatures of the 3D printed constructions, especially when combined with microfabrication technologies (Matai et al., 2020).

10.6.3 Cancer

To enable the development of cutting-edge cancer medicines, 3D printed models help to increase our understanding of the pathophysiology and metastasis of cancer. Simulating a cancer microenvironment is challenging, but 3D printing makes it possible by using materials like Pluronic F-127 to create sacrificial templates. This enables creating vessel-like structures that can be cellularized and perfused, thus increasing nutrient availability (Albritton et al., 2017).

Richards et al. described a vascular network can be printed using various techniques, but extrusion-based bioprinting can print complicated networks (Richards et al., 2017). According to Zervantonakis et al. (2012), 3D breast tumor bioprinted models linked to microfluidics can reproduce alterations in the endothelium barrier by interactions between tumor and endothelial cells and replicate intravasation. Discovering the mechanisms behind metastasis continues to be a serious obstacle in advancing the battle against cancer. One approach to studying metastasis is the Boyden chamber, currently the most popular 3D technique, in which cells flow across a physical barrier with pores (migration) or merge with a protein coating, frequently Matrigel, to simulate the ECM (invasion) (Guy et al., 2017).

10.7 THE EFFECTS OF 3D BIOPRINTING ON QUALITY OF LIFE

Transplant therapy is expanding noticeably worldwide, significantly extending life spans and improving patients' overall quality of life. Critical medical needs have been met through successful transplants of the cornea, liver, kidneys, lungs, heart, spleen, thyroid, and pancreas (Thongprayoon et al., 2022).

Because 3D bioprinting uses novel techniques to create artificial organs, it can solve the ongoing problem of organ donor scarcity by offering personalized part production and reducing the probability of rejection risks after transplantation. Moreover, it decreases wait times for organs. Important developments in bioprinting have opened the door for hopeful improvements in replacing internal organs, blood vessels, and bones in human patients (Yaneva et al., 2023). As 3D bioprinting is taking us toward personalized medicine, it will also resolve the issue of gender differences in transplantation; for instance, there is a difference between the anatomy and physiology of male versus female kidneys that can impact transplantation. Moreover, the ability of students to examine without risking patients is one of the main benefits.

One of the main benefits of bioprinting is that it allows students to examine medical scenarios without risking patients. It can be used to store patient cases with rare health conditions, providing opportunities to train medical professionals in unusual pathologies and challenging conditions. Additionally, printed models can reduce the need for human body parts and address moral and legal concerns about using corpses as teaching tools (Shopova et al., 2023).

10.8 ETHICAL AND LEGAL REGULATIONS IN 3D BIOPRINTING

Bioprinting has progressed from eradicating skepticism by focusing on its advantages ("apologetics") to focusing on its medical applications ("evangelism"). However, the use of live cells necessitates strict ethical and legal standards. Traditional ethical issues include donor confidentiality, informed permission, the potential for invasive cell creation, and donor cell ownership. The medical ethics principle of *Primum non nocere* is still debated considering the morality of the activity, the agent's aim, the method of action, potential negative impacts, and the balance of positive and negative outcomes. Additionally, commercialization is hindered by legal constraints such as the Russian Civil Code's subclause 3 of Clause 4 of Article 1349, and using cells from specific animal species can offend religious beliefs, complicating the use of xenogeneic sources in 3D printing (Kirillova et al., 2020). Swierstra and Rip (2009) also developed a model they called Ethics of New and Emerging Science and Technology that discusses ethics at two levels. At the first level is meta-ethical issues, which includes utilitarianism or consequentialism, deontology, theories of justice, and conceptions of the good life, which are moral ethics (Vijayavenkataraman et al., 2016). In examining the ethics of bioprinting, it is important to consider the technique's reliance on a digital model. As printed organs made from digital models replace natural ones, they will replace nature. As a result, the issue of who should bear responsibility for creating and assessing 3D models arises: Should "designers," biologists, or programmers be in charge of translating anatomical images into digital form, and if so, to what extent? Will models be usable without a patient's permission? The concepts of informed consent and confidentiality are the two cornerstones of protecting human rights in biomedical research (Patuzzo et al., 2018).

10.9 LIMITATIONS AND FUTURE PERSPECTIVES

Many of the difficulties the 3D bioprinting industry is currently encountering are related to certain technical, material, and cellular components of the bioprinting process. Despite the field's early development, it has already been successful in producing a number of human-scale tissues that are almost functional enough for transplantation. However, it is necessary to improve bioprinter technology and categorize printers based on the optimal materials for each. Additionally, development is required to boost resolution and make 3D bioprinting accessible at a commercial level.

Likewise, we need new biomaterials that enable creating sophisticated gradient combinations (Murphy and Atala, 2014). After implantation, the printed structures must have sufficient mechanical strength to maintain their shape and withstand external stress. Additionally, bio-ink must maintain low viscosity to avoid clogging the delivery nozzles, although most hydrogels used in bioprinting systems have low mechanical properties (Mallick and Cox, 2013). In addition, 3D printers frequently have limited printing space, which limits the maximum size of 3D-printed tissues and the ability to print entire organs. Even assembling smaller 3D-printed tissue constructs into a larger model leads to assembly errors.

Printing a new generation of tissue structures will be possible with smart polymers that can produce programmable tissue scaffolds by adjusting their characteristics in response to stimuli like pH, temperature, or magnetic fields. This reaction could improve interactions between cells and printed structures. We expect 3D bioprinting to develop beyond structural similarities to functional performance (Gara et al., 2024).

REFERENCES

Albritton JL, Miller JS. 3D bioprinting: Improving in vitro models of metastasis with heterogeneous tumor microenvironments. Dis Models Mech. 2017 Jan 1;10(1):3–14.

Ashammakhi N, Ahadian S, Xu C, Montazerian H, Ko H, Nasiri R, Barros N, Khademhosseini A. Bioinks and bioprinting technologies to make heterogeneous and biomimetic tissue constructs. Mater Today Bio. 2019 Jan 1;1:100008.

Axpe E, Oyen ML. Applications of alginate-based bioinks in 3D bioprinting. Int J Mol Sci. 2016 Nov 25;17(12):1976.

Balhaddad AA, Garcia IM, Mokeem L, Alsahafi R, Majeed-Saidan A, Albagami HH, Khan AS, Ahmad S, Collares FM, Della Bona A, Melo MA. Three-dimensional (3D) printing in dental practice: Applications, areas of interest, and level of evidence. Clin Oral Investig. 2023 Jun;27(6):2465–2481.

Benwood C, Chrenek J, Kirsch RL, Masri NZ, Richards H, Teetzen K, Willerth SM. Natural biomaterials and their use as bioinks for printing tissues. Bioengineering. 2021 Feb 20;8(2):27.

Bhatt T, Dey R, Hegde A, Ketkar AA, Pulianmackal AJ, Deb AP, Rampalli S, Jamora C. Initiation of wound healing is regulated by the convergence of mechanical and epigenetic cues. PLoS Biol. 2022 Sep 16;20(9):e3001777.

Billiet T, Vandenhaute M, Schelfhout J, Van Vlierberghe S, Dubruel P. A review of trends and limitations in hydrogel-rapid prototyping for tissue engineering. Biomaterials. 2012 Sep 1;33(26):6020–6041.

Blaeser A, Duarte Campos DF, Puster U, Richtering W, Stevens MM, Fischer H. Controlling shear stress in 3D bioprinting is a key factor to balance printing resolution and stem cell integrity. Adv Healthc Mater. 2016 Feb;5(3):326–333.

Brachet A, Bełżek A, Furtak D, Geworgjan Z, Tulej D, Kulczycka K, Karpiński R, Maciejewski M, Baj J. Application of 3D printing in bone grafts. Cells. 2023 Mar 10;12(6):859.

Cao X, Maharjan S, Ashfaq R, Shin J, Zhang YS. Bioprinting of small-diameter blood vessels. Engineering. 2021 Jun 1;7(6):832–844.

Chen S, Wang H, Mainardi VL, Talò G, McCarthy A, John JV, Teusink MJ, Hong L, Xie J. Biomaterials with structural hierarchy and controlled 3D nanotopography guide endogenous bone regeneration. Sci Adv. 2021 Jul 28;7(31):eabg3089.

Chia HN, Wu BM. Recent advances in 3D printing of biomaterials. J Biol Eng. 2015 Dec;9(1):1–4.

Datta P, Barui A, Wu Y, Ozbolat V, Moncal KK, Ozbolat IT. Essential steps in bioprinting: From pre-to post-bioprinting. Biotechnol Adv. 2018 Sep 1;36(5):1481–1504.

de Melo BA, Jodat YA, Cruz EM, Benincasa JC, Shin SR, Porcionatto MA. Strategies to use fibrinogen as bioink for 3D bioprinting fibrin-based soft and hard tissues. Acta Biomaterialia. 2020 Nov 1;117:60–76.

Decante G, Costa JB, Silva-Correia J, Collins MN, Reis RL, Oliveira JM. Engineering bioinks for 3D bioprinting. Biofabrication. 2021 Apr 8;13(3):032001.

Derakhshanfar S, Mbeleck R, Xu K, Zhang X, Zhong W, Xing M. 3D bioprinting for biomedical devices and tissue engineering: A review of recent trends and advances. Bioact Mater. 2018 Jun 1;3(2):144–156.

Dravid A, McCaughey-Chapman A, Raos B, O'Carroll SJ, Connor B, Svirskis D. Development of agarose–gelatin bioinks for extrusion-based bioprinting and cell encapsulation. Biomed Mater. 2022 Jun 15;17(5):055001.

Feng Y, Zhu S, Mei D, Li J, Zhang J, Yang S, Guan S. Application of 3D printing technology in bone tissue engineering: A review. Curr Drug Deliv. 2021 Aug 1;18(7):847–861.

Gara DK, Gujjala R, Prasad PS, Madaboosi N, Ojha S. 4D bioprinting: A review on smart bio-adaptable technology to print stimuli-responsive materials. Prog Addit Manuf. 2024 Dec 27:1–43.

Gaspar DA, Gomide V, Monteiro FJ. The role of perfusion bioreactors in bone tissue engineering. Biomatter. 2012 Oct 1;2(4):167–175.

Gillispie G, Prim P, Copus J, Fisher J, Mikos AG, Yoo JJ, Atala A, Lee SJ. Assessment methodologies for extrusion-based bioink printability. Biofabrication. 2020 Feb 19;12(2):022003.

Groll J, Boland T, Blunk T, Burdick JA, Cho DW, Dalton PD, Derby B, Forgacs G, Li Q, Mironov VA, Moroni L. Biofabrication: Reappraising the definition of an evolving field. Biofabrication. 2016 Jan 8;8(1):013001.

Gu Y, Forget A, Shastri VP. Biobridge: An outlook on translational bioinks for 3D bioprinting. Adv Sci. 2022 Jan;9(3):2103469.

Gungor-Ozkerim PS, Inci I, Zhang YS, Khademhosseini A, Dokmeci MR. Bioinks for 3D bioprinting: An overview. Biomater Sci. 2018;6(5):915–946.

Guy JB, Espenel S, Vallard A, Battiston-Montagne P, Wozny AS, Ardail D, Alphonse G, Rancoule C, Rodriguez-Lafrasse C, Magne N. Evaluation of the cell invasion and migration process: A comparison of the video microscope-based scratch wound assay and the boyden chamber assay. JoVE (J Visual Exp). 2017 Nov 17;129:e56337.

Haleem A, Javaid M, Khan RH, Suman R. 3D printing applications in bone tissue engineering. J Clin Orthopaed Trauma. 2020 Feb 1;11:S118–S124.

Ji S, Guvendiren M. Recent advances in bioink design for 3D bioprinting of tissues and organs. Front Bioeng Biotechnol. 2017 Apr 5;5:23.

Kačarević ŽP, Rider PM, Alkildani S, Retnasingh S, Smeets R, Jung O, Ivanišević Z, Barbeck M. An introduction to 3D bioprinting: Possibilities, challenges and future aspects. Materials. 2018 Nov 6;11(11):2199.

Kang MS, Jang J, Jo HJ, Kim WH, Kim B, Chun HJ, Lim D, Han DW. Advances and innovations of 3D bioprinting skin. Biomolecules. 2022 Dec 27;13(1):55.

Khoeini R, Nosrati H, Akbarzadeh A, Eftekhari A, Kavetskyy T, Khalilov R, Ahmadian E, Nasibova A, Datta P, Roshangar L, Deluca DC. Natural and synthetic bioinks for 3D bioprinting. Adv NanoBiomed Res. 2021 Aug;1(8):2000097.

Kirillova A, Bushev S, Abubakirov A, Sukikh G. Bioethical and legal issues in 3D bioprinting. Int J Bioprint. 2020;6(3).

Koch L, Gruene M, Unger C, Chichkov B. Laser assisted cell printing. Curr Pharmaceut Biotechnol. 2013 Jan 1;14(1):91–97.

Leonov DV, Spirina YA, Yatsenko AA, Kushnarev VA, Ustinov EM, Barannikov SV. Advanced 3D bioprinting technologies. Cell Tissue Biol. 2021 Nov;15:616–627.

Li J, Chen M, Fan X, Zhou H. Recent advances in bioprinting techniques: Approaches, applications and future prospects. J Transl Med. 2016 Dec;14(1):1–5.

Li X, Liu B, Pei B, Chen J, Zhou D, Peng J, Zhang X, Jia W, Xu T. Inkjet bioprinting of biomaterials. Chem Rev. 2020 Sep 9;120(19):10793–10833.

Loai S, Kingston BR, Wang Z, Philpott DN, Tao M, Cheng HL. Clinical perspectives on 3D bioprinting paradigms for regenerative medicine. Regen Med Front. 2019 Jul 23;1(1).

Mallick KK, Cox SC. Biomaterial scaffolds for tissue engineering. Front Biosci Elite. 2013 Jan 1;5(1):341–360.

Mandrycky C, Wang Z, Kim K, Kim DH. 3D bioprinting for engineering complex tissues. Biotechnol Adv. 2016 Jul 1;34(4):422–434.

Mangano F, Savkova N. 3D printing in dentistry. In: *Handbook of Surgical Planning and 3D Printing*. Academic Press, 2023 Jan 1, pp. 171–202.

Massai D, Cerino G, Gallo D, Pennella F, Deriu MA, Rodriguez A, Montevecchi FM, Bignardi C, Audenino A, Morbiducci U. Bioreactors as engineering support to treat cardiac muscle and vascular disease. J Healthc Eng. 2013 Aug;4:329–370.

Matai I, Kaur G, Seyedsalehi A, McClinton A, Laurencin CT. Progress in 3D bioprinting technology for tissue/organ regenerative engineering. Biomaterials. 2020 Jan 1;226:119536.

Midha S, Dalela M, Sybil D, Patra P, Mohanty S. Advances in three-dimensional bioprinting of bone: Progress and challenges. J Tissue Eng Regen Med. 2019 Jun;13(6):925–945.

Mohammad AM, Rashad HH, Habeeb QS, Rashad BH, Saeed SY. Demographic, clinical and angiographic profile of coronary artery disease in Kurdistan region of Iraq. Am J Cardiovasc Dis. 2021;11(1):39.

Mori H, Shimizu K, Hara M. Dynamic viscoelastic properties of collagen gels with high mechanical strength. Mater Sci Eng C. 2013 Aug 1;33(6):3230–3236.

Murphy SV, Atala A. 3D bioprinting of tissues and organs. Nat Biotechnol. 2014 Aug;32(8):773–785.

Murphy SV, De Coppi P, Atala A. Opportunities and challenges of translational 3D bioprinting. Nat Biomed Eng. 2020 Apr;4(4):370–380.

Norotte C, Marga FS, Niklason LE, Forgacs G. Scaffold-free vascular tissue engineering using bioprinting. Biomaterials. 2009 Oct 1;30(30):5910–5917.

Olejnik A, Semba JA, Kulpa A, Danczak-Pazdrowska A, Rybka JD, Gornowicz-Porowska J. 3D bioprinting in skin related research: Recent achievements and application perspectives. ACS Synth Biol. 2021 Dec 30;11(1):26–38.

Ozbolat IT, Peng W, Ozbolat V. Application areas of 3D bioprinting. Drug Discov Today. 2016 Aug 1;21(8):1257–1271.

Patuzzo S, Goracci G, Gasperini L, Ciliberti R. 3D bioprinting technology: Scientific aspects and ethical issues. Sci Eng Ethics. 2018 Apr;24:335–348.

Pekkanen AM, Mondschein RJ, Williams CB, Long TE. 3D printing polymers with supramolecular functionality for biological applications. Biomacromolecules. 2017 Sep 11;18(9):2669–2687.

Raggatt LJ, Partridge NC. Cellular and molecular mechanisms of bone remodeling. J Biol Chem. 2010 Aug 13;285(33):25103–25108.

Rajabi N, Rezaei A, Kharaziha M, Bakhsheshi-Rad HR, Luo H, RamaKrishna S, Berto F. Recent advances on bioprinted gelatin methacrylate-based hydrogels for tissue repair. Tissue Eng Part A. 2021 Jun 1;27(11–12):679–702.

Ramadan Q, Zourob M. 3D bioprinting at the frontier of regenerative medicine, pharmaceutical, and food industries. Front Med Technol. 2021 Jan 28;2:607648.

Ramasamy M, Raja R, Narendrakumar R. Implant surgical guides: From the past to the present. J Pharm Bioallied Sci. 2013 Jun;5(Suppl 1):S98.

Reddy VS, Ramasubramanian B, Telrandhe VM, Ramakrishna S. Contemporary standpoint and future of 3D bioprinting in tissue/organs printing. Curr Opin Biomed Eng. 2023 Apr 11;27:100461.

Richards D, Jia J, Yost M, Markwald R, Mei Y. 3D bioprinting for vascularized tissue fabrication. Ann Biomed Eng. 2017 Jan;45:132–147.

Rijal G. Bioinks of natural biomaterials for printing tissues. Bioengineering. 2023 Jun 10;10(6):705.

Rosser J, Thomas DJ. Bioreactor processes for maturation of 3D bioprinted tissue. In: *3D Bioprinting for Reconstructive Surgery*. Woodhead Publishing, 2018 Jan 1, pp. 191–215.

Saini G, Segaran N, Mayer JL, Saini A, Albadawi H, Oklu R. Applications of 3D bioprinting in tissue engineering and regenerative medicine. J Clin Med. 2021 Oct 26;10(21):4966.

Shopova D, Yaneva A, Bakova D, Mihaylova A, Kasnakova P, Hristozova M, Sbirkov Y, Sarafian V, Semerdzhieva M. (Bio) printing in personalized medicine: Opportunities and potential benefits. Bioengineering. 2023 Feb 23;10(3):287.

Singh M, Haverinen HM, Dhagat P, Jabbour GE. Inkjet printing: Process and its applications. Adv Mater. 2010 Feb 9;22(6):673–685.

Swierstra T, Rip A. NEST-ethics: Patterns of moral argumentation about new and emerging science and technology. Read Philos Technol. 2009 Oct 16:208.

Takebe T, Kobayashi S, Kan H, Suzuki H, Yabuki Y, Mizuno M, Adegawa T, Yoshioka T, Tanaka J, Maegawa J, Taniguchi H. Human elastic cartilage engineering from cartilage progenitor cells using rotating wall vessel bioreactor. Transplant Proc. 2012 May 1;44(4):1158–1161. Elsevier.

Thongprayoon C, Kaewput W, Pattharanitima P, Cheungpasitporn W. Progress and recent advances in solid organ transplantation. J Clin Med. 2022 Apr 11;11(8):2112.

van Lenthe GH, Hagenmüller H, Bohner M, Hollister SJ, Meinel L, Müller R. Nondestructive micro-computed tomography for biological imaging and quantification of scaffold–bone interaction in vivo. Biomaterials. 2007 May 1;28(15):2479–2490.

Vijayavenkataraman S, Lu WF, Fuh JY. 3D bioprinting: An ethical, legal and social aspects (ELSA) framework. Bioprinting. 2016 Mar 1;1:11–21.

Wang Z, Abdulla R, Parker B, Samanipour R, Ghosh S, Kim K. A simple and high-resolution stereolithography-based 3D bioprinting system using visible light crosslinkable bioinks. Biofabrication. 2015 Dec 22;7(4):045009.

Xing F, Xiang Z, Rommens PM, Ritz U. 3D bioprinting for vascularized tissue-engineered bone fabrication. Materials. 2020 May 15;13(10):2278.

Yan WC, Davoodi P, Vijayavenkataraman S, Tian Y, Ng WC, Fuh JY, Robinson KS, Wang CH. 3D bioprinting of skin tissue: From pre-processing to final product evaluation. Adv Drug Deliv Rev. 2018 Jul 1;132:270–295.

Yaneva A, Shopova D, Bakova D, Mihaylova A, Kasnakova P, Hristozova M, Semerdjieva M. The progress in bioprinting and its potential impact on health-related quality of life. Bioengineering. 2023 Aug 1;10(8):910.

Yi HG, Kim H, Kwon J, Choi YJ, Jang J, Cho DW. Application of 3D bioprinting in the prevention and the therapy for human diseases. Signal Transduct Target Ther. 2021 May 14;6(1):177.

Yu Y, Moncal KK, Li J, Peng W, Rivero I, Martin JA, Ozbolat IT. Three-dimensional bioprinting using self-assembling scalable scaffold-free "tissue strands" as a new bioink. Scient Rep. 2016 Jun 27;6(1):28714.

Zeng M, Jin S, Ye K. *Tissue and Organ 3D Bioprinting*. SLAS TECHNOLOGY: Translating Life Sciences Innovation, 2018.

Zervantonakis IK, Hughes-Alford SK, Charest JL, Condeelis JS, Gertler FB, Kamm RD. Three-dimensional microfluidic model for tumor cell intravasation and endothelial barrier function. Proc Natl Acad Sci. 2012 Aug 21;109(34): 13515–13520.

Zhang J, Wehrle E, Rubert M, Müller R. 3D Bioprinting of human tissues: Biofabrication, bioinks, and bioreactors. Int J Mol Sci. 2021 Apr 12;22(8):3971.

11 Scaffold-Based 3D Cultures in Cancer Research

Insights on Colorectal Cancer

Keerthi Nethaji,[1] *Amit Dey,*[1] *Akanksha Esmeeta,*[2]
Alakesh Das,[1] *Atil Bisgin,*[3] *Antara Banerjee,*[1] *and Surajit Pathak*[1]*
[1] Department of Medical Biotechnology, Faculty of Allied Health Sciences, Chettinad Academy of Research and Education, Chettinad Hospital and Research Institute, Tamil Nadu, India
[2] Amity Institute of Biotechnology, Amity University, Uttar Pradesh, India
[3] VariantGen Genetic Diagnosis, Treatment and Healthcare Center and InfoGenom RD Laboratories, Cukurova University Technopolis, Adana, Turkiye

11.1 INTRODUCTION

In vitro models are used in mammalian cell biology studies to grow cells in the lab and study specific biological mechanisms or processes in a variety of test environments. Such models mimic the behavior of cells in real tissues, as to how well the tumor cells cope with local signals and cues to evolve and develop the traits in accordance with the surroundings (Knight and Przyborski 2015). Two-dimensional (2D) cell cultures have been a major tool in cellular biology due to their reliable nature, simplicity, and reproducibility; however, their use degrades the *in vivo* environment of the living cells. Thus, researchers have developed 3D scaffolds and biomaterials to increase the capability to closely emulate the *in vitro* domain of the cells (Campuzano and Pelling 2019). For analysis, 3D cultures of cells are preferred over 2D because of the different topographies and cell behaviors (Greiner et al. 2012). The cell–cell and cell–matrix associations of cells that are raised on 2D tissue culture substrates differs greatly due to the shape (Ghibaudo et al. 2009; Kumar et al. 2011; Bisgin and Mujde 2022).

In contrast, the modeling systems may have a more distinct behavior that includes the function of the cells in the culture (Boxberger and Meyer 1994), differentiation (Pampaloni et al. 2010), drug metabolism (Loessner et al. 2010), gene expression (Ridky et al. 2010), and protein synthesis (Li et al. 2008). Tumour models built in 3D provide precise and error-free results on the aspects of drug screening and testing, metabolic profiling, and research (Jensen and Teng 2020). Due to the limitations of 3D cell culture, new advancements like microfabrication and microfluidics are being explored. This review sheds light on the evolution of cell culture techniques from 2D to 3D and applications in the field of research.

11.2 SCAFFOLD-BASED 3D CELL CULTURES

Three-dimensional scaffolds are made of permeable, porous, and fibrous biomaterials and facilitate the dispatch of gases and body fluids; they promote cell–cell and cell–matrix interactions, viability, and extracellular matrix (ECM) deposition in cell culture. With a better understanding of tumor tissue microenvironments and the associated cell phenotypes, numerous 3D tissue culture models have been utilized in tumor biomedical studies to evaluate cell signaling, gene expression, small-molecule interactions, and medication sensitivities (David et al. 2008).

The 3D tissue culture model offers tumor cells a foundation for restoring defects of nerves, bones, muscles, and vessels as well as for proliferation, ECM generation, and cell attachment (Nikolova and Chavali 2019). Because of their properties, the 3D scaffolds enable cancer cells to multiply heterogeneously, adopt different phenotypes, develop treatment resistance, and metastasize like human malignancies; the scaffolds that are formulated on synthetic polymers or selected ECM components can enhance cancer cells with signaling and proliferation (Unnikrishnan et al. 2021). The 3D scaffold-based tumor models regulate the biochemical and mechanical signals that promote cell–cell and cell–ECM interactions and hypoxic conditions in the original tumor microenvironment (Efraim et al. 2019; Li et al. 2018; Bisgin and Mujde 2022). Moreover, 3D microengineering provides a better idea about the complex architecture of cancer cells and it's *in vivo* properties. Because of the availability of a broad range of biocompatible materials, it has become easier for researchers to manipulate patterns and templates, with more precise control over properties like compositions, mechanics, geometry, cell–cell interaction, and diffusion (Li and Kilian 2015). The engineering of scaffolds as a part of the tumor microenvironment provides

* Corresponding author: drsurajitpathak@care.edu.in

DOI: 10.1201/9781032660592-13

cells with essential signals and growth factors that help them retain their functions (Girigoswami et al. 2021; Bisgin and Mujde 2022).

11.3 3D OVER 2D CELL CULTURE MODELS

Cell cultures in 2D play an important role in research and have been widely used for culturing cells since the early 1990s (Ferreira et al. 2018). Even though 2D culture has been traditionally applied in cancer research and many biological studies, it has limitations related to their inaccuracy in representing tissue cells *in vitro* (Mahmud et al. 2009). The 2D cultures have low maintenance but do not accurately replicate vitality, drug metabolism, or other cellular functions. Moreover, extracting cells from live tissues and plating them in a 2D environment changes the overall cell morphology and functions (Croce et al. 2016). Due to adverse changes, cells in a monolayer adherent arrangement lose polarity when they interact with the external environment. This changes their responses to external stimuli and phenomena like apoptosis, leading to inconsistent results in laboratory experiments (Weaver et al. 2002; Meyers et al. 2006; Bisgin and Mujde 2022). In contrast, the 2D cell culture models do not effectively mimic tumor microenvironments, and therefore most drug testing carried out in clinical trials fails (Unnikrishnan et al 2021).

The advancements in 3D model cell culture systems have led to significant breakthroughs in biological research and drug development. These 3D models have been advanced in numerous investigations of biological mechanisms, including differentiation, proliferation, cell number monitoring, migration, response to stimuli, and invasion of tumor cells into neighboring tissues, drug metabolism, stimulation of angiogenesis, and gene expression. Additionally, 3D models have proved to be physiologically relevant (Antoni et al. 2015). Embedding and cultivating tumor cells in a collagen-based scaffold to replicate the morphology of a tumor in *in vivo* microenvironment allows the 3D model system to facilitate drug testing (Le et al. 2016). However, growing cells in 2D monolayers does not accurately replicate the complex structure of intricate tumors and stroma (Valkenburg et al. 2018).

The 3D-based scaffolds facilitate cell–cell and cell–matrix interactions by influencing biochemical and mechanical signals and mimic the hypoxic and nutrient-deprived conditions of a native tumor environment (Efraim et al. 2019; Li et al. 2018). The cells are subjected to extreme nutrition and oxygenation. Due to the alterations in the alignment and grouping of the surface receptors, the synthesis of ECM proteins is also significantly altered, and the cells are unable to receive the right signals resulting from the natural ECM arrangement (Hutmacher et al. 2009; Castells-Sala et al. 2013). With numerous passages, tumor cells that grow rapidly in 2D cell cultures can initiate an immortalization mechanism. As a result, these cells inaccurately represent the tumor because they become more vulnerable to treatments that target quickly dividing cells (Burdett et al. 2010). Tumour cells frequently develop dorsal and ventral surfaces throughout the proliferation in 2D preparations, which results in an unnatural polarity. In contrast, 3D can reestablish tumor shape without regard to the stiffness or content of the matrix. Anticancer medications must be proven effective in preclinical studies that mimic the *in vivo* tumor microenvironment. The 3D tumor models show higher drug resistance to anticancer medications, making them desirable *in vitro* models for drug testing (Polonio-Alcalá et al. 2019).

The transition from 2D to 3D screening assays is being widely accepted in many science fields. Recent *in vitro* screening techniques have been developed to simulate the 3D microenvironment, resulting in culture models that better resemble *in vivo* physiology (Kimlin et al. 2013). Biomimetic scaffolds have demonstrated a crucial role in growing mammalian cells in complex 3D environments. Many tools have been devised to mimic the environment provided by the ECM. However, only structural scaffolds made of biopolymers can mimic the ECM appropriately. Along with it, 3D scaffolds are also an effective base for carrying out complex tissue engineering experiments including organ culturing as shown in Figure 11.1.

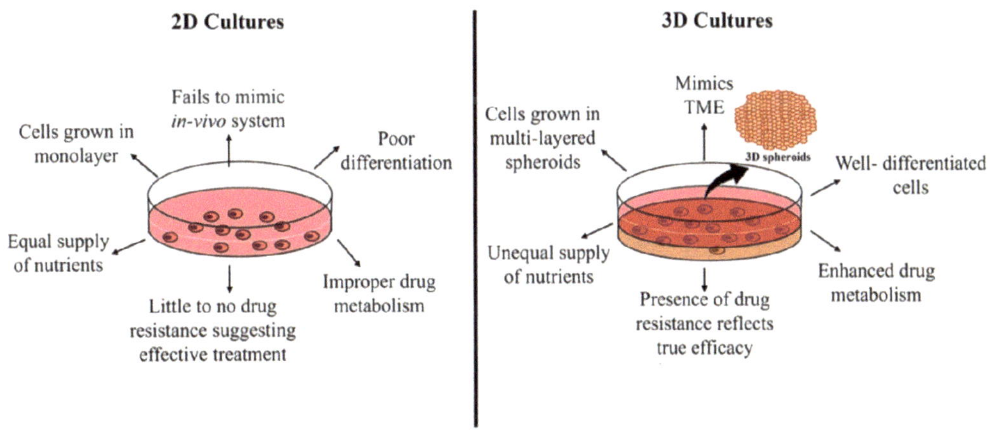

FIGURE 11.1 2D versus 3D cell cultures.

11.4 PROPERTIES OF 3D SCAFFOLDS

Studying cellular responses in tumor models using biochemical and physical cues is effective but still faces barriers. Current models lack a comprehensive understanding of how cells integrate and respond to physical and chemical stimuli. The physical properties of the scaffolds used in research are crucial, and adding growth factors, signaling, and adhesion molecules helps with extensively interpreting cell behavior and differentiation (DuFort et al. 2011; Ross et al. 2012).

Cells grown in 3D cultures exhibit improved glycolysis and upregulation in growth factors, the proteins involved in angiogenesis, and some important chemoattractant, in contrast to 2D cell cultures (Rijal et al. 2017). Depending on the research requirements, 3D scaffolds are fabricated using natural compounds like algae and collagen or synthetic polymers such as PCL, PEG, and polylactic-co-glycolic acid (PLGA) (Dhandayuthapani et al. 2011; Rijal and Li 2016). The use of 3D scaffolds in tissue engineering is essential for cancer research. The ideal scaffolds have elasticity that resembles that of native tissues and are biodegradable, permeable, biocompatible, repeatable, and resilient to support cells temporarily (Safran et al. 2008).

Cells cultured in 3D make major contributions in clinical research because they accurately replicate tumor microenvironment and gene and protein expression; furthermore, 3D cultured cells make it easy to study other processes such as invasion, metastasis, migration, angiogenesis, and tumor metabolism. Scaffold characteristics are determined by their architecture and materials, which in turn affect the engineering of the tissue. Biologists studying cartilage regeneration produce novel elastomeric compounds that were mechanically durable for soft tissue engineering: poly(1,8-octanediol-co-citrate), poly(glycerolsebacate), and polycaprolactone fumarate (Chen et al. 2008).

11.5 BIOMATERIALS USED IN 3D SCAFFOLDS

Hydrogels are a class of biomaterials that vary from synthetic molecules like PEG and polyamides to native proteins like fibrin and collagen and have proved to be well adapted for use in 3D scaffolds in cancer research (Aurand et al. 2012). Some of the commonly used scaffold biomaterials include ceramics like bioactive glasses and hydroxyapatite that can induce specific biological activity. Other materials like synthetic and natural polymers are also used in 3D scaffolds (Lasprilla et al. 2012; Saini et al. 2016).

11.5.1 Natural Polymers

Animals, plants, and microbes that generate sustainable sources of energy are used to produce natural polymers; they show good gelation and water binding. Natural polymers positively influence cells via distinct recognition because they interact with cells in the host organism. Therefore, hybrid materials are used to mimic the ECM. By combining two or three natural polymers and conjugating them, it is possible to create a material that chemically resembles the ECM.

11.5.1.1 Collagen

Collagen types I–V are used as raw biomaterials in 3D scaffolds for tissue engineering because of its bioavailability in the ECMs of various tissues. Collagen I is the most effective in extraction and shows high versatility in scaffolds (Balasubramanian et al. 2013). Collagen scaffolds support the tissue healing process. Additionally, culturing neuronal cancer stem cells on a 3D porous collagen scaffold greatly increases neurite outgrowth, and these models for breast cancer tumor progression have shown promising signs as a dynamic 3D framework with properties crucial for molecular signaling, shape, differentiation, proliferation, and motility (Rijal and Li 2016; Subramaniam et al. 2019).

In different work, striated fibrillar collagens played a direct role in tumor growth, and type III collagen, which serves as the structural foundation of the basement membranes of various solid organs, made the cells more invasive than type IV and V (David et al. 2008). Cancer cells adhere to the collagen matrix, allowing for effective proliferation, survival, migration, and invasion (Pickup et al. 2014; Banerjee et al. 2017; Malayaperumal et al. 2021; Bisgin and Mujde 2022). Collagen is also known to enhance the accumulation and tight organization of the matrix proteins that are part of the tumor environment, allowing the 3D tumor model to be more flexible and stable (Banerjee et al. 2019).

11.5.1.2 Alginate

Alginate is a natural polysaccharide that is widely used in biomedicine and regenerative biology owing to its biocompatibility and biodegradability (Sun and Tan 2013). It is rich in integrin and is often used to research integrin signaling in ECM. It is a polymer derived from brown seaweeds, and it is used as a 3D culture matrix because it facilitates cell integration and provides a foundation for cellular growth. Alginate's exceptional qualities include its biocompatibility, non-toxicity, ease of processing, and easy gelation; sodium alginate, oxidized alginate, and a few other derivatives are extensively used in tissue engineering, wound healing, and drug delivery. Alginate is readily converted into 3D scaffolds for cell development made of microspheres, fibers, sponges, hydrogels, microcapsules, and foams (Sahoo and Biswal 2021).

11.5.1.3 Hyaluronic Acid

Hyaluronic acid (HA) is a glycosaminoglycan that is seen in extracellular tissues in several regions of the human body. It is bioactive in promoting proliferation and the adhesion of complete and degraded endothelial cells, as interactions between HA and endothelial cells are mediated by CD44, ICAM-1, and RHAMM cell receptors to promote growth and tissue repair. These characteristics support the use of this naturally occurring compound as the main component in new biomaterials for vascular grafts (Zhu et al. 2014).

HA supports tissues in resisting external stress and compression as part of the extracellular space in the 3D scaffold. It also modulates the cell behavior in the tumor microenvironment (Girigoswami et al. 2021). HA modifies the biological activity of cancer cells by enhancing the activity of the TGF-β and FAK pathways and other important signaling pathways that contribute to cancer progression; these signaling pathways interact with cell adhesion receptors like CD44 and receptors for RHAMM, and these changes affect cells' responses to chemotherapeutic drugs. Consequently, 3D hydrogel matrices based on HA that have been covalently linked are appealing candidates for *in vitro* research into new treatments that target osteoblastic prostate cancer cells (Gurski et al. 2009).

11.5.2 Synthetic Polymers

11.5.2.1 PEG

PEG is a synthetic polyether with a linear and branching form that is hydrophilic, and biocompatible, non-immunogenic, and it has received US FDA approval for use in biomedicine as a viable option for cell encapsulation. PEG hydrogels are not biodegradable by nature; however, it can be modified to promote decomposition by adding degradable segments. The three most widely utilised hydrolytic blocks are PCL, polyglycolide or poly(glycolic acid) (PGA), and PLA (Liu and Wang 2020). PEG hydrogels are soft polymers, and their size, structure, and other characteristics can be easily changed; they can be used as drug carriers to differentially affect biological parameters *in vivo*. PEG hydrogels were employed as substrates for maintaining stem cell growth following damage, promoting cell migration, proliferation, and differentiation (Kong et al. 2017; Bisgin and Mujde 2022).

11.5.2.2 PCL

PCL is renowned for its excellent biocompatibility due to its slow rate of degradation and its durability; it also outperforms other polyesters in resisting acidic breakdown. Due to its hydrophobic nature, PCL inhibits cell attachment and cell growth, and it has been extensively utilized in 3D scaffolds in bone tissue engineering. When incorporated with other hydrogels and polymers, PCL displayed a unique ability to upgrade its characteristics and even form new combinations. It has also been approved by FDA for use in load-bearing applications (Liu and Wang 2020).

11.5.2.3 PLA

Polycondensation of lactic acid or polymerization of lactide generates PLA. specifically needs the requirement of These synthetic polymers are essential for scaffold fabrication and are the most suitable for this purpose. Its mechanical features and controlled degradation rate make it highly analogous to the proteins used in tissue engineering. Water and CO_2 are two PLA degradation products that are neither toxic nor cancerous to humans. This compound can be found in a variety of forms, such as poly(L-lactic acid), poly(D-lactic acid), and poly(D,L-lactic acid), that have different applications (Polonio-Alcalá et al. 2019).

11.5.2.4 PLGA

Cancer cells grown in 3D PLGA scaffolds showed vastly better morphological characteristics than cells grown in 2D PLGA.

11.6 GENE AND PROTEIN EXPRESSION IN 3D SCAFFOLDS

Gene and protein expression have been broadly analyzed in cancer research studies in which 3D scaffolds are being used to investigate the signaling pathways, stemness, epithelial–mesenchymal transition, migration, malignancy, and some induced functional and behavioral changes in cancer cells. Because of their poor functionality and performance, 2D cultured cells are not suitable for such gene expression studies (Fallica et al. 2011).

In contrast, 3D cell culture scaffolds show better and more accurate results in multiple cancers (Jiang et al. 2015; Sakuma et al. 2016). The impacts of 3D scaffolds on gene expression are influenced by differential intracellular contact, interactions with ECM, and several other factors that can be tested on 3D cultures. These factors are fundamental crucial for researching signaling pathways involved in cancer progression and migration, and 3D cultures have also assisted in determining molecular markers responsible for morphometric analyses.

Using computational techniques in conjunction with 3D cell cultures, researchers identified genes responsible for tumor morphology and aggregation. They also found that gene expression shifts in 3D spheroids mimic those in *in vivo* tumors, highlighting their functional importance. Extensive research showed that 3D culture could be used to investigate protein expression in cancer cell lines (Gaedtke et al. 2007). Such results aid in better understanding oncogenesis and tumor growth, as well as in producing lipid rafts on tumor cells that govern specific signaling pathways. For instance, when the microenvironment changes, cellular gene expression fluctuates significantly. However, cells are remarkably adaptable and can grow and multiply in changing conditions. Extensive experiments in 3D models have shown high repeatability with minimal changes in gene expression.

11.7 APPLICATIONS OF 3D SCAFFOLDS IN CANCER RESEARCH

As we highlighted in previous research, cultured 3D spheroids are popular for producing tumor models, for instance in studying breast acinar structures (Rijal and Li 2016; Chwalek et al. 2014). Hydrogels are extensively used in tissue engineering and drug delivery because they can hold large amounts of water. This capacity derives from their 3D networks of polymer chains that crosslink to form matrices (Zhang and Khademhosseini 2017). Cancer studies involving hydrogels have made it evident that the scaffolds respond to environmental stimuli such as temperature, pH, electric fields, ionic strength, and radiation (Godugu et al. 2013), and smart hydrogels also function as reservoirs for drugs, support natural tissue repair, serve as tissue barriers, and facilitate the delivery of bioactive molecules (Rosser et al. 2015).

Researchers found that the synthetic polymer PEG significantly improved the circulation half-life of drugs in clinical trials. Additionally, it reduced drug accumulation in the liver while enhancing the biocompatibility of the material (Fenton et al. 2018). Apart from the astounding biocompatibility and responsiveness of 3D biomaterial platforms to stimuli, they exhibit many other features like porous structures, stability, flexibility, volume, and elasticity. For instance, with their ability to adapt shapes and structures, 3D biomaterial platforms fit irregular lesions better than other compounds (Han and Wu 2022). Alginate-based 3D hydrogels were used in exploring therapeutics for neuroblastoma immunization and susceptibility (Marrella et al. 2019; Sitarski et al. 2018; Bisgin and Mujde 2022), and cell-laden hydrogel-based scaffolds were beneficial for the sustained co-culture of lymphocytes and prostate cancer cells. These scaffolds reinforced the *ex vivo* interaction of tumor cells with some T lymphocytes, NK cells, and B cells (Marrella et al. 2019).

Synthetic polymers like PLA, PCL, and blends of the two are being widely used for bioprinting 3D scaffold models as seedbeds for cancer cells of interest. The FDA approved PCL to be a precise biocompatible and biodegradable scaffold for bone grafts and other such applications (Sekita et al. 2017; Hung et al. 2016; Temple et al. 2014). It is possible to perform culture experiments on hydrophilic glass fiber with primary, secondary, and other cell lines. Moreover, hydrophilic glass fiber is beneficial for cancer biology experiments as it supports culturing advanced 3D tumor models for lengthy experiments *in vitro* (Matei et al. 2018). Researchers have performed different experiments including mixed cultures, sol-gel suspension co-culture methods, and stack and culture experiments (Rosser et al. 2015) and have established that PCL blended with PLA effectively mimics bone architecture.

Indeed, biomimetic synthetic polymers are a potential tool for constructively modeling the bone niche's architecture and biomechanical properties; these hard polymeric substances require 2D cell cultures to work with and are highly reproducible (Patricio et al. 2013). This material not only promotes tissue regeneration in cartilage, bone, vascular muscle, and ligaments but shows promising drug screening results when tested on the tumoroids grown on it (Rabionet et al. 2017; Matei et al. 2018; Carpenter et al. 2018; Bisgin and Mujde 2022). Bioprinting alginate/fibrinogen hydrogels with HeLa cells also supported the expansion of 3D tumor models in cervical cancer cells, and there was significant improvement in the behavioral similarity to the native cancer cells when HeLa cells were cultured in a 3D model (Zhao et al. 2014, 2020).

Natural hydrogels such as collagen and HA are bioactive and stimulate the native structure of ECM, and therefore, they are used most often, although synthetic hydrogels like PEG are mechanically tunable and support the distribution of bioactive signals. Hydrogels are extremely replicable, but their limited mechanical strength is a flaw. This issue can be often improved by crosslinking after printing the polymers (Ji and Guvendiren 2017). Another important technique in cell culture is magnetic levitation, which allows cell aggregation and promotes cell–cell interaction. This method includes cell loading with magnetic nanoparticles that are later subjected to an external magnetic field that causes the cells to aggregate and form spheroids (Haisler et al. 2013; Bisgin and Mujde 2022). The main advantages of this technique are that it does not require any specific medium and it works as easily as a 2D cell culture. Moreover, it is apt for a wide range of cell types and is not limited to 96-well plate. One of the best advantages of this technique is it forms a spheroid within 16 hours and does not require artificial protein substrate as the formation can take place during the development of ECM (Caballero et al. 2020).

Spheroids can also be formed and tested using a microplate with an ultra-low attachment; the ultra-low attachment coating forms more consistent spheroids. Microplates are typically made of polystyrene and receive a treatment of hydrophilic and hydrophobic coatings. Additionally, the bottom of the plates has a V shape to ensure consistency in the spheroid formation (Chen et al. 2019). One of the advantages of spheroid microplating is that it eliminates the step of spheroid transfer. Human breast cancer cells on microplates mimicked the conditions of hypoxia, resisted apoptosis, and demonstrated drug resistance. Furthermore, researchers effectively used neutrospheres to examine growth kinetics and drug toxicity (Hofmann et al. 2022).

11.8 ADVANTAGES OF 3D SCAFFOLD-BASED CULTURES

Biomaterial-based 3D scaffolds are much more feasible and selective than other techniques (Fenton et al. 2018). They respond to metabolic signals and environmental factors like pressure, temperature, pH, enzymes, and glucose and other stimuli (Lu et al. 2016). Numerous benefits like low immunogenicity, ECM biomimetic features or high cell affinity are present in naturally produced polymers (Chee and Nugent 2019). Through interactions with the cell matrix, it influences the tumor microenvironment, particularly in signaling, differentiation, and migration of breast cancer. Collagen fibrils arrange themselves in multiple terms, depending on the kind of tissue, to create collagen fibers appropriate for the distinct roles and characteristics of tissues. By altering the concentrations and preparation guidelines, collagen hydrogels' mechanical characteristics, architecture, and biodegradability can be precisely modified (Insua-Rodríguez and Oskarsson 2016; El-Sherbiny and Yacoub 2013).

One of the major categories of structural macromolecular elements of ECM is HA. It provides cellular mechanisms with hydration and defense (Sawicki et al. 2018). HA hydrogel can be chemically modified with acrylate or thiol groups to imitate native tumor tissue, leveraging its natural polysaccharide structure (Pluen et al. 2001). High cell viability, regulated proliferation or differentiation, and cell phenotypes resembling that are observed *in vivo* are found to be the outcomes of using hydrogels prepared from natural sources, like Matrigel, collagen, or fibril, support high cell viability, regulate proliferation or differentiation, and produce cell phenotypes resembling those observed *in vivo*. These hydrogels possess natural adhesive characteristics (Glowacki and Mizuno 2008).

Several cancer cell lines of breast, lung, and prostate were found to exhibit a very distinctive and coherent drug response when it was made to grow on scaffolds containing collagen (Stock et al. 2016). Synthetic polymers have the convenience of constant and customizable material characteristics, allowing for repeatable outcomes. Synthetic hydrogels like PEG or vinyl monomer polymers enhance the chemical and mechanical characteristics of cancer cells. Studies conducted with breast cancer cell lines demonstrated high viability and rigidity in shape and size of colonies within scaffolds made from a PEG–fibrinogen hydrogen model (Pradhan et al. 2017; Unnikrishnan et al. 2021). Additionally, PEG, polyamide, PGA, and other artificial polymer hydrogels are affordable and have predictable material characteristics that are usually straight forward to modify by crosslinking (Zhang and Khademhosseini 2017). Spheroid-based 3D models are one of the most widely applied methods to model tumors and mimic cell–cell and cell–ECM interactions because they are inexpensive and easy to use. Such benefits render them useful tools in tumor biology studies and preclinical drug screening. Nevertheless, they lack self-assembly and regeneration properties, hence being less sophisticated than organoid models (Baiao et al. 2022).

Breast cancer cells grown on PLA scaffolds displayed higher proliferation than those grown on 2D cultures (Polonio-Alcalá et al. 2019). The mechanical strength of scaffolds was maintained during cancer-associated fibroblast formation, decellularization, and primary breast cancer cell growth, as the initial breast cancer cells in PCL exhibited delayed disintegration kinetics (Nayak et al. 2019). Alginate hydrogel-grown ovarian tumor cell lines showed chemoresistance to paclitaxel, and alginate in 3D culture favored the retention of stemness features in hepatocellular carcinomas by supporting the Wnt/β-catenin signaling cascade (Takai et al. 2016). Fibroblasts cultured in 3D PLGA scaffolds were more porous and enhanced the spatial interactions between the neighboring fibroblast cells (Hakkinen et al. 2011).

11.9 LIMITATIONS OF 3D SCAFFOLD-BASED CULTURES

PCL has poor tensile strength and very high elongation at breakage (Gunatillake et al. 2006). A significant challenge with alginate gels is their inadequate long-term durability in physiological settings. Exchange reactions with monovalent cations can release divalent ions from these gels, leading to their deterioration. Additionally, alginate gels act as a matrix for the assembly of platelets and red blood cells, while the discharge of calcium ions from the matrix promotes homeostasis. Depending on the context, these characteristics can be advantageous or detrimental, but concerns about these biological processes, along with broader issues with ionically cross linked gels have driven interest in covalently crosslinked hydrogels composed of alginate (Suzuki et al. 1998).

The need for synthetic polymers arises because natural polymers have limitations along with their multiple advantages, in particular rapid *in vivo* degradation, potential risk of infections, and low stability in physiological conditions. Collagen has poor mechanical strength, which makes electrospinning or electrospray processing difficult (Khorshidi et al. 2016; Bisgin and Mujde 2022), and PLG-based scaffolds are demonstrably brittle and fragile (Leach et al. 2019). Natural scaffold systems disintegrate due to MMP-responsive cleavage, which hinders tissue regeneration.

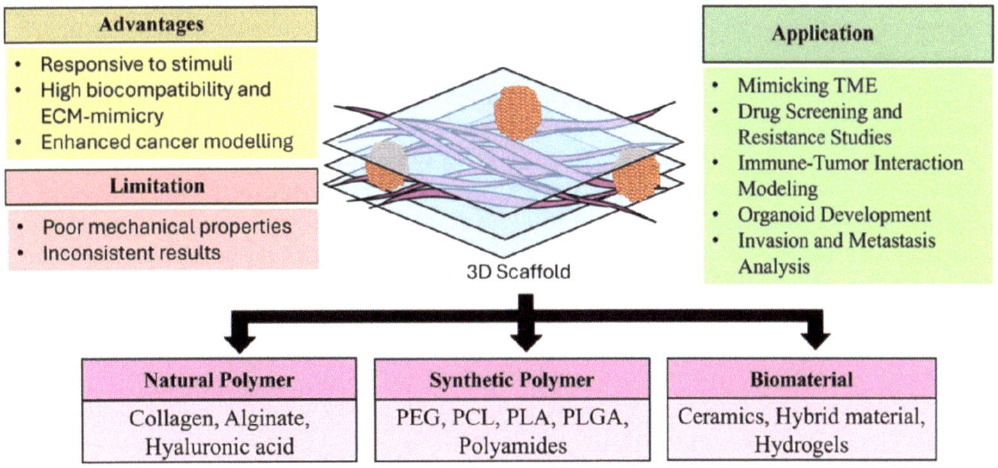

FIGURE 11.2 Overview of 3D scaffold-based models in cancer research.

While synthetic polymers such as PCL and PLA can replicate the structural characteristics of natural ECM, they fail to generate the biochemical signals essential for cell–ECM communication. Similarly, alginate, a plant derivative, lacks the native chemicals sites that bond to mammalian cells, but its inertness minimizes the impact of bioactive media on cell fate (Florczyk et al. 2012).

In addition to the limitations, normal 3D culture technique has shown poor results in many experiments due to lack of expensive coating on the plates. Many scientists have proven that most cell aggregation is due to a result of cell movement in the medium (Krishnamurthy and Nör 2013; Bisgin and Mujde 2022). Researchers have highlighted challenges in obtaining spheres for certain cell lines and with the lack of repeatability of experiments. To overcome such limitations, investigators developed a biomimetic scaffold that presented its own drawbacks including the flattening of cells under adherent conditions. Additionally, the scaffold material significantly influences cell behavior and growth patterns (Lee et al. 2008). Another major concern is restriction of cell observation and cell extraction in some analyses, which makes it difficult to interpret the result (Glicklis et al. 2000).

11.10 3D SCAFFOLD MODELS IN COLORECTAL CANCER

3D scaffold-based models are a new generation of bioengineered models used for anticancer drug screening in colorectal cancer (CRC). Collagen type I hydrogels are most frequently employed for the development of 3D models of CRC because they most closely mimic the ECM microenvironment (Magdeldin et al. 2017; Baker et al. 2013). Hydrogels are usually formed by changing the pH or temperature of the acidic collagen solution, forming a matrix that is cell-supportive in 3D and facilitates cell–cell and cell–ECM interaction. Such scaffolds permit structural support necessary, enabling cancer cells to behave more physiologically in the 3D culture. In a recent study, researchers have successfully created 3D CRC spheroids in an embedded stromal composition. Quantitative imaging techniques in conjugation with analysis of invasive markers and angiogenic factors were utilized in the study to reveal the degree of stromal complexity in the spheroids that significantly had an influence in CRC aggressiveness. Such type of studies emphasizes the need to incorporate stromal components in 3D scaffold models in an effort to enhance the research of tumor microenvironment (Pape et al. 2019).

CRC organoids represent another advanced form of 3D culture that offers significant advantages over traditional 2D models. In the recent times, scientists have made a successful generation of self-organizing intestinal organoids from intestinal stem cells. Such organoids display distinct crypt-like and villi-like structures that closely mimic the spatial organization of the intestine observed *in vivo* I (Pauli et al. 2017; Mahla 2016). Studies confirm that CRC organoids retain around 90% of the somatic mutations of the original tumors that have the ability to demonstrate their utility for precise disease modeling and personalized drug screening (Weeber et al. 2015). Bauleth-Ramos et al. (2020) set up a co-culture spheroid model of CRC by employing a blend of colon cancer cells, monocytes, and human intestinal fibroblasts in an agarose mold. It was applied to assess biocompatibility of nanoparticles and chemoimmunotherapy approaches for CRC treatment (Bauleth-Ramos et al. 2020). These models have shortcomings, for instance, they cannot exactly duplicate the intricacy of native human tissue structure and composition of extracellular matrix. To overcome such shortcomings, scaffold-based 3D systems were created as an artificial, acellular matrix that provides support for seeding, growth, and cell proliferation in a more physiological relevant environment (Langhans 2018).

Matano et al. were the first to use the clustered regularly interspaced short palindromic repeats–CRISPR-associated protein 9 (CRISPR–Cas9) system in human intestinal organoids. This method allowed for the introduction of several oncogenic mutations into organoids from normal colon tissue (Matano et al. 2015). Although, this was an essential step towards futuristic research, the present patient-derived organoid cultures fall short in accurately replicating the key aspects of TME. A deeper understanding of how the microenvironment drives tumorigenesis might lead to more predictive and biologically relevant experimental models that incorporate heterogeneous cell types and better capture the physical and cellular heterogeneity of tumors (Boehnke et al. 2016).

11.11 HIGH-THROUGHPUT SCREENING (HTS)

Drug development, including anticancer drugs, is typically time-consuming and expensive. The expense of advanced treatment candidates has greatly decreased with the development of cell screening techniques based on high-throughput methods, including the use of patient-derived cancer organoids, organs-on-a-chip, and other prognostic models (Jajoriya 2021).

The knowledge of human development and disease has been altered by 3D organoids, including organoid models of the breast, kidney, retina, brain, stomach, and other organs using embryonic and pluripotent stem cells (Lancaster and Knoblich 2014). Organoids replicate certain elements of organ function, and they contain important characteristics including a variety of cell types, including stem and progenitor cells, and the unique capacity to self-assemble (Ranga et al. 2014). These models are designed to support a variety of HTS technologies including micropatterning, automated robotic liquid handling, detection systems, and microfluidics, as they have proven to be effective for the discovery of cancer drugs (LaBarbera et al. 2012). A promising method for 3D imaging, especially with multicellular tumor spheroids MCTS models, is light sheet-based fluorescence microscopy, which implants fluorescence optical sectioning with respect to tomography. Additionally, new analysis techniques involve dissecting 3D cell culture-related components such morphological profiles and biomarker expression (Pampaloni et al. 2013). For instance, a tumor breast cancer cell line was uniformly arranged in 384-well plates and used in a recent HTS test model of breast cancer for finding drugs that target dormant nonproliferating cells inside the spheroid (Wenzel et al. 2014).

11.11.1 HTS in Colorectal cancer

3D scaffolding-based HTS has proven to be a groundbreaking strategy in the field of CRC studies, providing closely TME-mimicking models. Such an advancement helps to increase the predictive validity of preclinical testing for drugs and supports the identification of tailored treatments.

- *Patient-Derived Organoids (PDOs)*
  PDOs are developed from CRC tissues and maintain the original tumors' genetic and phenotypic traits. They function well as successful drug screening platforms and platforms for personal medicine. Research has revealed that PDOs were capable of predicting chemotherapy responses in patients and that increasing organoid scores had lower rates of tumor regression as well as poorer progression-free survival. In addition, PDOs have been utilized to assess off-label therapeutics, promoting the discovery of treatment alternatives for disease progress patients (Cho et al. 2022).
- *Improved Drug Response Predictability*
  Scaffold-based models are more representative of *in vivo* drug diffusion and metabolic gradients. In CRC, this means more reliable evaluations of chemotherapeutic activity, especially for drugs that affect quiescent or hypoxic tumor areas poorly represented in 2D culture. One study showed that CRC cells grown on 3D electrospun scaffolds showed different sensitivity to 5-FU, the significance of 3D environment in drug response (Breslin and O'Driscoll 2013).
- *Compatibility with Automation and HTS Platforms*
  Scaffold-based constructs are being adapted to HTS compatibility with formats like 96- or 384-well plates, coupled with robotic processing and automated imaging devices. This enables the parallel screening of extensive drug libraries on CRC cells without disrupting the structural integrity and biological accuracy of the 3D constructs. Furthermore, reduced-sized collagen- or alginate-based scaffolds have been modified for HTS with CRC patient-derived organoids for greater scalability of personalized drug screening (Tuveson and Clevers 2019).

11.12 FUTURE PERSPECTIVES

Decades of research have led to breakthroughs in 3D culture scaffolds in cancer studies. For instance, 3D scaffolds can be enhanced with drug compounds and responsive agents that support dual or triple approaches to tumor immunotherapy. This technology is still under rapid exploration, but advancements in durable 3D scaffolds hold promise for broader clinical applications. Microfabrication techniques such as photolithography, microcontact printing, and replica molding enable creating microscale structures that allow for precise control over cell and tissue shape and size to produce microenvironments for 3D cultures (Park and Shuler 2003).

Similarly, tissues-on-biochip model is a complex microfluidics model for performing 3D cell culture that offers controlled microenvironments that are easy to manipulate and that allow long-term cell differentiation. Additionally, microfluidics ensures regular and consistent flow of oxygen to the cells, thereby stabilizing their growth (Russo et al. 2022). In addition to these models, a more complex approach to 3D cell culture involves *in vitro* models that replicate tissue–tissue interfaces and mimic organ microarchitecture. One of the major benefits of these organ-level structures is that they simplify the study of chemical signals and provide dynamic (Huh et al. 2011).

There is still a long way to go to effectively address the limitations of these compounds such as their sometimes-poor biocompatibility. Overcoming these constraints requires more detailed understanding of cancer immunology, which will allow for more personalized, effective treatment. Enhancing tumor growth in ECM-like settings will be critical for expanding the outcomes of such clinical studies. And despite the fact that 3D cultures and bioprinting are ongoing, 4D bioprinting is emerging that possesses even more functionality and durability. In CRC, 3D scaffolds have been exceptionally valuable for examining tumor heterogeneity, drug resistance, and EMT, which are frequently not well recapitulated in standard 2D cultures. Stable scaffold materials such as collagen, polycaprolactone (PCL), and gelatin-methacryloyl (GelMA) provide ECM-like settings that support long-term CRC cell development and tissue arrangement (Chen et al. 2020).

11.13 CONCLUSION

Tumor models cultured in 3D are perfect for research because they accurately mimic the *in vivo* tumor microenvironment. The 3D culture scaffolds used in cancer research and drug development are also crucial for replacing time-consuming preclinical animal models; they can meet biochemical and functional needs by combining biomaterials with complimentary qualities to support local regeneration. However, synthetic polymers have shown disadvantages over natural biomaterial because they failed to provide the biochemical signals required for cell–ECM communication. Both 2D and 3D cultures are critical for research, but 3D scaffolds have the capacity to completely revolutionize the way innovative pharmacological treatments are tested, diseases are modeled, stem cells are employed, and organs are transplanted. That is, tailored 3D cell culture scaffolds with biofunctionalization can induce signal transduction cascades, sort cell types, and modulate stem cell proliferation.

REFERENCES

Antoni D, Burckel H, Josset E et al (2015) Three-dimensional cell culture: a breakthrough in vivo. Int J Mol. Sci 16(3): 5517–27.

Aurand ER, Lampe KJ, Bjugstad KB (2012) Defining and designing polymers and hydrogels for neural tissue engineering. Neurosci Res 72(3):199–213.

Baiao A, Dias S, Soares AF, Pereira CL, Oliveira C, Sarmento B (2022) Advances in the use of 3D colorectal cancer models for novel drug discovery. Expert Opin Drug Discov 17(6):569–80.

Baker AM, Bird D, Lang G, Cox TR, Erler JT (2013) Lysyl oxidase enzymatic function increases stiffness to drive colorectal cancer progression through FAK. Oncogene 32:1863–8.

Balasubramanian P, Prabhakaran MP, Sireesha M, Ramakrishna S (2013) Collagen in human tissues: structure, function, and biomedical implications from a tissue engineering perspective. In Polymer Composites–Polyolefin Fractionation–Polymeric Peptidomimetics–Collagens (pp. 173–206). Springer.

Banerjee A, Chabria Y, Kanna NR R, Gopi J, Rowlo P, Sun XF, Pathak S (2019) Role of tumor specific niche in colon cancer progression and emerging therapies by targeting tumor microenvironment. In Cell Biology and Translational Medicine, Volume 13: Stem Cells in Development and Disease (pp. 177–92). Springer.

Banerjee A, Pathak S, Subramanium VD et al (2017) Strategies for targeted drug delivery in treatment of colon cancer: current trends and future perspectives. Drug Discov Today 22(8):1224–32.

Bauleth-Ramos T, Feijão T, Gonçalves A, Shahbazi MA, Liu Z, Barrias C, Oliveira MJ, Granja P, Santos HA, Sarmento B (2020) Colorectal cancer triple co-culture spheroid model to assess the biocompatibility and anticancer properties of polymeric nanoparticles. J Control Release 323:398–411.

Bisgin A, Mujde C (2022) Bioreactor-based tissue models as an alternative approach in cancer research. In Handbook of Animal Models and Its Uses in Cancer Research (pp. 1–16). Singapore: Springer Nature Singapore.

Boehnke K, Iversen PW, Schumacher D, et al (2016) Assay establishment and validation of a high-throughput screening platform for three-dimensional patient-derived colon cancer organoid cultures. J Biomol Screen 21(9):931–41.

Boxberger HJ, Meyer TF (1994) A new method for the 3-D in vitro growth of human RT112bladder carcinoma cells using the alginate culture technique. Biol Cell 82(2–3):109–19.

Breslin S, O'Driscoll, L (2013) Three-dimensional cell culture: the missing link in drug discovery. Drug Discov Today 18(5–6):240–49.

Burdett E, Kasper FK, Mikos AG et al (2010) Engineering tumors: a tissue engineering perspective in cancer biology. Tissue Eng Part B Rev 16(3):351–9.

Caballero D, Reis RL, Kundu SC (2020) Trends in biomaterials for three-dimensional cancer modeling. In Biomaterials for 3D Tumor Modeling (pp. 3–41). Elsevier.

Campuzano S, Pelling AE (2019) Scaffolds for 3D cell culture and cellular agriculture applications derived from non-animal sources. Front Sustain Food Syst 3:38.

Carpenter RA, Kwak JG, Peyton SR et al (2018) Implantable pre-metastatic niches for the study of the microenvironmental regulation of disseminated human tumour cells. Nat Biomed Eng 2(12):915–29.

Castells-Sala C, Alemany-Ribes M, Fernández-Muiños T et al (2013) Current applications of tissue engineering in biomedicine. J Biochips Tiss Chips (S2):1.

Chee BS, Nugent M (2019) Electrospun natural polysaccharide for biomedical application. In Natural Polysaccharides in Drug Delivery and Biomedical Applications (pp. 589–615). Academic Press.

Chen B, Wu Y, Ao Z et al (2019) High-throughput acoustofluidic fabrication of tumor spheroids. Lab. Chip 19(10):1755–63.

Chen H, Cheng Y, Wang X, Wang J, Shi X, Li X, Tan W, Tan Z (2020) 3D printed in vitro tumor tissue model of colorectal cancer. Theranostics 10(26):12127.

Chen QZ, Bismarck A, Hansen U et al (2008) Characterisation of a soft elastomer poly (glycerol sebacate) designed to match the mechanical properties of myocardial tissue. Biomaterials 29(1):47–57.

Cho YW, Min DW, Kim HP, An Y, Kim S, Youk J, Chun J, Im JP, Song SH, Ju YS, Han SW (2022) Patient-derived organoids as a preclinical platform for precision medicine in colorectal cancer. Mol Oncol 16(12):2396–412.

Chwalek K, Bray LJ, Werner C (2014) Tissue-engineered 3D tumor angiogenesis models: potential technologies for anticancer drug discovery. Adv Drug Deliv. Rev 79:30–9.

Croce S, Wei Q, D'Imporzano G et al (2016) Anaerobic digestion of straw and corn stover: the effect of biological process optimization and pre-treatment on total bio-methane yield and energy performance. Biotechnol Adv 34(8):1289–304.

David L, Dulong V, Coquerel B et al (2008) Collagens, stromal cell-derived factor-1α and basic fibroblast growth factor increase cancer cell invasiveness in a hyaluronan hydrogel. Cell Prolif 41(2):348–64.

Dhandayuthapani B, Yoshida Y, Maekawa T, Kumar DS (2011) Polymeric scaffolds in tissue engineering application: a review. Int J Polym Sci 2011(1):290602.

DuFort CC, Paszek MJ, Weaver VM (2011) Balancing forces: architectural control of mechanotransduction. Nat Rev Mol Cell Biol 12(5):308–19.

Efraim Y, Schoen B, Zahran S et al (2019) 3D structure and processing methods direct the biological attributes of ECM-based cardiac scaffolds. Sci Rep 9(1):5578.

El-Sherbiny IM, Yacoub MH (2013) Hydrogel scaffolds for tissue engineering: progress and challenges. Glob Cardiol Sci Pract 2013(3):38.

Fallica B, Makin G, Zaman MH (2011) Bioengineering approaches to study multidrug resistance in tumor cells. Integr Biol 3(5):529–39.

Fenton OS, Olafson KN, Pillai PS et al (2018) Advances in biomaterials for drug delivery. Adv Mater 30(29):1705328.

Ferreira LP, Gaspar VM, Mano JF (2018) Design of spherically structured 3D in vitro tumor models-advances and prospects. Acta Biomater 75:11–34.

Florczyk SJ, Liu G, Kievit FM et al (2012) 3D porous chitosan–alginate scaffolds: a new matrix for studying prostate cancer cell–lymphocyte interactions in vitro. Adv Healthc Mater 1(5):590–9.

Gaedtke L, Thoenes L, Culmsee C et al (2007) Proteomic analysis reveals differences in protein expression in spheroid versus monolayer cultures of low-passage colon carcinoma cells. J Proteome Res 6(11):4111–8.

Ghibaudo M, Trichet L, Le Digabel J et al (2009) Substrate topography induces a crossover from 2D to 3D behavior in fibroblast migration. Biophys J 97(1):357–68.

Girigoswami K, Saini D, Girigoswami A (2021) Extracellular matrix remodeling and development of cancer. Stem Cell Rev Rep 17:739–47.

Glicklis R, Shapiro L, Agbaria R et al (2000) Hepatocyte behavior within three-dimensional porous alginate scaffolds. Biotechnol Bioeng 67(3):344–53.

Glowacki J, Mizuno S (2008) Collagen scaffolds for tissue engineering. Biopolymers 89(5):338–44.

Godugu C, Patel AR, Desai U et al (2013) AlgiMatrix™ based 3D cell culture system as an in-vitro tumor model for anticancer studies. PLoS One 8(1):e53708.

Greiner AM, Richter B, Bastmeyer M (2012) Micro-engineered 3D scaffolds for cell culture studies. Macromol. Biosci 12(10):1301–14.

Gunatillake P, Mayadunne R, Adhikari R (2006) Recent developments in biodegradable synthetic polymers. Biotechnol Annu Rev 12:301–47.

Gurski LA, Jha AK, Zhang C et al (2009) Hyaluronic acid-based hydrogels as 3D matrices for in vitro evaluation of chemotherapeutic drugs using poorly adherent prostate cancer cells. Biomater 30(30):6076–85.

Haisler WL, Timm DM, Gage JA et al (2013) Three-dimensional cell culturing by magnetic levitation. Nat Protoc 8(10): 1940–9.

Hakkinen KM, Harunaga JS, Doyle AD et al (2011) Direct comparisons of the morphology, migration, cell adhesions, and actin cytoskeleton of fibroblasts in four different three-dimensional extracellular matrices. Tissue Eng Part A 17(5–6):713–24.

Han S, Wu J (2022) Three-dimensional (3D) scaffolds as powerful weapons for tumor immunotherapy. Bioact Mater 17:300–19.

Hofmann S, Cohen-Harazi R, Maizels Y et al (2022) Patient-derived tumor spheroid cultures as a promising tool to assist personalized therapeutic decisions in breast cancer. Transl Cancer Res 11(1):134.

Huh D, Hamilton GA, Ingber DE (2011) From 3D cell culture to organs-on-chips. Trends Cell Biol 21(12):745–54.

Hung BP, Naved BA, Nyberg EL et al (2016) Three-dimensional printing of bone extracellular matrix for craniofacial regeneration. ACS Biomater Sci Eng 2(10):1806–16.

Hutmacher DW, Horch RE, Loessner D et al (2009) Translating tissue engineering technology platforms into cancer research. J Cell Mol Med 13(8a):1417–27.

Insua-Rodríguez J, Oskarsson T (2016) The extracellular matrix in breast cancer. Adv Drug Deliv Rev 97:41–55.

Jajoriya AK (2021) High throughput cell-based screening methods for cancer drug discovery. In Protocol Handbook for Cancer Biology (pp. 169–82). Academic Press.

Jensen C, Teng Y (2020) Is it time to start transitioning from 2D to 3D cell culture? Front Mol Biosci 7:33.

Ji S, Guvendiren M (2017) Recent advances in bioink design for 3D bioprinting of tissues and organs. Front Bioeng Biotechnol 5:23.

Jiang Z, Han B, Li H et al (2015) Preparation and antitumor metastasis of carboxymethyl chitosan. Carbohydr Polym 125:53–60.

Khorshidi S, Solouk A, Mazinani S et al (2016) A review on different approaches for improving cell infiltration in electrospun nanofibrous scaffolds. Pathobiol Res 18(4):1–22.

Kimlin L, Kassis J, Virador V (2013) 3D in vitro tissue models and their potential for drug screening. Expert Opin. Drug Discov 8(12):1455–66.

Knight E, Przyborski S (2015) Advances in 3D cell culture technologies enabling tissue-like structures to be created in vitro. J Anat 227(6):746–56.

Kong XB, Tang QY, Chen XY et al (2017). Polyethylene glycol as a promising synthetic material for repair of spinal cord injury. Neural Regen Res 12(6):1003.

Krishnamurthy S, Nör JE (2013) Orosphere assay: a method for propagation of head and neck cancer stem cells. Head Neck 35(7):1015–21.

Kumar G, Tison CK, Chatterjee K et al (2011) The determination of stem cell fate by 3D scaffold structures through the control of cell shape. Biomaterials 32(35):9188–96.

LaBarbera DV, Reid BG, Yoo BH (2012) The multicellular tumor spheroid model for high-throughput cancer drug discovery. Expert Opin Drug Discov 7(9):819–30.

Lancaster MA, Knoblich JA (2014) Organogenesis in a dish: modeling development and disease using organoid technologies. Science 345(6194):1247125.

Langhans SA (2018) Three-dimensional in vitro cell culture models in drug discovery and drug repositioning. Front. Pharmacol 9:6.

Lasprilla AJ, Martinez GA, Lunelli BH et al (2012) Poly-lactic acid synthesis for application in biomedical devices – a review. Biotechnol Adv 30(1):321–8.

Le VM, Lang MD, Shi WB et al (2016) A collagen-based multicellular tumor spheroid model for evaluation of the efficiency of nanoparticle drug delivery. Artif Cells Nanomed Biotechnol 44(2):540–4.

Leach DG, Young S, Hartgerink JD et al (2019) Advances in immunotherapy delivery from implantable and injectable biomaterials. Acta Biomater 88:15–31.

Lee J, Cuddihy MJ, Kotov NA (2008) Three-dimensional cell culture matrices: state of the art. Tissue Eng Part B Rev 14(1):61–86.

Li CL, Tian TA, Nan KJ et al (2008) Survival advantages of multicellular spheroids vs. monolayers of HepG2 cells in vitro. Oncol Rep 20(6):1465–71.

Li W, Hu X, Yang S et al (2018) A novel tissue-engineered 3D tumor model for anticancer drug discovery. Biofabrication 11(1):015004.

Li Y, Kilian KA (2015) Bridging the gap: from 2D cell culture to 3D microengineered extracellular matrices. Adv Healthc Mater 4(18):2780–96.

Liu F, Wang X (2020) Synthetic polymers for organ 3D printing. Polymers 12(8):1765.

Loessner D, Stok KS, Lutolf MP et al (2010) Bioengineered 3D platform to explore cell–ECM interactions and drug resistance of epithelial ovarian cancer cells. Biomaterials 31(32):8494–506.

Lu Y, Aimetti AA, Langer R et al (2016) Bioresponsive materials. Nat Rev Mater 2(1):1–7.

Magdeldin T, López-Dávila V, Pape J, Cameron GW, Emberton M, Loizidou M, Cheema U (2017) Engineering a vascularised 3D in vitro model of cancer progression. Sci Rep;7(1):44045.

Mahla RS (2016) Stem cells applications in regenerative medicine and disease therapeutics. Int J Cell Biol 2016, 6940283.

Mahmud G, Campbell CJ, Bishop KJ et al (2009) Directing cell motions on micropatterned ratchets. Nat Phys 5(8):606–12.

Malayaperumal S, Sriramulu S, Jothimani G et al (2021) A review of AEG-1 oncogene regulating microRNA expression in colon cancer progression. EMIDDT 21(1):27–34.

Marrella A, Dondero A, Aiello M et al (2019) Cell-laden hydrogel as a clinical-relevant 3D model for analyzing neuroblastoma growth, immunophenotype, and susceptibility to therapies. Front Immunol 10:1876.

Matano M, Date S, Shimokawa M, et al (2015) Modeling colorectal cancer using CRISPR-Cas9-mediated engineering of human intestinal organoids. Nat Med 21:256–62.

Matei I, Rampersaud S, Lyden D (2018) Engineered niches model the onset of metastasis. Nat Biomed Eng 2(12):885–7.

Meyers J, Craig J, Odde DJ (2006) Potential for control of signaling pathways via cell size and shape. Curr Biol 16(17): 1685–93.

Nayak B, Balachander GM, Manjunath S et al (2019) Tissue mimetic 3D scaffold for breast tumor-derived organoid

culture toward personalized chemotherapy. Colloids Surf. B Biointerfaces 180:334–43.
Nikolova MP, Chavali MS (2019) Recent advances in biomaterials for 3D scaffolds: a review. Bioact Mater 4:271–92.
Pampaloni F, Ansari N, Stelzer EH (2013) High-resolution deep imaging of live cellular spheroids with light-sheet-based fluorescence microscopy. Cell Tissue Res 352:161–77.
Pampaloni F, Stelzer EH, Leicht S et al. (2010) Madin–Darby canine kidney cells are increased in aerobic glycolysis when cultured on flat and stiff collagen-coated surfaces rather than in physiological 3-D cultures. Proteomics 10(19):3394–413.
Pape J, Magdeldin T, Ali M, Walsh C, Lythgoe M, Emberton M, Cheema U (2019) Cancer invasion regulates vascular complexity in a three-dimensional biomimetic model. Eur J Cancer 119:179–93.
Park TH, Shuler ML (2003) Integration of cell culture and microfabrication technology. Biotechnol Prog 19(2):243–53.
Patricio T, Domingos M, Gloria A et al (2013) Characterisation of PCL and PCL/PLA scaffolds for tissue engineering. Procedia CIRP 5:110–4.
Pauli C, Hopkins BD, Prandi D, Shaw R, Fedrizzi T, Sboner A, Sailer V, Augello M, Puca L, Rosati, R, et al (2017) Personalized in vitro and in vivo cancer models to guide precision medicine. Cancer Discov 7:462–77.
Pickup MW, Mouw JK, Weaver VM (2014) The extracellular matrix modulates the hallmarks of cancer. EMBO Rep 15(12):1243–53.
Pluen A, Boucher Y, Ramanujan S et al (2001) Role of tumor–host interactions in interstitial diffusion of macromolecules: cranial vs. subcutaneous tumors. PNAS 98(8):4628–33.
Polonio-Alcalá E, Rabionet M, Gallardo X et al (2019) PLA electrospun scaffolds for three-dimensional triple-negative breast cancer cell culture. Polymers 11(5):916.
Pradhan S, Hassani I, Seeto WJ et al (2017) PEG-fibrinogen hydrogels for three-dimensional breast cancer cell culture. J Biomed Mater 105(1):236–52.
Rabionet M, Yeste M, Puig T et al (2017) Electrospinning PCL scaffolds manufacture for three-dimensional breast cancer cell culture. Polymers 9(8):328.
Ranga A, Gjorevski N, Lutolf MP (2014) Drug discovery through stem cell-based organoid models. Adv Drug Deliv Rev 69:19–28.
Ridky TW, Chow JM, Wong DJ et al (2010) Invasive three-dimensional organotypic neoplasia from multiple normal human epithelia. Nat Med 16(12):1450–5.
Rijal G, Bathula C, Li W (2017) Application of synthetic polymeric scaffolds in breast cancer 3D tissue cultures and animal tumor models. Int J Biomater 2017(1):8074890.
Rijal G, Li W (2016) 3D scaffolds in breast cancer research. Biomaterials 81:135–56.
Ross AM, Jiang Z, Bastmeyer M et al (2012) Physical aspects of cell culture substrates: topography, roughness, and elasticity. Small 8(3):336–55.
Rosser J, Calvo IO, Schlager M et al (2015) Recent advances of biologically inspired 3D microfluidic hydrogel cell culture systems. J Cell Biol Cell Metable 2(5):1–4.
Russo M, Cejas CM, Pitingolo G (2022) Advances in microfluidic 3D cell culture for preclinical drug development. Prog Mol Biol Transl Sci 187(1):163–204.
Safran MR, Kim H, Zaffagnini S (2008) The use of scaffolds in the management of articular cartilage injury. J Am Acad Orthop Surg 16(6):306–11.
Sahoo DR, Biswal T (2021) Alginate and its application to tissue engineering. SN Appl Sci 3(1):30.
Saini P, Arora M, Kumar MR (2016) Poly (lactic acid) blends in biomedical applications. Adv Drug Deliv Rev 107:47–59.
Sakuma K, Tanaka A, Mataga I (2016) Collagen gel droplet-embedded culture drug sensitivity testing in squamous cell carcinoma cell lines derived from human oral cancers: optimal contact concentrations of cisplatin and fluorouracil. Oncol Lett 12(6):4643–50.
Sawicki LA, Choe LH, Wiley KL et al (2018) Isolation and identification of proteins secreted by cells cultured within synthetic hydrogel-based matrices. ACS Biomater Sci Eng 4(3):836–45.
Sekita A, Matsugaki A, Nakano T (2017) Disruption of collagen/apatite alignment impairs bone mechanical function in osteoblastic metastasis induced by prostate cancer. Bone 97:83–93.
Sitarski AM, Fairfield H, Falank C et al (2018) 3D tissue engineered in vitro models of cancer in bone. ACS Biomater Sci Eng 4(2):324–36.
Stock K, Estrada MF, Vidic S et al (2016) Capturing tumor complexity in vitro: comparative analysis of 2D and 3D tumor models for drug discovery. Sci Rep 6(1):28951.
Subramaniam VD, Prasad SV, Banerjee A et al (2019) Health hazards of nanoparticles: understanding the toxicity mechanism of nanosized ZnO in cosmetic products. Drug Chem Toxicol 42(1):84–93.
Sun J, Tan H (2013) Alginate-based biomaterials for regenerative medicine applications. Materials 6(4):1285–309.
Suzuki Y, Nishimura Y, Tanihara M et al (1998) Evaluation of a novel alginate gel dressing: cytotoxicity to fibroblasts in vitro and foreign-body reaction in pig skin in vivo. J Biomed Mater Res B Appl Biomater 39(2):317–22.
Takai A, Fako V, Dang H et al (2016) Three-dimensional organotypic culture models of human hepatocellular carcinoma. Sci Rep 6(1):1–1.
Temple JP, Hutton DL, Hung BP et al (2014) Engineering anatomically shaped vascularized bone grafts with hASCs and 3D-printed PCL scaffolds. J Biomed Mater Res A 102(12):4317–25.
Tuveson D, Clevers H (2019) Cancer modeling meets human organoid technology. Science 364(6444):952–5.
Unnikrishnan K, Thomas LV, Ram Kumar RM (2021) Advancement of scaffold-based 3D cellular models in cancer tissue engineering: an update. Front Oncol 4468.
Valkenburg KC, De Groot AE, Pienta KJ (2018) Targeting the tumour stroma to improve cancer therapy. Nat Rev Clin Oncol 15(6):366–81.
Weaver VM, Lelièvre S, Lakins JN et al (2002) β4 integrin-dependent formation of polarized three-dimensional architecture confers resistance to apoptosis in normal and malignant mammary epithelium. Cancer Cell 2(3):205–16.
Weeber F, van de Wetering M, Hoogstraat M, Dijkstra KK, Krijgsman O, Kuilman T, Gadellaa-van Hooijdonk, CG, van der Velden DL, Peeper DS, Cuppen EP, et al (2015) Preserved genetic diversity in organoids cultured from biopsies of human colorectal cancer metastases. Proc Natl Acad Sci USA 112:13308–11.
Wenzel C, Riefke B, Gründemann S et al (2014) 3D high-content screening for the identification of compounds that target cells in dormant tumor spheroid regions. Exp Cell Res 323(1):131–43.
Zhang YS, Khademhosseini A (2017) Advances in engineering hydrogels. Science 356(6337):eaaf3627.
Zhao S, Zuo WJ, Shao ZM et al (2020) Molecular subtypes and precision treatment of triple-negative breast cancer. Ann Transl Med 8(7).
Zhao Y, Yao R, Ouyang L et al (2014) Three-dimensional printing of Hela cells for cervical tumor model in vitro. Biofabrication 6(3):035001.
Zhu C, Fan D, Wang Y (2014) Human-like collagen/hyaluronic acid 3D scaffolds for vascular tissue engineering. Mater Sci Eng C 34:393–401.

Section III

Distinct Aspects and Management of Diseases of Different Organs

12 Biomedical Innovations in Respiratory Ailments

*Manickam Subramanian**

Professor, Department of Anatomy, PSP Medical College Hospital & Research Institute, Oragadam, Kanchipuram, Tamil Nadu, India

LIST OF ABBREVIATIONS

CFTR	Cystic fibrosis transmembrane conductance regulator
COPD	Chronic obstructive pulmonary disease
CRISPR	Clustered regularly interspaced short palindromic repeats
CT	Computerized tomography
IL	Interleukins
mRNA	Messenger RNA
MSCs	Mesenchymal stem cells
SARS-CoV-2	Severe acute respiratory syndrome related coronavirus 2

12.1 INTRODUCTION

Breathing is the fundamental function of all living systems. In the vast realm of medical science, the field of respiratory medicine stands as a cornerstone of healthcare, addressing the intricate mechanisms that sustain life itself. The significance of respiratory health cannot be overstated, as the intricate network of airways and lungs serves as the primary interface between the human body and its environment. From the rhythmic exchange of oxygen and carbon dioxide to the intricate orchestration of immune defenses, the respiratory system plays a pivotal role in maintaining overall wellbeing.

Respiratory disorders affect 544.9 million individuals globally and rank as the third leading cause of death, following cardiovascular diseases and cancer, accounting for 20% of total deaths. Ailments like lung cancer, chronic obstructive pulmonary disease (COPD), cystic fibrosis, and asthma have considerable financial and emotional impact on both patients and their caregivers (1). Among respiratory diseases, COPD has the highest prevalence followed by asthma. Their prevalence rose at the alarming rate of 39.8% between 1990 and 2017 in tandem with environmental and lifestyle changes. Smoking and household pollution from solid fuels are the predominant risk factors for chronic respiratory diseases (2).

Despite significant advancements in respiratory medicine, many currently available treatments for respiratory ailments suffer from inadequacies. Traditional therapies such as bronchodilators and corticosteroids for conditions like asthma and COPD often fail to address underlying inflammatory processes adequately, leading to disease exacerbations and symptom progression over time (3). Similarly, antibiotics used to treat respiratory infections face antimicrobial resistance and limited effectiveness against evolving pathogens (4). In cystic fibrosis, while treatments like airway clearance and pancreatic enzyme replacement alleviate symptoms, they do not address the underlying genetic defect and fail to provide a cure (5). Additionally, therapies for respiratory cancers, such as surgery, chemotherapy, and radiation, frequently result in notable adverse effects and offer limited effectiveness during advance stages of disease. In summary, the present state of respiratory treatments emphasizes the urgent need for innovative therapeutic strategies to overcome the deficiencies of current interventions and enhance patient outcomes.

Amid these challenges, the field of respiratory medicine has witnessed a remarkable surge of innovations aimed at revolutionizing diagnosis, treatment, and management strategies. From cutting-edge biologicals targeting specific molecular pathways to the transformative potential of gene therapy and stem cell interventions, the arsenal of therapeutic modalities continues to expand. In parallel, advancements in nanotechnology have unlocked novel avenues for drug delivery and targeted therapies, while telemedicine platforms bridge geographical barriers to enhance patient care and monitoring. Artificial intelligence emerges as a game-changer, offering unparalleled insights into disease progression, treatment optimization, and personalized patient management through data driven algorithms. Furthermore, the dawn of precision medicine heralds a new era of tailored interventions in which individualized genetic, molecular, and clinical profiles guide therapeutic decisions. Vaccines, long hailed as triumphs of preventive medicine, continue to evolve, offering hope in the fight against infectious respiratory diseases.

In this chapter, we delve into the multifaceted landscape of innovations in respiratory medicine, exploring the transformative potential of biologicals, gene therapy, stem cell therapy, nanotechnology, telemedicine, AI, precision medicine, and vaccines (Figure 12.1). Through a comprehensive examination of these advancements, we illuminate the path toward a future where respiratory health is optimized and the burdens of respiratory ailments are alleviated for individuals across the globe.

* Corresponding author: gandhiayu@hotmail.com

DOI: 10.1201/9781032660592-15

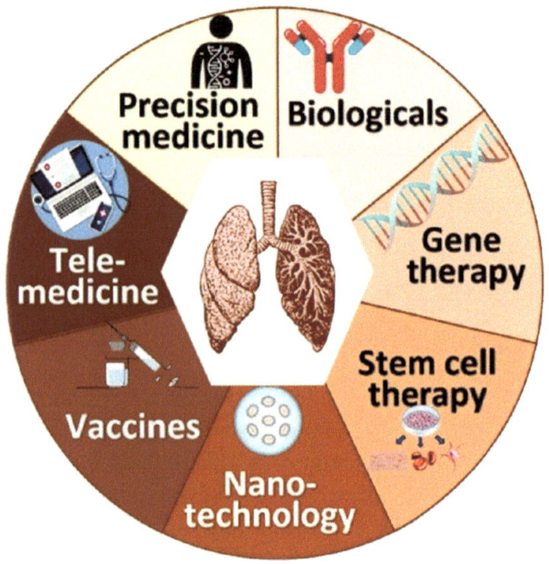

FIGURE 12.1 Biomedical innovations in respiratory care.

12.2 BIOLOGICALS AND TARGETED THERAPY

Respiratory diseases like asthma and COPD have long posed significant challenges in the realm of healthcare. Conventional treatments, though effective to some extent, often come with limitations and adverse effects (6). In recent years, a promising avenue for innovation has emerged in the form of biologicals and targeted therapies, offering a more precise and personalized approach to managing respiratory ailments.

Biologicals, a class of drugs derived from living organisms, and targeted therapies, which aim to interfere with specific molecules involved in the disease process, have revolutionized the treatment of asthma, idiopathic pulmonary fibrosis, and COPD. These therapeutic approaches address the underlying causes of respiratory conditions, targeting pathways and mechanisms that traditional treatments might overlook.

One notable area of focus for biologicals and targeted therapies is the treatment of severe asthma. Asthma phenotypes are based on observable traits resulting from hereditary and environmental influence, whereas endotypes are based on the cellular and molecular pathophysiological mechanisms that make individuals respond differently to the same treatment. Based on the inflammatory mechanism involved, asthmatics are classified into endotypes as type 2 high or low inflammation, with type 2 high being the more common; it is characterized by Th2 cytokine response (IL-4, IL-5, and IL-13). Traditional asthma management relies heavily on inhaled corticosteroids and bronchodilators. However, a small percentage of patients with severe asthma suffer from persistent symptoms despite maximal treatment. This is where monoclonal antibodies, a type of biological, have shown remarkable success. Monoclonal antibodies target specific proteins, such as interleukins or immunoglobulins, involved in the inflammatory cascade of asthma. By doing so, they effectively modulate the immune

TABLE 12.1
Monoclonal Antibodies for Treating Respiratory Diseases

Respiratory Disease	Target	Drug Name	Research Outcome
Asthma (7)	IgE	Omalizumab	Improved lung function and reduced use of oral corticosteroids
	IL-4	Pascolizumab	Animal studies show significant outcome. Not effective clinically.
	IL-5	Mepolizumab Reslizumab	Ameliorates eosinophil-driven type 2 endotype
	IL-13	Lebrikizumab Tralokinumab	Reduce IgE levels and eosinophil counts with no effect on clinical symptoms
	IL-23	Lectin-8 Dexpramipexole Risankizumab	
	IL-33	Texepelumab Itepekimab	Reduce serum IgE and peripheral eosinophils suggesting their role in type 2 high and low endotypes
Idiopathic pulmonary fibrosis (8)	IL-11	9MW3811	Reduces TGF-β1 and tissue fibrosis
	PDGF		Controls abnormal fibroblast proliferation and collagen production
	Wnt/–β catenin		Prevents epithelial-mesenchymal transition and fibrosis
	MMPs		Prevents epithelial-mesenchymal transition and fibrosis
Cystic fibrosis (9)	CFTR	Elexacaftor Tezacaftor Ivacaftor Lumacaftor	Improves lung function and quality of life, reduces antibiotic use

Abbreviations: IL: Interleukin; PDGF: Platelet-derived growth factor; MMPs: Matrix metalloproteinases; CFTR: Cystic fibrosis transmembrane conductance regulator.

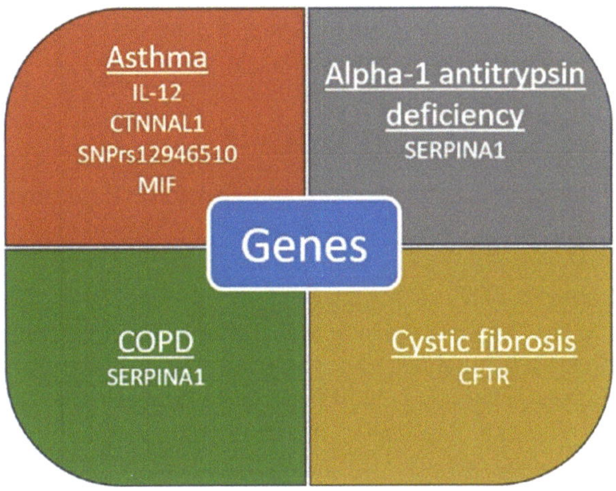

FIGURE 12.2 Gene therapy for pulmonary diseases.

response and reduce the frequency and severity of asthma exacerbations. Table 12.1 displays the currently available or researched monoclonal antibodies for various respiratory diseases.

In the realm of COPD, which is characterized by progressive airflow limitation, biologicals and targeted therapies are also showing promise. One notable example is bronchodilators that selectively target specific receptors, such as long-acting muscarinic antagonists and long-acting beta-agonists. These drugs provide more sustained relief with fewer side effects than with traditional bronchodilators.

Despite the promise of biologics and targeted therapies, challenges remain. Only a few aforesaid biologics are approved for human use, and others are in in various stages of clinical trials and may not be clinically effective due to their inability to inhibit the other inflammatory cytokines, or autoimmune reactions. Further, access to these advanced treatments, their cost, and the need for further research into long-term safety are significant considerations. Additionally, the identification of biomarkers for certain respiratory conditions is an evolving field that requires ongoing research.

12.3 GENE THERAPY

Gene therapy emerges as a promising avenue in the treatment landscape of respiratory illnesses, encompassing conditions like pulmonary hypertension, alpha-1 antitrypsin deficiency, asthma, and cystic fibrosis. This innovative therapeutic strategy involves the administration of genetic material aimed at either substituting or complementing defective genes, inhibiting the expression of harmful genes, or introducing novel genetic elements into cellular structures. In contrast with conventional pharmacological interventions, gene therapy holds the potential for sustained efficacy.

Diverse gene therapy methodologies have been devised to address respiratory diseases, including viral-vector-based delivery systems and CRISPR-associated protein 9 (CRISPR/Cas9). Viral vectors, notably adeno-associated viruses (AAVs) and lentiviruses, serve as common vehicles for transporting therapeutic genes to targeted cells (10, 11). Clinical trials have exhibited promising outcomes with AAVs in treating cystic fibrosis and other hereditary lung disorders, and CRISPR/Cas9 has enabled precise gene editing within living cells, showcasing its potential in rectifying mutations observed in animal models afflicted with cystic fibrosis and alpha-1 antitrypsin deficiency (12) (Figure 12.2). However, challenges persist, including immune reactions to viral vectors and the prospect of off-target effects during genome editing (8).

Significant strides have been made in recent years in the realm of gene therapy for respiratory ailments. Noteworthy advancements include clinical trials employing AAV gene therapy to target CFTR for cystic fibrosis, resulting in notable enhancements in lung functionality and overall quality of life for patients. Furthermore, encouraging preclinical findings have surfaced through CRISPR/Cas9 gene editing in cystic fibrosis and other respiratory disorders. Nevertheless, it remains imperative to acknowledge existing challenges within this domain, such as the need for refined delivery methodologies and comprehensive exploration of the associated risks linked with genome editing (13, 14).

12.4 STEM CELL THERAPY AND TISSUE ENGINEERING

Respiratory diseases ranging from chronic conditions like COPD to interstitial lung diseases pose significant challenges to conventional treatments. While traditional therapies aim to manage symptoms, stem cell therapy offers the potential to repair damaged lung tissue and transform the treatment landscape for respiratory ailments.

Stem cells possess an extraordinary capability to transform into diverse cell types and facilitate tissue regeneration, making them indispensable in biomedical research. Within the respiratory system, adult stem cells are strategically situated along the basal layer and submucosal glands of the upper airway, as well as in the alveolar epithelium of the lower airway. Notably, alveolar type II cells and club cells in the respiratory tract harbor a unique capacity for self-renewal and differentiation into alveolar type I cells and ciliated cells, respectively, thus presenting promising avenues for restoring damaged respiratory epithelium (15). However, the aging process inevitably diminishes the renewing potential of these adult stem cells, necessitating exploration into alternative sources such as embryonic stem cells, mesenchymal stem cells (MSCs), hemopoietic stem cells, and induced pluripotent stem cells (16).

Among these options, MSCs have emerged as a favored choice in pulmonary medicine due to their multipotent nature and reduced risk of immune rejection. Derived from various sources namely umbilical cord, amniotic fluid, placenta, bone marrow, and adipose tissue, these versatile cells can be introduced into the patient's body through various routes such as intravenous or intraperitoneal injection or intratracheal instillation.

The burgeoning interest in stem cell therapy is evidenced by the 271 clinical trials registered at clinicaltrials.gov, focusing on elucidating the role of stem cells in an array of pulmonary ailments including asthma, pulmonary hypertension, COPD, non-small cell lung cancer, acute lung injury, cystic fibrosis, and pulmonary fibrosis (15, 17). These investigations span different phases of clinical trials and bestow promising results by virtue of their paracrine function and anti-inflammatory actions.

One of the potential applications of stem cell therapy is in the treatment of COPD, a progressive condition characterized by airflow limitation and chronic inflammation. MSCs modulate immune response, reduce inflammation, and stimulate tissue repair, improving lung function and quality of life for COPD patients. Moreover, stem cell therapy holds potential for treating idiopathic pulmonary fibrosis and scarring of lung tissue that led to impaired function. Stem cells mitigate fibrosis by restoring normal lung architecture and function (17).

MSCs have emerged as pivotal therapeutic agents in the management of influenza and acute respiratory distress syndrome, demonstrating remarkable efficacy in dampening immune cell-mediated inflammatory cascades and mitigating further lung injury. With the outbreak of COVID-19, the spotlight intensified on stem cell-based interventions, evidenced by the registration of approximately 94 clinical trials on clinicaltrials.gov targeting COVID-19 treatment. Despite this surge, only a handful of trials have advanced beyond phase II evaluations. Stem cell therapies have exhibited promising outcomes by attenuating the inflammatory milieu implicated in lung fibrosis and contributing to the recovery from pulmonary damage incurred by COVID-19 (18). Notably, research indicates that MSC transplantation exerts regulatory effects on the expression of angiotensin-converting enzyme 2 and transmembrane serine protease 2 genes, pivotal for the viral entry into host cells, thus offering a potential avenue for combating COVID-19 pathogenesis (19, 20).

Organoids are multicellular clusters produced by stem cell culture related to a specific organ. Recently, various organoid models of respiratory system are being developed for disease modeling, regeneration and pharmacological studies and has enormous potential in the research of lung cancer treatment (21).

Bioengineered trachea transplantation was performed successfully in lung cancer patients; the lung-on-a-chip model recapitulated the structure and functional elements of human lung, mimicking the mucociliary barrier, alveola–capillary interface, and inflammatory response that are critical in disease modeling and drug testing (15). However, despite the promise of stem cell therapy, challenges persist as the field is still in its early stage. Optimizing the delivery of stem cells to the lungs, ensuring their survival and integration into existing lung tissue, and addressing potential safety concerns are ongoing areas of research. Additionally, ethical considerations surrounding the use of embryonic stem cells necessitate careful navigation (15, 22).

12.5 NANOTECHNOLOGY

The mucus lining of the respiratory passage is negatively charged, and positively charged particles effectively interact with the epithelium and increase the retention time and cellular uptake of particles (23). This property enables researchers to develop drug delivery systems with charged particles of interest using nanotechnology. Aerosol drugs developed using nanotechnology that can be delivered directly to lungs have opened a new era in the treatment of lung diseases such as cystic fibrosis, asthma, lung cancer, and COPD and emerging infectious diseases like COVID-19.

12.5.1 Disease Diagnosis

Nanobased biosensors are used in the diagnosis of lung cancer, asthma, and bacterial lung infections. Recently, researchers fabricated a high-sensitivity nano-biosensor using graphene-based nanosheets coated with SARS-CoV-2 spike proteins. Researchers developed gold nanoparticles to shield DNA for the colorimetric diagnosis of Middle East respiratory syndrome coronavirus (MERS-CoV) and created a nanodisk mass spectrometry assay and a gold–copper nanoshell immunoassay, both using early secreted antigenic target 6 kDa and 10 kDa culture filtrate protein (CFP-10) antibodies, for the rapid and highly sensitive diagnosis of tuberculosis (24). Additionally, they developed a nanosensor-based electronic nose to diagnose COPD, lung cancer, and asthma by analyzing the volatile organic compounds present in the patient's exhaled air. The electronic nose also identified the four asthma phenotypes (24).

12.5.2 Targeted Drug Delivery

Nanoparticles encapsulate and conjugate with a variety of compounds such as peptides, DNAs, and RNAs, facilitating their targeted delivery to organs with reduced systemic toxicity. Targeted drug delivery utilizing nanotechnology focuses on two primary methods: targeted and stimuli-responsive nanocarriers. Targeted nanocarriers are equipped with surface-coated ligands that bind specifically to targeted cell types, enhancing precision in drug delivery. In contrast, stimuli-responsive nanocarriers release the encapsulated drug in response to specific stimuli such as pH, enzymes, and ultrasound. Recent advancements have enabled targeting intracellular organelles, exemplified by recent *in vivo* research on a breast cancer lung metastasis mouse model utilizing paclitaxel encapsulated in a nanocarrier with a mitochondria-targeting triphenylphosphine head group (24). Clinically available nanomedicines for the treatment of specific lung diseases are outlined in Table 12.2.

TABLE 12.2
Clinically Available Nanomedicines for Lung Diseases (25)

Disease	Pharmacological Agent	Nanoparticle Type
Lung cancer	Doxorubicin	PEGylated liposome
	Paclitaxel	Micellar
	Paclitaxel	Polymeric
Cystic fibrosis	Amikacin	Liposome
Pneumonia	Amikacin	Liposome

12.5.2.1 Cancer

Doxorubicin-PEGylated liposome, irinotecan nanoparticles (etirinotecan pegol, Onivyde), and paclitaxel nanoparticles (Genexol-PM, Abraxane) are among the nano-formulations currently available in the market or undergoing various stages of clinical trials for the treatment of small cell and non-small cell lung cancers. A novel immunotherapy nano-platform involves gold particle-coated polymers carrying small interfering RNA, which effectively downregulates the expression of human programmed death ligand 1 in lung cancer (25, 26).

12.5.2.2 Cystic Fibrosis

Mutations in the CFTR gene causes cystic fibrosis. Clinical studies show that PEG nanoparticles encapsulating plasmid DNA and cationic liposomal nonviral vectors carrying the CFTR gene significantly improved forced expiratory volume in 1 second (FEV1). Although efforts to correct the genetic defect in cystic fibrosis have been ongoing, clinically proven applications remain limited. Recently, lipid-based nanoparticles containing modified cystic fibrosis mRNA (messenger RNA) were clinically tested to reduce chloride secretion and increase FEV1 (25).

12.5.2.3 COPD

PEG or chitosan encapsulated amikacin or cisplatin, mediated by black phosphorus quantum dots, have shown the ability to penetrate the lung mucosal barrier, adhere to epithelial cells, and significantly alleviate symptoms of COPD in animal models. In COPD, the reduction of histone deacetylase 2 (HDAC2) by reactive oxygen species leads to glucocorticoid resistance. Research is currently underway to address this defect by applying lipid-polymer nanoparticles carrying HDAC2-encoding plasmid DNA (25).

12.5.2.4 Fungal Infections

In vivo studies involving aerosol administration of amphotericin B-polymethacrylic acid nanoparticles against *Aspergillus*-induced fungal lung infections demonstrated the ability to prevent fungal growth and reduce lung inflammation. Similarly, oral administration of itraconazole–chitosan nano-preparation effectively eliminated *Cryptococcus neoformans* and provided protection to infected mice (25).

12.5.2.5 Viral Infections

Preclinical research has focused on developing a pulmonary surfactant-GAMP adjuvanted H1N1 vaccine and a nanoparticle vaccine delivering the interferon gene (encoding for S protein) to protect against influenza virus, MERS-CoV, and SARS-CoV-2 (25).

12.5.2.6 Asthma

Nasal administration of beclomethasone dipropionate and salbutamol as nanoparticles improved drug penetration, enhanced bioavailability, and offered better therapeutic effects. Phytochemicals such as bavachinin and andrographolide encapsulated with nanoparticles were tested for their anti-asthmatic effects. Additionally, gene therapy in combination with nanotechnology has been explored for asthma treatment. Murine models demonstrated a significant reduction in airway hyperresponsiveness to allergic asthma induced by ovalbumin through the use of chitosan-interferon-γ nanoparticles (25).

Nanotechnology offers an opportunity to load drug combinations and improve the stability and bioavailability of packaged drugs. A dual-loaded nanodrug delivery system containing paclitaxel and cisplatin demonstrated enhanced efficacy against lung cancer. Peptide nanoparticle-loaded etoposide exhibited controlled drug release, improved pharmacokinetics, and showed superior anticancer activity. In an *in vivo* study, weekly administration of nano-encapsulated rifampicin, pyrazinamide, and isoniazid combination proved to be equally effective as daily oral administration. Nebulized ethionamide, a second-line anti-tubercular drug encapsulated in polymeric nanoparticles, significantly reduced bacterial load in experimental tuberculosis mice without causing systemic side effects. Experimental inhalation of lumacaftor and ivacaftor encapsulated in lipid-based nanoparticles effectively mitigated tissue damage in cystic fibrosis mice (24).

12.5.3 Vaccine Development

NUX-CoV2373, a protein subunit vaccine developed using SARS-CoV-2 spike glycoprotein nanoparticles in a vector with an adjuvant, showed significant immune response and is utilized in COVID-19 prevention (27). Chitosan, polymers derived from chitosan, and cationic lipids serve as mucoadhesive, enhancing vaccine permeation and retention in the respiratory mucosa. Polymeric and lipid-based nanoparticles are commonly utilized in pulmonary vaccines, and they serve as adjuvants that help to overcome the respiratory barrier and improves the vaccine immunogenicity and efficacy (28).

12.5.4 Theranostics

In the realm of theranostics, which involves the application of nanocarriers in the diagnosis and therapeutics of targeted disease organs, remarkable advancements have been made.

A ruthenium-loaded nanodrug delivery system enhanced with green fluorescent protein, specifically fuses with non-small cell lung carcinoma cells. Real-time monitoring displays the accumulation of nanoparticles in tumor cells and a two-fold inhibition of tumor growth in an *in vivo* model. Similarly, the theranostic application of doxorubicin and vinorelbine in a liposomal drug delivery system showed significant penetration and growth inhibition in a mouse lung cancer model (24).

Although nanotechnology is widely used in disease imaging and drug delivery, its effectiveness in clinical practice is limited due to various constraints such as instability, poor reproducibility, challenges in large-scale production, and high production costs. Ongoing research efforts are expected to overcome these limitations, thereby improving the management of lung diseases in the future.

12.6 VACCINES

12.6.1 COVID-19

One of the most remarkable innovations in COVID-19 vaccine development has been the utilization of mRNA. Although scientists began exploring mRNA vaccines as early as 1990, their progress was hindered by challenges such as instability, excessive immunogenic response, and ineffective delivery systems. However, the emergence of the COVID-19 pandemic, caused by the novel coronavirus SARS-CoV-2, catalyzed an unprecedented global effort to develop effective vaccines, with 183 vaccines in clinical development and 199 in preclinical development (29). This urgency spurred groundbreaking innovations in vaccine preparation, marking a historic chapter in immunology.

Messenger RNA vaccines emerged as promising candidates due to their ability to degrade without metabolic toxicity and their non-involvement in nuclear transformation of DNA structure. Moreover, their development required only the virus gene sequence, allowing for easy modification even in the face of viral mutations. This streamlined the vaccine development process during the pandemic, simplifying vaccine production and rendering it cost-effective. Vaccines mRNA-1273, BNT162b1, and BNT162b2 are lipid nanoparticle-encapsulated mRNA vaccines encoding the full-length spike protein receptor binding domain of SARS-CoV-2. In experimental studies, these vaccines induced robust neutralizing IgG antibodies against SARS-CoV-2 without causing lung tissue damage. Additionally, they do not elicit antibody-dependent enhancement in the vaccinated animals' serum (30).

In addition to mRNA vaccines, other vaccine types have been developed, including non-replicating viral vector vaccines, replicating viral vector vaccines, inactivated virus vaccines, protein subunit vaccines, virus-like particles, live attenuated virus vaccines, and bacterial antigen-spore expression vector vaccines (Table 12.3). Many of these vaccines, following preclinical evaluation, have undergone clinical trials and have been approved for human usage. Among the 11 vaccines approved by the WHO, four are adenovirus-based non-replicating viral vector vaccines, offering advantages such as stability, purity, economic viability, large-scale production, and multiple routes of administration.

Unprecedented cooperation between global organizations, governments, pharmaceutical firms, and research institutions in exchanging data, resources, and knowledge has expedited the development of vaccines within a short duration. Initiatives like COVAX led by the Vaccine Alliance, the Coalition for Epidemic Preparedness Innovations (CEPI), the WHO, and UNICEF have ensured fair and equitable access to vaccines worldwide. In addition to research and collaboration, the importance of public trust and vaccine acceptance was recognized early, with priority given to community engagement and transparent communication. These collective efforts have led to the remarkable coverage of 13 billion people under COVID-19 vaccination programs, saving the lives of 14.4 million worldwide (29).

The rapid development and approval of COVID-19 vaccines were not without challenges and ethical considerations. Issues such as vaccine distribution disparities, vaccine hesitancy, and the ethical implications of accelerated clinical trials have raised concerns. Striking a balance between speed and thorough evaluation has become a central challenge, emphasizing the importance of ethical frameworks in vaccine development.

12.6.2 Other Respiratory Diseases

Studies are being conducted on polymer-based delivery of mRNA vaccines to access the nasal-associated lymphoid tissue, crossing the nasal epithelial barrier. Similarly, chitosan-involved nanoparticle delivery systems are being explored for influenza virus mRNA vaccines to deliver naked RNA across the cellular membrane. Lundstrom has provided an account of mRNA vaccines being studied

TABLE 12.3
Candidate Vaccines Investigated against COVID-19 (27, 29)

Vaccine Type	No. of Candidate Vaccines (Percentage)	Representative Vaccines
mRNA	43 (24%)	mRNA-1273 BNT162b1 BNT162b2
Nonreplicating viral vector	25 (14%)	Ad5-nCoV Covishield (AZD1222) Sputnik JNJ-78436735
Inactivated virus	22 (12%)	CoronaVac Sinopharm Covaxin BBIBP-CorV
Protein subunit	59 (32%)	NUX-CoV2373

against influenza, human parainfluenza infection, human immunodeficiency virus, MERS-CoV, respiratory syncytial virus, and human metapneumovirus (30, 31).

12.6.3 Intranasal Vaccines

All vaccines developed against SARS-CoV-2 are administered intramuscularly and do not offer sterilizing immunity; this means they cannot completely block the virus from infecting and multiplying, which limits their ability to stop its spread. Respiratory vaccines eliminate the need for trained personnel, maintain the cold chain, and improve patient compliance. Therefore, efforts are underway to develop respiratory vaccines for COVID-19, with two authorized for human use: ChAdOx1 nCOV-19 and ChAd-SARS-CoV-2-S, which are adenovirus-based nonreplicating viral vector vaccines, and COVI-VAC and MV-014–212, live-attenuated vaccines being developed for intranasal application. There are two approved vaccines for influenza, namely Fluenz and Nasovac, that are administered intranasally. Extensive research is ongoing to develop intranasal vaccines for pneumonia, tuberculosis, and measles.

12.7 TELEMEDICINE AND REMOTE MONITORING

Respiratory diseases, such as asthma, cystic fibrosis, COPD, and interstitial lung diseases, have long posed challenges in effective management and monitoring. However, in the age of digital health, innovative solutions are emerging to revolutionize respiratory care. From wearable devices to mobile applications, these digital health solutions offer real-time monitoring, personalized insights, and improved patient engagement, marking a transformative era in the management of respiratory conditions. Telemedicine, envisioned for offering healthcare where distance is a critical factor, has been in use for many years. However, the COVID-19 lockdown led to its widespread implementation across all medical specialties, irrespective of distance.

12.7.1 Continuous Patient Monitoring

Wearable devices such as accelerometers, fitness trackers, and smartphone applications have been designed to continuously monitor and generate data. These devices track metrics such as respiratory rate, oxygen saturation, and physical activity, offering a comprehensive view of a patient's respiratory health over time. Smart inhalers, for instance, have garnered attention for their ability to track medication usage, inhalation techniques, and environmental factors influencing respiratory health. The data generated by these wearables can be transmitted to healthcare providers in real time through mobile applications, enabling timely interventions and personalized adjustments to treatment plans for better outcomes. Tailor-made interventions based on a patient's specific needs optimize the efficacy of the treatment and contribute to the concept of "precision medicine" (32)

12.7.2 Patient Compliance

Mobile applications play a crucial role in enhancing patient engagement and self-management. Asthma and COPD management apps, for example, provide features like symptom tracking, medication reminders, and peak flow monitoring. Patients can input daily symptoms, receive medication adherence reminders, and share data with their healthcare providers, fostering a collaborative approach to respiratory care. Additionally, these apps often incorporate educational resources to empower patients with information about their conditions and self-care strategies. Digital electronic adherence recordings (medication possession ratio and prescription refill data), data tracking nebulizers, and web-based adherence monitoring systems make data visible to both patients and healthcare providers, motivating improved adherence among patients (33).

Project Fizzyo is a trial project that aims to promote physiotherapy exercise among patients via a gaming approach. This project involves an airway clearance technique device coupled with breath pressure sensors and activity trackers, making daily physiotherapy an enjoyable game. A cohort study with the MyCyFAPP application among cystic fibrosis children in Europe, which keeps a diary of food consumed and provides advice about the dosage of pancreatic enzymes to be consumed, showed improvement in quality of life and abdominal symptoms at the end of six months. Furthermore, psychological support offered through telemedicine was shown to reduce depression by 38% as experts spent adequate time addressing the patients' concerns (33).

12.7.3 Access to Health Resources

Telemedicine, facilitated by digital health solutions, has become increasingly integral in respiratory care. Every medical consultation or treatment involves waiting. The Cystic Fibrosis Trust report mentions that patients spent around 137 minutes during treatment days in the United Kingdom. Virtual consultations allow patients to connect with healthcare providers remotely, eliminating barriers to access and promoting regular check-ins, thus saving valuable time. Respiratory therapists can assess inhaler techniques, review symptoms, and adjust treatment plans, all from the comfort of the patient's home. Telemedicine not only enhances accessibility but also reduces the burden on healthcare systems, particularly in managing chronic respiratory conditions. In 2020, CLEW, an artificial intelligence-based tele-ICU, was implemented in Israel, where intensivists remotely monitored and managed critically ill COVID-19 patients. This was a good example of better management of healthcare human resources, where doctors can offer their clinical expertise to a large group regardless of their location (34).

Despite the numerous advantages, challenges exist in the widespread adoption of digital health solutions for respiratory care. Issues such as lack of physical examination, delays in early detection of diseases, data security, interoperability, and the need for standardized protocols require careful consideration. Additionally, addressing the digital divide to ensure equitable access to these technologies is essential to prevent disparities in healthcare delivery.

12.8 ARTIFICIAL INTELLIGENCE

The development of artificial intelligence has brought transformative breakthroughs in the field of medicine. Techniques such as natural language processing, deep learning, decision trees, and machine learning have brought revolutionary changes in the diagnosis, prognosis, and management of specific diseases. While these newer technologies cannot replace clinical knowledge, they can be a valuable support in clinical decision-making.

12.8.1 Synthesizing Meaningful Data from Large Datasets

A recent study analyzing the electronic health records of 100,000 bronchial asthma patients undergoing systemic corticosteroid treatment through natural language processing found that these patients were older with a prevalence of diabetes, hypertension, dyslipidemia, and depression. Furthermore, the study revealed that systemic corticosteroid usage is associated with higher adverse events such as osteoporosis and Cushing's syndrome, raising serious concerns about frequent systemic corticosteroid usage (35).

Big data analysis of 50,000 COPD patients over 40 years showed that current strategies and guidelines have a marginal impact on disease management (36). Such huge datasets are monitored in real time, yielding unbiased information and reducing the dependency on registries and audits, which require significant time and human resources.

Large datasets can be exploited faster than conventional clinical studies, as evidenced in the recent COVID-19 pandemic. Using digitized datasets, researchers were able to determine that COVID-19 patients with comorbidities like COPD and asthma faced a higher risk of mortality. However, asthma patients treated with biologicals and inhaled corticosteroids had a reduced risk of hospitalization (37, 38). Machine learning was used to identify factors that predicted the intensive care unit admissions of COVID-19 patients.

12.8.2 Improving Diagnosis and Prognosis

Radiologically, the lung is a tissue with high contrast, well suited for automated image analysis. Convolutional neural networks, a deep learning architecture that can recognize a learned pattern anywhere in an image irrespective of its position and orientation, are used in image classification. For instance, nnUNet, an image segmentation framework, enables the recognition and localization of objects in images. Radiology has adopted many such machine learning algorithms in the diagnosis, monitoring, and prognostication of various diseases (39).

Radiomics is an emerging field that involves segmenting medical images and extracting and identifying features that contribute additional insight into a patient's condition. The concept of computer-aided diagnosis is being extensively used in radiologically diagnosing diseases, aiming to match human diagnostic abilities. Computer-aided diagnosis involving AI was used by Walsh et al. (40) and Christe et al. (41) for fibrotic lung disease classification from high-resolution computerized tomography (CT); the algorithm had accuracy of 73.3% in diagnosing the condition and produced results similar to those from a radiologist. Researchers are applying AI to analyze disease prognosis and progression. For instance, Siemens has integrated AI into their CT image software, which can identify and quantify lung lesions automatically to diagnose pneumonia and COVID-19 (39).

Computer-aided diagnosis enhances the predictive value of chest CT in detecting lung cancer in 5–8 mm nodules. AI-RAD, a convolutional neural network prototype, can automatically and accurately detect pulmonary nodules from low-dose CT. SVM-LASSO, a radiomics model, was able to detect malignant nodules with accuracy of 84.6%. Another deep-learning model, SD-Unet, performed lung nodule segmentation better, enabling the determination of nodule composition and their potential to turn malignant. The use of predictive models incorporating lung cancer biomarkers, scan images, and AI can enhance the accuracy and specificity of lung cancer diagnosis. Including lung cancer biomarker proteins like gastrin, CA125, β2-microglobulin with artificial neural networks increased the prediction rate of lung cancer by 85%. Inclusion of clinical parameters, circulating abnormal cells, and Pulmonary Nodule Artificial Intelligence Diagnostic System increased the prediction rate by 87.3%, 94.1%, and 96.1%, respectively. Similarly, convolutional neural networks were able to classify squamous cell carcinoma, small cell lung cancer, and adenocarcinoma from histopathological images with accuracies of 60%, 70%, and 89%, respectively. AI used in image analysis helps in precise staging of tumors, prevents misdiagnosis, and averts unnecessary surgery in benign nodules (42).

Combining imaging data, clinical information, results of lung function tests, and laboratory investigations, AI tools can assess the progression of clinical disease. Regression tasks and proportional hazards models are the currently used tools to assess the prognosis and progression of the disease. Fibrosis-Net and FVC-Net are deep learning architectures that combine CT image data with clinical and demographic data to predict forced vital capacity in interstitial lung diseases through regression analysis. These studies have provided evidence that the model can be applied in various diseases like COVID–19. Deep learning using clinical and CT imaging data was able to predict survival or mortality in idiopathic pulmonary fibrosis patients. Despite the prediction of disease progression using proportional hazard models being in early stages, the results are very encouraging and need further research. Furthermore, the Airway Transfer Network developed using generative models uses CT airway measurements to synthesize realistic airways and predict mortality. In the COVID-19 pandemic, chest imaging played a central role in diagnosis and prognosis, spurring the development of algorithms by various research groups that provided remarkable results for the diagnosis and prognosis of pneumonia (39).

12.8.3 Treatment

There are many FDA-approved AI applications that help predict outcomes and recurrence following targeted therapies for lung cancer patients. QUANIC utilizes various types of data collected over time to create personalized immunotherapy models and to understand resistance to treatment agents. Additionally, these applications can optimize radiation beam angles, monitor radiation levels, and aid in radiotherapy decision-making (42).

Recently, Chinese researchers developed a computational framework and machine learning algorithms to analyze the bulk transcriptome and single-cell sequencing data of immune and lung adenocarcinoma cells. Based on the analysis, key tumor-infiltrating immune cell signatures were established as predictive markers for lung adenocarcinoma patients' response to immunotherapy (43). Watson for Oncology, developed by IBM, is an AI system that extracts information from cancer patients' electronic medical records and suggests relevant treatment options. Currently, the usage of this system has limitations due to variation in genetic mutation, treatment protocols, and drug availability across different countries. Incorporation of region-specific data and personalized patient information would widen the application of this system with improved consistency.

Currently, these innovations are at the research level and lack large datasets. Additionally, they cannot provide additional information like risk factors favoring the progression of a disease. Radiological images, demographic and clinical data, and data from pulmonary function tests and laboratory investigations can be combined into algorithms to generate much more meaningful data. Furthermore, implementing these technologies in clinical practice requires financial resources and trained healthcare professionals. Ethical concerns like patient safety and data security are also to be taken care of.

12.9 PRECISION MEDICINE

In the pursuit of more effective and personalized treatment strategies, precision medicine has emerged as a revolutionary approach in respiratory care. By leveraging individualized patient data, genetic information, and advanced diagnostic tools, precision medicine aims to tailor interventions to the unique characteristics of each patient, marking a paradigm shift in the way respiratory conditions are understood and managed.

One of the fundamental principles of precision medicine in respiratory care is the identification of specific biomarkers associated with individual patients. Biomarkers can include genetic variations, protein levels, and other molecular indicators that provide insights into the underlying mechanisms of respiratory diseases. For instance, in asthma, identifying biomarkers related to airway inflammation and hyperreactivity allows healthcare providers to categorize patients into subtypes with distinct treatment responses (8, 44).

Pharmacogenomics, a subset of precision medicine, examines how an individual's genetic composition affects their reaction to drugs. In respiratory medicine, knowledge of a patient's genetic characteristics can assist in selecting the right medications and dosages. This method reduces the likelihood of adverse drug reactions and enhances treatment results, ensuring that patients receive the most effective treatments tailored to their needs.

Furthermore, precision medicine utilizes advanced diagnostic technologies to assess disease severity and progression. Imaging techniques, such as high-resolution CT scans, provide detailed insights into lung structure and function. This allows healthcare providers to tailor treatment plans based on a comprehensive understanding of the specific anatomical and functional changes in the patient's respiratory system.

The advent of digital health solutions contributes to the precision medicine approach by enabling continuous monitoring of patients' respiratory parameters. Wearable devices and mobile applications track real-time data on lung function, symptoms, and environmental factors. By analyzing this information, healthcare providers can identify patterns, trigger factors, and individual responses to treatment, facilitating a more personalized and proactive approach to respiratory care.

Despite the promising advancements, challenges exist in the widespread implementation of precision medicine in respiratory care. Access to advanced diagnostic technologies, the standardization of biomarker identification, and the incorporation of genetic information into clinical practice require careful consideration (45). Additionally, ethical considerations related to patient privacy and the responsible use of genetic data must be addressed.

12.10 CONCLUSION

Innovations in respiratory medicine have ushered in a new era of possibilities, offering hope for improved outcomes and quality of life for patients with respiratory conditions. From biologicals to gene therapy, stem cell therapy, nanotechnology, telemedicine, artificial intelligence, and precision medicine, these advancements hold tremendous promise for revolutionizing respiratory healthcare. However, amid the excitement surrounding these innovations, it is essential to acknowledge their limitations and consider their future role in shaping the landscape of respiratory medicine.

Biologicals have emerged as potent therapeutic agents for severe asthma and COPD, offering targeted treatment options for patients' refractory to conventional therapies. However, the high cost of biologics and the need for long-term monitoring of treatment response pose challenges to their widespread adoption in clinical practice. Additionally, concerns about immunogenicity and the development of treatment resistance underscore the need for ongoing research to optimize the use of biological agents in respiratory conditions.

Gene therapy holds promise for treating genetic respiratory disorders, but challenges such as the delivery of therapeutic genes to target cells and the potential for off-target effects necessitate further refinement of gene editing technologies. Moreover, ethical considerations surrounding gene editing and the long-term safety of genetically modified cells raise important questions about the future of gene therapy in respiratory medicine.

Stem cell therapy offers exciting possibilities for regenerating damaged lung tissue, but significant hurdles remain in translating preclinical findings into clinically viable treatments. Challenges such as immune rejection of transplanted cells, tumorigenicity, and ethical concerns about the use of embryonic stem cells underscore the need for rigorous preclinical and clinical studies to evaluate the safety and efficacy of stem cell-based interventions.

Nanotechnology has revolutionized drug delivery in respiratory medicine, but challenges such as nanoparticle toxicity, limited drug loading capacity, and the potential for off-target effects necessitate careful consideration of the safety and efficacy of nano-formulations. Moreover, the scalability and cost-effectiveness of nanomedicine pose challenges to its widespread implementation in clinical practice.

Telemedicine has emerged as a valuable tool for remote monitoring and management of respiratory conditions, but issues such as the digital divide, patient privacy concerns, and the lack of regulatory frameworks for telehealth services underscore the need for further development and standardization of telemedicine platforms. Moreover, the challenges of ensuring equitable access to telemedicine services in underserved communities and addressing disparities in digital literacy highlight the importance of addressing these issues in the future of respiratory healthcare.

Artificial intelligence holds great promise for transforming respiratory medicine through its ability to analyze complex clinical data and facilitate personalized treatment approaches. However, challenges such as data privacy concerns, algorithm bias, and the need for robust validation of AI-driven diagnostic and prognostic models underscore the importance of addressing these issues to maximize the potential of artificial intelligence in respiratory healthcare.

Precision medicine offers personalized treatment strategies tailored to the unique needs of each patient, but challenges such as the high cost of genetic testing, limited availability of targeted therapies, and the complexity of integrating genetic and molecular data into clinical practice pose barriers to its widespread adoption. Moreover, ethical considerations surrounding patient consent, data privacy, and the equitable distribution of precision medicine technologies highlight the need for ongoing dialogue and collaboration among stakeholders in respiratory healthcare.

While innovations in respiratory medicine offer tremendous promise for improving patient outcomes and transforming the delivery of respiratory care, it is essential to address the limitations and challenges associated with these advancements. By acknowledging these challenges and working collaboratively to overcome them, we can harness the full potential of innovations such as biologicals, gene therapy, stem cell therapy, nanotechnology, telemedicine, artificial intelligence, and precision medicine to shape the future of respiratory healthcare and improve the lives of patients worldwide. The success of COVID-19 vaccine development has showcased the ability of the scientific community to adapt, collaborate, and deliver solutions at an unprecedented pace, undoubtedly aiding in overcoming formidable public health challenges in the future.

REFERENCES

1. Chang LH, Rivera MP. Respiratory diseases: Meeting the challenges of screening, prevention, and treatment. N C Med J. 2013;74(5):385–92.
2. GBD Chronic Respiratory Disease Collaborators. Prevalence and attributable health burden of chronic respiratory diseases, 1990–2017: A systematic analysis for the Global Burden of Disease Study 2017. Lancet Respir Med. 2020;8(6):585–96.
3. Patel VH, Thannir S, Dhanani M, Augustine I, Sandeep SL, Mehadi A, et al. Current limitations and recent advances in the management of asthma. Dis Mon. 2023;69(7):101483.
4. Drug-resistant TB: Treatment enrolment, coverage and outcomes. WHO; 2023.
5. Burgess JK. Current and future models for understanding airway diseases. Pulm Pharmacol Ther. 2011;24(5):443.
6. Lommatzsch M, Mohme SN, Stoll P, Virchow JC. Response to various biologics in patients with both asthma and chronic obstructive pulmonary disease. Respiration. 2023;102(12):986–90.
7. Salter B, Lacy P, Mukherjee M. Biologics in asthma: A molecular perspective to precision medicine. Front Pharmacol. 2021;12:793409.
8. Wang J, Wang P, Shao Y, He D. Advancing treatment strategies: A comprehensive review of drug delivery innovations for chronic inflammatory respiratory diseases. Pharmaceutics. 2023;15(8):2151.
9. Elborn JS, Blasi F, Burgel PR, Peckham D. Role of inhaled antibiotics in the era of highly effective CFTR modulators. Eur Respir Rev. 2023;32(167):220154.
10. Alton E, Boyd AC, Davies JC, Gill DR, Griesenbach U, Harman TE, et al. Gene therapy for respiratory diseases: Progress and a changing context. Hum Gene Ther. 2020;31(17–18):911–6.
11. Lundstrom K. Viral vectors engineered for gene therapy. Int Rev Cell Mol Biol. 2023;379:1–41.
12. Chiuchiolo MJ, Crystal RG. Gene therapy for alpha-1 antitrypsin deficiency lung disease. Ann Am Thorac Soc. 2016;13(Suppl 4):S352–69.

13. McLachlan G, Alton E, Boyd AC, Clarke NK, Davies JC, Gill DR, et al. Progress in respiratory gene therapy. Hum Gene Ther. 2022;33(17–18):893–912.
14. Vu A, McCray Pb, Jr. New directions in pulmonary gene therapy. Hum Gene Ther. 2020;31(17–18):921–39.
15. Jungebluth P, Macchiarini P. Stem cell-based therapy and regenerative approaches to diseases of the respiratory system. Br Med Bull. 2011;99:169–87.
16. Hosseinirad H, Rashidi M, Moghaddam MM, Tebyanian H, Nouraei S, Mirhosseini SA, et al. Stem cell therapy for lung diseases: From fundamental aspects to clinical applications. Cell Mol Biol (Noisy-le-grand). 2018;64(10):92–101.
17. Cheng W, Zeng Y, Wang D. Stem cell-based therapy for pulmonary fibrosis. Stem Cell Res Ther. 2022;13(1):492.
18. Shi L, Wang L, Xu R, Zhang C, Xie Y, Liu K, et al. Mesenchymal stem cell therapy for severe COVID–19. Signal Transduct Target Ther. 2021;6(1):339.
19. Li Z, Niu S, Guo B, Gao T, Wang L, Wang Y, et al. Stem cell therapy for COVID-19, ARDS and pulmonary fibrosis. Cell Prolif. 2020;53(12):e12939.
20. Dey A, Vaishak K, Deka D, Radhakrishnan AK, Paul S, Shanmugam P, et al. Epigenetic perspectives associated with COVID-19 infection and related cytokine storm: An updated review. Infection. 2023;51(6):1603–18.
21. Wang J, Li X, Chen H. Organoid models in lung regeneration and cancer. Cancer Lett. 2020;475:129–35.
22. Ntolios P, Steiropoulos P, Karpathiou G, Anevlavis S, Karampitsakos T, Bouros E, et al. Cell therapy for idiopathic pulmonary fibrosis: Rationale and progress to date. BioDrugs. 2020;34(5):543–56.
23. Foged C, Brodin B, Frokjaer S, Sundblad A. Particle size and surface charge affect particle uptake by human dendritic cells in an in vitro model. Int J Pharm. 2005;298(2):315–22.
24. Doroudian M, O'Neill A, Mac Loughlin R, Prina-Mello A, Volkov Y, Donnelly SC. Nanotechnology in pulmonary medicine. Curr Opin Pharmacol. 2021;56:85–92.
25. Zhong W, Zhang X, Zeng Y, Lin D, Wu J. Recent applications and strategies in nanotechnology for lung diseases. Nano Res. 2021;14(7):2067–89.
26. Girigoswami A, Girigoswami K. Potential applications of nanoparticles in improving the outcome of lung cancer treatment. Genes (Basel). 2023;14(7):1370.
27. Beeraka NM, Tulimilli SV, Karnik M, Sadhu SP, Pragada RR, Aliev G, et al. The current status and challenges in the development of vaccines and drugs against devere acute respiratory syndrome-corona virus-2 (SARS-CoV-2). Biomed Res Int. 2021;2021:8160860.
28. He X, Chen X, Wang H, Du G, Sun X. Recent advances in respiratory immunization: A focus on COVID-19 vaccines. J Control Release. 2023;355:655–74.
29. Increasing COVID-19 vaccination uptake. World Health Organization; 2023.
30. Wang Y, Zhang Z, Luo J, Han X, Wei Y, Wei X. mRNA vaccine: A potential therapeutic strategy. Mol Cancer. 2021;20(1):33.
31. Lundstrom K. Replicon RNA viral vectors as vaccines. Vaccines (Basel). 2016;4(4):39.
32. Honkoop P, Usmani O, Bonini M. The current and future role of technology in respiratory care. Pulm Ther. 2022;8(2):167–79.
33. Fainardi V, Capoferri G, Tornesello M, Pisi G, Esposito S. Telemedicine and its application in cystic fibrosis. J Pers Med. 2023;13(7):1041.
34. Sharma S, Rawal R, Shah D. Addressing the challenges of AI-based telemedicine: Best practices and lessons learned. J Educ Health Promot. 2023;12:338.
35. Izquierdo JL, Almonacid C, Campos C, Morena D, Benavent M, González-de-Olano D, et al. Systemic corticosteroids in patients with bronchial asthma: A real-life study. J Investig Allergol Clin Immunol. 2023;33(1):30–6.
36. Izquierdo JL, Morena D, González Y, Paredero JM, Pérez B, Graziani D, et al. Clinical management of COPD in a real-world setting. A big data analysis. Arch Bronconeumol (Engl Ed). 2021;57(2):94–100.
37. Izquierdo JL, Almonacid C, González Y, Del Rio-Bermudez C, Ancochea J, Cárdenas R, et al. The impact of COVID-19 on patients with asthma. Eur Respir J. 2021;57(3):2003142.
38. Graziani D, Soriano JB, Del Rio-Bermudez C, Morena D, Díaz T, Castillo M, et al. Characteristics and prognosis of COVID-19 in patients with COPD. J Clin Med. 2020;9(10):3259.
39. Dack E, Christe A, Fontanellaz M, Brigato L, Heverhagen JT, Peters AA, et al. Artificial intelligence and interstitial lung disease: Diagnosis and prognosis. Invest Radiol. 2023;58(8):602–9.
40. Walsh SLF, Calandriello L, Silva M, Sverzellati N. Deep learning for classifying fibrotic lung disease on high-resolution computed tomography: A case-cohort study. Lancet Respir Med. 2018;6(11):837–45.
41. Christe A, Peters AA, Drakopoulos D, Heverhagen JT, Geiser T, Stathopoulou T, et al. Computer-aided diagnosis of pulmonary fibrosis using deep learning and CT images. Invest Radiol. 2019;54(10):627–32.
42. Gandhi Z, Gurram P, Amgai B, Lekkala SP, Lokhandwala A, Manne S, et al. Artificial intelligence and lung cancer: Impact on improving patient outcomes. Cancers (Basel). 2023;15(21):5236.
43. Zhang N, Zhang H, Liu Z, Dai Z, Wu W, Zhou R, et al. An artificial intelligence network-guided signature for predicting outcome and immunotherapy response in lung adenocarcinoma patients based on 26 machine learning algorithms. Cell Prolif. 2023;56(4):e13409.
44. Louis R, Bureau F, Desmet CJ. Advances toward precision medicine for asthma. Biochem Pharmacol. 2020;179:114081.
45. Johnson KB, Wei WQ, Weeraratne D, Frisse ME, Misulis K, Rhee K, et al. Precision medicine, AI, and the future of personalized healthcare. Clin Transl Sci. 2021;14(1):86–93.

13 Biomarkers in Lung Cancer Detection

Sayani Jati,[1] *Varshith Reddy R,*[1] *Kanthesh M Basalingappa,*[1*] *and Gopenath TS*[2]

[1] Division of Molecular Biology, School of Life Sciences, JSS AHER, Mysuru, India

[2] Department of Biotechnology and Bioinformatics, School of Life Sciences, JSS AHER, Mysuru, India

13.1 INTRODUCTION

Every year, more than 11 million people are diagnosed with cancer, and as of 2020, there were an additional 16 million cases [1]. Cancer is characterized by uncontrolled cell proliferation that creates tumor masses that seldom respond to traditional therapy. Tumor cells have the potential to move via blood arteries and lymph nodes, a process known as "invasion," which happens when cancer cells migrate and establish themselves in lymph nodes or other distant organs. Tumor progression is a complex process that relies on the coordination of numerous proteins, cell types, and signaling pathways. It is usually assumed that cancer is untreatable once it has spread. However, as cancer progresses, changes in cell signatures can aid in the preliminary identification and diagnosis of certain malignancies with careful analysis [2].

According to the WHO, a biomarker is a measurable element, pattern, or activity in the body or its outputs that indicates or predicts the likelihood of a specific event such as a disease. Karpetsky, Humphrey, and Levy coined the term biomarker in April 1977. They reported in the *Journal of the National Cancer Institute* that blood RNase levels were not a predictor of the presence or size of plasma cell tumors.

Biomarkers can be biochemical, cellular, molecular, genetic, proteomic, or epigenetic changes. According to the NIH Working Group, they are used to monitor diverse biological and pathological processes or responses to pharmacological interventions. These are major targets for cancer detection and risk evaluation in the early stages. Biomarkers that can be measured in serum, urine, or sputum are advantageous because they can be derived from tissues, cells, and fluids, that boost the utility and lower the price of screening [3]. Pepe presented a five-step procedure for producing, analysing, and verifying biomarkers, as well as steering the results of biomarker evolution to the clinical fields from the research laboratory. First is comparing gene or gene product expression profiles between damaged and healthy tissues to identify those that are either up- or down-regulated in malignant cells, followed by medical testing on noninvasive samples. The third phase is retrospective trials to see if a biomarker can detect preclinical disease utilizing a large sample size of people who appear to be healthy but are being tracked for cancer development. In the fourth phase, asymptomatic or higher-risk patients are checked for a specific biomarker, and others who meet criteria are usually followed up to see if they acquire malignancy, allowing researchers to assess the efficacy of a particular biomarker in early diagnosis programs, and in the fifth phase, researchers determine if screening a broad group of people reduces disease-related morbidity and mortality [4]. Replacing clinical outcomes with biomarkers in early-phase clinical trials can save time and money. This approach also aids in redefining illnesses and their therapies by shifting the focus from the traditional reliance on morphology and symptoms toward a more objective molecular framework [5].

Biomarkers are measurable parameters that serve as reliable indicators for evaluating a disease and assessing the effectiveness of therapeutic treatments [6]. Indeed, they perform a variety of functions, for instance in early cancer detection and determining the absence or existence of cancer [9, 10]. Prognostic biomarkers are used to estimate a patient's survivability or to predict how the tumor will behave by detecting the aggressive phenotype; predictive biomarkers determine if a treatment will be beneficial or track treatment recovery; and target biomarkers identify the molecular targets for new medicines, as well as how therapy influenced the expression of molecular markers [11, 12].

13.2 GENETICS OF LUNG CANCER

Lung carcinoma is the leading cause of cancer-related mortality worldwide, making it a significant public health concern. The disease that results is either small cell or non-small cell lung cancer (NSCLC), which includes squamous cell carcinoma, adenocarcinoma, and large cell carcinoma and accounts for approximately 75% of all lung tumors [7, 8].

Adenocarcinoma is increasing around the world, possibly because contemporary cigarettes carry greater doses of particular carcinogens [13]. Researchers have demonstrated that environmental and genetic factors cause lung cancer, including cigarette smoking, asbestos, and ionizing radiation (radon) exposure. Cigarette smoking accounts for 80%–90% of lung cancer cases, and environmental (asbestos, radon) and hereditary factors account for 20% of all lung cancer cases [14]. Because several carcinogenic chemicals require metabolic activation before they can interact with biological components, cross-variation in carcinogen metabolic activities influences their effects on cancer development [15, 16].

* Corresponding author: kantheshmb@jssuni.edu.in

DOI: 10.1201/9781032660592-16

Current diagnostic techniques, on the other hand, are either too costly or not sensitive enough to detect lung cancer early, which affect the disease's outcome [17, 18].

Research links several genetic loci to sporadic lung cancers. Molecular abnormalities in oncogenes that promote tumor growth and in tumor suppressor genes play a role, with mutations in KRAS, an oncogene, found in 30% of lung adenocarcinomas [19, 20]. MYC, Cyclin D1, and EGFR15 are elevated and upregulated in 2.5%–10%, 5%, and 6%, respectively, of NSCLC patients. C-erbB2 (Her-2/neu) or BCL2 overexpression is seen in 25% of cases [20]. Oncogene resequencing discovered unique BRAF mutations, which are seen in about 2% of adenocarcinoma patients and are only found in tumors that do not have KRAS mutations [21, 22]. Following that, abnormalities in the EGFR gene were discovered, and their presence was linked to sensitivity to small-molecule kinase inhibitors (likely gefitinib or erlotinib) [23, 24].

The transforming EML4-ALK fusion gene is expressed in some NSCLC patients [25]. Novel potential oncogenes have been discovered using high-resolution copy-number alterations screens (e.g., IAP1/2) or a method that combines expression and genomic data [26]. In one recent study, changes in transcriptional factors during lung development are more common in lung cancer than initially; tumor suppressor genes associated with this type of cancer include p53, p16, and RB [27]. Another potential tumor suppressor gene for lung cancer, p34, was recently discovered [28]. CYGB, previously associated only with sporadic head and neck cancers, is now linked to tumor suppressor genes in pathogenic cancer. Additionally, tyrosine kinases like EGFR homolog ERBB4 and various Ephrin receptor genes including EPHA3, VEGFR2 (KDR), and NTKR are also often altered in cancer [29]. These study findings on critical molecular pathways involved in lung carcinogenesis can support developing more targeted drugs.

Preneoplastic lesions show changes in gene expression and chromosome structure that appear to correlate with increasing recurrence rates, greater atypia, hyperproliferation, and the loss of cell cycle regulator. Abnormalities in the RAS genes, the p53 pathway, and genes in the genomic region of 3p14.2; methylation of the promoter of a gene that is abnormal; reduced expression of retinoic acid and retinoid X receptor; enhanced vascular development; changed extracellular matrix; and altered protein expressions are just a few of the changes found in preneoplastic lesions [30–33].

13.3 LUNG CANCER BIOMARKERS

The obvious first option for biomarker candidates for lung cancer screening is blood. Biomarkers based on blood give a full view of the patient's whole body, including the main tumor, metastatic sickness, immunological response, and peritumoral stroma. Sputum, bronchial lavage or aspirate samples, exhaled breath, and airway epithelium sampling, on the other hand, are exclusively accessible in lung and other respiratory tract cancers as alternative biomarkers. These might give information on genetic alterations that are closer to the cancer cells and their environs physically, and consequently more useful for treatment decision-making in screened patients with early-stage sickness. Urine and saliva were also collected as prospective biomarker sources. In a metabolomics-based approach, the former is very important.

13.3.1 DNA-Based Biomarkers

Recent technological breakthroughs in next-generation sequencing (NGS) have driven extensive study of circulating tumor DNA (ctDNA) in recent years. Additionally, NGS enables the cost-effective examination of customized gene panels [35]. However, although ctDNA has been successful in assessing late lung cancer, its application in the early identification of tumors is limited; its accuracy in the early stage of NSCLC varies from 15% to 50% [34–37]. Recent findings indicate that the absence of necrotic regions in confined small cancer lesions reduces ctDNA shedding; this limits the possibilities of catching nucleotide changes in stage I ctDNA. Researchers used personalized screening by DNA sequencing (CAPP-seq) to overcome the limited sensitivity of the traditional techniques in liquid samples as an alternative [34]. CAPP-seq provided a basic bioinformatics strategy for identifying target genes in tumors with recurrently mutated areas. Despite the promising results for tracking residual molecular illness following lung cancer treatment, CAPP-seq showed poor sensitivity in diagnosing early-stage lung cancer [38, 39]. To avoid the constraints of customized gene panels, researchers sequenced whole-exome or whole-genome ctDNA including the whole set of known human genes [40, 41].

Some ctDNA detection challenges stem from clonal hematopoiesis, an age-related, nontumor-derived abnormality in which somatic abnormalities in hematopoietic stem cells and progenitor cells lead to the clonal expansion of mutant hematopoietic cells. To address this, sequencing both white blood cell DNA and cfDNA can help determine tumor-specific cfDNA mutation ratios [42]. Other researchers discovered methylation analysis of cfDNA as a reservoir of novel, less invasive malignancy indications in addition to detecting ctDNA changes. Although cancer cells exhibit a global hypomethylation of DNA, hypermethylated regions correlating with the promoters of tumor suppressor genes with CpG regions have been used to identify ctDNA [43]. Measuring ctDNA fractions in cfDNA through specific methylation signatures represents a promising approach for both the diagnosis and prognosis of lung cancer [44, 45].

13.3.2 RNA-Based Biomarkers

Various circulating RNA species (miRNA, microRNA, piwi-interacting RNAs, piRNA, transfer RNAs, tRNA; small nucleolar RNAs; small nuclear RNAs) were found in human serum [46]. Because of their extraordinary stability in hostile conditions and resistance to circulating RNAses, circulating microRNAs (c-miRNAs) have been widely explored in the literature as perfect biomarkers [47].

Virtually all human cells are designed to release C-miRNAs through passive mechanisms such as apoptotic bodies, which are complexed with AGO protein, and active mechanisms such as exosomes and microvesicles [48, 49], as well as to regulate tissue homeostasis via paracrine signaling [49] or pathogenic mechanisms such as neoplastic transformation and tumor progression [50, 51]. MiRNAs are released into the microenvironment by cancer-associated fibroblasts, cancer cells, and blood cells, and assessing miRNA species and their levels in the blood could be a viable method of early lung cancer detection; c-miRNA has been thoroughly verified as a less invasive biomarker for lung cancer early detection in a few studies [49, 52].

13.3.3 Complement Fragments

The classical complement pathway may be utilized by the cancer to triggers in the complementary cascade activation. C4d, a downstream split product of this pathway, is discovered in greater amounts in the body fluids of the patients. The International Early Lung Cancer Action Program included 190 asymptomatic people, including 32 individuals with cancer discovered through screening, plasma C4d levels were associated to an enhanced lung cancer risk. Potential factor like emphysema did not change plasma levels of C4d [53, 93].

Unfortunately, in screening trials, samples from the MILD CT study could not be utilized as indicators for identifying high-risk individuals [54]. Patient's plasma samples with malignant nodules in two distinct cohorts exhibited much larger levels of C4d than those from persons with benign nodules. In people with intermediate-sized pulmonary nodules, C4d plasma levels suggested benign lung nodules with an 84% negative predictive value and a 54% positive predictive value [93], but later they showed high specificity (89%) but poor sensitivity (44%) [54, 93].

13.3.4 DNA Methylation

Tumor tissue displays hypomethylation across the board, as well as hypermethylation of particular CpG islands in the promoter regions of tumor suppressor genes [55]. Twenty years ago, hypermethylation of at least one of the four genes investigated was seen in fifteen of twenty-two NSCLC tumours (68%) but not in any matched normal lung tissue sample. In the methylated primary tumours, eleven of fifteen samples (73%) included aberrant methylation DNA in the matching blood sample [56]. In lung adenocarcinoma, DNA methylation changes are common, and they may play a crucial role in tumorigenesis, detection, and prognosis. As useful markers, the status of four CpG loci methylation were highlighted in homeobox A9 or HOXA9, cyclin D1 or CCND1, keratin-associated protein 8–1 or KRTAP8–1, and tubby-like protein 2 or TULP2. The methylation of HOXA9 was dramatically elevated in these lung adenocarcinoma patients, but the methylation of KRTAP8–1, CCND1, and TULP2 was clearly downregulated in tumor samples compared with neighboring tissues [57, 94].

13.3.5 Epidermal Growth Factor Receptor

With the introduction of novel targeted lung cancer medicines, researchers are seeking for more precise molecular markers to predict response and prognosis. The role of the epidermal growth factor receptor (EGFR) and its intracellular tyrosine kinase in the biology of NSCLC has been investigated, and currently it is being targeted. There are small medicines that inhibit tyrosine kinase, such as gefitinib and erlotinib, also some monoclonal antibodies targeting extracellular EGFR (most likely cetuximab and panitumumab), are among them. The existence of a mutant EGFR gene has been linked to the response to tyrosine kinase inhibitors. Other critical traits, like copy numbers of epidermal growth factor receptor and expression of mRNA, may also be essential, according to current studies but In biopsy specimens, above mentioned both may be detected, it might assist the clinician in the selection of therapeutic drugs [58].

13.3.6 Autoantibodies

Autoantibodies occur in certain persons with lung cancer in response to an abnormal tumor antigen, frequently in the preclinical period prior to the onset of symptoms or the possibility of imaging-based detection. They have been found in lung cancer of all histological varieties and stages but are also frequently absent or present only in low concentrations not only in people without cancer but also in many cancer patients. As a consequence, autoantibodies panels will almost certainly be sensitive but not specific. Researchers used a well-validated autoantibodies panel as a risk management approach for lung cancer in several screening groups [59–66].

13.3.7 Gene Hypermethylation

Multiple tumor suppressor genes have their promoters methylation in the beginning of the lung cancer development. Transcriptional silencing is linked to hypermethylation of gene promoter regions. Gene silence caused by hypermethylation has an impact on genes involved in every aspect of healthy cell function. It also acts as primary trigger for malignant transformation and progression [67].

Investigators used polymerase chain reaction to determine methylation using bisulfite to transmute all unmethylated cytosine residues to uracil residues and then amplified DNA between primers specific for methylated and unmethylated sequences. Other researchers have investigated methylation and identified abnormal methylation of p16INK4a, APC, TMS1, RASSF1, and DAPK in lung cancer [68–72]. Also in lung cancer, abnormal promoter methylation was seen in PRSS3, a serine protease and putative tumor suppressor gene; human DAB2 association protein gene; and apoptosis-associated speck-lick protein containing a CARD [73–75]. Hypermethylation of p16 and FHIT has been linked to a higher risk of recurring lung cancer after treatment [76, 77].

RASSF1A, p16, H-cadherin, and RARb methylation have been connected to the frequency of lung carcinoma in bronchoalveolar lavage, while FHIT methylation has been linked to cigarette usage [78]. Hypermethylation of an RASSF1A promoter was identified in 88% of SCLC and 28% of NSCLC utilizing bronchial aspirates. In persons with mild pulmonary disease, no hypermethylation was found [79], and previous researchers discovered abnormal promoter APC methylation [80], p16, and RARB2 [81] in bronchial aspirates from lung cancer patients; with this combination of genes, they identified lung cancer with a sensitivity of 69% and specificity of 87%.

13.3.8 Gene and Protein Expression

Researchers are constantly looking into different blood proteins and sputum proteins as potential lung cancer biomarkers, for instance hnRNPB1, an RNA-binding protein associated in the mRNA transport and RNA associated mutagenesis, that is typically produced by squamous cell carcinoma of the lung. In certain populations, sputum expression was overexpressed in the early stages of lung cancer. Researchers discovered that hnRNPB1 levels in lung cancer patients' plasma were high in 46% of the overall of cancer patients compared with 12% of controls [82].

Dysfunction is a major factor in the onset and progression of cancer. In humans, telomeres cover the ends of chromosomes and human telomerase (hTERT) catalytic component is known to be higher in cancers, while EGFR is commonly amplified in lung cancer [83]. Survivin is a protein molecule inhibits apoptosis while promoting mitosis. The presence of nuclear survivin in tumor tissue indicated reappearance and poor survival in patients with resected NSCLC [84, 85], and overexpression of osteopontin was identified in lung cancer patients' tissue and serum and associated with more aggressive illness [86]. The Fas-associated death domain (Fas) inactivates the NF-kB signaling pathway and is associated with cyclin D1 and B1 overexpression by changing cell cycle regulation. In lung cancer, higher levels of phosphorylated Fas-associated death domain mRNA and protein were related to poorer prognosis [87], and matrix metalloproteinase-9 functional polymorphisms have also been associated with lung cancer risk and recurrence [88]. Lung cancer patients also showed more soluble E-cadherin [89].

13.3.9 Other Molecular Markers

A biomarker for genomic instability is the cytokinesis-block micronucleus test. The quantity of micronuclei, nucleoplasmic bridges, and nuclear buds may be exploited to predict cancer risk. This test is acutely susceptible to NNK-induced genomic damage, which is known to be a tobacco carcinogen and might be used to predict the risk of lung cancer [90]. Many carcinogens induce cancer by generating DNA adducts in the presence of metabolically active intermediates. The existence of these adducts in human tissues may be exploited as a tool in cancer risk studies [91]. Mitochondrial DNA has been considered as a biomarker for cancer. Scientists have developed a technique for identifying sequence changes in mitochondrial DNA that is both quick and high throughput [92].

REFERENCES

1. Dalton WS, Friend SH. Cancer biomarkers an invitation to the table. Science. 2006;12:1165–1168.
2. Klausner RD. The nation's investment in cancer research: a plan and budget proposal for fiscal year 2003. Bethesda, MD: The National Cancer Institute; 2002.
3. Verma M, Kagan J, Sidransky D, Srivastava S. Proteomic analysis of cancer cell mitochondria. Nat Rev Cancer. 2003;3:789–795.
4. Pepe MS, Etzioni R, Feng Z, et al. Phases of biomarker development for early detection of cancer. J Natl Cancer I. 2001;93:1054–1061.
5. Manne U, Srivastava R-G, Srivastava S. Recent advances in biomarkers for cancer diagnosis and treatment. DDT. 2005;10.
6. Atkinson, AJ, Colburn, WA, DeGruttola, VG, DeMets, DL, Downing, GJ, Hoth, DF, Oates, JA, Peck, CC, Schooley, RT, Spilker, BA, Woodcock, J, & Zeger, SL. NCI-FDA Biomarkers Definitions Working Group; Biomarkers and surrogate endpoints: preferred definitions and conceptual framework. Clin Pharmacol Ther. 2001;69(3):89–95.
7. Ginsberg RJ, Vokes EE, Raben A. Section 2 non-small cell lung cancer. In chapter 30: Cancer of the lung. In: DeVita VT Jr, Hellman S, Rosenberg SA, eds. Cancer: principles and practice of oncology, 5th ed. Philadelphia, PA: Lippincott-Raven. 1997:858–865.
8. Marby M, Nelkin BD, Baylin SB. Chapter 41 lung cancer. In: Vogelstein B, Kinzler KW, eds. The genetic basis of human cancer, 2nd ed. New York: The McGraw-Hill Companies; 1998:671–679.
9. Stellman SD, Muscat JE, Thompson S, Hoffmann D, Wynder EL. Risk of squamous cell carcinoma and adenocarcinoma of the lung in relation to lifetime filter cigarette smoking. Cancer. 1997;80:382–388.
10. Lee PN. Relation between exposure to asbestos and smoking jointly and the risk of lung cancer. Occup Environ Med. 2001;58:145–153.
11. Alberg AJ, Ford JG, Samet JM. Epidemiology of lung cancer: ACCP evidence-based clinical practice guidelines (2nd edition). Chest. 2007;132:29S–55S.
12. Khuder SA. Effect of cigarette smoking on major histological types of lung cancer: a meta-analysis. Lung Cancer. 2001;31:139–148.
13. Mattson ME, Pollack ES, Cullen JW. What are the odds that smoking will kill you? Am J Public Health. 1987;77:425–431.
14. Peto R, Darby S, Deo H, Silcocks P, Whitley E, Doll R. Smoking, smoking cessation, and lung cancer in the UK since 1950: combination of national statistics with two case-control studies. BMJ. 2000;321:323–329.
15. Bartsch H, Nair U, Risch A, Rojas M, Wikman H, Alexandrov K. Genetic polymorphism of CYP genes, alone or in combination, as a risk modifier of tobacco-related cancers. Cancer Epidemiol Biomarkers Prev. 2009;9:3–28.
16. Schwartz AG, Prysak GM, Bock CH, Cote ML. The molecular epidemiology of lung cancer. Carcinogenesis. 2007;28:507–518.
17. Henschke CI, Yankelevitz DF, Libby DM, Pasmantier MW, Smith JP, Miettinen OS. Survival of patients with stage I lung cancer detected on CT screening. N Engl J Med. 2006;355:1763–1771.

18. Markowitz SB, Miller A, Miller J, Manowitz A, Kieding S, Sider L, Morabia A. Ability of low-dose helical CT to distinguish between benign and malignant noncalcified lung nodules. Chest. 2007;131:1028–1034.
19. Sekido Y, Fong KM, Minna JD. Progress in understanding the molecular pathogenesis of human lung cancer. Biochim Biophys Acta. 1998;1378:F21–F59.
20. Salgia R, Skarin AT. Molecular abnormalities in lung cancer. J Clin Oncol. 1998;16:1207–1217.
21. Naoki K, Chen TH, Richards WG, Sugarbaker DJ, Meyerson M. Missense mutations of the BRAF gene in human lung adenocarcinoma. Cancer Res. 2002;62:7001–7003.
22. Brose MS, Volpe P, Feldman M, et al. BRAF and RAS mutations in human lung cancer and melanoma. Cancer Res. 2002;62:6997–7000.
23. Lynch TJ, Bell DW, Sordella R, Gurubhagavatula S, Okimoto RA, Brannigan BW, Harris PL, Haserlat SM, Supko JG, Haluska FG, Louis DN, Christiani DC, Settleman J, Haber DA. Activating mutations in the epidermal growth factor receptor underlying responsiveness of non-small-cell lung cancer to gefitinib. N Engl J Med. 2004 May 20;350(21):2129–39.
24. Paez JG, Janne PA, Lee JC, et al. EGFR mutations in lung cancer: correlation with clinical response to gefitinib therapy. Science. 2004;304:1497–1500.
25. Soda M, Choi YL, Enomoto M, et al. Identification of the transforming EML4-ALK fusion gene in non-small-cell lung cancer. Nature. 2007;448:561–566.
26. Dai Z, Zhu WG, Morrison CD, et al. A comprehensive search for DNA amplification in lung cancer identifies inhibitors of apoptosis cIAP1 and cIAP2 as candidate oncogenes. Hum Mol Genet. 2003;12:791–801.
27. Kendall J, Liu Q, Bakleh A, et al. Oncogenic cooperation and coamplification of developmental transcription factor genes in lung cancer. Proc Natl Acad Sci USA. 2007;104:16663–16668.
28. Khuder SA. Effect of cigarette smoking on major histological types of lung cancer: a meta-analysis. Lung Cancer. 2001;31:158–168.
29. Xinarianos G, McRonald FE, Risk JM, Bowers NL, Nikolaidis G, Field JK, Liloglou T. Frequent genetic and epigenetic abnormalities contribute to the deregulation of cytoglobin in non-small cell lung cancer. Hum Mol Genet. 2006;15:2038–2044.
30. Travis WD Colby TV, Corrin B, in collaboration with pathologists from 14 countries. Histological typing of lung and pleural tumors. New York: Springer; 1999.
31. Greenberg AK, Yee H, Rom WN. Preneoplastic lesions of the lung. Respir Res. 2002;3:20.
32. Belinsky S. Gene-promoter hypermethylation as a biomarker in lung cancer. Nat Rev Cancer. 2004;4:707–717.
33. Brabender J, Metzger R, Salonga D, et al. Comprehensive expression analysis of retinoic acid receptors and retinoid X receptors in nonsmall cell lung cancer: implications for tumor development and prognosis. Carcinogenesis. 2002;26:525–530.
34. Newman AM, Bratman SV, To J, et al. An ultrasensitive method for quantitating circulating tumor DNA with broad patient coverage. Nat Med. 2014;20:548–554.
35. van Nimwegen KJM, van Soest RA, Veltman JA, Nelen MR, van der Wilt GJ, Vissers LELM, Grutters JPC. Is the $1000 genome as near as we think? A cost analysis of next-generation sequencing. Clin Chem. 2016;62:1458–1464.
36. Cohen JD, Li L, Wang Y, et al. Detection and localization of surgically resectable cancers with a multi-analyte blood test. Science. 2018;359:926–930.
37. Abbosh C, Birkbak NJ, Wilson GA, et al. Phylogenetic CtDNA analysis depicts early-stage lung cancer evolution. Nature. 2017;545:446–451.
38. Chabon JJ, Hamilton EG, Kurtz DM, et al. Integrating genomic features for non-invasive early lung cancer detection. Nature. 2020;580:245–251.
39. Chaudhuri AA, Chabon JJ, Lovejoy AF, et al. Early detection of molecular residual disease in localized lung cancer by circulating tumor DNA profiling. Cancer Discov. 2017;7: 1394–1403.
40. Giroux Leprieur E, Hélias-Rodzewicz Z, Takam Kamga P, et al. Sequential CtDNA whole-exome sequencing in advanced lung adenocarcinoma with initial durable tumor response on immune checkpoint inhibitor and late progression. J Immunother Cancer. 2020;8.
41. Keller L, Belloum Y, Wikman H, Pantel K. Clinical relevance of blood-based ctDNA analysis: mutation detection and beyond. Br J Cancer. 2021;124:345–358.
42. Chan HT, Chin YM, Nakamura Y, Low S-K. Clonal hematopoiesis in liquid biopsy: from biological noise to valuable clinical implications. Cancers. 2020;12.
43. Ehrlich M. DNA hypomethylation in cancer cells. Epigenomics. 2009;1:239–259.
44. Ooki A, Maleki Z, Tsay J-CJ, et al. A panel of novel detection and prognostic methylated DNA markers in primary non-small cell lung cancer and serum DNA. Clin Cancer Res. 2017;23:7141–7152.
45. Hulbert A, Jusue-Torres I, Stark A, et al. Early detection of lung cancer using DNA promoter hypermethylation in plasma and sputum. Clin Cancer Res. 2017;23:1998–2005.
46. Umu SU, Langseth H, Bucher-Johannessen C, et al. A comprehensive profile of circulating RNAs in human serum. RNA Biol. 2018;15:242–250.
47. Chen X, Ba Y, Ma L, et al. Characterization of microRNAs in serum: a novel class of biomarkers for diagnosis of cancer and other diseases. Cell Res. 2008;18:997–1006.
48. Chevillet JR, Kang Q, Ruf IK, et al. Quantitative and stoichiometric analysis of the microRNA content of exosomes. Proc Natl Acad Sci USA. 2014;111:14888–14893.
49. Turchinovich A, Weiz L, Burwinkel B. Extracellular MiRNAs: the mystery of their origin and function. Trends Biochem Sci. 2012;37:460–465.
50. Le MTN, Hamar P, Guo C, Basar E, Perdigão-Henriques R, Balaj L, Lieberman J. MiR-200-containing extracellular vesicles promote breast cancer cell metastasis. J Clin Invest. 2014;124:5109–5128.
51. Melo SA, Sugimoto H, O'Connell JT, et al. Cancer exosomes perform cell-independent microRNA biogenesis and promote tumorigenesis. Cancer Cell. 2014;26:707–721.
52. Pritchard CC, Kroh E, Wood B, et al. Blood cell origin of circulating microRNAs: a cautionary note for cancer biomarker studies. Cancer Prev Res. 2012;5:492–497.
53. Ajona D, Pajares MJ, Corrales L, et al. Investigation of complement activation product C4d as a diagnostic and prognostic biomarker for lung cancer. J Natl Cancer Inst. 2013;105:1385–1395.
54. Ajona D, Okrój M, Pajares MJ, et al. Complement C4dspecific antibodies for the diagnosis of lung cancer. Oncotarget. 2018;9:6346–6355.
55. Ehrlich M. DNA hypomethylation in cancer cells. Epigenomics. 2009;1:239–259.
56. Esteller M, Sanchez-Cespedes M, Resell R, Sidransky D, Baylin SB, Herman JG. Detection of aberrant promoter hypermethylation of tumor suppressor genes in serum DNA from non-small cell lung cancer patients. Cancer Res. 1999;59:67–70.
57. Shen N, Du J, Zhou H, Chen N, Pan Y, Hoheisel JD, Jiang Z, Xiao L, Tao Y and Mo X (2019) A Diagnostic Panel of DNA Methylation Biomarkers for Lung Adenocarcinoma. Front Oncol. 9:1281.

58. Dziadziuszko R, Witta SE, Cappuzzo F, et al. Epidermal growth factor receptor messenger RNA expression, gene dosage, and gefitinib sensitivity in nonsmall cell lung cancer. Clin Cancer Res. 2006;12:3078–3084.
59. Lam S, Boyle P, Healey GF, et al. EarlyCDT-Lung: an immunobiomarker test as an aid to early detection of lung cancer. Cancer Prev Res. 2011;4:1126–1134.
60. Macdonald IK, Murray A, Healey GF, et al. Application of a high throughput method of biomarker discovery to improvement of the EarlyCDT®-Lung test. PLoS One. 2012;7: e51002.
61. Chapman CJ, Healey GF, Murray A, et al. EarlyCDT®-lung test: improved clinical utility through additional autoantibody assays. Tumor Biol. 2012;33:1319–1326.
62. Healey GF, Lam S, Boyle P, Hamilton-Fairley G, Peek LJ, Robertson JFR. Signal stratification of autoantibody levels in serum samples and its application to the early detection of lung cancer. J Thorac Dis. 2013;5:618–625.
63. Jett JR, Peek LJ, Fredericks L, Jewell W, Pingleton WW, Robertson JFR. Audit of the autoantibody test, EarlyCDT®-lung, in 1600 patients: an evaluation of its performance in routine clinical practice. Lung Cancer. 2014;83:51–55.
64. Massion PP, Healey GF, Peek LJ, et al. Autoantibody signature enhances the positive predictive power of computed tomography and nodule-based risk models for detection of lung cancer. J Thorac Oncol. 2017;12:578–584.
65. Sullivan FM, Farmer E, Mair FS, et al. Detection in blood of autoantibodies to tumour antigens as a case-finding method in lung cancer using the EarlyCDT®-Lung Test (ECLS): study protocol for a randomized controlled trial. BMC Cancer. 2017;17:187.
66. Edelsberg J, Weycker D, Atwood M, Hamilton-Fairley G, Jett R. Cost-effectiveness of an autoantibody test (EarlyCDT-Lung) as an aid to early diagnosis of lung cancer in patients with incidentally detected pulmonary nodules. PLoS One. 2018 May 22;13(5):e0197826.
67. Belinsky S. Gene-promoter hypermethylation as a biomarker in lung cancer. Nat Rev Cancer. 2004;4:707–717.
68. Xue X, Zhu YM, Woll PJ. Circulating DNA and lung cancer. Ann N Y Acad Sci. 2006;1075:154–164.
69. Usadel H, Brabender J, Danenberg KD, et al. Quantitative adenomatous polyposis coli promoter methylation analysis in tumor tissue, serum, and plasma DNA of patients with lung cancer. Cancer Res. 2002;62:371–375.
70. Ramirez JL, Sarries C, De Castro PL, et al. Methylation patterns and K-ras mutations in tumor and paired serum of resected nonsmall-cell lung cancer patients. Cancer Lett. 2003;193: 207–216.
71. Esteller M, Sanchez-Cespedes M, Rosell R, et al. Detection of aberrant promoter hypermethylation of tumor suppressor genes in serum DNA from nonsmall cell lung cancer patients. Cancer Res. 1999;59:67–70.
72. Belinsky SA, Klinge DM, Dekker JD, et al. Gene promoter methylation in plasma and sputum increases with lung cancer risk. Clin Cancer Res. 2005;11:6505–6511.
73. Marsit CJ, Okpukpara C, Danaee H, Kelsey KT. Epigenetic silencing of the PRSS3 putative tumor suppressor gene in nonsmall cell lung cancer. Mol Carcinog. 2005;44:146–150.
74. Yano M, Toyooka S, Tsukuda K, et al. Aberrant promoter methylation of human DAB2 interactive protein (hDAB2IP) gene in lung cancers. Int J Cancer. 2005;113:59–66.
75. Zhang Z, Tan S, Zhang L. Prognostic value of apoptosis-associated speck-like protein containing a CARD gene promoter methylation in resectable nonsmall cell lung cancer. Clin Lung Cancer. 2006;8:62–65.
76. Maruyama R, Sugio K, Yoshino I, et al. Hypermethylation of FHIT as a prognostic marker in nonsmall cell lung carcinoma. Cancer. 2004;100:1472–1477.
77. Kim JS, Kim JW, Han J, et al. Cohypermethylation of p16 and FHIT promoters as prognostic factor of recurrence in surgically resected stage I nonsmall cell lung cancer. Cancer Res. 2004;66:4049–4054.
78. Kim H, Kwon YM, Kim JS, et al. Tumor-specific methylation in bronchial lavage for the early detection of nonsmall cell lung cancer. J Clin Oncol. 2004;22:2363–2370.
79. Grote HJ, Schmiemann V, Geddert H, et al. Methylation of RAS association domain family protein A as a biomarker of lung cancer. Cancer. 2006;108:129–134.
80. Grote HJ, Schmiemann V, Kiel S, et al. Aberrant methylation of the adenomatous polyposis coli promoter 1A in bronchial aspirates from patients with suspected lung cancer. Int J Cancer. 2004;110:751–755.
81. Grote HJ, Schmiemann V, Geddert H, et al. Aberrant promoter methylation of p16INK4a, RARB2 and SEMA3B in bronchial aspirates from patients with suspected lung cancer. Int J Cancer. 2005;116:720–725.
82. Sueoka E, Sueoka N, Iwanaga K, et al. Detection of plasma hnRNP B1 mRNA, a new cancer biomarker, in lung cancer patients by quantitative real-time polymerase chain reaction. Lung Cancer. 2005;48:77–83.
83. Miura N, Nakamura H, Sato R, et al. Clinical usefulness of serum telomerase reverse transcriptase (hTERT) mRNA and epidermal growth factor receptor (EGFR) mRNA as a novel tumor marker for lung cancer. Cancer Sci. 2006;97: 1366–1373.
84. Lu B, Gonzalez A, Massion PP, et al. Nuclear survivin as a biomarker for nonsmall lung cancer. Br J Cancer. 2004;91: 537–540.
85. Shinohara ET, Gonzalez A, Massion PP, et al. Nuclear survivin predicts recurrence and poor survival in patients with resected nonsmall lung carcinoma. Cancer. 2005;103:1685–1692.
86. Hu Z, Lin D, Yuan J, et al. Overexpresson of osteopontin is associated with more aggressive phenotype in human nonsmall lung cancer. Clin Cancer Res. 2005;11:4646–4652.
87. Chen G, Bhojani MS, Heaford AC, et al. Phosphorylated FADD induces NF-kappaB, perturbs cell cycle, and is associated with poor outcome in lung adenocarcinomas. Proc Natl Acad Sci USA. 2005;102:12507–12512.
88. Hu Z, Huo X, Luo D, et al. Functional polymorphisms of matrix metalloproteinase-9 are associated with risk of occurrence and metastasis of lung cancer. Clin Cancer Res. 2005;11:5433–5439.
89. Charalabopoulos K, Gogali A, Dalavaga Y, et al. The clinical significance of soluble E-cadherin in nonsmall cell lung cancer. Exp Oncol. 2006;28:83–85.
90. El-Zein RA, Schabath MB, Etzel CJ, et al. Cytokinesis-blocked micronucleus assay as a novel biomarker for lung cancer risk. Cancer Res. 2006;66:6449–6456.
91. Phillips DH. DNA adducts as markers of exposure and risk. Mutat Res. 2005;577:284–292.
92. Jakupciak JP, Wang W, Markowitz ME, et al. Mitochondrial DNA as a cancer biomarker. J Mol Diagn. 2005;7: 258–267.
93. Seijo LM, Peled N, Ajona D, et al. Biomarkers in lung cancer screening: achievements, promises, and challenges. J Thorac Oncol. 2019;14:343–357.
94. Shen N, Du J, Zhou H, et al. A diagnostic panel of DNA methylation biomarkers for lung adenocarcinoma. Front Oncol. 2019;9:1281.

14 Evolution of Molecular Markers in Prostate Cancer

Rohit Chauhan, *Aditya Prakash Sharma, Ravimohan S Mavuduru, and Uttam K Mete**
Department of Urology, Postgraduate Institute of Medical Education and Research, Chandigarh, India

14.1 INTRODUCTION

The landscape of prostate cancer diagnosis and management has witnessed a transformative evolution, marked by the integration of molecular markers. Over the years, advancements in our understanding of the molecular underpinnings of prostate cancer have paved the way for novel diagnostic, prognostic, and therapeutic strategies. This journey has been fueled by a desire to move beyond traditional methods and embrace precision medicine tailored to individual patient profiles.

Early detection and accurate risk stratification are crucial in guiding clinical decisions for prostate cancer patients. Molecular markers offer a promising avenue, providing insights into the intricate molecular alterations driving prostate cancer initiation, progression, and response to treatment. As we delve into the evolution of molecular markers in prostate cancer, it becomes evident that these biomarkers hold the key to refining diagnosis, predicting outcomes, and personalizing therapeutic approaches for improved patient care. This chapter encompasses the milestones achieved, current challenges, and the promising future that lies ahead in the dynamic realm of prostate cancer research.

Prostate cancer ranks as the sixth leading cause of cancer-related deaths in men globally, resulting in approximately 358,989 deaths in 2018. In 2016, prostate cancer held the fifth position in terms of incidence rates among males in India, registering at 4.8 per 100,000. Notably, there was a substantial 29.8% rise in the age-standardized incidence rate from 1990 to 2016 [1]. The traditional belief that the prevalence of prostate cancer in India is lower than in Western nations is evolving, as population-based cancer registries reveal a consistent and rapid upswing. According to the Indian Council of Medical Research, the reported incidence rate stands at 9–10 cases per 100,000 population. While this rate is higher than that observed in several Asian and African countries, it remains lower than the rates reported in the United States and Europe [2–4].

In the field of oncology, the rationale for choosing a biomarker hinge on its capacity to forecast early detection, gauge therapeutic response, and determine staging. Additionally, the chosen biomarker should contribute to minimizing over-diagnosis, differentiating between lesion types to identify subclasses of prostate cancer, aiding in the selection of diverse treatment options for patients, enhancing life expectancy, and ultimately improving the overall quality of life for the individual. Biomarkers are typically categorized based on their intended use. Screening or early detection biomarkers are designed to forecast the occurrence of the disease during the asymptomatic phase of cancer progression. Diagnostic biomarkers can predict cancer in individuals suspected of having a disease, while prognostic biomarkers anticipate the course of disease progression, and predictive biomarkers can estimate the risk of disease onset or progression or forecast the response to a specific therapy within a subgroup of the patient population. Finally, surrogate biomarkers serve as substitutes for clinical endpoints, aiding in the measurement of clinical benefit or harm to the patient (Figure 14.1).

A perfect diagnostic biomarker possesses notable specificity (accurate identification of those without the disease; true negative rate), considerable sensitivity (precise identification of those with the disease; true positive rate), simplicity in application, reproducibility, easily interpretable results for clinicians, cost-effectiveness, and quantifiable measures derived from readily accessible biological specimens. This chapter is grounded in the current understanding of existing diagnostic and prognostic molecular markers in prostate cancer, along with those in the developmental stage, and evaluates their practicality in clinical settings.

14.1.1 Blood-Based Markers

Prostate-specific antigen (PSA) is a 33 kDa serine protease encoded by the Kallikrein 3 gene situated on chromosome 19q 13.3–13.4. Originating from the ductal and acinar epithelium of the prostate, PSA's discovery stands out as a milestone in 20th-century scientific exploration. Initially recognized as a marker of human semen, PSA exhibits qualitative and quantitative variations in serum levels that allow differentiation between prostate carcinoma and benign inflammatory conditions. Approximately 65% to 95% of total PSA (tPSA) forms complexes with protease inhibitors like -1-antichymotrypsin (PSA-ACT). On the other hand,

* Corresponding authors: urologypgimer@gmail.com; rcsobti@pu.ac.in

DOI: 10.1201/9781032660592-17

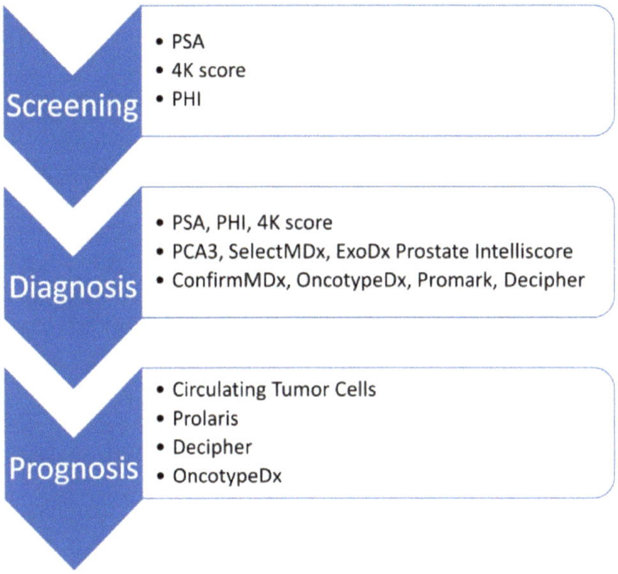

FIGURE 14.1 Biomarkers for prostate cancer.

free PSA (fPSA) represents the molecular form of PSA not bound to protein complexes. A low ratio of fPSA to tPSA serves as a characteristic of prostate cancer. However, due to differences in the stabilities of PSA molecular forms, caution is advised regarding the primary use of PSA as a screening method in clinical settings [5].

The FDA initially sanctioned the PSA test in 1986 to monitor the progression of prostate cancer in men already diagnosed with the disease. Subsequently, in 1994, the FDA granted approval for diagnostic screening of prostate cancer in conjunction with a digital rectal exam (DRE). The integration of PSA screening and early detection strategies has notably contributed to the reduction in prostate cancer mortality over the past thirty years. Remarkably, during the early phases of PSA testing, African American (AA) men with newly diagnosed prostate cancer exhibited higher serum PSA levels at initial diagnosis than Caucasian American (CA) men as well as larger tumor volumes. The tumor volume disparity could be linked to screening behavior and/or biological distinctions in prostate cancer among black men [6].

Limited data exists on race-specific biomarkers in prostate cancer, as most biomarker studies focus on CA men. Therefore, evaluating the utility of existing markers and exploring new biomarkers in the context of race is crucial. While a serum PSA threshold of 4 ng/mL is still utilized for patients with an abnormal DRE, rising PSA values over time, even below 4 ng/mL, have proven to be a more significant diagnostic parameter. Persistently elevated PSA levels prompt biopsy; however, high PSA remains an imprecise marker for cancer risk, as higher PSA levels do not consistently indicate the presence of cancer [7]. Non-cancerous conditions like prostatitis and benign prostatic hyperplasia can also elevate serum PSA, potentially leading to unnecessary biopsies [8, 9].

Complications arising from unnecessary biopsies due to abnormal PSA levels have resulted in various issues for patients, with erectile dysfunction and urinary incontinence being the most prevalent [10]. Subsequently, ongoing debates have ensued regarding the pros and cons of PSA-based screening for prostate cancer. The revised C grade recommendation from the USPSTF, as of May 2018, suggests that the decision for periodic PSA-based screening in men aged 55–69 should be individualized. Before opting for screening, men should engage in discussions with their clinicians to comprehend the potential benefits and drawbacks, allowing them to incorporate their values and preferences into the decision-making process. However, the USPSTF maintains its recommendation against PSA-based screening for prostate cancer in men aged 70 years and older [11].

The Prostate Health Index (phi; Beckman Coulter, Brea, CA, USA) is a blood-based test that evaluates levels of fPSA, tPSA, and [–2]proPSA (p2PSA) to assess the risk of Gleason 7 disease during biopsy [12]. FDA-approved in 2012, the phi assay distinguishes prostate cancer from benign conditions in men aged 50 with nonsuspicious DRE findings and serum tPSA levels of 4 to 10 ng/mL. In a multicenter study on 892 men, phi significantly improved specificity for prostate cancer detection in the 4 to 10 ng/mL PSA range, with a score of 27.0 demonstrating 90% sensitivity and 31.1% specificity for predicting prostate cancer at biopsy [13]. Recent research involving 658 men with PSA levels of 4 to 10 ng/mL, revealed phi's superiority over free and total PSA, enhancing the prediction of high-grade, clinically significant prostate cancer with a high AUC (0.698) [14]. Despite the 2015 NCCN guidelines recommending phi for early prostate cancer detection, its use in first-line screening remains limited due to limited prospective analyses in US populations, although a phi above 35 provides insight into the probability of prostate cancer, especially in biopsy-naive or previously negative biopsy patients [15].

The 4Kscore® Test (OPKO Lab, Nashville, TN, USA) delivers a composite score based on four kallikrein proteins: tPSA, fPSA, intact PSA, and hK2. Since 2015, the NCCN Prostate Cancer Early Detection guidelines have incorporated the 4Kscore [16]. Alongside clinical data like age and prior negative biopsy history, the 4Kscore estimates the likelihood of a positive biopsy for prostate cancer. In a meta-analysis, the 4Kscore exhibited an 8% to 10% improvement in biopsy predictive accuracy, potentially avoiding 48% to 56% of unnecessary biopsies [17]. Notably, it can predict the risk of distant prostate cancer metastases up to two decades later from a blood sample in men with PSA >2 ng/mL. A Swedish study with 12,542 men illustrated that the 4K model significantly outperformed total PSA in predicting distant metastasis, especially in those with elevated PSA at ages 50 and 60 [18]. The 4Kscore, associated with a low risk, predicted a minimal chance of metastatic prostate cancer development, showcasing its potential in risk stratification. The multifaceted utility and predictive capabilities underscore the promising role of 4Kscore in prostate cancer assessment.

Circulating tumor cells (CTCs) have garnered substantial attention as biomarkers in advanced prostate cancer.

Numerous investigations have demonstrated the presence of tumor cells in the blood and bone marrow of individuals with prostate cancer [19, 20]. The integration of next-generation sequencing (NGS) and highly sensitive CTC detection assays, conducted in the blood plasma of cancer patients, is introducing CTCs and circulating cell-free tumor DNAs (ct-DNA) as valuable tools for liquid biopsy in oncology. These assays provide minimally invasive methods for monitoring tumor burden, as well as tumor genome and biology in pre- and post-treatment scenarios.

The FDA-approved CellSearchTM platform (Janssen Diagnostics, Raritan, NJ, USA) is a notable system for prognostic use in prostate cancers. It employs immunomagnetic capture and fluorescence imaging technology for the detection and enumeration of CTCs. The CellSearchTM CTC test stands as an independent predictor of overall and progression-free survival in metastatic prostate cancer [21]. Ongoing CTC research is concentrated on molecular characterization to identify biomarkers for treatment response prediction. Molecular analysis from CTCs has shown comparability with that of primary tumor tissue and/or metastasis, offering real-time insight into metastatic prostate cancer [22, 23]. Various studies exploring the diagnostic and therapeutic potential of CTCs underscore their role as promising prognostic and predictive biomarkers for clinical outcomes and treatment responses in prostate cancer [24, 25].

An alternative to CTCs is ct-DNAs or cell-free (cf) DNA (cf-DNAs), which comprise small nucleic acid fragments released from tumor cells. Pantel and colleagues reported the potential use of ct-DNA in the blood-plasma of prostate cancer patients as a biomarker [26]. Azad et al. highlighted AR amplification in ct-DNAs of castration-resistant prostate cancer (CRPC) [27]. Wyatt et al. demonstrated that genomic alterations in, including AR amplification, AR mutation, and RB1 loss, strongly associated with enzalutamide resistance in CRPC patients [28]. While emerging technologies offer clear advantages over existing ones, they also present limitations. Challenges include the extremely low abundance of ct-DNA or CTC, which may lead to overdiagnosis/treatment and impose psychological and unnecessary financial burdens. Additionally, the correlation of CTC/ct-DNA with the age/sex of patients is not well established. Despite liquid biopsy finding its niche in oncology, particularly during the treatment of metastatic breast and lung cancers, its utility in other cancers, including CRPC, requires extensive evaluation.

14.1.2 Urine-Based Markers

The discovery of prostate cancer antigen 3 (PCA3) stemmed from transcriptome evaluations conducted on both normal and tumor prostate tissues. The initial findings reported the presence of prostate tissue-specific and prostate cancer-associated long non-coding RNAs, namely differential display 3/(PCA3) and a prostate cancer gene expression marker 1 [29]. However, it was observed that PCA3 was overexpressed in virtually all prostate cancers [30].

Indeed, the diagnostic effectiveness of PCA3 RNA in post-DRE urine has undergone thorough evaluation [31]. The Progensa PCA3 assay (Progensa Test Kit, Hologic, Marlborough, MA, USA) is an FDA-approved diagnostic test for prostate cancer recommended for men aged 50 years or older with elevated serum PSA and prior negative biopsy results [32]. The European Association of Urology guidelines include the Progensa PCA3 assay for decision-making regarding repeat biopsy. PCA3 scores indicate the ratio of PCA3 RNA molecules to PSA RNA molecules detected in a patient's urinary specimen after a DRE. The sensitivity, specificity, positive predictive value (PPV), and negative predictive value (NPV) of PCA3 vary depending on the selected cutoff score (25 or 35). A PCA3 score <25 is linked to a reduced likelihood of prostate cancer on subsequent repeat biopsy. A PCA3 score of 35 demonstrated sensitivity ranging from 58% to 82%, specificity from 58% to 76%, PPV of 67% to 69%, NPV of 87%, and an AUC of 0.68 to 0.87 (95% CI for 0.87: 0.81–0.92) [33]. Several studies on large patient cohorts indicate that urine PCA3 outperforms serum PSA in diagnosing prostate cancer [34]. It has also been observed that urinary PCA3 before radical prostatectomy could predict extracapsular extension and tumor volume; however, the predictive value of PCA3 for aggressive disease has been inconsistent [35].

The SelectMDx (MDxHealth, Irvine, CA, USA) qRT-PCR assay offers the probability of detecting prostate cancer through biopsy, along with the likelihood of high-grade versus low-grade disease, achieving an AUC of 0.89 (95% CI: 0.86–0.92). It is conducted on post-DRE urine specimens from patients with clinical risk factors for prostate cancer under consideration for biopsy, assessing the mRNA levels of DLX1 and HOXC6 utilizing Kallikrein 3 expression as an internal reference to select patients for prostate biopsy. A three-gene panel comprising HOXC6, TDRD1, and DLX1 significantly enhanced the detection of clinically significant prostate cancer, boasting an AUC of 0.77 in contrast with 0.72 for PSA and 0.68 for PCA3 [36].

Genomic rearrangements resulting in the fusion of the AR-regulated TMPRSS2 gene promoter and the N-terminally deleted ERG coding sequence constitute the most prevalent prostate cancer-specific driver gene alteration [37]. The oncogenic activation of ERG through these gene fusions is identified in 50% to 65% of prostate cancer patients of European ancestry [38]. A comprehensive multicenter study involving TMPRSS2:ERG (T2-ERG) and PCA3 in post-DRE urine from 1312 men demonstrated a substantial improvement over serum PSA for detecting clinically relevant cancer during biopsy [39]. In a test developed by the University of Michigan, MLabs, for post-DRE urine known as the Mi-Prostate Score, T2-ERG displayed low sensitivity of 24.3% to 37% but high specificity of 93%, with PPV of 94%. Moreover, when combined with serum PSA (cut-off of 10 ng/mL) and urinary PCA3, T2-ERG enhanced the accuracy of prostate cancer diagnosis, achieving a sensitivity and specificity of 80% and 90%, respectively. This test also provides a risk assessment for aggressive disease [40].

In one study, T2-ERG gene fusions exhibited a strong association with a Gleason score of 7 or higher and prostate cancer-related death. The combination of PCA3 and T2-ERG, along with the Prostate Cancer Prevention Trial risk calculator, assisted in determining the urgency of biopsy after an elevated serum PSA [41]. In an analysis of ERG mRNA in post-DRE urine from 237 men, predictive accuracy was 0.80 for prostate cancer diagnosis in Caucasian men with PSA less than 4.0 ng/mL. Recently, three studies further underscored the diagnostic utility of T2-ERG and PCA3 in urine. Collectively, data from these studies show that a combination of urine PCA3 and T2-ERG, along with serum PSA, significantly improves the detection of aggressive disease (Gleason score greater than or equal to 7) during the initial biopsy and reduces unnecessary biopsies by 42% [42–44].

The ExoDx Prostate Intelliscore from Exosome Diagnostics Inc., based in Cambridge, MA, USA, is a standard urine (non-DRE) exosome-based assay that evaluates ERG and PCA3 alongside SPDEF as an internal reference. This test is recommended for men aged over 50 with PSA ranging from 2 to 10 ng/mL who are being considered for an initial prostate biopsy [45]. It is utilized in combination with the current standard of care (SOC) variables, including PSA, age, race, and family history, to assess the risk of Gleason 6 or 7 and benign disease during the initial biopsy. An assessment of the ExoDx score (EXO106) in 195 patients during the diagnostic biopsy stage revealed that EXO106 exhibited strong clinical performance in predicting biopsy results for both any cancer and high-grade disease. In the case of high-grade disease, it significantly ($p = 0.0009$) enhanced the predictive performance of SOC (EXO106 and SOC, AUC 0.80; SOC alone, AUC 0.67) [46]. In a multicentric study involving 774 patients (255 in the training cohort and 519 in the validation cohort) from twenty-two community practice and academic urologic sites in the US, the ExoDx assay, when combined with SOC features, outperformed the ExoDx assay or SOC data alone in predicting the presence of Gleason grade ≥7 cancer and negative biopsy results. For high-grade prostate cancer, utilizing a predefined cut point, 27% of biopsies could have been avoided, missing only 5% of patients with dominant pattern 4 high-risk GS7 disease [47].

14.1.2.1 Tissue-Based Markers

ConfirmMDx, developed by MDxHealth, Inc., based in Irvine, CA, USA, is a DNA methylation assay that relies on prostate tissue biopsy. This multiplexed, epigenetic test assesses the methylation status of glutathione S-transferase Pi 1 (GSTP1), adenomatous polyposis coli (APC), and Ras association (RalGDS/AF-6) domain family member 2 (RASSF2). The methylation of APC and RASSF2 genes, in addition to GSTP1, is commonly observed in prostate cancer. Interestingly, a positive ConfirmMDx test for pathology-determined cancer in a negative biopsy suggests the possibility of cancer cells or regions being missed during the biopsy procedure. Commercially available, this test aims to distinguish true negative prostate biopsies from those that may harbor occult cancer [48]. It is incorporated into the National Comprehensive Cancer Network (NCCN) guidelines for men with at least one prior negative biopsy. ConfirmMDx was validated in two studies: Methylation Analysis to Locate Occult Cancer trial, where the assay demonstrated a significant association with patient outcomes with an odds ratio of 3.17, and Detection Of Cancer Using Methylated Events in Negative Tissue, where the assay independently correlated with prostate cancer detection in a repeat biopsy, achieving an 88% negative predictive value [49, 50].

The Oncotype DX GPS test, developed by Genomic Health in Redwood City, CA, USA, is a genomic test based on biopsy samples. This test assesses the mRNA expression of 17 genes crucial for the growth and survival of tumor cells. The initial phase involved the study of 4,500 patients, using reverse transcription PCR (RT-PCR) and formalin-fixed paraffin-embedded radical prostatectomy (RP) specimens from 441 patients with low to intermediate risk prostate cancer [51]. Out of the 732 selected genes, 288 were identified to predict clinical recurrence (local recurrence or distant metastasis), and 198 were associated with aggressive disease after adjusting for PSA, Gleason score (GS), and clinical stage [52]. Subsequently, a 17-gene prognostic expression signature known as GPS was defined. In a validation study involving 395 needle biopsies from patients under active surveillance (AS), GPS accurately predicted high-grade (odds ratio (OR) per 20 GPS units: 2.3; 95% confidence interval (CI), 1.5–3.7; $p < 0.001$) and high-stage (OR per 20 GPS units: 1.9; 95% CI, 1.3–3.0; $p = 0.003$) disease at surgical pathology. Furthermore, GPS predicted high-grade and/or high-stage disease, even after considering established clinical factors (age, PSA, clinical stage, biopsy GS), resulting in an OR of 2.1 (95% CI, 1.4–3.2) when adjusting for the Cancer of the Prostate Risk Assessment score (CAPRA).

The Promark, developed by Metamark in Cambridge, MA, USA, is a protein-based prognostic assay designed to predict cancer aggressiveness in patients with biopsy Gleason scores of 3 + 3 and 3 + 4. This assay assesses eight protein markers through an automated, quantitative, and multiplex immunofluorescence assay conducted on formalin-fixed paraffin-embedded tissues [53]. The protein marker panel generates a score ranging from 0 to 1, predicting adverse pathology at the time of radical prostatectomy (RP) for patients falling under NCCN very low-risk or low-risk categories who are considering active surveillance. In an initial study involving 381 patient biopsies with corresponding prostatectomy pathology, the risk score for favorable and unfavorable pathology was established [54]. A risk score below 0.33 demonstrated predictive values for favorable pathology in the NCCN very low-risk and low-risk groups, as well as the low-risk D'Amico group, at 95%, 81.5%, and 87.2%, respectively. These values surpassed the predictive capabilities of the current risk classification groups themselves, which were 80.3%, 63.8%, and 70.6%,

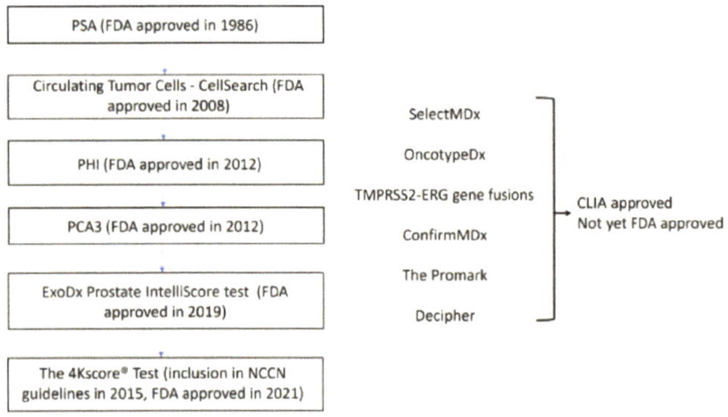

FIGURE 14.2 Timeline of molecular markers of prostate cancer.

respectively. Conversely, at a risk score above 0.8, the predictive value for unfavorable pathology was 76.9% across all risk groups.

The Decipher test, developed by GenomeDx in San Diego, CA, USA, is a genomics-driven assay that evaluates RNA expression levels of 22 different genes. The selection of these genes was based on distinctive patterns of differential expression observed in a retrospective, nested case-control study comparing 192 early metastatic cases (within 5 years of rising PSA) to 271 controls [55]. These genes play crucial roles in various biological pathways, including cell cycle progression, cell proliferation, differentiation, adhesion, and immune response-regulating signaling pathways.

The twenty-two-gene signature known as the Decipher genomic classifier (GC) is applicable to both RP and prostate biopsy specimens, represented as a score ranging from 0 to 1.0 (with a higher score indicating a higher probability of clinical metastasis). NCCN guidelines recommend it for patients exhibiting adverse pathology after RP [56]. The Decipher test calculates the likelihood of clinical metastasis within five years of RP and ten-year prostate cancer-specific mortality in men with high-risk pathology or high-risk clinical features post-RP. In validation studies, the test demonstrated excellent discrimination in predicting clinical metastasis (AUC = 0.75–0.83) and cancer-specific mortality (AUC = 0.78), surpassing the performance of available clinicopathologic characteristics (AUC = 0.69) [57]. Recently, the Decipher biopsy test was validated, providing risk assessments at RP for pathologic grade upgrading (Gleason pattern 4 or 5), five-year development of metastasis, and ten-year prostate cancer-specific mortality [58]. Additionally, Decipher scores can guide the timing of postoperative radiotherapy (adjuvant vs. salvage) and serve as the sole independent predictor of clinical metastasis in patients experiencing biochemical recurrence after surgery [59]. Figure 14.2 depicts the timeline of molecular markers in prostate cancer.

14.1.3 The Role of Molecular Biomarkers in Clinical Decision-Making

Molecular biomarkers for prostate cancer exhibit significant promise in enhancing risk assessment, minimizing unnecessary treatments, and facilitating more targeted therapeutic interventions for individuals with high-risk disease. The phi, 4Kscore, and PSA tests offer valuable insights to identify patients warranting biopsy referrals. Both suspected prostate cancer patients requiring initial biopsy and those necessitating re-biopsy following an initial negative biopsy can benefit from information provided by assays such as PSA, 4Kscore, phi, SelectMDx, PCA3, TMPRSS2-ERG, ExoDxIntelliscore, SelectMDx, and ConfirmMDx. Prognostic biomarker tests, notably Prolaris and Oncotype DX, contribute to risk stratification, aiding in decision-making for treatment post a positive biopsy for prostate cancer. Furthermore, prognostic markers offer valuable guidance for treatment post-surgery, as observed in the case of Decipher.

14.2 CONCLUSION

The molecular biomarkers for prostate cancer we highlighted in this review are becoming pivotal in enhancing diagnosis and treatment. While a promising array of biomarkers has emerged, clinicians and reimbursement agencies face the challenge of evaluating and prioritizing these new markers. These biomarkers show potential in guiding biopsy decisions, especially in the PSA gray zone (4.0 to 10.0 ng/mL), reducing unnecessary procedures. They also aid in distinguishing clinically significant prostate cancer, influencing treatment strategies. Emerging technologies, including liquid biopsy, multi-genomics, and advanced imaging, offer additional prospects. The integration of robust biomarkers and advanced imaging, like multiparametric MRI, holds promise for personalized risk assessment in prostate cancer diagnosis and management.

REFERENCES

1. Bakshi G, Tongaonkar H, Addla S, et al. Expert survey on management of prostate cancer in India: Real-world insights into practice patterns. Indian J Cancer. 2022;59(Supplement): S19–S45. doi: 10.4103/ijc.ijc_1145_21
2. Hassanipour S, Delam H, Arab-Zozani M, et al. Survival rate of prostate cancer in Asian countries: A systematic review and meta-analysis. Ann Glob Health. 2020;86:2.
3. Emanuele C. European Commission. Epidemiology of prostate cancer in Europe. 2015. Available from: https://ec.europa.eu/jrc/en/publication/epidemiology-prostate-cancer-europe [Last accessed on 2021 Jan 5].
4. American Cancer Society. Survival rates for prostate cancer. 2020. Available from: https://www.cancer.org/cancer/prostate-cancer/detection-diagnosis staging/survival-rates.html [Last accessed on 2021 Jan 5].
5. Khan MA, Sokoll LJ, Chan DW, et al. Clinical utility of proPSA and "benign" PSA when percent free PSA is less than 15%. Urology. 2004;64(6):1160–1164. doi: 10.1016/j.urology.2004.06.033
6. Moul JW, Sesterhenn IA, Connelly RR, et al. Prostate-specific antigen values at the time of prostate cancer diagnosis in African-American men. JAMA. 1995;274(16):1277–1281.
7. Thompson IM, Pauler DK, Goodman PJ, et al. Prevalence of prostate cancer among men with a prostate-specific antigen level < or = 4.0 ng per milliliter [published correction appears

in N Engl J Med. 2004 Sep 30;351(14):1470]. N Engl J Med. 2004;350(22):2239–2246. doi: 10.1056/NEJMoa031918
8. Djavan B, Zlotta A, Remzi M, et al. Optimal predictors of prostate cancer on repeat prostate biopsy: A prospective study of 1,051 men. J Urol. 2000;163(4):1144–1149.
9. Vickers AJ, Cronin AM, Aus G, et al. A panel of kallikrein markers can reduce unnecessary biopsy for prostate cancer: Data from the European Randomized Study of Prostate Cancer Screening in Göteborg, Sweden. BMC Med. 2008;6:19. Published 2008 Jul 8. doi: 10.1186/1741-7015-6-19
10. Loeb S, Vellekoop A, Ahmed HU, et al. Systematic review of complications of prostate biopsy. Eur Urol. 2013;64(6): 876–892. doi: 10.1016/j.eururo.2013.05.049
11. Bibbins-Domingo K, Grossman DC, Curry SJ. The US Preventive Services Task Force 2017 draft recommendation statement on screening for prostate cancer: An invitation to review and comment. JAMA. 2017;317(19):1949–1950. doi: 10.1001/jama.2017.4413
12. Beckman Coulter, I. FDA approves new blood test to improve prostate cancer detection. 2012. Available from: https://www.prnewswire.com/news-releases/fda-approves-new-blood-test-to-improveprostate-cancer-detection-160267195.html [Last accessed on 2019 Apr 10].
13. Catalona WJ, Partin AW, Sanda MG, et al. A multicenter study of [-2]pro-prostate specific antigen combined with prostate specific antigen and free prostate specific antigen for prostate cancer detection in the 2.0 to 10.0 ng/ml prostate specific antigen range [published correction appears in J Urol. 2011 Jul;186(1):354]. J Urol. 2011;185(5):1650–1655. doi: 10.1016/j.juro.2010.12.032
14. Loeb S, Catalona WJ. The Prostate Health Index: A new test for the detection of prostate cancer. Ther Adv Urol. 2014;6(2):74–77. doi: 10.1177/1756287213513488
15. Oncology NCCNCPGI Prostate Cancer Early Detection Version 1. 2015. Available from: http://www.nccn.org/professionals/physician_gls/pdf/prostate_detection.pdf [Last accessed on 2015 Jun 30].
16. Carroll PR, Parsons JK, Andriole G, et al. NCCN guidelines insights: Prostate cancer early detection, Version 2.2016. J Natl Compr Canc Netw. 2016;14(5):509–519. doi: 10.6004/jnccn.2016.0060
17. Voigt JD, Zappala SM, Vaughan ED, et al. The Kallikrein Panel for prostate cancer screening: Its economic impact. Prostate. 2014;74(3):250–259. doi: 10.1002/pros.22746
18. Stattin P, Vickers AJ, Sjoberg DD, et al. Improving the specificity of screening for lethal prostate cancer using prostate-specific antigen and a panel of kallikrein markers: A nested case-control study. Eur Urol. 2015;68(2):207–213. doi: 10.1016/j.eururo.2015.01.009
19. Bettegowda C, Sausen M, Leary RJ, et al. Detection of circulating tumor DNA in early- and late-stage human malignancies. Sci Transl Med. 2014;6(224):224ra24. doi: 10.1126/scitranslmed.3007094
20. Morgan TM, Lange PH, Vessella RL. Detection and characterization of circulating and disseminated prostate cancer cells. Front Biosci. 2007;12:3000–3009. Published 2007 May 1. doi: 10.2741/2290
21. de Bono JS, Scher HI, Montgomery RB, et al. Circulating tumor cells predict survival benefit from treatment in metastatic castration-resistant prostate cancer [published correction appears in Clin Cancer Res. 2009 Feb 15;15(4):1506]. Clin Cancer Res. 2008;14(19):6302–6309. doi: 10.1158/1078-0432.CCR-08-0872
22. Punnoose EA, Ferraldeschi R, Szafer-Glusman E, et al. PTEN loss in circulating tumour cells correlates with PTEN loss in fresh tumour tissue from castration-resistant prostate cancer patients. Br J Cancer. 2015;113(8):1225–1233. doi: 10.1038/bjc.2015.332
23. Krebs MG, Metcalf RL, Carter L, et al. Molecular analysis of circulating tumour cells-biology and biomarkers. Nat Rev Clin Oncol. 2014;11(3):129–144. doi: 10.1038/nrclinonc.2013.253
24. Antonarakis ES, Lu C, Wang H, et al. AR-V7 and resistance to enzalutamide and abiraterone in prostate cancer. N Engl J Med. 2014;371(11):1028–1038. doi: 10.1056/NEJMoa1315815
25. Thalgott M, Rack B, Maurer T, et al. Detection of circulating tumor cells in different stages of prostate cancer. J Cancer Res Clin Oncol. 2013;139(5):755–763. doi: 10.1007/s00432-013-1377-5
26. Schwarzenbach H, Alix-Panabières C, Müller I, et al. Cell-free tumor DNA in blood plasma as a marker for circulating tumor cells in prostate cancer. Clin Cancer Res. 2009;15(3):1032–1038. doi: 10.1158/1078-0432.CCR-08-1910
27. Azad AA, Volik SV, Wyatt AW, et al. Androgen receptor gene aberrations in circulating cell-free DNA: Biomarkers of therapeutic resistance in castration-resistant prostate cancer. Clin Cancer Res. 2015;21(10):2315–2324. doi: 10.1158/1078-0432.CCR-14-2666
28. Wyatt AW, Azad AA, Volik SV, et al. Genomic alterations in cell-free DNA and enzalutamide resistance in castration-resistant prostate cancer. JAMA Oncol. 2016;2(12):1598–1606. doi: 10.1001/jamaoncol.2016.0494
29. Srikantan V, Zou Z, Petrovics G, et al. PCGEM1, a prostate-specific gene, is overexpressed in prostate cancer. Proc Natl Acad Sci U S A. 2000;97(22):12216–12221. doi: 10.1073/pnas.97.22.12216
30. Petrovics G, Liu A, Shaheduzzaman S, et al. Frequent overexpression of ETS-related gene-1 (ERG1) in prostate cancer transcriptome [published correction appears in Oncogene. 2007 Oct 11;26(46):6684. Furasato, Bungo [corrected to Furusato, Bungo]]. Oncogene. 2005;24(23):3847–3852. doi: 10.1038/sj.onc.1208518
31. Nakanishi H, Groskopf J, Fritsche HA, et al. PCA3 molecular urine assay correlates with prostate cancer tumor volume: Implication in selecting candidates for active surveillance. J Urol. 2008;179(5):1804–1810. doi: 10.1016/j.juro.2008.01.013
32. Groskopf J, Aubin SM, Deras IL, et al. APTIMA PCA3 molecular urine test: Development of a method to aid in the diagnosis of prostate cancer. Clin Chem. 2006;52(6):1089–1095. doi: 10.1373/clinchem.2005.063289
33. Leyten GH, Hessels D, Jannink SA, et al. Prospective multicentre evaluation of PCA3 and TMPRSS2-ERG gene fusions as diagnostic and prognostic urinary biomarkers for prostate cancer. Eur Urol. 2014;65(3):534–542. doi: 10.1016/j.eururo.2012.11.014
34. Wang T, Qu X, Jiang J, et al. Diagnostic significance of urinary long non-coding PCA3 RNA in prostate cancer. Oncotarget. 2017;8(35):58577–58586. Published 2017 Apr 20. doi: 10.18632/oncotarget.17272
35. Hessels D, Schalken JA. Urinary biomarkers for prostate cancer: A review. Asian J Androl. 2013;15(3):333–339. doi: 10.1038/aja.2013.6
36. Leyten GH, Hessels D, Smit FP, et al. Identification of a candidate gene panel for the early diagnosis of prostate cancer. Clin Cancer Res. 2015;21(13):3061–3070. doi: 10.1158/1078-0432.CCR-14-3334
37. Sanguedolce F, Cormio A, Brunelli M, et al. Urine TMPRSS2: ERG fusion transcript as a biomarker for prostate cancer: Literature review. Clin Genitourin Cancer. 2016;14(2): 117–121. doi: 10.1016/j.clgc.2015.12.001
38. Sreenath TL, Dobi A, Petrovics G, et al. Oncogenic activation of ERG: A predominant mechanism in prostate cancer. J Carcinog. 2011;10:37. doi: 10.4103/1477-3163.91122

39. Rice KR, Chen Y, Ali A, et al. Evaluation of the ETS-related gene mRNA in urine for the detection of prostate cancer. Clin Cancer Res. 2010;16(5):1572–1576. doi: 10.1158/1078-0432.CCR-09-2191
40. Yang Z, Yu L, Wang Z. PCA3 and TMPRSS2-ERG gene fusions as diagnostic biomarkers for prostate cancer. Chin J Cancer Res. 2016;28(1):65–71. doi: 10.3978/j.issn.1000-9604.2016.01.05
41. Tomlins SA, Aubin SM, Siddiqui J, et al. Urine TMPRSS2: ERG fusion transcript stratifies prostate cancer risk in men with elevated serum PSA. Sci Transl Med. 2011;3(94): 94ra72. doi: 10.1126/scitranslmed.3001970
42. Sanda MG, Feng Z, Howard DH, et al. Association between combined TMPRSS2:ERG and PCA3 RNA urinary testing and detection of aggressive prostate cancer. JAMA Oncol. 2017;3(8):1085–1093. doi: 10.1001/jamaoncol.2017.0177
43. Tomlins SA, Day JR, Lonigro RJ, et al. Urine TMPRSS2:ERG Plus PCA3 for individualized prostate cancer risk assessment. Eur Urol. 2016;70(1):45–53. doi: 10.1016/j.eururo.2015.04.039
44. McLeod DG, Petrovics G. Re: Association between combined TMPRSS2:ERG and PCA3 RNA urinary testing and detection of aggressive prostate cancer. Eur Urol. 2018;73(2):301–302. doi: 10.1016/j.eururo.2017.09.035
45. McKiernan J, Donovan MJ, O'Neill V, et al. A novel urine exosome gene expression assay to predict high-grade prostate cancer at initial biopsy. JAMA Oncol. 2016;2(7):882–889. doi: 10.1001/jamaoncol.2016.0097
46. Donovan MJ, Noerholm M, Bentink S, et al. A molecular signature of PCA3 and ERG exosomal RNA from non-DRE urine is predictive of initial prostate biopsy result. Prostate Cancer Prostatic Dis. 2015;18(4):370–375. doi: 10.1038/pcan.2015.40
47. McKiernan J, Donovan MJ, Margolis E, et al. A prospective adaptive utility trial to validate performance of a novel urine exosome gene expression assay to predict high-grade prostate cancer in patients with prostate-specific Antigen 2–10ng/ml at initial biopsy. Eur Urol. 2018;74(6):731–738. doi: 10.1016/j.eururo.2018.08.019
48. Van Neste L, Partin AW, Stewart GD, et al. Risk score predicts high-grade prostate cancer in DNA-methylation positive, histopathologically negative biopsies. Prostate. 2016;76(12):1078–1087. doi: 10.1002/pros.23191
49. Stewart GD, Van Neste L, Delvenne P, et al. Clinical utility of an epigenetic assay to detect occult prostate cancer in histopathologically negative biopsies: Results of the MATLOC study. J Urol. 2013;189(3):1110–1116. doi: 10.1016/j.juro.2012.08.219
50. Partin AW, Van Neste L, Klein EA, et al. Clinical validation of an epigenetic assay to predict negative histopathological results in repeat prostate biopsies. J Urol. 2014;192(4):1081–1087. doi: 10.1016/j.juro.2014.04.013
51. Van Den Eeden SK, Lu R, Zhang N, et al. A biopsy-based 17-gene genomic prostate score as a predictor of metastases and prostate cancer death in surgically treated men with clinically localized disease. Eur Urol. 2018;73(1):129–138. doi: 10.1016/j.eururo.2017.09.013
52. Klein EA, Cooperberg MR, Magi-Galluzzi C, et al. A 17-gene assay to predict prostate cancer aggressiveness in the context of Gleason grade heterogeneity, tumor multifocality, and biopsy undersampling. Eur Urol. 2014;66(3):550–560. doi: 10.1016/j.eururo.2014.05.004
53. Shipitsin M, Small C, Choudhury S, et al. Identification of proteomic biomarkers predicting prostate cancer aggressiveness and lethality despite biopsy-sampling error. Br J Cancer. 2014;111(6):1201–1212. doi: 10.1038/bjc.2014.396
54. Blume-Jensen P, Berman DM, Rimm DL, et al. Development and clinical validation of an in situ biopsy-based multimarker assay for risk stratification in prostate cancer. Clin Cancer Res. 2015;21(11):2591–2600. doi: 10.1158/1078-0432.CCR-14-2603
55. Karnes RJ, Bergstralh EJ, Davicioni E, et al. Validation of a genomic classifier that predicts metastasis following radical prostatectomy in an at risk patient population. J Urol. 2013;190(6):2047–2053. doi: 10.1016/j.juro.2013.06.017
56. Mohler JL, Armstrong AJ, Bahnson RR, et al. Prostate cancer, Version 1.2016. J Natl Compr Canc Netw. 2016;14(1):19–30. doi: 10.6004/jnccn.2016.0004
57. Den RB, Yousefi K, Trabulsi EJ, et al. Genomic classifier identifies men with adverse pathology after radical prostatectomy who benefit from adjuvant radiation therapy [published correction appears in J Clin Oncol. 2015 Apr 20;33(12):1416]. J Clin Oncol. 2015;33(8):944–951. doi: 10.1200/JCO.2014.59.0026
58. Klein EA, Haddad Z, Yousefi K, et al. Decipher genomic classifier measured on prostate biopsy predicts metastasis risk. Urology. 2016;90:148–152. doi: 10.1016/j.urology.2016.01.012
59. Ross AE, Feng FY, Ghadessi M, et al. A genomic classifier predicting metastatic disease progression in men with biochemical recurrence after prostatectomy. Prostate Cancer Prostatic Dis. 2014;17(1):64–69. doi: 10.1038/pcan.2013.49

15 Developments and Trends in Cancer Research and Healthcare Interventions

Monika Kadian,[1] *Garima Sharma,*[1] *Sandeep Kaur,*[2] *and Anil Kumar*[1]*
[1] Pharmacology Division, University Institute of Pharmaceutical Sciences, UGC Centre of Advanced Study (UGC-CAS), Panjab University, Chandigarh, India
[2] Department of Zoology, Panjab University, Chandigarh, India

15.1 INTRODUCTION

Cancer is unregulated cell division that results in a malignant growth (or tumor). It is epidemic in nature (Sajjad et al., 2021; Bekisz and Geris, 2020). According to recent statistics data, human malignant growth is becoming the leading cause of death worldwide (Sajjad et al., 2021) and accounts for about 8 million fatalities per year worldwide (Bekisz and Geris, 2020). Cancer is becoming more prevalent and is now considered to be epidemic in scope. In 2018, there were more than 18 million new instances of cancer reported globally, and 9.6 million people died of cancer (WHO, 2018). Due to population increase and aging, unfortunately this cancer burden is projected to remain unchanged in the future. Researching the growth, comprehending the invasion, and monitoring the spread of cancer tumors has been significantly hampered in recent years by the lack of in-depth knowledge of cancer biology (Vargo-Gogola and Rosen, 2007; Wang et al., 2014).

Oncology research, like research on other diseases, is extremely dependent on a valid and adequate model paradigm, but cancer is not uniform; rather, it is a heterogeneous system with a wide range of characteristics, so selecting the model that best represents the given tumor system is one of the key challenges for cancer investigation (Breitenbach and Hoffmann, 2018). Even though our understanding of this class of illnesses is constantly expanding, certain cancers are still fatal and untreatable while others are treatable to some extent, and patients can live for years longer. Strategies for treating these malignant variations are frequently futile because of their complexity and heterogeneity (Stanta and Bonin, 2018).

Both naturally occurring and artificial cancer models have characteristics similar to actual tumors. *In vitro* cancer models are incapable of completely replicating the diversity of human cancer cells, their microenvironment, and the stromal compartment, which hampers the in-depth comprehension of tumor progression, therapeutic responses, and adverse effects (Voskoglou-Nomikos et al., 2003). Tumor cell lines, 3D model organoids, and species like *Drosophila melanogaster*, zebrafish, transgenic mouse model, pigs, patient-derived xenografts, and computational cancer models are some of the systems employed for investigating human cancer. These models serve as the foundation for research into the biochemical and genetic processes and pathogenesis of cancer. The cumulative data from cancer models aids in a more thorough knowledge of the nuances of cancer development (Schachtschneider et al., 2017).

In order to achieve a proven response to treatment and to use clinically appropriate dosages of curative drugs to evaluate survival, it is necessary to obtain roughly a 50% delay in tumor progression. Furthermore, it is crucial to determine whether tumor regeneration occurs after the medicine is stopped. If this is the case, it is also important to determine whether the regrowth is quick when therapy is delayed compared with before treatment started. All cancer models attempt to replicate at least some aspects of human cancer, but since no model is flawless, it must be determined how to access our data while staying inside the confines of the test's constraints (Richmond and Su, 2008).

15.2 HISTORICAL CANCER MODELING: HOW DID IT START?

The initial phase of cancer modeling is to identify the problem or an issue, for instance the symptoms of scrotal cancer to uniquely afflict English chimney sweepers by Percivall Pott in the 18th century (Brown and Thornton, 1957). Pott observed multiple risk factors for cancer based on certain aspects of the chimney sweeping work profile. Specifically, chimney soot residue accumulated on the workers' lower scrotums, where Pott also observed cancerous lesions, which led to the conclusion that some compound in chimney residues can induce skin cancer.

Once the problem is identified, the next step is to build the models. This step is divided into three parts: choosing an experimental structure that is physiologically and pathologically pertinent to the research, identifying the system's propelling and capping factors, and assessing the system's identifiable and/or quantifiable parameters. Followed by Pott's historical research, Yamagiwa hypothesized in the 20th century that continuous exposure to coal tar (present in chimney residue) induced metastasis in skin cells. He validated this hypothesis by applying coal tar on the ears of rabbit, which was the first animal model

* Corresponding author: kumaruips@pu.ac.in

DOI: 10.1201/9781032660592-18

published in the field of cancer studies (Yamagiwa and Ichikawa, 1918).

While testing the models, Yamagiwa established that both control and treated groups must be included in the inputs and that reproducibility is essential to the effectiveness of the experimental setup. The result should make it possible to compare any deviations from the modeled phenomenon with clinical data or results. By adjusting the amount of coal tar applied and the length of exposure, Yamagiwa was able to modify his model and determine the elements necessary for coal tar to cause cancer in rabbit models. He came to the conclusion that coal tar causes cancer and offered the first evidence in favor of chemical carcinogenesis (Yamagiwa and Ichikawa, 1918).

Depending on the hypothesis, various factors can be examined as the output for productive evaluation, such as carcinoma cases, histology, growth dynamics, or therapy response, but for the model to be useful, the modeled subject must always be clinically relevant. In Yamagiwa's investigation, he provided histological proof that the tumor in his model developed similarly to skin carcinoma in humans. Over the years, polycyclic aromatic hydrocarbons extracted from coal tar were studied in different mouse models and proved efficacious as carcinogens (Lippman and Hawk, 2009).

Although mouse models best integrate applicability and economics, there are alternative models available with varying benefits and drawbacks, such as *in vitro* yeast, worm, and fly models (Brumby and Richardson, 2005; Shoemaker, 2006; Xu and Kim, 2011). One example of modeling chemotherapy is from late 1950, when methotrexate was the go-to treatment for choriocarcinoma, but it only produced a transient response. Min Chiu Li hypothesized that even after remission, patients may still have residual illness and eventually lead to recurrence using human chorionic gonadotropin (hCG) as a tumor indicator. He kept treating patients until hCG levels returned to normal and no signs of recurrence were seen. These findings revealed that in order to avoid recurrence and decrease toxicity, cyclic treatment was necessary (DeVita and Chu, 2008; Hertz et al., 1963).

Based on these findings, Howard Skipper proposed in the 1960s that the impact of chemotherapy on tumor growth kinetics, particularly remnant disease, impacted patient survival. He used syngeneic mice with the murine leukemia L1210 implanted intraperitoneally as a model for drug-induced cell death and used the average percentage increase in survival, net log10 cell-kill, and long-term relicts to measure the effectiveness of the treatment. By changing the model's parameters, he learned that leukemia cells grow exponentially and destroy the host once they reach the limit of tumor burden; effective therapy must completely cure any remaining disease; and chemotherapy kills leukemia cells by a constant fraction rather than by a constant number, a response known as log-kill (Skipper, 1965; Skipper et al., 1964; Skipper and Perry, 1970). Following clinical findings, these principles were subsequently verified in human leukemias as well (Berenbaum, 1972). These results provided one of the pillars of conventional chemotherapy for cancer therapy.

15.3 WHAT MAKES AN EFFECTIVE CANCER MODEL?

Cancer is a complicated disease. Each cell's high-dimensional nonlinear dynamic system interacts with countless of neighboring cells and across the body via many chemical and physical pathways. Cancer cells differ genetically from one another, which affects and complicates intricate connections between them and host cells of all kinds, especially in advanced adult-onset malignancies. This poses significant hurdles for both cancer studies and clinical therapeutics. Therapies aim to maximize patient survival by eliminating this entire complex system when possible and controlling it indefinitely when not. Any effective strategy must take into account disease's extreme complexity between and among individuals. Because of this heterogenicity, typical exploratory experimental procedures are impractical in terms of both time and resources. To select and identify relevant hypotheses for experimental and clinical testing, we must use mathematical and computational tools.

Because cells and molecules can communicate in unanticipated manners, mathematics gives a rational tool for organizing cognition about this unexpected heterogenicity, and it can produce hypotheses or make projections that intuition alone cannot. Efforts are currently underway to control cancer through the use of models that capture the core aspects of cancer heterogeneity while combining useful precision with appropriate ease, although it is necessary to determine how we strike a balance.

Further, mathematical modeling's value lies in its ability to reveal these ideas and put them to the test by comparing results with projected outcomes and experiments. As a result, in order for a biological or mathematical model's hypotheses to be evaluated and challenged, it is crucial to identify and publicly describe as many of them as possible. If anticipated outcomes match observations, additional predictions may be generated to drive future experiments; if hypotheses do not match data, a gap in knowledge has been recognized, and assumptions must be addressed. The variables to be included (state variables) and the guidelines that control how they change and interact with one another are both significant topics covered by the hypotheses (Beckman et al., 2020).

Furthermore, descriptive and conceptual methodologies are available for mathematical modeling of biological processes. Descriptive models, which are similarly crucial for fundamental research, aim to account for every component of the foundation biology and ideally produce a virtual replication of the relevant biological system. Researchers used these models with some effectiveness in preclinical drug research in SimCyp (Jamei et al., 2009; Abduljalil et al., 2016) and used Gastroplus to predict drug metabolism throughout the course of drug development (Gobeau et al., 2016; Naylor et al., 2006).

15.3.1 Conceptual Models

Conceptual models concentrate on a small number of biological systems and are essential for fundamental research. For instance, PKPD (pharmacokinetic/pharmacodynamic) models track the dynamics of a drug's concentration across various bodily compartments (the PK component) as well as the effects of the medication on cells (the PD part). The PK component mandates that compartments be chosen, such as particular organ systems, and that rules for drug transit, degradation, and excretion be established. Based on binding and response dynamics, the PD component tracks alterations in cells or tissues, such as altered protein levels or drug-dependent alterations in tumor growth, mortality, mutation, and metastasis (Beckman et al., 2020).

15.3.2 Agent-Based Models

Agent-based models (ABMs) can be conceptual or descriptive: To anticipate long-term dynamics with or without therapy, a conceptual ABM comprises few cell types and cell communications (Norton et al., 2017; Wu et al., 2019), whereas a descriptive ABM attempts to encompass all known cell types and their interactions (Gong et al., 2017; Norton et al., 2017).

The main distinction between equation- and agent-based models is that with equation-based models, a clear mathematical explanation of how variables interact with each other is provided, and then numeric answers are created to these equations that represent the phenomena of interest. ABMs execute computer simulations that track each individual cell and define the rules by which agents such as cells communicate; they enable simulations to include spatial components along with high chance variation around average behavior. Because outcome is treated as experimental data, it must be reproduced in order to gain enough statistical power. As such, ABMs act as a bridge between equation-based models and research lab trials (Beckman et al., 2020).

15.3.3 Deterministic Models

Deterministic models consistently produce the same result each time; in contrast, stochastic models include random number generation and represent chance fluctuations around the average behavior, providing varied outputs in repeat runs. As with any modeling choice, there is a trade-off between the mathematical simplicity and viability of deterministic models and the greater complexity and potential for realism of stochastic models. Small groupings of cells during the beginning of carcinogenesis are an example of a smaller system in which chance mutations have a greater impact than they do in bigger systems (Beckman et al., 2020).

15.3.4 Machine Learning

Machine learning, ML, offers a profoundly different quantitative method for data comprehension. Rather than defining specific mechanisms, ML algorithms use data to discover links and uncover correlations. Within ML, there are two main approaches: supervised and unsupervised. Supervised learning included sorting data into already-defined categories; the hypothesis is applied on data for which an answer key is provided, with the purpose of automated data classification that lacks such set of responses into proper categories.

Assessing the characteristics of drugs that became effective pharmaceuticals for forecasting valuable novel compounds at the preliminary phase is an example of supervised ML in drug discovery; this process sorts results into successful or unsuccessful drugs). Unsupervised ML, on the other hand, entails discovering data categories, for instance exploring the typical qualities of successful drugs. ML algorithms, for example, can be incredibly effective in selecting patients who will react well to a certain medication, but they can only do so currently within the spectrum of current patients. Mechanistic models offer two significant advantages. First, they allow for linking errors from predictions to specific assumptions that can be addressed, resulting in better prediction as well as deeper knowledge. Second, proven models can be expanded outside the range of known data to create further testable predictions (Beckman et al., 2020).

15.4 WHAT CAN MODELS ADDRESS?

15.4.1 Tumor Growth Models

The tumor itself was the earliest measurable quantity, of course, and the most fundamental models focus on describing the dynamics of tumor formation. The dynamics are effectively represented by logistic or Gompertz designs, although controversy about which one provides a more accurate mechanistic justification over the dynamics keeps going. Tumor growth curves are generally shaped by *in vitro* and *in vivo* studies, displaying the initial exponential expansion that over time either decreases or saturates as the overall population hits certain thresholds (Benzekry et al., 2014; Kareva and Karev, 2018).

Further, the Simeoni design accurately explains tumor progression in xenograft models in pharmaceutical modeling. This model forecasts early expansion at an exponential rate followed by gradual expansion (Simeoni et al., 2004, 2013), and the plateau phase is frequently not reached *in vivo* and so is not required. Large lesions in humans are frequently surgically removed or irradiated if present in isolation, whereas mouse models include one tumor that continues to develop until an animal is euthanized (thus deliberately linking surviving to the development of a single large lesion). Death in the metastatic phase is typically associated with many micrometastases contaminating the crucial organs. Given their small size, these micrometastases are most likely still growing in an exponential fashion. Radiological data from human studies also includes huge lesions apparent in CT, and these, like large lesions in mouse models, exhibit distinct growth qualities from those in common tiny tumors that are expected to have a greater influence on survival. Because oxygen and nutrients easily diffuse in, they are less likely to exhibit comparable dynamics among tumor cells (Beckman and Loeb, 2020).

that although all medications were capable of directly killing cancer cells, only one crossed the *in vitro* blood–brain barrier and killed tumor cells in an *in vivo*-like scenario. This study supports the use of organs on chips to replicate each organ's distinct microenvironmental settings for improved drug screening.

15.5.4.4 Organoids

Finally, organoids are a new and promising *in vitro* cancer research approach. Organoids are 3D constructions made from embryonic stem cells, induced pluripotent stem cells, and somatic stem cells (Xu et al., 2018). These cells can develop into a variety of cell types, organizing them similarly to their *in vivo* arrangements and exhibiting organ-specific activities (Lancaster and Knoblich, 2014). Sato et al. (2009) made the first breakthrough in 2009 when they cultivated an organoid of the intestines in 3D. In brief, they were able to develop 3D mice crypt architecture for eight months, resulting in villi-like epithelial structures including all differentiated cell types reported *in vivo*. Researchers have effectively replicated organoids such as the liver and stomach that mimic the genotype, phenotype, and cellular activities of parent tissues (Xu et al., 2018), and their normal genomes remain stable throughout time (Weeber et al., 2017). Along with ability to produce organoids, malignant cells can also be used to create models of healthy tissues in order to investigate diseased tissues.

Disease-free and malignant organoids are continuously being used in *in vitro* cancer research. Drost et al. (2017), for example, created disease-free colon organoids as models to determine the genesis of mutation markers in cancer. They used CRISPR/Cas9 to knock down a mismatch DNA repair gene that has been linked to colorectal cancer in the organoids. This enabled them to investigate the impact of a single interruption of the DNA repair gene in an otherwise healthy genetic background. They discovered that deleting this gene caused a mutation in the organoids similar to that seen in mismatch repair-deficient colorectal cancer following subcloning and two-month passage of the organoids. Drost et al. studied the lesser-known involvement of base excision repair genes in cancer formation using this unique technique to track the origin of mutations and found that altering one of these genes results in a mutational signature similar to that reported in breast cancer. Thus, organoids can be utilized to unravel cancer formation and rigorously investigate the involvement of particular genes in disease onset. Organoids are also employed to learn more about how microorganisms cause cancer beginning in a predisposed subset of patients in India. First, Scanu et al. (2015) created gallbladder organoids with a mutational landscape to replicate these patients' genetic propensity. Second, they infected gallbladder organoids with *Salmonella*, which resulted in neoplastic cellular transformation along with aberrant development in settings with limited growth factors. Table 15.3 compares cancer models.

TABLE 15.3

Advantages and Disadvantages of Different Cancer Models

Cancer Model	Cancer Type	Strengths	Weaknesses
Pig	Adenomatous polyposis, spontaneous osteosarcomas, leukemia, lymphoma, soft tissue, pancreatic ductal adenocarcinoma, HCC	Lower price, effective cancer evolution recapitulation, and effective modeling of chromosomal rearrangement displaying clinically significant histological and genotypic tumor characteristics	Inability to demonstrate tumor–stroma interaction, inadequate for immune integration, higher housing requirements, prolonged intervals of generation, biosafety concerns
Organoid	Prostate cancer, pancreatic cancer, colorectal cancer, breast cancer	Sustain the tumor's expressing pattern, including copy number alterations, transcriptional landscape, and mutation status In the organoid cancer model, genetic stability lasted a long time It is possible to investigate the tumor microenvironment	Drug sensitivity, gene expression, and signaling pathways are all negatively affected by growth enhancers and suppressors in patient-derived organoids
Cancer cell line	Chronic myeloid leukemia, prostate adenocarcinoma, malignant melanoma	Convenient maintenance, low cost, immortal, reduced heterogeneity at cellular levels, and high growth rates Similar to a human tumor in terms of gene expression, patterns and CNA	Cross-contamination, mislabeling, antiproliferative medication proliferation, absent in stromal components Serial transmission resulted in genotypic and phenotypic diversity
Drosophila	Multiple endocrine neoplasia Type 2	Asymmetric division, centrosome malfunctioning, instability in genome, drug metabolism, and unscheduled genetic expression are investigated Genetic resemblance with humans, centrosomal anomalies, ease of care, and a distinct lymphatic system from that in humans	When compared with mammals, rudimentary hematopoietic systems and distinct lymphatic systems inhibited tumor spread potential

(Continued)

TABLE 15.3
(Continued)

Cancer Model	Cancer Type	Strengths	Weaknesses
Genetically engineered mouse model	Lymphomas, osteosarcomas, and hemangiosarcoma, pancreatic cancer, breast cancer, and prostate cancer	Cost efficacy along with enhanced precision for initial diagnosis after exposing to genotoxic and nongenotoxic carcinogens; tumor formation in different organs/tissues allows for targeted tissue development with a lower chance of false-positive results, and genetic mutations can be analyzed for driving tumor-initiating and signaling pathways	Only malignancies driven by certain signaling pathways can be produced; drug safety and efficacy need periodic dose–response evaluation; and increased biosafety issues due to possible susceptibility to grow a tumor in acute durations
Patient-derived xenograft	Ovarian cancer, colorectal cancer, prostate cancer, gastric cancer, renal cell carcinoma, non-small cell lung cancer	The tumor's heterogeneity is preserved, allowing for histological and genetic characterization; tailored therapy with effectual agent response monitoring; and simulation of the physiological, hormonal, and oxygen circumstances in the patient's primary tumor-originating site using enlarged PDXs Co-clinical research, drug screening and biomarker, and precision medicine development	When standard PDX is employed, metastatic implant is impossible; human tumor microenvironment communications aren't well documented; and forming PDX is tedious and highly demanding Engraftment rejection, long development time, and cost; immune-modulating research is challenging
Zebrafish	HCC, lung adenocarcinoma, non-small cell lung carcinoma, liver cancer, breast cancer, melanoma and squamous cell carcinoma, skin cancer	Characterization and visualization of a single cancerous cell, tumor detection, diagnosis, followed by preclinical research conducted early on, the testing system is established through positive controls, tumorigenesis, and the impact of inflammation were convincingly demonstrated Cell imaging is used for high-throughput drug screening, and genes involved with lineage-specific pathways can be analyzed	Rejection of a tumor transplant can reveal heterogeneity and tumor evolution; strategies used to prevent engraftment rejection may lower survival rates; specialized equipment is needed; and invasive drug administration techniques, which can impair drug absorption, metabolism, and excretion, are rarely employed
Computational cancer models	Lung cancer, breast cancer, pancreatic cancer	Starting from any sample, unbiased examination of the DNA, RNA, and protein landscape. Data creation is both quick and robust. Data repositories are created that can be utilized for future research or providing validity by other researchers. New biomarkers, drivers, and therapeutic targets were identified Specific mutations associated with medication responsiveness were identified To create a model for toxicity pathway evaluation, molecular techniques were used Many chemical and physical characteristics of molecules that are related to their environmental fate and movement were determined	High costs for handling sample and beginning amount, instrumentation, and data analysis and integration time The lack of coherence amongst -omics methods (for example, genomics, transcriptomics, proteomics, and metabolomics) Single-cell analysis has immense promise but is still in its early stages Tumors are diverse, therefore -omics data from one biopsy site might not be indicative of the complete tumor

15.6 CONCLUSION

Ultimately, the goal of any model is to provide outcomes that exhibit beneficial effect on patient wellbeing. In preclinical testing, the combined use of *in vitro*, *in vivo*, and computer models enhances the cost effectiveness of medication development and approval. The higher growth rate and simplicity of care of tumor cell lines provide it a benefit over the pig's lengthier generation period and massive housing requirements. Tumor graft successfully mimics the tumor *in vivo*, which cell lines are unable to replicate accurately. Organoids provide a novel cancer model for studying tumor heterogeneity and microenvironment. Human cancer mouse models are important resources for oncology study. Patient-derived xenografts (PDXs) are easy to use, repeatable, and reasonably priced, but they have the disadvantage that unlike genetically engineered mouse models (GEMMs), they don't imitate the genetics and histology of living tissue tumors because of the absence of a full immune system.

This limitation is surpassed by distinct GEMMs strains designed to generate cancer that mimics the phenotypic, biochemical, and genetic aspects of certain human cancers. In the future, GEMMs are extremely promising for assessing targeted therapies. Although *Drosophila* has explained the molecular basis of carcinogenesis due to genetic similarities with *Homo sapiens*, morphological and physiological distinctions with humans have restricted its usage. Oncopig, which shows genetic, epigenomic, and chromosomal homology with humans, has bypassed this shortcoming. Zebrafish have numerous advantages over standard cell lines; they are used in the laboratory for a variety of purposes since they produce accurate findings for a range of phenotypes, although they do have some limitations. Zebrafish could be a great model for predicting novel biological markers if accurate procedures and methodologies are applied.

Computational models alleviate the shortcomings of *in vitro* and *in vivo* models. Mechanistic computer models are created using -omics data from patients and are used to analyze treatment effectiveness and side effects. However, the current computational model doesn't adequately replicate the intricacy of the human tumor under development. Computational models improve understanding of molecular changes in disease-related pathways and enable competent prescreening for essential patient selection. It has the potential to increase knowledge of cancer evolution and response against drug treatment. Cancer models, whether *in vitro*, *in vivo*, or computational, allow us to undertake experiments on patients that would be difficult due to economic, moral, and welfare issues. Patients benefit from our temporarily or permanently collecting data and information from these models.

REFERENCES

Abduljalil K, Edwards D, Barnett A, Rose RH, Cain T, Jamei M (2016) A tutorial on pharmacodynamic scripting facility in Simcyp. CPT Pharmacometrics Syst Pharmacol 5(9): 455–465. doi: 10.1002/psp4.12102. Epub 2016 Aug 29. PMID: 27393710; PMCID: PMC5036420.

Basanta D, Gatenby RA, Anderson AR (2012) Exploiting evolution to treat drug resistance: combination therapy and the double bind. Mol Pharm 9(4):914–921. doi:10.1021/mp200458e.

Beckman RA (2010) Efficiency of carcinogenesis: is the mutator phenotype inevitable? Semin Cancer Biol 20(5):340–352. doi: 10.1016/j.semcancer.2010.10.004. https://pubmed.ncbi.nlm.nih.gov/20934514/

Beckman RA, Kareva I, Adler FR (2020) How should cancer models be constructed? Cancer Control 27(1):1073274820962008. doi: 10.1177/1073274820962008. https://www.ncbi.nlm.nih.gov/pmc/articles/PMC7791451/

Beckman RA, Loeb LA (2006) Efficiency of carcinogenesis with and without a mutator mutation. Proc Natl Acad Sci USA 103(38):14140–14145. doi: 10.1073/pnas.0606271103. https://www.pnas.org/doi/10.1073/pnas.0606271103

Beckman RA, Loeb LA (2020) Rare mutations in cancer drug resistance and implications for therapy. Clin Pharmacol Ther 108(3):437–439. doi: 10.1002/cpt.1938. https://www.ncbi.nlm.nih.gov/pmc/articles/PMC7484911/

Beckman RA, Schemmann GS, Yeang CH (2012) Impact of genetic dynamics and single-cell heterogeneity on development of nonstandard personalized medicine strategies for cancer. Proc Natl Acad Sci USA 109(36):14586–14591. doi: 10.1073/pnas.1203559109. https://pubmed.ncbi.nlm.nih.gov/22891318/

Beckman RA, Weiner LM, Davis HM (2007) Antibody constructs in cancer therapy: protein engineering strategies to improve exposure in solid tumors. Cancer 109(2):170–179. doi: 10.1002/cncr.22402. https://pubmed.ncbi.nlm.nih.gov/17154393/

Beckman RA, Yeang CH (2014) Nonstandard personalized medicine strategies for cancer may lead to improved patient outcomes. Per Med 11(7):705–719. doi: 10.2217/pme.14.57. https://pubmed.ncbi.nlm.nih.gov/29764056/

Bekisz S, Geris L (2020) Cancer modeling: from mechanistic to data-driven approaches, and from fundamental insights to clinical applications. J Comput Sci 46:101198.

Benzekry S, Lamont C, Beheshti A, Tracz A, Ebos JM, Hlatky L, Hahnfeldt P (2014) Classical mathematical models for description and prediction of experimental tumor growth. PLoS Comput Biol 10(8):e1003800. doi: 10.1371/journal.pcbi.1003800. https://pubmed.ncbi.nlm.nih.gov/25167199/

Berenbaum MC (1972) In vivo determination of the fractional kill of human tumor cells by chemotherapeutic agents. Cancer Chemother Rep 56(5):563–571. https://pubmed.ncbi.nlm.nih.gov/4652587/

Bersini S, Jeon JS, Dubini G, Arrigoni C, Chung S, Charest JL, Moretti M, Kamm RD (2014) A microfluidic 3D in vitro model for specificity of breast cancer metastasis to bone. Biomaterials 35(8):2454–2461. doi: 10.1016/j.biomaterials.2013.11.050. https://pubmed.ncbi.nlm.nih.gov/24388382/

Breitenbach M, Hoffmann J (2018) Editorial: cancer models. Front Oncol 8:401. doi: 10.3389/fonc.2018.00401. https://pubmed.ncbi.nlm.nih.gov/30338241/

Brown JR, Thornton JL (1957) Percivall Pott (1714–1788) and chimney sweepers' cancer of the scrotum. Br J Ind Med 14(1):68–70. doi: 10.1136/oem.14.1.68. https://www.ncbi.nlm.nih.gov/pmc/articles/PMC1037746/

Brumby AM, Richardson HE (2005) Using Drosophila melanogaster to map human cancer pathways. Nat Rev Cancer 5(8):626–639. doi: 10.1038/nrc1671. https://pubmed.ncbi.nlm.nih.gov/16034367/

Campbell BB, Light N, Fabrizio D, Zatzman M, Fuligni F, de Borja R, Davidson S, Edwards M, Elvin JA, Hodel KP, Zahurancik WJ, Suo Z, Lipman T, Wimmer K, Kratz CP, Bowers DC, et al. (2017) Comprehensive analysis of hypermutation in human cancer. Cell 171(5):1042–1056.e10. doi: 10.1016/j.cell.2017.09.048. https://pubmed.ncbi.nlm.nih.gov/29056344/

Chandarlapaty S, Sawai A, Scaltriti M, Rodrik-Outmezguine V, Grbovic-Huezo O, Serra V, Majumder PK, Baselga J, Rosen N (2011) AKT inhibition relieves feedback suppression of receptor tyrosine kinase expression and activity. Cancer Cell 19(1):58–71. doi: 10.1016/j.ccr.2010.10.031. https://pubmed.ncbi.nlm.nih.gov/21215704/

Chelliah V, Lazarou G, Bhatnagar S, Gibbs JP, Nijsen M, Ray A, Stoll B, Thompson RA, Gulati A, Soukharev S, Yamada A, Weddell J, Sayama H, Oishi M, Wittemer-Rump S, Patel C, et al. (2021) Quantitative systems pharmacology approaches for immuno-oncology: adding virtual patients to the development paradigm. Clin Pharmacol Ther 109(3):605–618. doi: 10.1002/cpt.1987. https://pubmed.ncbi.nlm.nih.gov/32686076/

Chen Z, He S, Zilberberg J, Lee W (2019) Pumpless platform for high-throughput dynamic multicellular culture and chemosensitivity evaluation. Lab Chip 19(2):254–261. doi: 10.1039/c8lc00872h. https://www.ncbi.nlm.nih.gov/pmc/articles/PMC6333476/

DeVita VT Jr, Chu E (2008) A history of cancer chemotherapy. Cancer Res 68(21):8643–8653. doi: 10.1158/0008-5472.CAN-07-6611. https://pubmed.ncbi.nlm.nih.gov/18974103/

Domcke S, Sinha R, Levine DA, Sander C, Schultz N (2013) Evaluating cell lines as tumour models by comparison of genomic profiles. Nat Commun 4:2126. doi: 10.1038/ncomms3126. https://pubmed.ncbi.nlm.nih.gov/23839242/

Drost J, van Boxtel R, Blokzijl F, Mizutani T, Sasaki N, Sasselli V, de Ligt J, Behjati S, Grolleman JE, van Wezel T, Nik-Zainal S, Kuiper RP, Cuppen E, Clevers H (2017) Use of CRISPR-modified human stem cell organoids to study the origin of mutational signatures in cancer. Science 358(6360):234–238. doi: 10.1126/science.aao3130. https://pubmed.ncbi.nlm.nih.gov/28912133/

Du Q, Bian XL, Xu XL, Zhu B, Yu B, Zhai Q (2013) Role of mitochondrial permeability transition in human hepatocellular carcinoma Hep-G2 cell death induced by rhein. Fitoterapia 91:68–73. doi: 10.1016/j.fitote.2013.08.008. https://pubmed.ncbi.nlm.nih.gov/23994628/

Ehsan SM, Welch-Reardon KM, Waterman ML, Hughes CC, George SC (2014) A three-dimensional in vitro model of tumor cell intravasation. Integr Biol (Camb) 6(6):603–610. doi: 10.1039/c3ib40170g. https://pubmed.ncbi.nlm.nih.gov/24763498/

Erkan M, Reiser-Erkan C, Michalski CW, Kleeff J (2010) Tumor microenvironment and progression of pancreatic cancer. Exp Oncol 32(3):128–131. https://pubmed.ncbi.nlm.nih.gov/21403605/

Ferreira D, Adega F, Chaves R (2013) The Importance of Cancer Cell Lines as in Vitro Models in Cancer Methylome Analysis and Anticancer Drugs Testing. In: Lopez-Camarillo C, editor. Oncogenomics and Cancer Proteomics – Novel Approaches in Biomarkers Discovery and Therapeutic Targets in Cancer. InTech; London, pp. 139–166. doi: 10.5772/1745

Gatenby RA, Silva AS, Gillies RJ, Frieden BR (2009) Adaptive therapy. Cancer Res 69(11):4894–4903. doi: 10.1158/0008-5472.CAN-08-3658. https://pubmed.ncbi.nlm.nih.gov/19487300/

Gerlowski LE, Jain RK (1983) Physiologically based pharmacokinetic modeling: principles and applications. J Pharm Sci 72(10):1103–1127. doi: 10.1002/jps.2600721003. https://pubmed.ncbi.nlm.nih.gov/6358460/

Ghajar CM, Peinado H, Mori H, Matei IR, Evason KJ, Brazier H, Almeida D, Koller A, Hajjar KA, Stainier DY, Chen EI, Lyden D, Bissell MJ (2013) The perivascular niche regulates breast tumour dormancy. Nat Cell Biol 15(7):807–817. doi: 10.1038/ncb2767. https://www.ncbi.nlm.nih.gov/pmc/articles/PMC3826912/

Gobeau N, Stringer R, De Buck S, Tuntland T, Faller B (2016) Evaluation of the GastroPlus™ Advanced Compartmental and Transit (ACAT) model in early discovery. Pharm Res 33(9):2126–2139. doi:10.1007/s11095-016-1951-z.

Gong C, Milberg O, Wang B, Vicini P, Narwal R, Roskos L, Popel AS (2017) A computational multiscale agent-based model for simulating spatio-temporal tumour immune response to PD1 and PDL1 inhibition. J R Soc Interface 14(134):20170320. doi: 10.1098/rsif.2017.0320. https://pubmed.ncbi.nlm.nih.gov/28931635/

Goodspeed A, Heiser LM, Gray JW, Costello JC (2016) Tumor-derived cell lines as molecular models of cancer pharmacogenomics. Mol Cancer Res 14(1):3–13. doi: 10.1158/1541-7786.MCR-15-0189. https://pubmed.ncbi.nlm.nih.gov/26248648/

Haase K, Kamm RD (2017) Advances in on-chip vascularization. Regen Med 12(3):285–302. doi:10.2217/rme-2016-0152.

Helmlinger G, Sokolov V, Peskov K, Hallow KM, Kosinsky Y, Voronova V, Chu L, Yakovleva T, Azarov I, Kaschek D, Dolgun A, Schmidt H, Boulton DW, Penland RC (2019) Quantitative systems pharmacology: an exemplar model-building workflow with applications in cardiovascular, metabolic, and oncology drug development. CPT Pharmacometrics Syst Pharmacol 8(6):380–395. doi: 10.1002/psp4.12426. https://pubmed.ncbi.nlm.nih.gov/31087533/

Hertz R, Ross GT, Lipsett MB (1963) Primary chemotherapy of nonmetastatic trophoblastic disease in women. Am J Obstet Gynecol 86:808–814. doi: 10.1016/s0002-9378(16)35198-5. https://pubmed.ncbi.nlm.nih.gov/13954160/

Hu R, Kim BR, Chen C, Hebbar V, Kong AN (2003) The roles of JNK and apoptotic signaling pathways in PEITC-mediated responses in human HT-29 colon adenocarcinoma cells. Carcinogenesis 24(8):1361–1367. doi: 10.1093/carcin/bgg092. https://pubmed.ncbi.nlm.nih.gov/12819185/

Huang XH, Zheng QS (2010) Pharmacokinetic and pharmacodynamic data analysis: concepts and applications. Am J Pharm Educ 74(3):53b. https://www.ncbi.nlm.nih.gov/pmc/articles/PMC2865421/

Huh D, Matthews BD, Mammoto A, Montoya-Zavala M, Hsin HY, Ingber DE (2010) Reconstituting organ-level lung functions on a chip. Science 328(5986):1662–1668. doi: 10.1126/science.1188302. https://pubmed.ncbi.nlm.nih.gov/20576885/

Ito M, Ishii G, Nagai K, Maeda R, Nakano Y, Ochiai A (2012) Prognostic impact of cancer-associated stromal cells in patients with stage I lung adenocarcinoma. Chest 142(1):151–158. doi: 10.1378/chest.11-2458. https://pubmed.ncbi.nlm.nih.gov/22302300/

Jamei M, Turner D, Yang J, et al (2009) Population-based mechanistic prediction of oral drug absorption. AAPS J 11(2):225–237. doi:10.1208/s12248-009-9099-y.

Kareva I, Berezovskaya F (2015) Cancer immunoediting: a process driven by metabolic competition as a predator–prey–shared resource type model. J Theor Biol 380:463–472. https://doi.org/10.1016/j.jtbi.2015.06.007 https://europepmc.org/article/med/26116366

Kareva I, Berezovskaya F, Castillo-Chavez C (2010) Myeloid cells in tumor–immune interactions. J Biol Dyn 4(4):315–327. doi: 10.1080/17513750903261281. https://www.tandfonline.com/doi/full/10.1080/17513750903261281

Kareva I, Karev G (2018) From experiment to theory: what can we learn from growth curves? Bull Math Biol 80(1):151–174. doi: 10.1007/s11538-017-0347-5. https://pubmed.ncbi.nlm.nih.gov/29150759/

Kim HJ, Huh D, Hamilton G, Ingber DE (2012) Human gut-on-a-chip inhabited by microbial flora that experiences intestinal peristalsis-like motions and flow. Lab Chip 12(12):2165–2174. doi: 10.1039/c2lc40074j. https://pubmed.ncbi.nlm.nih.gov/22434367/

Kim S, Lee H, Chung M, Jeon NL (2013) Engineering of functional, perfusable 3D microvascular networks on a chip. Lab Chip 13(8):1489–1500. doi: 10.1039/c3lc41320a. https://pubmed.ncbi.nlm.nih.gov/23440068/

Knudsen ES, Balaji U, Mannakee B, Vail P, Eslinger C, Moxom C, Mansour J, Witkiewicz AK (2018) Pancreatic cancer cell lines as patient-derived avatars: genetic characterisation and functional utility. Gut 67(3):508–520. doi: 10.1136/gutjnl-2016-313133. https://www.ncbi.nlm.nih.gov/pmc/articles/PMC5868284/

Koparal AT, Zeytinoglu M (2003) Effects of carvacrol on a human Non-Small Cell Lung Cancer (NSCLC) cell line, A549. Cytotechnology 43(1–3):149–154. doi: 10.1023/b:cyto.0000039917.60348.45. https://pubmed.ncbi.nlm.nih.gov/19003220/

Lancaster MA, Knoblich JA (2014) Organogenesis in a dish: modeling development and disease using organoid technologies. Science 345(6194):1247125. doi: 10.1126/science.1247125. https://pubmed.ncbi.nlm.nih.gov/25035496/

Li XM, Luo XG, He JF, Wang N, Zhou H, Yang PL, Zhang TC (2018) Induction of apoptosis in human cervical carcinoma HeLa cells by active compounds from Hypericum ascyron L. Oncol Lett 15(3):3944–3950. doi: 10.3892/ol.2018.7812. https://pubmed.ncbi.nlm.nih.gov/29556280/

Lippman SM, Hawk ET (2009) Cancer prevention: from 1727 to milestones of the past 100 years. Cancer Res 69(13):5269–5284. https://pubmed.ncbi.nlm.nih.gov/19491253/

Lloyd MC, Alfarouk KO, Verduzco D, Bui MM, Gillies RJ, Ibrahim ME, Brown JS, Gatenby RA (2014) Vascular measurements correlate with estrogen receptor status. BMC Cancer 14:279. https://pubmed.ncbi.nlm.nih.gov/24755315/

Loeb LA, Kohrn BF, Loubet-Senear KJ, Dunn YJ, Ahn EH, O'Sullivan JN, Salk JJ, Bronner MP, Beckman RA (2019) Extensive subclonal mutational diversity in human colorectal cancer and its significance. Proc Natl Acad Sci 116(52):26863–26872. https://pubmed.ncbi.nlm.nih.gov/31806761/

Mabrey FL, Chien SS, Martins TS, Annis J, Sekizaki TS, Dai J, Beckman RA, Loeb LA, Carson A, Patay B, Blau CA (2018) High throughput drug screening of leukemia stem cells reveals resistance to standard therapies and sensitivity to other agents in acute myeloid leukemia. Blood 132:180.

Messenheimer DJ, Jensen SM, Afentoulis ME, Wegmann KW, Feng Z, Friedman DJ, Gough MJ, Urba WJ, Fox BA (2017) Timing of PD-1 blockade is critical to effective combination immunotherapy with anti-OX40. Clin Cancer Res 23(20):6165–6177. https://pubmed.ncbi.nlm.nih.gov/28855348/

Moore H (2018) How to mathematically optimize drug regimens using optimal control. J Pharmacokinet Pharmacodyn 45(1):127–137. https://www.ncbi.nlm.nih.gov/pmc/articles/PMC5847021/

Moore H, Strauss L, Ledzewicz U (2018) Optimization of combination therapy for chronic myeloid leukemia with dosing constraints. J Math Biol 77(5):1533–1561. https://pubmed.ncbi.nlm.nih.gov/29992481/

Naylor TA, Connolly PC, Martini LG, Elder DP, Minekus M, Havenaar R, Zeijdner E (2006) Use of a gastro-intestinal model and Gastroplus™ for the prediction of in vivo performance. J Appl Ther Res 6(1):15.

Norton KA, Wallace T, Pandey NB, Popel AS (2017) An agent-based model of triple-negative breast cancer: the interplay between chemokine receptor CCR5 expression, cancer stem cells, and hypoxia. BMC Syst Biol 11(1):68. https://pubmed.ncbi.nlm.nih.gov/28693495/

Orcutt KD, Adams GP, Wu AM, Silva MD, Harwell C, Hoppin J, Matsumura M, Kotsuma M, Greenberg J, Scott AM, Beckman RA (2017) Molecular simulation of receptor occupancy and tumor penetration of an antibody and smaller scaffolds: application to molecular imaging. Mol Imaging Biol 19(5):656–664. https://pubmed.ncbi.nlm.nih.gov/28213834/

Orlando PA, Gatenby RA, Brown JS (2012) Cancer treatment as a game: integrating evolutionary game theory into the optimal control of chemotherapy. Phys Biol 9(6):065007. https://pubmed.ncbi.nlm.nih.gov/23197192/

Pardoll DM (2012) The blockade of immune checkpoints in cancer immunotherapy. Nat Rev Cancer 12(4):252–264. https://pubmed.ncbi.nlm.nih.gov/22437870/

Prahallad A, Sun C, Huang S, Di Nicolantonio F, Salazar R, Zecchin D, Beijersbergen RL, Bardelli A, Bernards R (2012) Unresponsiveness of colon cancer to BRAF(V600E) inhibition through feedback activation of EGFR. Nature 483(7387):100–103. https://pubmed.ncbi.nlm.nih.gov/22281684/

Reddy M, Yang RS, Andersen ME, Clewell HJ III (2005) Physiologically Based Pharmacokinetic Modeling: Science and Applications. John Wiley & Sons.

Richmond A, Su Y (2008) Mouse xenograft models vs GEM models for human cancer therapeutics. Dis Model Mech 1(2–3):78–82. https://pubmed.ncbi.nlm.nih.gov/19048064/

Rocchetti M, Poggesi I, Germani M, Fiorentini F, Pellizzoni C, Zugnoni P, Pesenti E, Simeoni M, De Nicolao G (2005) A pharmacokinetic-pharmacodynamic model for predicting tumour growth inhibition in mice: a useful tool in oncology drug development. Basic Clin Pharmacol Toxicol 96(3):265–268. https://pubmed.ncbi.nlm.nih.gov/15733231/

Rommerskirch W, Graeber I, Grässmann M, Grässmann A (1988) Homologous recombination of SV40 DNA in COS7 cells occurs with high frequency in a gene dose independent fashion. Nucleic Acids Res 16(3):941–952. https://pubmed.ncbi.nlm.nih.gov/2830596/

Sajjad H, Imtiaz S, Noor T, Siddiqui YH, Sajjad A, Zia M (2021) Cancer models in preclinical research: a chronicle review of advancement in effective cancer research. Animal Model Exp Med 4(2):87–103. https://pubmed.ncbi.nlm.nih.gov/34179717/

Sato T, Vries RG, Snippert HJ, van de Wetering M, Barker N, Stange DE, van Es JH, Abo A, Kujala P, Peters PJ, Clevers H (2009) Single Lgr5 stem cells build crypt-villus structures in vitro without a mesenchymal niche. Nature 459(7244):262–265. https://pubmed.ncbi.nlm.nih.gov/19329995/

Scanu T, Spaapen RM, Bakker JM, Pratap CB, Wu LE, Hofland I, Broeks A, Shukla VK, Kumar M, Janssen H, Song JY, Neefjes-Borst EA, te Riele H, Holden DW, Nath G, Neefjes J (2015) Salmonella manipulation of host signaling pathways provokes cellular transformation associated with gallbladder carcinoma. Cell Host Microbe 17(6):763–774. https://pubmed.ncbi.nlm.nih.gov/26028364/

Schachtschneider KM, Schwind RM, Newson J, Kinachtchouk N, Rizko M, Mendoza-Elias N, Grippo P, Principe DR, Park A, Overgaard NH, Jungersen G, Garcia KD, Maker AV, Rund LA, Ozer H, Gaba RC, Schook LB (2017) The oncopig cancer model: an innovative large animal translational oncology platform. Front Oncol 7:190. https://pubmed.ncbi.nlm.nih.gov/28879168/

Shah NP, Skaggs BJ, Branford S, Hughes TP, Nicoll JM, Paquette RL, Sawyers CL (2007) Sequential ABL kinase inhibitor therapy selects for compound drug-resistant BCR-ABL mutations with altered oncogenic potency. J Clin Invest 117(9):2562–2569. https://pubmed.ncbi.nlm.nih.gov/17710227/

Sharabi AB, Lim M, DeWeese TL, Drake CG (2015) Radiation and checkpoint blockade immunotherapy: radiosensitisation and potential mechanisms of synergy. Lancet Oncol 16(13):e498–e509. https://pubmed.ncbi.nlm.nih.gov/26433823/

Sharma SV, Haber DA, Settleman J (2010) Cell line-based platforms to evaluate the therapeutic efficacy of candidate anticancer agents. Nat Rev Cancer 10(4):241–253. https://pubmed.ncbi.nlm.nih.gov/20300105/

Shoemaker RH (2006) The NCI60 human tumour cell line anticancer drug screen. Nat Rev Cancer 6(10):813–823. https://pubmed.ncbi.nlm.nih.gov/16990858/

Shrimali RK, Ahmad S, Verma V, Zeng P, Ananth S, Gaur P, Gittelman RM, Yusko E, Sanders C, Robins H, Hammond SA, Janik JE, Mkrtichyan M, Gupta S, Khleif SN (2017) Concurrent PD-1 Blockade negates the effects of OX40 agonist antibody in combination immunotherapy through inducing T-cell apoptosis. Cancer Immunol Res 5(9):755–766. https://pubmed.ncbi.nlm.nih.gov/28848055/

Silva AS, Kam Y, Khin ZP, Minton SE, Gillies RJ, Gatenby RA (2012) Evolutionary approaches to prolong progression-free survival in breast cancer. Cancer Res 72(24):6362–6370. https://pubmed.ncbi.nlm.nih.gov/23066036/

Simeoni M, De Nicolao G, Magni P, Rocchetti M, Poggesi I (2013) Modeling of human tumor xenografts and dose rationale in oncology. Drug Discov Today Technol 10(3):e365–e372. https://pubmed.ncbi.nlm.nih.gov/24050133/

Simeoni M, Magni P, Cammia C, De Nicolao G, Croci V, Pesenti E, Germani M, Poggesi I, Rocchetti M (2004) Predictive pharmacokinetic-pharmacodynamic modeling of tumor growth kinetics in xenograft models after administration of anticancer agents. Cancer Res 64(3):1094–1101. https://pubmed.ncbi.nlm.nih.gov/14871843/

Skipper HE (1965) The effects of chemotherapy on the kinetics of leukemic cell behavior. Cancer Res 25(9):1544–1550. https://pubmed.ncbi.nlm.nih.gov/5861078/

Skipper HE, Perry S (1970) Kinetics of normal and leukemic leukocyte populations and relevance to chemotherapy. Cancer Res 30(6):1883–1897. https://pubmed.ncbi.nlm.nih.gov/4917694/

Skipper HE, Schabel FM Jr, Wilcox WS (1964) Experimental evaluation of potential anticancer agents. Xiii. On the criteria and kinetics associated with "curability" of experimental leukemia. Cancer Chemother Rep 35:1–111. https://pubmed.ncbi.nlm.nih.gov/14117037/

Spill F, Reynolds DS, Kamm RD, Zaman MH (2016) Impact of the physical microenvironment on tumor progression and metastasis. Curr Opin Biotechnol 40:41–48. https://pubmed.ncbi.nlm.nih.gov/26938687/

Stanta G, Bonin S (2018) Overview on clinical relevance of intratumor heterogeneity. Front Med (Lausanne) 5:85. https://pubmed.ncbi.nlm.nih.gov/29682505/

Tran L, Xiao JF, Agarwal N, Duex JE, Theodorescu D (2021) Advances in bladder cancer biology and therapy. Nat Rev Cancer 21(2):104–121. doi:10.1038/s41568-020-00313-1.

Vargo-Gogola T, Rosen JM (2007) Modelling breast cancer: one size does not fit all. Nat Rev Cancer 7(9):659–672. https://pubmed.ncbi.nlm.nih.gov/17721431/

Voskoglou-Nomikos T, Pater JL, Seymour L (2003) Clinical predictive value of the in vitro cell line, human xenograft, and mouse allograft preclinical cancer models. Clin Cancer Res 9(11):4227–4239. https://pubmed.ncbi.nlm.nih.gov/14519650/

Wang C, Tang Z, Zhao Y, Yao R, Li L, Sun W (2014) Three-dimensional in vitro cancer models: a short review. Biofabrication 022001. https://pubmed.ncbi.nlm.nih.gov/24727833/

Weeber F, Ooft SN, Dijkstra KK, Voest EE (2017) Tumor organoids as a pre-clinical cancer model for drug discovery. Cell Chem Biol 24(9):1092–1100. https://pubmed.ncbi.nlm.nih.gov/28757181/

Whisler JA, Chen MB, Kamm RD (2014) Control of perfusable microvascular network morphology using a multiculture microfluidic system. Tissue Eng Part C Methods 20(7):543–552. https://pubmed.ncbi.nlm.nih.gov/24151838/

World Health Organization (2018) Cancer. https://www.who.int/health-topics/cancer

Wu DJ, Aktipis A, Pepper JW (2019) Energy oversupply to tissues: a single mechanism possibly underlying multiple cancer risk factors. Evol Med Public Health 2019(1):9–16. https://pubmed.ncbi.nlm.nih.gov/31893122/

Xu H, Li Z, Yu Y, Sizdahkhani S, Ho WS, Yin F, Wang L, Zhu G, Zhang M, Jiang L, Zhuang Z, Qin J (2016) A dynamic in vivo-like organotypic blood-brain barrier model to probe metastatic brain tumors. Sci Rep 6:36670. https://pubmed.ncbi.nlm.nih.gov/27830712/

Xu H, Lyu X, Yi M, Zhao W, Song Y, Wu K (2018) Organoid technology and applications in cancer research. J Hematol Oncol 11(1):116. https://pubmed.ncbi.nlm.nih.gov/30219074/

Xu X, Kim SK (2011) The early bird catches the worm: new technologies for the Caenorhabditis elegans toolkit. Nat Rev Genet 12(11):793–801. https://pubmed.ncbi.nlm.nih.gov/21969037/

Yamagiwa K, Ichikawa K (1918) Experimental study of the pathogenesis of carcinoma. J Cancer Res 3(1):1–29.

Zhang J, Benavente CA, McEvoy J, Flores-Otero J, Ding L, Chen X, Ulyanov A, Wu G, Wilson M, Wang J, Brennan R, Rusch M, Manning AL, Ma J, et al. (2012) A novel retinoblastoma therapy from genomic and epigenetic analyses. Nature 481(7381):329–334. https://pubmed.ncbi.nlm.nih.gov/22237022/

Zhao P, Zhang L, Grillo JA, Liu Q, Bullock JM, Moon YJ, Song P, Brar SS, Madabushi R, Wu TC, Booth BP, Rahman NA, Reynolds KS, Gil Berglund E, Lesko LJ, Huang SM (2011) Applications of physiologically based pharmacokinetic (PBPK) modeling and simulation during regulatory review. Clin Pharmacol Ther (2):259–267. https://pubmed.ncbi.nlm.nih.gov/21191381/

Section IV

Transformative Technologies in Reproductive Health and Neonatology

16 Reproductive Health and Biotechnological Innovations

*Sheena Mariam Thomas and Ramakrishnan Veerabathiran**
Human Cytogenetics and Genomics Laboratory, Faculty of Allied Health Sciences, Chettinad Hospital and Research Institute, Chettinad Academy of Research and Education, Kelambakkam, Tamil Nadu 603103, India

16.1 INTRODUCTION

The state of being physically, mentally, and socially associated with one's reproductive system is referred to as reproductive health (Natarajan and Pandiyan, 2018). It involves possessing the ability to reproduce and the liberty to decide whether, when, and how often to exercise that choice. It also requires having a fulfilling and safe sexual life and access to knowledge and instruction for making informed reproductive health decisions. Treatments for infertility, maternal health, and contraception will all be completely transformed by biotechnological advancements in reproductive health. These developments are meant to fill gaps in the field and increase the access to and effectiveness of reproductive health technologies (Callahan et al., 2020).

Assisted reproductive technologies (ARTs) have revolutionized the treatment of infertility; the chances of successful conception have increased considerably risen because of procedures like preimplantation genetic testing (PGT), intracytoplasmic sperm injection (ICSI), and *in vitro* fertilization (IVF). Genetic screening and testing are now possible because of genomic technologies, which have also enhanced our understanding of the genetic variables affecting reproductive health (Gayatri et al., 2016). The gene editing technique CRISPR/Cas9 shows promise in repairing genetic defects before pregnancy (Rubeis and Steger, 2018). Biotechnological developments have resulted in the creation of innovative and highly efficacious contraceptives, including hormonal implants, long-acting reversible contraceptives (LARCs), and contraceptive patches (Stoddard et al., 2011). The novel contraceptive products, improved IVF, and microfluidics and nanofluidics to streamline conventional reproductive treatments are examples of recent biotechnology advances in reproductive health (Choudhary et al., 2016). Furthermore, the development of next-generation contraceptives, including once-monthly pills, injectable contraceptives lasting six months, and covert microarray patches, along with the use of artificial intelligence (AI) in ultrasounds, are revolutionizing the field of reproductive health technologies. The ultimate goal of these advances is to improve reproductive health outcomes by giving men and women more options that suit their tastes and circumstances (Edmonds, 2007).

Many tactics are being used to enhance reproductive health. The goal of research and development (R&D) funds and programs like Ananya is to decrease maternal and child mortality, increase infection resilience in mothers and their infants, and encourage improvements in nutrition and health. The organization is dedicated to creating cutting-edge contraceptive methods that overcome obstacles like cost and accessibility while also catering to the desires of women and girls in low- and middle-income nations (Darmstadt et al., 2020). Additionally, efforts are being made to boost financing and focus on reproductive health research and development for women's health issues, with an emphasis on developing new and improved reproductive health alternatives based on a thorough understanding of women and girls' desires. Making educated decisions about one's sexual and reproductive health begins with comprehensive sex education. Education programs give people the information and abilities they need while promoting a culture of accountability, consent, and open communication. Everyone must have access to safe and legal abortion services, infertility therapies, family planning, and prenatal care. By guaranteeing early identification and treatment, this lessens the burden associated with reproductive health problems. Enhancing women's decision-making skills, access to education, and economic opportunity is essential for improving reproductive health outcomes. Policies that are inclusive of all genders help create a climate that is favorable to reproductive health. Campaigns for public awareness are essential to eradicating stigma and encouraging appropriate sexual behavior. A more aware and proactive society is fostered by campaigns that highlight family planning, STI prevention, and contraceptive techniques (Stover et al., 2016).

16.2 INFERTILITY TREATMENT

The clinical definition of infertility is the inability to achieve a healthy pregnancy after engaging in at least 12 months of unprotected sexual activity. Infertility can be attributed to factors related to females (20% to 40% of couples), males (35% to 40% of couples), a combination of both (20% to 30% of couples), or may remain undetermined. In women, it is often associated with conditions such as endometriosis, tubal blockages, and irregular menstruation. In men, it

* Corresponding author: rkgenes@gmail.com

DOI: 10.1201/9781032660592-20

frequently results from obstructions in the sperm ducts or abnormalities in sperm production and function. Monitoring and addressing infertility can involve various approaches, including fertility drugs, lifestyle modifications, medical and surgical treatments for underlying conditions, fertility counselling, and the use of ARTs such as intrauterine insemination (IUI) and IVF or ICSI. Recent research has focused on optimizing outcomes for both mothers and newborns in IVF/ICSI, including factors such as the number of embryos transferred and the choice between fresh or frozen sperm, eggs, or embryos (Nardelli et al., 2014).

16.2.1 IVF

IVF is a complex process aimed at overcoming infertility and producing a live birth as a direct result of the intervention; it entails retrieving an oocyte from ovarian follicles and activating the ovaries with a mix of fertility drugs. The recovered eggs can either be cryopreserved for later use or fertilized *in vitro* to produce embryos. The embryos that are produced are then placed inside the uterus. An IVF cycle is the period during which these processes usually take place, and it lasts roughly two weeks (Weng, 2019).

Timely retrieval of oocytes for fertilization is a prerequisite for successful induction. Problems in the laboratory can include sperm–oocyte membrane fusion, sperm penetration failure, failed nuclei fusion, and unsuccessful embryo cleavage. Implantation and transfer can also be challenging. IVF aims to replicate natural processes as precisely as possible within an artificial framework (DeCherney, 1986) and is depicted in Figure 16.1.

16.2.2 IUI

The process of IUI involves the direct insertion of sperm into a woman's uterus to enhance fertility, especially in cases where natural conception proves challenging (Joseph and Salai, 2021). In comparison with more invasive and costly IVF, IUI is less invasive and more cost-effective. Semen preparation techniques have evolved alongside advancements in ARTs like IVF with uterine embryo transfer. These techniques aim to separate morphologically normal, motile sperm from seminal plasma, which contains dead sperm, bacteria, and white blood cells that produce oxygen radicals detrimental to healthy sperm function (Jonge, 2002). A randomized controlled experiment focusing on male subfertility revealed that semen preparation significantly increased the likelihood of conception following IUI (Goldenberg et al., 1992).

Various sperm preparation techniques are available, broadly falling into three categories. The swim-up technique selects sperm based on their swimming ability. In this method, liquid semen is covered with a culture medium, allowing mobile sperm to swim into the culture. The top portion of the layered media containing the selected sperm is then carefully removed for further use. The second technique involves density gradients, where sperm are separated

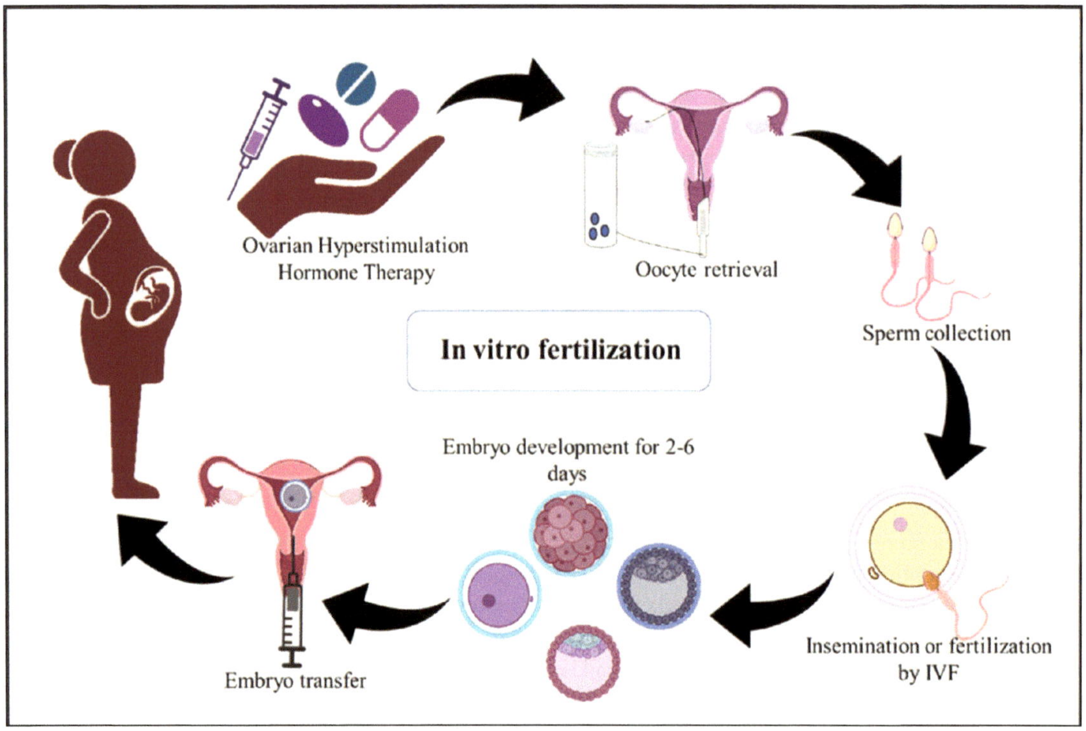

FIGURE 16.1 In IVF, the woman is administered hormones to stimulate the ovary, and the oocyte is retrieved to fertilize with healthy sperm cells in laboratory conditions. The zygote is then allowed to divide in the laboratory, and the fully maturated embryo is transferred into the uterus.

based on their density through centrifugation (Edition, 2010). The third approach combines centrifugation with the traditional wash method, particularly useful in cases of male factor infertility. The selection of a sperm preparation method is contingent upon the characteristics of the semen sample. The swim-up technique is commonly chosen when semen samples are deemed largely normal, whereas density gradient centrifugation is preferred in instances of male factor infertility because it retrieves more motile spermatozoa (Henkel and Schill, 2003).

The insemination procedure involves inserting a small catheter to deliver a 0.2–0.5 mL sperm solution into the uterus, typically without the assistance of imaging guidance. Fallopian tube sperm perfusion uses a larger 4 mL inseminate that allows the fluid to fill the uterine cavity, Fallopian tubes, and possibly the peritoneal cavity (Kahn et al., 1992). Depending on ovulation timing, insemination may occur once or multiple times, with the common time frame being 32–36 hours after human chorionic gonadotropin (hCG) administration (Ragni et al., 2004).

16.2.3 Intracytoplasmic Sperm Injection

ICSI offers greater prospects for couples with minimal or no success using traditional IVF methods and other assisted fertilization procedures like partial zona dissection, zona drilling, or sub-zonal insemination. For each metaphase II oocyte to be injected, one motile spermatozoon must be present for ICSI to be successful. Injecting a motile sperm cell can result in elevated rates of fertilization and successful pregnancies. When spermatozoa that are either immotile or potentially non-vital are injected, rates decrease. The success rates of ICSI are not impacted by the concentration, morphology, or elevated levels of anti-sperm antibodies in semen (Neri et al., 2014).

It is feasible to implant oocytes that have successfully undergone the extrusion of the first polar body. Other oocyte morphological features, including the cytoplasm, perivitelline space, and zona pellucida do not show an increased rate of success of ICSI. The outcome of an ICSI cycle might also be influenced by several technical factors. Executing the injection method, an inverted microscope equipped with manipulators and injectors is essential. A straightforward injection procedure necessitates accurately crafted microneedles possessing appropriate characteristics, and these needles must be correctly positioned on the inverted microscope. Because of these features, injecting an oocyte might be somewhat challenging. The manner the oolemma is ruptured and the zona pellucida penetrated can affect how the injection process turns out. Typically, morphological examinations are unable to identify these oocyte traits; instead, they are only visible during the injection process (Palermo et al., 2017).

ICSI involves carefully selecting and immobilizing morphologically normal and motile sperm. The sperm is captured into an ICSI pipette filled with a perivitelline space solution. The procedure required precision in using an oil–hydraulic injection system. The selected sperm was inserted into an oocyte held by gentle suction, ensuring that the sperm was deposited intracytoplasmically. This meticulous process aimed to minimize oocyte damage, with an incidence of less than 10% per injected oocyte. After ICSI, oocytes were incubated in the EBSS-serum medium, and pronuclei formation was assessed 16 to 18 hours later. Normal fertilization was confirmed by the presence of two distinct pronuclei. Embryo cleavage was evaluated 24 hours later, and on the second day, the morphologically best embryos were transferred to the patient's uterine cavity. Extra embryos were cryopreserved. Embryo transfer was facilitated using tight difficult transfer, and luteal phase support was provided through injections or vaginal capsules. Pregnancy was confirmed by urinary hCG tests, and clinical pregnancy was verified through vaginal ultrasound at six to seven weeks of gestation (Svalander et al., 1995).

16.2.4 Surrogacy

IVF has made surrogacy a prominent fertility treatment. By utilizing an embryo either formed by the surrogate or obtained from a donor and then transferred to the uterus of a gestational carrier, women facing challenges in conceiving, such as those with uterine anomalies or serious medical conditions, now have the opportunity to become mothers (Söderström-Anttila et al., 2016).

The primary indicator for surrogacy is the total absence of a uterus. This condition may arise from Mayer–Rokitansky–Kuster–Hauser syndrome (Lindenman et al., 1997) or hysterectomy, but other indicators are a significant anatomical abnormalities like a small unicornuate uterus, a T-shaped uterus, or multiple fibroids, coupled with unsuccessful fertility treatments. The additional criteria for surrogacy are women with serious medical issues (diseases of the heart or kidneys) that make pregnancy unsafe. When all other avenues for self-pregnancy have been explored and a patient has recurrent miscarriages and implantation failure, surrogacy may be the final option (Aflatoonian et al., 2013).

There are two main forms of surrogacy: traditional and gestational. In traditional (genetic/partial/straight) surrogacy, the surrogate mother undergoes artificial insemination using the intended father's sperm, resulting in the intended father and the surrogate becoming the genetic parents. In gestational surrogacy, also referred to as host or complete surrogacy, an embryo from the intended parents or donated oocytes or sperm is transferred to the surrogate's uterus. The mother carrying the child through gestational surrogacy is not genetically related to the child.

India's draft Assisted Reproductive Technology (Regulation) Bill of 2014 states that a surrogate is typically a married woman aged between 23 and 35 (or 25 and 35 in the 2016 Surrogacy Bill). She should have at least one child who is a minimum of three years old, and the age gap between deliveries should not exceed two years. Additionally, the surrogate must obtain consent from her spouse to become a surrogate mother. In-depth background

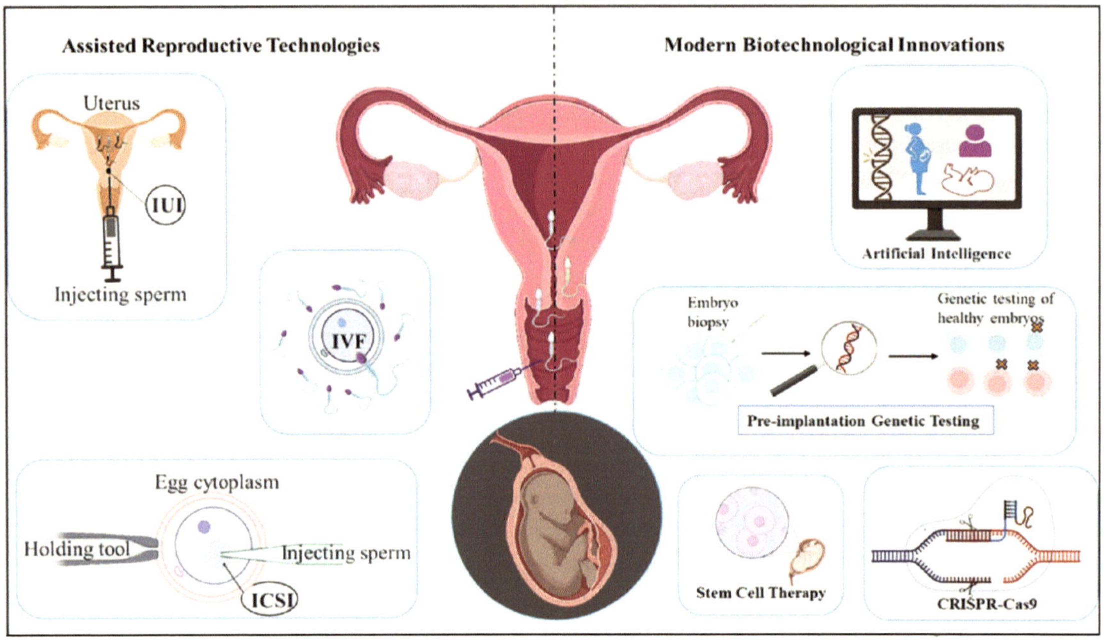

FIGURE 16.2 Technological innovations in fertility treatments.

investigations on criminal and financial history are usually included in the screening procedure, along with a comprehensive physical and psychiatric evaluation. In addition to testing for the hepatitis C virus, hepatitis B surface antigen, and human immunodeficiency virus, routine blood tests are performed; Pap smears, mammograms, and electrocardiograms are advised. To rule out any anatomical abnormalities, she will also have a comprehensive pelvic and abdominal ultrasonography (Patel et al., 2018).

16.3 MODERN BIOTECHNOLOGICAL INNOVATIONS IN REPRODUCTIVE HEALTH

Human fertility and contraception have advanced significantly as a result of contemporary biotechnology discoveries in reproductive health. Reproductive biotechnologies, contraceptive research and development, and revolutionary advancements in reproductive, maternal, infant, and child health are just a few of the many advancements that fall under this broad category of breakthroughs. Reproductive biotechnologies are beginning to take the place of traditional methods, opening up new avenues for assisted reproductive technologies and fertility treatments (Choudhary et al., 2016). Preimplantation genetic testing (PGT) and time-lapse imaging improve genetic screening and embryo selection in ARTs like IVF (Ly et al., 2011).

By repairing genetic mutations, CRISPR/Cas9 gene editing has the potential to prevent inherited disorders. Cancer patients can preserve their fertility with ovarian tissue transplantation, and family planning flexibility can be achieved by egg storage, even for social reasons. Infertility issues are addressed by technologies such as IUI and ICSI. In reproductive medicine, 3D printing is used to create customised implants, and stem cell treatment regenerates the endometrium (Royal Society et al., 2020). Reproductive health monitoring is aided by wearable technology and digital health, including fertility tracking applications. Furthermore, remarkable developments in biology, chemistry, and medicine have resulted in the creation of novel contraceptive products outside of the market (Goldenberg et al., 1992). The availability of drugs and the widespread use of contraception are two examples of transformative innovations that have improved morbidity and mortality in the fields of reproductive, maternal, newborn, and child health (Engmann et al., 2016). Figure 16.2 depicts the technologies that have been carried out in modern biotechnological innovations used to treat infertility cases.

16.3.1 Preimplantation Genetic Testing

PGT involves extracting a cellular biopsy from a developing human oocyte or embryo during an IVF cycle, examining its genetic composition, and utilizing the findings to determine the most suitable embryos for later placement in the uterus. When the sex of cleavage stage embryos in two couples – one of which had X-linked disease – was discovered in 1990, PG testing was first reported (Brezina et al., 2012). Chromosome aneuploidy, a prevalent issue in human gametes and preimplantation embryos, affects less than 0.3% of births but significantly contributes to IVF failures, miscarriages, and stillbirths. Aneuploidies, which are often a result of errors in maternal meiosis, particularly prevalent in women in their late 30s and early 40s, account for a significant proportion of genetic abnormalities. This age-related increase in aneuploidies is associated with a heightened risk of miscarriage and reduced live birth rates following IVF. If genetic testing methods were flawless, there would be no clinical rationale for transferring aneuploid embryos in infertility treatments aimed at achieving successful pregnancies. However, accurately distinguishing between euploid and aneuploid embryos without compromising treatment poses a substantial challenge. The term "preimplantation genetic testing for aneuploidy" (PGT-A) has replaced the previous term "preimplantation genetic screening". Initial techniques involved single-cell analysis using fluorescent *in situ* hybridization, polar body or cleavage stage biopsy, and a limited number of chromosome-specific probes (typically 7–9 chromosomes). However, several randomized control studies (RCTs) failed to demonstrate significant advantages. A substantial trial in the Netherlands suggested a decrease in the number of live births. Over the past decade, advancements in whole-genome amplification and copy number analysis for all 24 chromosomes have occurred, transitioning from tests based on microarray comparative genomic hybridization to those utilizing next-generation sequencing (NGS). Progress has also been made in vitrification for cryopreservation and blastocyst culture. The combination of blastocyst culture, trophectoderm biopsy, PGT-A, vitrification, and subsequent transfer of euploid blastocysts in unstimulated cycles has demonstrated high implantation rates and live births per transfer. Various randomized control trials, employing diverse technologies and patient populations, have documented improved clinical outcomes with PGT-A. Notably, randomization often occurred at the blastocyst stage, with a preference for selecting the most morphologically healthy embryos (Rosenwaks and Handyside, 2018).

16.3.2 CRISPR/Cas9

The most precise, dependable, and effective method for editing the genome of a cell in any living organism involves the utilization of CRISPR (clustered regularly interspaced short palindromic repeats) and its associated protein, Cas9. This technology finds applications across various disciplines. Three primary phases are involved in CRISPR/Cas9 genome editing: recognition, cleavage, and repair. By using complementary base pairing to identify the target gene sequence, the engineered single guide RNA (sgRNA) directs the Cas9 protein to it. Without sgRNA, Cas9 is not active. Double-strand breaks (DSBs) are produced at a location just before the Protospacer Adjacent Motif (PAM) sequence, a brief conserved region located downstream of the cut site, by the Cas9 nuclease. The PAM sequence is identified by the widely used Cas9 protein as 5′-NGG-3′. After locating the target location, Cas9 creates an RNA–DNA hybrid by causing local DNA melting. Uncertainty surrounds the precise mechanism by which Cas9 melts the target DNA. The complementary strand of the DNA is then cut by the activated Cas9 utilizing its HNH domain, and the non-complementary strand by its RuvC domain. The resulting DSBs are primarily blunt-ended. The host's cellular mechanism repairs the DSB in the last stage (Asmamaw and Zawdie, 2021).

Depending on whether the target population is gametes, embryos, or stem cells, different factors need to be taken into account when using germline gene editing. When dealing with embryos, the editing system is directly introduced into the cytoplasm or pronuclei of zygotes. Consequently, additional screening is necessary to identify embryos with accurately modified genomes and without any unintended genetic alterations. However, the effectiveness of genomic editing in embryos is impeded by the creation of mosaic embryos resulting from ineffective nuclease cutting and erroneous DNA repair before cleavage. In non-human primates, modified offspring were born when Cas9 or TALENs were microinjected into zygotes. The efficiency of TALENs or Cas9 in mammalian zygotes varies from 0.5% to 40.9% per injected zygote. Recent CRISPR/Cas9 editing of human zygotes aimed to assess specificity and fidelity. Out of the 86 injected 3PN zygotes, successful testing was conducted on 54 DNA samples, revealing 28 genome-edited zygotes (5.6% of the total) with the correct genetic material added through homologous recombination. However, edited embryos displayed mosaicism, consistent with observations in other systems. Several 'off-target' mutations were found, which could have been caused by intrinsic defects in 3PN zygote-derived embryos or by CRISPR/Cas9 operating elsewhere in the genome. The fact that off-target mutations were only confirmed in a subset of the genome (exome) suggests that the frequency of these changes may be underestimated. PGD cannot reliably anticipate outcomes due to the mosaic nature of gene repairs (Vassena et al., 2016).

Another strategy that replaces the zygote approach is the application of gene changes during gametogenesis. The CRISPR/Cas9 system can be employed in developing immature oocytes or sperm to generate gene-corrected mature sperm or oocytes, which can subsequently be utilized in ART. The objective of this approach is to address genetic disorders in the succeeding generation. In cases of infertility attributed to maturation arrest, spermatogonial stem cells are considered a more viable target, as mature post-meiotic sperm in the male germ cell line may not be readily susceptible to genetic modifications. The feasibility of this approach is increasing with the advancement of *in vitro* spermatogonia stem cell production. Manipulating the oocyte in the female germ line genetically is comparatively more straightforward. However, there are issues because each patient only receives a small number of mature oocytes, and the CRISPR procedure is not very effective. The germinal vesicle stage during *in vitro* egg maturation offers a potential window for required modifications, although further research is needed to evaluate the effectiveness and safety of CRISPR/Cas9 during meiosis (Nickkholgh et al., 2014).

16.3.3 Mitochondrial Replacement Therapy

Mitochondrial replacement therapy (MRT) or mitochondrial donation refers to an embryo with the parents' nuclear

DNA undergoing IVF to acquire the donor female's mitochondrial DNA. MRT attempts to treat infertility and give women a reasonable opportunity to have biologically related, healthy children in this age of cutting-edge scientific technology. Maternal spindle transfer (MST) and pronuclear transfer (PNT) are the two basic methods used in MRT. However, the third option of polar body genome transfer has also been introduced (Craven et al., 2017).

PNT involves raising two zygotes *in vitro* following fertilization, one from the donor, with healthy mitochondria and faulty pronuclei, and the other from the biological parents. By employing electric pulses or the inactivated Japanese hemagglutinating virus, the pronuclei from a donor zygote that has eliminated pronuclei are transferred into the recipient's zygote containing healthy mitochondria, and the mother's womb receives the modified zygote. The carryover ranges from being undetectable to less than 5%. The successful birth of the world's second three-parent baby in Ukraine, without encountering any clinical complications, fully addressed the involved parties' safety concerns regarding the procedure (Sharma et al., 2020).

MST is a form of selective reproduction conducted before fertilization that is akin to prenatal diagnostic and preimplantation genetic diagnosis. In this process, the damaged egg from the mother is utilized to extract the maternal spindle complex during the metaphase stage. Then, this combination is introduced into the perivitelline gap of the donor's enucleated egg, which harbors healthy mitochondria, and the genetically modified embryo is next implanted into the mother's uterus. This method is superior because the maternal spindle contains less cytoplasm, ultimately reducing the likelihood of mitochondrial DNA carryover and mutations (Costa-Borges et al., 2020).

Polar body genome transfer entails transferring polar bodies, two small structures generated during oogenesis in mammalian oocytes. These bodies possess varying amounts of cytoplasm and the same number of chromosomes as in the nucleus of the primordial oocyte. In primates, these bodies are diploid and haploid, respectively. The method's significance lies in the presence of rare mitochondria with limited cytoplasm, minimizing the likelihood of mitochondrial DNA carryover. Using the two ways mentioned above, all of the nuclear material contained within polar bodies becomes more potent and can produce oocytes and zygotes that have been reestablished (Christodoulaki et al., 2021).

16.3.4 Microfluidics

Treatment for infertility and ART have found a potential new tool in the realm of microfluidics technology. It provides exceptional chances to minimize human mistakes, lessen stress on gametes and embryos, and automate IVF procedures. Sperm sorting, oocyte modification, insemination, embryo culture, and sperm and embryo quality assessment have all been done using microfluidic devices. In the end, they help in the treatment of male infertility by facilitating the optimization of sperm sample preparation, analysis, and manipulation. The recovery of motile sperm from semen, sperm quality measurement, and the isolation of pure sperm populations for the treatment of male infertility have all been demonstrated to be improved using microfluidic systems for sperm sorting. These systems can increase sample preparation efficiency and consistency in cell/embryo culturing environments, and they can closely simulate *in vivo* circumstances of sperm selection (Nosrati et al., 2017).

Microfluidic devices offer benefits in the field of ART, including improved automation, scalability, and faster processing times for the diagnosis and treatment of male infertility. These tools, which are frequently made of polydimethylsiloxane (PDMS), are essential for managing gametes, developing oocytes, and cultivating embryos in environments that are quite similar to those found *in vivo*. A microfluidic biochip's primary tasks include expansion, restoration, sorting, and emulsion droplet creation. One such instance is a microfluidic device based on PDMS that is used in a mouse embryo system to expand a droplet containing an embryo and sort and dispose of another droplet that does not contain an embryo. It has been investigated how to control different phases of assisted reproductive programs using microfluidic systems in IVF. Oocytes are frequently unnecessarily damaged by traditional IVF methods because of the manipulation of di-electric force. These problems are addressed by microfluidic IVF, which uses chips with multiple layers of different materials, most commonly PDMS because it is inexpensive, gas-permeable, and biocompatible. Studies investigating the biocompatibility of materials utilized in microfluidic IVF, conducted with mouse embryos and pig oocytes, have revealed that materials such as PDMS, silicon oxide, silicon nitrate, borosilicate glass, gold, chromium, and titanium do not adversely impact embryo development (Alias et al., 2021).

16.3.5 Stem Cell Therapy

Gamete deficits resulting from genetic abnormalities induce infertility in certain couples, even with the recent advancements in ART. In these situations, ART might not be helpful, but many couples would rather find a less invasive and more affordable solution to their genetically connected problems. For the treatment of infertility in both experimental and clinical contexts, stem cells present a viable option for cell-based therapy. Stem cells are present in both adult tissues and embryos. They are distinguished by their undifferentiated state and capacity for self-renewal and differentiation. They support the capacity of differentiated organs to undergo repair. In the realm of infertility treatments, various types of stem cells have been explored, including induced pluripotent stem cells (iPSC), adult stem cells (such as mesenchymal stem cells or MSC), ovarian stem cells, spermatogonial stem cells (SSCs), and embryonic stem cells (ESCs) (Wu et al., 2022). The remarkable potential of embryonic stem cells (ESCs) to differentiate into all three embryonic lineages and self-renew indefinitely makes them highly promising in the field of regenerative medicine.

Human embryonic stem cells (hESC), derived from the inner cell mass of blastocysts, express the Oct 4 transcription factor. Research has indicated that both mouse and human embryonic stem cells (ESCs) can be induced to develop into primordial germ cells *in vitro*, undergo meiosis, and generate male and female gametes. In models of sperm-deficient mice, ESCs – especially those produced from cloned blastocysts employing gene repair technology – have shown the capacity to develop viable sperm. The extensive use of ESCs in cell replacement treatment is restricted by ethical issues, notwithstanding their efficaciousness. On the other hand, it has been discovered that hESCs are essential for endometrial damage repair and function restoration (Volarevic et al., 2014). The definition provided by the International Society for Cellular Therapy characterizes mesenchymal stem cells (MSCs) as cells expressing surface markers CD105, CD73, and CD90, possessing adhesive properties to plastic, and demonstrating the capability to differentiate into osteoblasts, adipocytes, and chondroblasts. MSCs can be categorized based on their origin, including bone marrow stromal cells, adipose-derived stem cells, menstrual blood-derived MSCs, umbilical cord-derived MSCs, amniotic fluid-derived MSCs, placental tissue-derived MSCs, salivary gland-derived MSCs, and dental pulp-derived MSCs. The therapeutic potential of MSCs in treating infertility associated with ovarian and endometrial dysfunction has been the subject of numerous investigations. Research has extensively explored the therapeutic capabilities of bone marrow mesenchymal stem cells, commonly known as BMSCs. BMSCs are useful in endometrial regeneration and the healing of intrauterine adhesions. They can differentiate into mesodermal lineages, endometrial, and follicular cells. Similar encouraging results have been noted with BMSC therapies for chemotherapy-induced ovarian dysfunction. Menstrual blood mesenchymal stem cells, or MB-MSCs, provide a simple, safe, and non-invasive source for therapeutic use. Research has demonstrated the protective effects of MB-MSCs against endometrial damage, cyclophosphamide-induced premature ovarian failure, busulfan-induced rat models of ovarian dysfunction, and epirubicin-induced ovarian damage (Saha et al., 2021).

16.3.6 Artificial Intelligence

Since the 1956 Dartmouth conference, artificial intelligence (AI) has been studied, covering the use of computers to simulate human mental processes. Appropriate hardware, such as graphics processing units (GPUs) from Nvidia, is essential for running AI software. Modern storage systems are required due to the daily generation of enormous amounts of electronic data. AI relies heavily on natural language processing (NLP), which enables computers to comprehend and analyse human language. NLP is especially useful for analysing unstructured data, such as electronic medical records (Raimundo and Cabrita, 2021). Empirical machine learning methods, which can be classified as unsupervised, supervised, or reinforcement learning, are the cornerstone of AI. AI has advanced medicine significantly, especially in areas like pathology and radiology, where it has decreased errors and allowed for real-time prediction. AI has been used in reproductive medicine since the late 20th century. In non-surgical reproductive medicine applications, supervised learning methods like decision trees and support vector machines are frequently employed. These methods aid in tasks such as determining optimal morpho-kinetic parameters for embryos and assessing the cost-effectiveness of cryopreserving human oocytes. Within the realm of image processing, unsupervised learning approaches like principal component analysis and K-means are utilized to predict pregnancy outcomes based on oocyte quality as depicted in Figure 16.3. In RM, Artificial neural networks (ANN) are utilized for tasks like IVF outcome prediction and embryo segmentation. Reinforcement AI machine learning is frequently used in robotic surgery and other robotic parts of RM. Even though AI in RM seems promising, more randomized controlled trials are required to confirm its effectiveness. Additional clarification is also needed about legal and ethical considerations. Enhancing data connectivity is essential for integrating AI into routine medical practice because the technology depends on massive datasets for training (Medenica et al., 2022).

16.4 COMPLICATIONS AND CHALLENGES

There are several complications and risk factors associated with reproductive and infertility treatments that cause adverse effects.

16.4.1 Complications Associated with Reproductive Techniques

Infertility treatments can have several complications, including increased risk of stroke: Compared with people who conceived naturally, those who underwent infertility treatment had a 66% higher risk of experiencing a stroke within a year of giving birth (Liang et al., 2022). Unfavorable prenatal and pregnancy outcomes are also possible: In an extensive study, advanced infertility treatments were associated with an elevated risk of adverse pregnancy and perinatal outcomes such as preterm birth, cesarean delivery, and placental abruption (Palomba et al., 2016). Multiple pregnancies are the most frequent side effect of infertility therapy, increasing the chance of early labor and delivery (Greil et al., 2010).

Additionally, compared with women who conceived naturally, those who underwent infertility treatments were more likely to experience acute kidney failure, irregular heartbeat, placental abruption, cesarean delivery, premature birth, and increased hospital expenses. The fact that preexisting medical disorders including high blood pressure, diabetes, and obesity are more prevalent in those who use infertility therapies should be noted as a potential contributing factor to some of the dangers related to these treatments (Hanson et al., 2017).

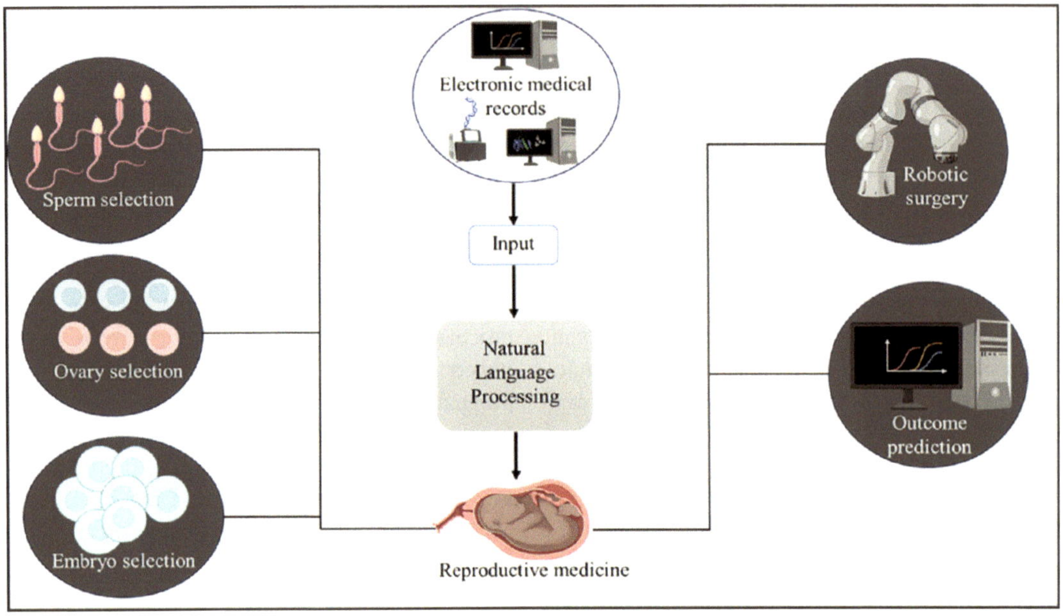

FIGURE 16.3 Applications of artificial intelligence in reproductive medicine. The medical records are processed into machine readable language and are used for selecting healthy sperm, ovary, and embryo in infertility treatment and the application of artificial intelligence is also widened in the field of robotic surgery and outcome prediction.

16.4.2 Ethical Concerns Related to Reproductive Health

Ensuring the autonomy, privacy, and wellbeing of individuals engaging in reproductive health decisions and research is contingent upon several critical concerns. First, encouraging women to take charge of their own reproductive decisions about sex, contraception, abortion, and the use of reproductive technologies can improve their health, but the right to procreate, the moral standing of the embryo, and the participation of outside parties in reproduction through surrogacy and genetic material donation provide ethical challenges (Schenker and Eisenberg, 1997).

Informed consent, upholding privacy and confidentiality, and adhering to communication principles are ethical considerations in research about sexual health, particularly when studies involve culturally taboo subjects and vulnerable participants like adolescents (Shirmohammadi et al., 2018), and in the field of reproductive health, legal and clinical safety for individuals undergoing self-managed abortions is a crucial ethical factor. The ability of mature minors to give informed consent is another ethical factor to take into account when undertaking sexual and reproductive health research with female adolescents involved in high-risk behaviors (Zhang et al., 2019). Finally, contraceptive regulations, intimate partner consent, and the appropriate participation of pregnant women in research studies are ethical considerations with including women as research participants. It is necessary to weigh the interests of the expectant mother with those of the fetus when pregnant women participate in scientific investigations (Douglass et al., 2020).

16.5 FUTURE ASPECTS AND CURRENT INNOVATIONS

Two types of medical innovation are recognized: clinical innovation, which involves physicians experimenting with novel treatments that have not yet been thoroughly studied in a research setting, and formal medical research. Clinical innovation avoids the strict guidelines of formal research, which could lead to patients being ignorant of the experimental nature and accompanying hazards; examples include innovative surgical methods and off-label drug use. Even while novel therapies are frequently published, they usually lack the systematic design required for reliable data on safety and efficacy, which could result in the acceptance of practices that are not sufficiently validated (Margo, 2001).

In contrast to the drawn-out process of controlled research, proponents contend that clinical innovation provides a quicker means of making breakthroughs available for patients. For instance, during the AIDS pandemic (Wilkes and Johns, 2008), several medications were used off-label to save lives. But there's a risk that these innovations won't be thoroughly tested before being incorporated into routine practice, which could have negative or disastrous effects. Preimplantation genetic screening was adopted prematurely in assisted reproduction, for instance, requiring a formal study to reevaluate the situation (Dondorp and Wert, 2019).

According to ethical declarations like the Helsinki Declaration and the Belmont Report, novel treatments should be investigated as soon as it is practically possible. It can be difficult to distinguish between genuine innovation and straightforward clinical modifications. The Belmont Report allows for subjective interpretation by distinguishing between "major innovations" and "radically new procedures." The success of assisted reproduction as a whole has historically been attributed to significant technologies like cryopreservation and ICSI without a thorough preclinical study phase. Recent assessments, however, point to possible hazards for children created only through IVF, such as an increased likelihood of unfavorable pregnancy outcomes and uncommon imprinting abnormalities. Although most IVF cases have resulted in excellent outcomes, questions have been raised regarding potential long-term health implications and epigenetic changes (Owen and Segars, 2009).

16.5.1 Personalized Fertility Treatment

Numerous academics have suggested that ideas and approaches from personalized medicine can be used in fertility treatments. One proposition is to leverage real-world data routinely gathered from patients visiting reproductive medicine and fertility clinics, including information from electronic health records, which are now widely employed. This data could be employed in analyses to investigate patient profiles, subgroups, and patterns, providing valuable insights into differences in fertility rates, responses to fertility-enhancing interventions, and other relevant factors. The outcomes of such analyses could subsequently influence the development of future treatment strategies for individuals dealing with reproductive challenges (Jungheim and Carson, 2018).

Additionally, there have been proposals to develop smartphone applications capable of providing personalized coaching information to enhance pregnancy outcomes within the realm of digital medicine. Genetic variations that have been shown to affect fertility may potentially be utilized to support personalized intervention plans or diagnoses. Lastly, it has been suggested that adaptive trial designs be utilized to evaluate the effectiveness of tailored strategies for increasing awareness regarding fertility and the time to conception (Laber et al., 2016).

Apart from the conventional methods of tailoring fertility treatments, several novel tactics are being developed to augment female fertility; currently, it is possible, for example, to freeze and preserve samples of a woman's ovarian tissue and oocytes for later reimplantation when she is ready to conceive. Given that this process utilizes the patient's cells and considers her individual reproductive preferences, it becomes highly personalized. Theoretically, gene editing could be employed to correct genetic flaws in the relevant cells of these tissues. However, the success of this process depends on the viability and integrity of the preserved tissues. An innovative and contentious personalized fertility intervention known as *in vitro* gametogenesis employs cell reprogramming to create sperm and egg cells from various cell types such as skin cells. These modified cells can then be used to produce *de novo* gametes for fertilization (Goetz and Schork, 2018).

16.5.2 New Contraceptive Methods

The process of creating novel contraceptive methods is drawn out and intricate, requiring clinical testing, regulatory approval, and breakthroughs in our understanding of basic reproductive biology. The best birth control option is determined by a person's preferences, general health, and ethical, social, cultural, and religious frameworks (Poltrock, 1989). Oral contraceptive pills, intrauterine devices, injectables, vaginal rings, patches, male and female sterilization, lactational amenorrhea, condoms, withdrawal, and fertility awareness-based approaches are just a few of the various forms of contraception available. New contraceptive methods that are acceptable, safe, and effective are needed because the current options do not satisfy the needs of all users (Maguire, 2003).

Table 16.1 describes the various contraceptive methods used in reproductive medicine along with their typical use percentage effectiveness. Encouraging advancements and possible innovations in contraceptive approaches include a contraceptive vaccine, extended-release implantable steroids, reversible methods for male and female sterilization, new spermicidal agents with antiviral properties, a monthly pill designed as a menstrual inducer, inventive strategies for ovulation prediction and self-detection, and techniques that interfere with male spermatogenesis. The WHO carries out leading research to increase access to and improve the provision of contraceptive information and services, and it also takes part in the development of new contraceptive methods (Madden et al., 2023).

Today, LARCs, oral contraceptive tablets, female sterilization, and male condoms are the most widely used forms of birth control. Female sterilization is the most widely

TABLE 16.1
Contemporary Contraceptive Methods

Method	Typical Use Effectiveness (%)	Description
Reversible implant	0.05	A small rod inserted under the skin of the upper arm
Intrauterine device	0.2–0.8	A small T-shaped device is inserted into the uterus
Injection	6	Hormonal injection
Pill	9	Daily oral contraceptives containing hormones
Condom	18	Barrier method also used to protect against STI
Vaginal ring	9	Flexible ring inserted into the vagina
Permanent sterilization	0.15–0.5	Surgical procedure to provide long-term contraception
Spermicide	28	Chemical substances inserted into the vagina before intercourse to kill sperm

used method, followed by male condoms, LARCs, and oral contraceptives (Darroch and Singh, 2013). According to the Guttmacher Institute, IUDs, tablets, and male condoms are the most often used forms of contraception for women. The male condom and female sterilization are the two most widely utilized techniques globally, according to the UN (Angdembe et al., 2022).

16.5.3 Improved Techniques for Reproductive Diseases

As circulating exosomes contain more material than other indicators, they provide a substantial advantage for genetic investigations. They have the potential to revolutionize the way that diseases, particularly reproductive abnormalities, are diagnosed in the future by eliminating the need for biopsies in the case of ailments like cancer. Nevertheless, research on exosomes and their molecular cargo in bodily fluids is still in its infancy, and before they are used as conventional biomarkers in clinical settings, more validation through clinical studies following established protocols is needed. For exosomes to be used as efficient diagnostic agents, precise extraction and characterization from bodily fluids is required, notwithstanding debates over their isolation and detection. Exosome isolation can be made easier by several commercial kits, such as ExoQuick, however standardizing procedures is essential to overcoming current obstacles (Helwa et al., 2017).

Cutting-edge technologies that improve exosomal collection and detection efficiency include NGS and nanoparticle-tracking analysis. These technologies also help find new exosomes associated with reproductive issues. A novel strategy for treating reproductive problems is targeted therapy, which aims to change key signaling pathway components associated with the advancement of the disease. Personalized medicine may be made possible by the identification of novel targets, which may be made easier by comprehending the function of exosomes in the reproductive system secretome. The treatment of reproductive problems using stem cell therapy is gaining traction because of its advantages, which include plentiful sources, self-renewal, differentiation, and low ethical considerations. Exosomal stem cells exhibit potential in treating a range of illnesses, including reproductive abnormalities, due to their low risk of aneuploidy and decreased likelihood of immunological rejection. With the transfer of RNAs and proteins, exosomes produced from stem cells can reprogram sick cells, potentially offering therapeutic benefits (Esfandyari et al., 2021).

16.6 CONCLUSION

Advancements in biotechnology have opened up new possibilities and options for individuals and couples, significantly influencing the realm of reproductive health. The capacity for reproduction and the freedom to decide when and how to exercise it, contributing to fulfilling and secure sexual lives, are crucial elements of reproductive health. Increasing the likelihood of a successful conception, assisted reproductive technologies like IVF, IUI, and ICSI have revolutionized infertility therapies. Pregnancy can potentially be prevented by addressing genetic abnormalities using genomic technologies, such as CRISPR/Cas9 gene editing. Biotechnological advancements have led to the development of highly effective contraceptives including hormonal implants, long-acting reversible contraceptives, and contraceptive patches.

Microfluidics and nano-fluidics are two emerging technologies that are simplifying conventional reproductive therapies, while artificial intelligence is finding its way into ultrasounds and improving reproductive health technology. Personalized approaches to fertility therapy that make use of digital and real-world data are another example of the current developments in reproductive health. Offering a variety of efficient birth control alternatives is the goal of the development of new contraceptive technologies, such as innovative spermicidal chemicals, implanted steroids, and contraceptive vaccinations. But these innovations also bring with them difficulties and moral dilemmas. Consideration must be given to the dangers of particular health problems, the complications connected with reproductive procedures, and the ethical difficulties around informed consent and confidentiality. Continued consideration is necessary for the ethical issues raised by the right to procreate, assisted reproduction, and the involvement of third parties in the reproductive procedure. Clinical innovation is seeing a rise in the use of digital tools and real-world data to support personalized approaches to reproductive therapy, which improve patient outcomes.

Global access to safer and more acceptable contraceptive choices is the goal of the development of novel contraceptive methods led by research and implementation activities. The field of reproductive health is constantly innovating, providing individuals and couples facing infertility difficulties with options, hope, and better results. To make sure that technological advancements improve people's lives and society as a whole, it is critical to address the problems and moral issues that come with them.

ACKNOWLEDGMENT

The authors thank the Chettinad Academy of Research Education for their constant support and encouragement.

REFERENCES

Aflatoonian, N., Eftekhar, M., Aflatoonian, B., Rahmani, E., & Aflatoonian, A. (2013). Surrogacy as a good option for treatment of repeated implantation failure: A case series. *Iranian Journal of Reproductive Medicine*, *11*(1), 77.

Alias, A. B., Huang, H. Y., & Yao, D. J. (2021). A review on microfluidics: An aid to assisted reproductive technology. *Molecules*, *26*(14), 4354.

Angdembe, M. R., Sigdel, A., Paudel, M., Adhikari, N., Bajracharya, K. T., & How, T. C. (2022). Modern contraceptive use among young women aged 15–24 years in selected municipalities of Western Nepal: Results from a cross-sectional survey in 2019. *BMJ Open*, *12*(3), e054369.

Asmamaw, M., & Zawdie, B. (2021). Mechanism and applications of CRISPR/Cas-9-mediated genome editing. *Biologics: Targets and Therapy*, 353–361.

Brezina, P. R., Brezina, D. S., & Kearns, W. G. (2012). Preimplantation genetic testing. *BMJ*, *345*.

Callahan, R. L., Mehta, N. J., Nanda, K., & Kopf, G. S. (2020). The new contraceptive revolution: Developing innovative products outside of industry. *Biology of Reproduction*, *103*(2), 157–166.

Choudhary, K. K., Kavya, K. M., Jerome, A., & Sharma, R. K. (2016). Advances in reproductive biotechnologies. *Veterinary World*, *9*(4), 388.

Christodoulaki, A., Boel, A., Tang, M., De Roo, C., Stoop, D., & Heindryckx, B. (2021). Prospects of germline nuclear transfer in women with diminished ovarian reserve. *Frontiers in Endocrinology*, *12*, 635370.

Costa-Borges, N., Spath, K., Miguel-Escalada, I., Mestres, E., Balmaseda, R., Serafín, A., & Calderón, G. (2020). Maternal spindle transfer overcomes embryo developmental arrest caused by ooplasmic defects in mice. *Elife*, *9*, e48591.

Craven, L., Tang, M. X., Gorman, G. S., De Sutter, P., & Heindryckx, B. (2017). Novel reproductive technologies to prevent mitochondrial disease. *Human Reproduction Update*, *23*(5), 501–519.

Darmstadt, G. L., Weng, Y., Pepper, K. T., Ward, V. C., Mehta, K. M., Borkum, E., . . . & Ananya Study Group. (2020). Impact of the Ananya program on reproductive, maternal, newborn and child health and nutrition in Bihar, India: Early results from a quasi-experimental study. *Journal of Global Health*, *10*(2).

Darroch, J. E., & Singh, S. (2013). Trends in contraceptive need and use in developing countries in 2003, 2008, and 2012: An analysis of national surveys. *Lancet (London, England)*, *381*(9879), 1756–1762.

DeCherney, A. H. (1986). In vitro fertilization and embryo transfer: A brief overview. *The Yale Journal of Biology and Medicine*, *59*(4), 409.

De Jonge, C. (2002). The clinical value of sperm nuclear DNA assessment. *Human Fertility*, *5*(2), 51–53.

Dondorp, W., & De Wert, G. (2019). Refining the ethics of preimplantation genetic diagnosis: A plea for contextualized proportionality. *Bioethics*, *33*(2), 294–301.

Douglass, K. M., Strobel, K. M., Richley, M., Mok, T., de St Maurice, A., Fajardo, V., . . . & Afshar, Y. (2020). Maternal-neonatal dyad outcomes of maternal COVID-19 requiring extracorporeal membrane support: A case series. *American Journal of Perinatology*, *38*(1), 82–87.

Edition, F. (2010). *Examination and processing of human semen*. Geneva: World Health Organization.

Edmonds, D. K. (Ed.). (2007). *Dewhurst's textbook of obstetrics & gynaecology* (p. 534). Blackwell Pub.

Engmann, C. M., Khan, S., Moyer, C. A., Coffey, P. S., & Bhutta, Z. A. (2016). Transformative innovations in reproductive, maternal, newborn, and child health over the next 20 years. *PLoS Medicine*, *13*(3), e1001969.

Esfandyari, S., Elkafas, H., Chugh, R. M., Park, H. S., Navarro, A., & Al-Hendy, A. (2021). Exosomes as biomarkers for female reproductive diseases diagnosis and therapy. *International Journal of Molecular Sciences*, *22*(4), 2165.

Gayatri, S. S. D., Pandiyan, N., & Pandiyan, R. (2016). Does the endometrial thickness really matter in assisted reproductive technology cycles – A retrospective analysis. *Chettinad Health City*, 13.

Goetz, L. H., & Schork, N. J. (2018). Personalized medicine: Motivation, challenges, and progress. *Fertility and Sterility*, *109*(6), 952–963.

Goldenberg, M., Rabinovici, J., Bider, D., Lunenfeld, B., Blankstein, J., & Weissenberg, R. (1992). Intra-uterine insemination with prepared sperm vs. unprepared first split ejaculates. A randomized study. *Andrologia*, *24*(3), 135–140.

Greil, A. L., Slauson-Blevins, K., & McQuillan, J. (2010). The experience of infertility: A review of recent literature. *Sociology of Health & Illness*, *32*(1), 140–162.

Hanson, B., Johnstone, E., Dorais, J., Silver, B., Peterson, C. M., & Hotaling, J. (2017). Female infertility, infertility-associated diagnoses, and comorbidities: A review. *Journal of Assisted Reproduction and Genetics*, *34*, 167–177.

Helwa, I., Cai, J., Drewry, M. D., Zimmerman, A., Dinkins, M. B., Khaled, M. L., Seremwe, M., Dismuke, W. M., Bieberich, E., Stamer, W. D., Hamrick, M. W., & Liu, Y. (2017). A comparative study of serum exosome isolation using differential ultracentrifugation and three commercial reagents. *PLoS One*, *12*(1), e0170628.

Henkel, R. R., & Schill, W. B. (2003). Sperm preparation for ART. *Reproductive Biology and Endocrinology*, *1*(1), 1–22.

Joseph, M., & Salai, G. (2021). Effect of acupressure on anxiety among women undergoing intrauterine insemination in a tertiary referral hospital, Tamil Nadu, India-A randomized controlled trial. *Annals of Tropical Medicine & Public Health*, *24*, 232–3109.

Jungheim, E. S., & Carson, K. R. (2018). Leveraging real-world data to move toward more personalized fertility treatment. *Fertility and Sterility*, *109*(4), 608–609.

Kahn, J. A., Von Düring, V., Sunde, A., Sørdal, T., & Molne, K. (1992). Fallopian tube sperm perfusion: First clinical experience. *Human Reproduction*, *7*(suppl_1), 19–24.

Laber, E. B., Zhao, Y. Q., Regh, T., Davidian, M., Tsiatis, A., Stanford, J. B., . . . & Kosorok, M. R. (2016). Using pilot data to size a two-arm randomized trial to find a nearly optimal personalized treatment strategy. *Statistics in Medicine*, *35*(8), 1245–1256.

Liang, C., Chung, H. F., Dobson, A. J., & Mishra, G. D. (2022). Infertility, miscarriage, stillbirth, and the risk of stroke among women: A systematic review and meta-analysis. *Stroke*, *53*(2), 328–337.

Lindenman, E., Shepard, M. K., & Pescovitz, O. H. (1997). Müllerian agenesis: An update. *Obstetrics & Gynecology*, *90*(2), 307–312.

Ly, K. D., Agarwal, A., & Nagy, Z. P. (2011). Preimplantation genetic screening: Does it help or hinder IVF treatment and what is the role of the embryo? *Journal of Assisted Reproduction and Genetics*, *28*, 833–849.

Madden, T., Cohen, S. Y., Paul, R., Hurley, E. G., Thomas, M. A., & Pauletti, G. (2023). Women's preferences for a new contraceptive under development: An exploratory study. *Frontiers in Global Women's Health*, *4*, 1095112.

Maguire, D. C. (Ed.). (2003). *Sacred rights: The case for contraception and abortion in world religions*. Oxford University Press.

Margo, C. E. (2001). When is surgery research? Towards an operational definition of human research. *Journal of Medical Ethics*, *27*(1), 40–43.

Medenica, S., Zivanovic, D., Batkoska, L., Marinelli, S., Basile, G., Perino, A., . . . & Zaami, S. (2022). The future is coming: Artificial intelligence in the treatment of infertility could improve assisted reproduction outcomes – The value of regulatory frameworks. *Diagnostics*, *12*(12), 2979.

Nardelli, A. A., Stafinski, T., Motan, T., Klein, K., & Menon, D. (2014). Assisted reproductive technologies (ARTs): Evaluation of evidence to support public policy development. *Reproductive Health*, *11*(1), 1–14.

Natarajan, P. (2018). Chettinad Health City Medical Journal is reproduction a fundamental right? A clinical, ethical and personal perspective. *Chettinad Health City Medical Journal*, *6*(4), 152–153.

Neri, Q. V., Lee, B., Rosenwaks, Z., Machaca, K., & Palermo, G. D. (2014). Understanding fertilization through intracytoplasmic sperm injection (ICSI). *Cell Calcium*, *55*(1), 24–37.

Nickkholgh, B., Mizrak, S. C., van Daalen, S. K., Korver, C. M., Sadri-Ardekani, H., Repping, S., & van Pelt, A. M. (2014). Genetic and epigenetic stability of human spermatogonial stem cells during long-term culture. *Fertility and Sterility*, *102*(6), 1700–1707.

Nosrati, R., Graham, P. J., Zhang, B., Riordon, J., Lagunov, A., Hannam, T. G., . . . & Sinton, D. (2017). Microfluidics for sperm analysis and selection. *Nature Reviews Urology*, *14*(12), 707–730.

Owen, C. M., & Segars, J. H. (2009, September). Imprinting disorders and assisted reproductive technology. In *Seminars in reproductive medicine* (Vol. 27, No. 5, pp. 417–428). Thieme Medical Publishers.

Palermo, G. D., O'Neill, C. L., Chow, S., Cheung, S., Parrella, A., Pereira, N., & Rosenwaks, Z. (2017). Intracytoplasmic sperm injection: State of the art in humans. *Reproduction*, *154*(6), F93–F110.

Palomba, S., Homburg, R., Santagni, S., La Sala, G. B., & Orvieto, R. (2016). Risk of adverse pregnancy and perinatal outcomes after high technology infertility treatment: A comprehensive systematic review. *Reproductive Biology and Endocrinology*, *14*(1), 1–25.

Patel, N. H., Jadeja, Y. D., Bhadarka, H. K., Patel, M. N., Patel, N. H., & Sodagar, N. R. (2018). Insight into different aspects of surrogacy practices. *Journal of Human Reproductive Sciences*, *11*(3), 212.

Poltrock, S. E. (1989, March). Innovation in user interface development: Obstacles and opportunities. In *Proceedings of the SIGCHI conference on human factors in computing systems* (pp. 191–195). Association for Computing Machinery, New York, United States.

Ragni, G., Somigliana, E., & Vegetti, W. (2004). Timing of intrauterine insemination: Where are we? *Fertility and Sterility*, *82*(1), 25–26.

Raimundo, J. M., & Cabrita, P. (2021). Artificial intelligence at assisted reproductive technology. *Procedia Computer Science*, *181*, 442–447.

Rosenwaks, Z., & Handyside, A. H. (2018). Is preimplantation genetic testing for aneuploidy an essential tool for embryo selection or a costly 'add-on' of no clinical benefit? *Fertility and Sterility*, *110*(3), 351–352.

Royal Society, National Academy of Sciences, National Academy of Medicine, & International Commission on the Clinical Use of Human Germline Genome Editing. (2020). *Heritable human genome editing*. National Academies Press (US).

Rubeis, G., & Steger, F. (2018). Risks and benefits of human germline genome editing: An ethical analysis. *Asian Bioethics Review*, *10*, 133–141.

Saha, S., Roy, P., Corbitt, C., & Kakar, S. S. (2021). Application of stem cell therapy for infertility. *Cells*, *10*(7), 1613.

Schenker, J. G., & Eisenberg, V. H. (1997). Ethical issues relating to reproduction control and women's health. *International Journal of Gynecology & Obstetrics*, *58*(1), 167–176.

Sharma, H., Singh, D., Mahant, A., Sohal, S. K., & Kesavan, A. K. (2020). Development of mitochondrial replacement therapy: A review. *Heliyon*, *6*(9).

Shirmohammadi, M., Kohan, S., Shamsi-Gooshki, E., & Shahriari, M. (2018). Ethical considerations in sexual health research: A narrative review. *Iranian Journal of Nursing and Midwifery Research*, *23*(3), 157.

Söderström-Anttila, V., Wennerholm, U. B., Loft, A., Pinborg, A., Aittomäki, K., Romundstad, L. B., & Bergh, C. (2016). Surrogacy: Outcomes for surrogate mothers, children and the resulting families – a systematic review. *Human Reproduction Update*, *22*(2), 260–276.

Stoddard, A., McNicholas, C., & Peipert, J. F. (2011). Efficacy and safety of long-acting reversible contraception. *Drugs*, *71*, 969–980.

Stover, J., Hardee, K., Ganatra, B., & García, C. (2016). Interventions to improve reproductive health. *Reproductive, Maternal, Newborn, and Child Health*, *95*.

Svalander, P., Forsberg, A. S., Jakobsson, A. H., & Wikland, M. (1995). Factors of importance for the establishment of a successful program of intracytoplasmic sperm injection treatment for male infertility. *Fertility and Sterility*, *63*(4), 828–837.

Vassena, R., Heindryckx, B., Peco, R., Pennings, G., Raya, A., Sermon, K., & Veiga, A. (2016). Genome engineering through CRISPR/Cas9 technology in the human germline and pluripotent stem cells. *Human Reproduction Update*, *22*(4), 411–419.

Volarevic, V., Bojic, S., Nurkovic, J., Volarevic, A., Ljujic, B., Arsenijevic, N., . . . & Stojkovic, M. (2014). Stem cells as new agents for the treatment of infertility: Current and future perspectives and challenges. *BioMed Research International*, *2014*.

Weng, L. (2019). IVF-on-a-chip: Recent advances in microfluidics technology for in vitro fertilization. *SLAS TECHNOLOGY: Translating Life Sciences Innovation*, *24*(4), 373–385.

Wilkes, M., & Johns, M. (2008). Informed consent and shared decision-making: A requirement to disclose to patients off-label prescriptions. *PLoS Medicine*, *5*(11), e223.

Wu, J. X., Xia, T., She, L. P., Lin, S., & Luo, X. M. (2022). Stem cell therapies for human infertility: Advantages and challenges. *Cell Transplantation*, *31*.

Zhang, X. D., Luk, S. C. Y., Han, Y. H., Chen, G. R., Zhang, G. L., Zhu, B. S., & Luchters, S. (2019). Ethical considerations for conducting sexual and reproductive health research with female adolescents engaged in high-risk behaviours in China. *Sexual & Reproductive Healthcare*, *20*, 46–53.

17 Prenatal Diagnosis and Genetic Screening

Rashmi Bagga[1]* *and Sujata Bhardwaj*[2]
[1] Department of Obstetrics and Gynaecology, PGIMER, Chandigarh, India
[2] High Risk Pregnancy and Perinatology, Department of Obstetrics and Gynaecology, PGIMER, Chandigarh, India

17.1 INTRODUCTION

Prenatal diagnosis is the science of identifying the chromosomal abnormalities, malformations, and other various genetic syndromes in the fetus.[1] It includes routine screening tests for aneuploidy and neural tube defects, invasive diagnostic tests, and diagnosis of structural malformations with specialized fetal imaging techniques. Chromosomal abnormalities arise when the normal karyotype of 46 chromosomes is altered, usually during cell division either due to addition or deletion of the chromosomes. Many of these chromosomal abnormalities are not compatible with life and often result in spontaneous miscarriage. Some being compatible produce distinct syndromes. The main goal of prenatal diagnosis is to provide information regarding prognosis, potential therapy, and recurrence chances to optimize fetal outcomes.[2] Testing for prenatal diagnosis encompasses both invasive and noninvasive tests as well as taking detailed family histories as well as history of any congenital or genetic disorders in previous childbirth. A pedigree analysis is always preferred if any presence of genetic disorder in the family.

17.2 PREGNANCY SCREENING

17.2.1 Screening Intervals

The various invasive and noninvasive tests are administered in either the first or the second trimester. First trimester screening takes place between 11 and 13 + 6 weeks of gestation corresponding to crown–rump length of 45–84 mm. The screening is guided by needs related to maternal age: Advanced maternal age increases the likelihood of a fetus with a chromosomal abnormality such as trisomy 13, 18, and 21, although maternal age is not associated risk of Turner syndrome.

The first trimester includes measuring nuchal translucency, the maximum thickness of the subcutaneous translucent area between skin and soft tissue over the spine at neck level, measured via ultrasound. Nuchal translucency is elevated in Downs syndrome and other chromosomal anomalies. The dual test is also administered: This entails measuring beta-hCG (a subunit of human chorionic gonadotrophin) and PAPP-A; both of which are produced by the placenta. In most fetuses with aneuploidy, beta-hCG is elevated and PAPP-A is low; additional ultrasound markers in the first trimester that have been associated with chromosomal abnormality include absent nasal bone, reversal of a wave in DV doppler, and tricuspid regurgitation. Today's most-used screening protocols combine all four measures in what's called a combined screen, and these have increased the detection rates for Downs syndrome and other aneuploidies from 79% to 87%–90% with a 5% false-positive rate.[3]

Second-trimester screening is carried out between 15 and 20 weeks of gestation. A triple serum screening tests for MSAFP, beta-hCG and unconjugated estriol (uE3) between 15 and 18 weeks of gestation. MSAFP and uE3 both are lower and beta-hCG is higher in trisomy 21, and all three are lower in trisomy 18. This test has a detection rate of 61%–70%.[4] For a quadruple test, inhibin A is added to these, and the quadruple test has a detection rate of 80% with 5% false positives; screening takes place at 16–20 weeks gestation. Comprehensive ultrasound, referred to as level 2 detailed anomaly scan, is carried out at 18–20 weeks of gestation to rule out the gross congenital malformations.

Many markers that can potentially be seen on second trimester ultrasound suggest a possible chromosomal anomaly and are referred to as soft markers. The various soft markers of clinical significance include increased nuchal thickness, echogenic bowel, echogenic foci in heart, absence or hypoplastic nasal bone, dilated renal pelvis, choroid plexus cyst, short femur or humerus, aberrant right subclavian artery and other cord abnormalities.[1] The presence of one or more of these markers at the anomaly scan should always prompt referral to a specialist in fetal medicine for further evaluation.

17.2.2 Invasive Tests

In the event of any screening abnormalities, such as any significantly deranged biochemical marker values or gross congenital malformations detected on ultrasound, diagnostic techniques are available to confirm the suspected diagnosis of chromosomal abnormalities. These diagnostic tests are invasive and carry a risk of miscarriage.

Chorionic villus sampling (CVS) entails obtaining chorionic cells under ultrasound guidance for karyotyping and is

* Corresponding author: rashmibagga@gmail.com

DOI: 10.1201/9781032660592-21

performed at 11–14 weeks of gestation; the sample is taken either transabdominally or transvaginally under aseptic conditions. The indications for CVS include for detecting chromosomal genetic disorders such as haemoglobinopathies and sickle cell anemia and inborn errors of metabolism. The procedure has the advantages of both earlier diagnosis and more DNA than can be extracted from uncultured amniocytes. Rapid karyotyping can produce preliminary results within next 48 hours; however, because of the risk of false positives, term cultures are for confirmation. The risks associated with CVS include fetal loss, bleeding, rupture of membranes, chorioamnionitis, cramping and isoimmunization of Rh-negative females.[5] It is never performed prior to nine weeks since it can lead to fetal limb defects.

Amniocentesis involves aspirating amniotic fluid from the amniotic sac under ultrasound guidance via needle inserted transabdominally; the fetal skin cells are obtained from amniotic fluid for culture to karyotype the fetus. It is also performed after 16 weeks of gestation[6] and is one of the most performed invasive procedures. Fetal risks include miscarriage, amniotic fluid leak, fetal trauma, and so on. The common diagnostic indicators for amniocentesis include the antenatal diagnosis of chromosomal abnormalities and genetic disorders, such as sex-linked disorders, and karyotyping to identify inborn errors of metabolism. Therapeutic indicators for amniocentesis include decompressing the uterus in unresponsive cases of chronic hydramnios and amnioinfusion in oligohydramnios.

Cordocentesis is indicated for the evaluation and treatment of fetal isoimmunization when there is a suspected need for in utero transfusion. At ≥20 weeks of gestational age, the blood sample is taken from the umbilical vein at the site where the cord attaches to the placenta. An increased risk of fetal loss (1.5% to 2%) has been observed during cordocentesis.[6]

17.2.3 DNA-Based Noninvasive Testing

The revolutionary technique of noninvasive prenatal testing (NIPT) centers on the cell-free fetal DNA (cfDNA) present in maternal circulation. Fetal DNA enters the maternal bloodstream through placental microparticles; the distinguishing factor is that fetal fragments are notably smaller (<200 base pairs) than maternal ones. Around 10%–15% of maternal plasma cfDNA is of placental origin after 10 weeks of gestation, sufficient for testing fetal disorders.[7] Analyzing specific DNA fragment lengths, methylation patterns, genetic polymorphisms, and quantitative differences in genomic representation aids in the analysis.

NIPT employs massively parallel shotgun sequencing (MPS), where all cfDNA is sequenced and mapped to the human reference genome. Increased fragments of a trisomic chromosome indicate fetal abnormalities. However, the technique requires sequencing numerous DNA fragments to ensure reliable statistical differences between normal and trisomic cases. This balance between sequencing enough fragments and managing costs is essential.

Targeted sequencing approaches like Digital Analysis of Selected Regions (DANSR) have emerged. DANSR involves selective amplification of specific chromosome regions, enabling more efficient sequencing and reducing costs. Combining DANSR with fetal fraction optimization improves accuracy for detecting trisomies 21 and 18 while reducing the number of reads and costs. Considerations affecting NIPS results include fetal DNA quantity, maternal age-based risk, multifetal gestations, and placental mosaicism. Clinical applications encompass fetal sex determination for X-linked disease risk assessment, blood type analysis, detection of microdeletion/duplication syndromes, chromosomal disorders, autosomal single-gene disorders, and copy number variants.

However, NIPT has limitations such as null calls, applicability only after the 10th week of gestation, inapplicability to obese individuals or recent blood transfusion recipients, and its screening nature. It cannot predict late pregnancy complications or screen for single-gene disorders. It also doesn't replace routine fetal anatomical screening or maternal serum α-fetoprotein testing for open neural tube defects.

NIPT leverages fetal DNA in maternal circulation for prenatal screening. It employs techniques like massively parallel shotgun sequencing and targeted sequencing, enabling the detection of a range of genetic abnormalities. While it offers valuable insights, certain limitations and considerations must be kept in mind during its clinical application. A variety of genetic tests can be performed on the fetal samples obtained (the amniotic fluid or the chorionic villi): karyotyping, chromosomal microarray analysis, fluorescent *in situ* hybridization (FISH), whole genome sequencing, and whole exome sequencing.

A standard karyotype takes approximately two weeks; cells need to be cultured and divided for analysis. Karyotyping includes testing numerical chromosomal abnormalities like aneuploidy or polyploidy, identifying balanced or unbalanced structural rearrangements. Karyotyping evaluates only dividing cells, and the growth of cultured cells affects how with which results are obtained. Amniotic fluid, which contains amniocytes, epithelial cells, and gastrointestinal mucosal cells yields results in 7–10 days. These dividing cells are arrested in metaphase, and their chromosomes are stained to reveal light and dark bands.

Chromosomal microarray analysis is 100 times more sensitive than karyotyping. It can detect microdeletions and microduplications as small as 50–100 kilobases. If performed on uncultured cells, direct CMA can yield results in three to seven days whereas if cultured cells are required then results are available in 10–14 days. Microarray uses one of two platforms, comparative genomic hybridization and single nucleotide polymorphism or it could be a combination of both.[8] These both platforms detect aneuploidy, unbalanced translocations, microdeletions and microduplications. FISH is used to rapidly identify specific chromosomal abnormalities such as trisomy 21, 18, 13, X or Y. It does not provide information on the entire chromosomal complement but merely the chromosomal or gene region of interest. FISH is typically selected for cases in which findings could alter pregnancy management because of its one- to two-day turnaround time. Finally, whole exome sequencing involves analyzing only the DNA coding regions (i.e., exons), which account for 1%–2% of the genome but contain 85% of the genetic anomalies[9]; whole genome sequencing meanwhile entails analyzing the entire genome.

In terms of clinical applications of noninvasive DNA testing, in pregnancies at increased risk for autosomal trisomy on aneuploidy screening, karyotyping, FISH, and CMA should be made available. Genetic counseling should include information regarding the benefits and limitations of both CMA and karyotyping. CMA can identify autosomal dominant genetic disorders that have not yet manifested in an affected parent. Whole exome sequencing is typically performed on the fetus and parents simultaneously. Testing can reveal an unrelated but medically actionable finding in one parent that could complicate counseling. Consequently, the clinical utility of this technology for prenatal cases is currently limited and is not recommended for routine clinical use.

REFERENCES

1. Bedei I, Wolter A, Weber A, Signore F, Axt-Fliedner R. Chances and Challenges of New Genetic Screening Technologies (NIPT) in Prenatal Medicine from a Clinical Perspective: A Narrative Review. Genes. 2021 Apr;12(4):501.
2. Rappaport VJ. Prenatal Diagnosis and Genetic Screening – Integration into Prenatal Care. Obstetrics and Gynecology Clinics of North America. 2008 Sep 1;35(3):435–58.
3. Carlson LM, Vora NL. Prenatal Diagnosis: Screening and Diagnostic Tools. Obstetrics and Gynecology Clinics. 2017 Jun 1;44(2):245–56.
4. Pös O, Budiš J, Szemes T. Recent Trends in Prenatal Genetic Screening and Testing. F1000Res. 2019 May 31;8:F1000 Faculty Rev–764.
5. Xu Y, Chen L, Liu Y, Hao Y, Xu Z, Deng L, et al. Screening, Prenatal Diagnosis, and Prenatal Decision for Sex Chromosome Aneuploidy. Expert Review of Molecular Diagnostics. 2019 Jun 3;19(6):537–42.
6. Wieacker P, Steinhard J. The Prenatal Diagnosis of Genetic Diseases. Deutsches Ärzteblatt International. 2010 Dec;107(48):857–62.
7. Christiaens L, Chitty LS, Langlois S. Current Controversies in Prenatal Diagnosis: Expanded NIPT That Includes Conditions Other Than Trisomies 13, 18, and 21 Should Be Offered. Prenatal Diagnosis. 2021;41(10):1316–23.
8. Levy B, Stosic M. Traditional Prenatal Diagnosis: Past to Present. In: Levy B, editor. Prenatal Diagnosis. New York, NY: Springer; 2019 [cited 2023 Aug 26]. pp. 3–22. (Methods in Molecular Biology). Available from: https://doi.org/10.1007/978-1-4939-8889-1_1
9. Kane SC, Willats E, Bezerra Maia e Holanda Moura S, Hyett J, da Silva Costa F. Pre-Implantation Genetic Screening Techniques: Implications for Clinical Prenatal Diagnosis. Fetal Diagnosis and Therapy. 2016 Sep 29;40(4):241–54.

18 Innovations in Biomedical Therapies for Male Reproductive Tract Disorders

Madhan Krishnan,[1] *Shyamaladevi Babu,*[1] *and Ahamed Basha Abdul Bari*[2]*

[1] Faculty of Research, Chettinad Hospital and Research Institute, Chettinad Academy of Research and Education, Kelambakkam, Tamil Nadu, India

[2] Department of Physiology, Chettinad Hospital and Research Institute, Chettinad Academy of Research and Education, Kelambakkam, Tamil Nadu, India

18.1 INTRODUCTION

The male reproductive tract (MRT) encompasses a complex network of organs and tissues responsible for the production, maturation, and transport of spermatozoa, as well as the synthesis and secretion of hormones essential for male sexual development and function. Disorders affecting the MRT can significantly impact fertility, sexual health, and overall quality of life (Pandiyan and Khan, 2017). Male factor infertility attributed to abnormalities in sperm production, function, or delivery contributes to nearly 50% of infertility cases. Biomedical advancements have led to sophisticated diagnostic techniques such as computer-assisted semen analysis of sperm parameters including motility, morphology, and concentration. Additionally, genetic testing is useful for identifying chromosomal abnormalities and gene mutations associated with male infertility, allowing for personalized treatment approaches and informed reproductive counseling (Leslie et al., 2024).

In the realm of therapeutic interventions, microsurgical procedures such as varicocelectomy and vasectomy reversal have offered hope to men seeking to restore fertility or alleviate obstructive reproductive tract conditions (Sengupta et al., 2024). Beyond infertility, biomedical innovations have also transformed the management of other MRT ailments including benign prostatic hyperplasia (BPH), erectile dysfunction (ED), and testicular cancer. Minimally invasive surgical techniques, targeted pharmacotherapies, and emerging regenerative medicine approaches hold promise for improving treatment outcomes and reducing the burden of these conditions on men's health (Franco et al., 2023). In this chapter, we explore the latest biomedical innovations driving advancements in the diagnosis, treatment, and prevention of ailments affecting the MRT. By highlighting recent developments and future directions in male reproductive health research and clinical practice, we aim to provide insights into the evolving landscape of male reproductive medicine and its implications for patient care and public health.

18.2 STRUCTURE AND FUNCTION OF THE MRT

The MRT is a finely tuned system consisting of intricate organs and structures that each play a pivotal role in the complex process of human reproduction. A fundamental understanding of the structure and function of these components is essential for appreciating the male reproductive system's normal functioning and its susceptibility to various disorders (O'Donnell et al., 2017).

18.2.1 The Testes

The testes, situated in the scrotum, are the principal male reproductive organs with a central role in sperm production through the seminiferous tubules of the testes, which ensure spermatogenesis. Beyond their reproductive function, the testes play a crucial endocrine role by secreting testosterone. This hormone is vital for the development of male sexual characteristics, influencing factors such as muscle mass, voice deepening, and growth of facial and body hair. This dual functionality of the testes underscores their significance in both the reproductive and hormonal aspects of male physiology (Smith and Walker, 2014).

18.2.1.1 Spermatogenesis

Spermatogenesis is a sophisticated and ongoing process that entails the generation of sperm cells, which mature as they navigate through the tubules, ultimately gaining the capacity to fertilize an egg. The seamless coordination of spermatogenesis is imperative for the continuation of human life, ensuring a constant supply of mature and viable sperm capable of participating in the fertilization process. This intricate biological dance within the testes highlights the significance of spermatogenesis in the broader context of human reproduction and emphasizes its vital role in maintaining the continuity of the human species (Suede et al., 2023).

18.2.1.2 The Epididymis

The epididymis is linked to the testes through a coiled tube. This structure plays a role in the maturation and storage of sperm, essential for successful reproduction. Upon release from the testes, sperm cells enter the epididymis, where they undergo transformation. The epididymis is crucial in preparing and storing sperm, contributing to the overall success of fertilization (James et al., 2020).

* Corresponding author: drahmedbasha@gmail.com

DOI: 10.1201/9781032660592-22

18.2.1.3 The Vas Deferens

The vas deferens, a pivotal component of the male reproductive system's ductal network, is a muscular tube that transports mature sperm from the epididymis to the urethra during ejaculation. This muscular conduit plays a crucial role in facilitating the journey of sperm from their maturation site to the point of delivery. By connecting the epididymis to the urethra, the vas deferens contributes significantly to the coordinated and efficient movement of sperm, highlighting its indispensable role in the male reproductive process (Gurung et al., 2023).

18.2.2 The Prostate

Nestled beneath the bladder and encircling the urethra, the prostate is a walnut-sized entity that holds considerable importance in the composition of semen, the fluid released during ejaculation. Semen serves a dual purpose by providing nourishment and protection to sperm, thus contributing significantly to their overall health and functionality. The prostate's walnut-sized structure belies its substantial impact on the reproductive process; the fluid produced by the prostate gland is integral to the sustenance and viability of sperm as they embark on their journey (InformedHealth.org, 2011). Understanding the functions of the prostate gland provides insights into the intricate mechanisms that support reproductive success in males (Verze et al., 2016).

18.2.3 The Seminal Vesicles

The seminal vesicles, a pair of diminutive glands integral to the male reproductive system, assume a crucial role by augmenting semen with additional fluid characterized by its abundance of fructose, enzymes, and prostaglandins that serve multiple functions that significantly impact the reproductive process. One of their key roles is to furnish energy to sperm, as the fructose acts as a vital energy source for the sperm cells. Furthermore, the enzymes facilitate various physiological processes that are essential for the sperm's functionality. The prostaglandins in the seminal vesicle fluid enhance sperm motility, thereby increasing the likelihood of successful fertilization (Gonzales, 2001).

18.2.4 The Penis and Urethra

The penis, with its dual functionality for both reproductive and urinary processes, undergoes erection during sexual arousal; this physiological response facilitates the introduction of sperm into the female reproductive tract. The urethra, extending through the penis, serves as a conduit for both urine and semen. In essence, the MRT's structure and functions underscore its significance in the perpetuation of human life. From the initiation of spermatogenesis in the testes to the culmination of the reproductive process during ejaculation, each component plays a unique and indispensable role. A comprehensive understanding of this, lays the foundation for addressing disorders, optimizing fertility, and promoting overall male reproductive health (Krassioukov and Elliott, 2017).

18.3 DISORDERS OF THE MRT

The MRT is a multifaceted system that is pivotal to human reproduction and encompasses organs and structures dedicated to sperm production, storage, and transport. This intricate system comprises key components, each with specific functions crucial to the overall reproductive process. A fundamental grasp of the anatomy and physiology of the MRT is essential for comprehending both its normal functioning and the diverse range of disorders that can impact it.

At the core of this system are the testes, responsible for sperm production through spermatogenesis and the secretion of testosterone. The epididymis, a coiled tube connected to the testes, facilitates sperm maturation and storage, playing a vital role in ensuring the sperm's readiness for fertilization. The vas deferens serves as the conduit for mature sperm and transporting them during ejaculation. Other integral components include the prostate gland, seminal vesicles, and the penis with its urethra (Gurung et al., 2023). Disorders of the MRT can profoundly affect fertility, sexual health, and overall wellbeing. Conditions such as testicular torsion, testicular cancer, and prostatitis highlight the vulnerability of this intricate system. Understanding these disorders and their respective treatments is crucial for healthcare professionals and individuals alike for ensuring prompt intervention and management (Dimitriadis et al., 2023).

Understanding the pathophysiology, symptoms, and potential treatments for these disorders is crucial for both medical professionals and individuals seeking insights into male reproductive health. In this chapter, we explore disorders such as cryptorchidism and testicular cancer, as well as issues related to sperm transport such as obstructive azoospermia and complications after vasectomy; we also look at disorders including prostatitis and ejaculatory duct obstruction (Revenig et al., 2014). By navigating the complex landscape of disorders affecting the MRT, this exploration aims to provide a comprehensive resource for healthcare professionals, researchers, and individuals seeking knowledge about reproductive health challenges in men. Insights gleaned from this analysis can foster a deeper understanding of these disorders, ultimately contributing to improved diagnostic approaches, treatment strategies, and overall reproductive wellbeing (Kaltsas et al., 2023).

18.3.1 The Testes

Testicular disorders encompass a spectrum of conditions that can significantly impact fertility, hormonal balance, and overall male reproductive health. They can manifest as abnormalities in structure, function, or hormonal regulation of the testes. Factors such as genetic predisposition, infections, and hormonal imbalances contribute to the development of these disorders

18.3.1.1 Cryptorchidism and Orchidopexy

Cryptorchidism is a prevalent congenital disorder characterized by undescended testes; it occurs when one or both testes fail to descend into the scrotum and instead remain within the abdominal cavity or inguinal canal, it affects newborn males, and it requires timely intervention to prevent complications and ensure optimal reproductive health. Cryptorchidism is often identified during routine physical examinations in newborns or infants, and diagnosis is crucial to initiate timely intervention and mitigate potential complications because the failure of testicular descent can impact both reproductive and general health. Undescended testes are vulnerable to temperature-related damage, potentially affecting sperm production and fertility in adulthood. Additionally, the condition is associated with an increased risk of testicular cancer (Nguyen et al., 2023).

Orchidopexy, a surgical procedure, is the standard treatment for repositioning the undescended testis into the scrotum; also known as orchiopexy, it involves making a small incision in the groin or scrotum, locating the undescended testis, and securing it in the scrotum. In cases of bilateral cryptorchidism, where both testes are affected, bilateral orchidopexy may be performed. Performing orchidopexy at an early age, preferably before two years old, is crucial for preventing potential complications, establishing normal testicular development, reducing infertility risk, and decreasing the likelihood of testicular cancer in adulthood. Orchidopexy is usuallly successful, with a high rate of testicular descent into the scrotum. However, postoperative follow-up is essential for monitoring long-term outcomes and complications (Moore et al., 2021).

18.3.1.2 Testicular Cancer

Testicular cancer, though relatively rare, is a significant health concern, although advances in medical research and technology have led to effective therapies. Testicular cancer is most prevalent among young and middle-aged men; the tumors are broadly categorized as seminomas and non-seminomas, each requiring tailored treatment approaches. Several risk factors contribute to the development of testicular cancer, including a family history of the disease, undescended testicles, and genetic conditions. Early diagnosis is crucial for successful treatment. Common diagnostic methods include physical examination, imaging studies (ultrasound), and tumor marker tests (AFP, hCG, LDH) (Al-Obaidy et al., 2022).

Testicle cancer management options (Chovanec and Cheng, 2022) include orchiectomy, chemotherapy, and radiation treatment among other options. With an orchiectomy, the primary treatment is the surgical removal of the affected testicle, which aids in both diagnosis and treatment. With chemotherapy, the combination of bleomycin, etoposide, and cisplatin (BEP) is both standard and effective against both seminomatous and non-seminomatous tumors. Radiation therapy is either adjuvantor salvage and is employed particularly for seminomas; adjuvant therapy targets potential residual cancer cells after surgery, while salvage therapy addresses recurrence.

High-dose chemotherapy and stem cell transplant are used in resistant cases, specifically high-dose chemotherapy followed by autologous stem cell transplant. After completion of initial treatment, regular surveillance is crucial to monitor for any signs of recurrence; this typically involves imaging studies and tumor marker tests. Ongoing research has led to advancements in targeted therapies and immunotherapy for testicular cancer. These approaches aim to enhance efficacy in the medical management.

A lesser-known disorder of the testicles is testicular torsion, a critical condition characterized by the twisting of the spermatic cord, resulting in diminished blood supply to the testes. This urological emergency demands immediate surgical intervention to prevent irreversible damage. Prompt diagnosis and treatment are crucial for preserving testicular function and mitigating potential complications.

18.3.2 The Epididymis

Epididymitis, characterized by inflammation of the epididymis, is often a result of infection; it manifests with pain and swelling in the affected area. Antibiotics are the primary course of treatment for infectious epididymitis, highlighting the importance of identifying the causative agent for targeted therapy. Separately, benign cysts in the epididymis can cause discomfort, and although they are generally non-threatening, symptomatic cases can necessitate drainage or removal. Understanding the nature of these cysts and their impact on reproductive health is crucial for determining appropriate management strategies (Nickel, 2003).

18.3.3 The Vas Deferens and Urethra

Often linked with cystic fibrosis, the congenital absence of the vas deferens can result in male infertility. Conception can require assisted reproductive technologies to overcome the impediment presented by this condition. Separately, urethral stricture, the narrowing of the urethra, can hinder the flow of urine and semen. Treatment options vary from dilation to surgical intervention depending on the severity of the stricture (de Souza et al., 2018).

18.3.4 The Prostate

Prostatitis, either acute or chronic, involves inflammation of the prostate that leads to pain and urinary symptoms; treatment involves antibiotics and anti-inflammatory medications. BPH can result in urinary symptoms, and management options range from medications to surgical intervention. Prostate cancer is the most prevalent cancer in men, and treatment modalities encompass surgery, radiation, and hormone therapy. Early detection through screening enables timely intervention, contributing to better prognosis and outcomes (Strauss and Dimitrakov, 2010).

18.3.5 The Penis and Seminal Vesicles

Erectile dysfunction, the inability of a penis to achieve erection, can stem from various causes, including vascular, neurological, or psychological factors. Treatment options span from oral medications to surgical interventions, with a tailored approach based on the underlying cause. Peyronie's disease, another penis disorder, is characterized by fibrous plaques in the penis leading to curvature and pain during erections. Early diagnosis and intervention can help manage symptoms and preserve sexual function, and treatment options include medications and surgery (Patel et al., 2012). Meanwhile, seminal vesiculitis, inflammation of the seminal vesicles often due to infection can impact semen quality; while often asymptomatic, the cysts can cause discomfort that requires treatment, and antibiotic therapy is typically prescribed (Dagur et al., 2016).

18.4 INFERTILITY AND REPRODUCTIVE HEALTH

Male infertility arises from hormonal imbalances, sperm abnormalities, or morphological issues. Diagnosis involves a thorough evaluation of semen quality and hormonal profiles. Treatment strategies encompass lifestyle modifications (Dupesh et al., 2020), medications, or assisted reproductive technologies to address specific causative factors. Each condition demands specific considerations for diagnosis, treatment, and ongoing management. By understanding the pathophysiology of these disorders, healthcare professionals and individuals alike can make informed decisions, leading to better outcomes and improved reproductive health (Babakhanzadeh et al., 2020).

18.5 BIOMEDICAL INNOVATIONS IN THERAPIES FOR MALE REPRODUCTIVE DISEASES

We have addressed a number of diseases that significantly impact fertility, sexual function, and overall reproductive health. In recent years, advancements in medical research and technology have led to the development of various therapeutic approaches for addressing male reproductive diseases (Stewart and Shipley, 2021) that we discuss in this section.

18.5.1 Erectile Dysfunction

ED is common in men and is marked by the incapacity to attain or sustain an erection adequate for sexual satisfactory performance. Understanding the multifaceted nature of ED is crucial for tailoring effective treatment approaches (Irwin, 2019).

Treatment for ED aims to improve erectile function, help circulatory health, and enhance the quality of life for men with ED (Kim et al., 2021) and includes a number of primary therapeutic options. For instance, lifestyle modifications such as smoking cessation, reduced alcohol intake, and increased physical activity can enhance erectile function. Via the oral route, phosphodiesterase type 5 inhibitors like tadalafil, sildenafil and vardenafil are commonly prescribed; these medications facilitate increased blood flow to the penis to support erections. Externally, vacuum-assisted erectile devices are noninvasive and utilize a vacuum to draw blood into the penis and induce an erection; intra urethral suppositories are small pellets that are inserted into the urethra to promote erection; with intra cavernosal injections, medication is injected directly into the penis to facilitate an erection; and with surgical implants, penile prostheses are surgically implanted to enable erections. In more recent advances, ongoing research is exploring external prosthetic devices, shock-wave therapy, and injecting stem cells or platelet-rich plasma for ED. These approaches are in the experimental phase, necessitating further investigation into their safety and effectiveness (Kim et al., 2021; Towe et al., 2022).

18.5.2 Male Infertility

Male infertility poses significant challenges for a considerable number of men, often manifesting as issues like poor sperm quality, low sperm count, or reproductive system blockages. Fortunately, a range of treatment options exists to address male infertility and enhance the likelihood of conception. These interventions entail lifestyle adjustments that support ARTs (Choy and Eisenberg, 2018). For instance, lifestyle modifications can have a notable impact on male fertility and are often recommended as initial measures to improve sperm quality and overall reproductive health (Taitson et al., 2022).

When male infertility is linked to hormonal imbalances, hormone therapy can improve sperm production, and other can increase sperm count and motility. ARTs encompass methodologies like IVF, ICSI, and IUI. These techniques involve manipulating sperm and eggs to facilitate fertilization and embryo development (Taitson et al., 2022). Surgical procedures such as varicocele repair, vasectomy reversal, and sperm retrieval can address anatomical abnormalities, restore sperm flow, or retrieve viable sperm for use in ART procedures. These interventions are tailored to the specific needs of each patient and can significantly impact male fertility outcomes (Taitson et al., 2022; American Urological Association, 2023). Finally, recent developments in male infertility involve advancements in stem cell therapy that can regenerate tissue.

18.5.3 BPH

BPH is common in aging men and is characterized by noncancerous prostate gland enlargement, leading to lower urinary tract symptoms of varying prevalence and severity (Langan, 2019). Treatment takes multiple forms depending on disease severity. For patients with mild symptoms, for instance, watchful waiting is often recommended as BPH is not a progressive disease for all men. This approach involves regular monitoring of symptoms and is suitable for those with minimal impact on their quality of life (Franco et al., 2023).

When drugs are called for, alpha-blockers help relax the muscles in the prostate and bladder, making it easier to urinate, and they are often used to improve urine flow and reduce symptoms, and 5-alpha-reductase inhibitors work by reducing the size of the prostate (Devlin et al., 2021). For necessary interventional therapy, transurethral resection removes the inner part of the prostate that is pressing against the urethra. It is considered the gold standard for surgical treatment of BPH (Sciacqua et al., 2023).

However, minimally invasive treatments are also available for BPH, such as laser therapy, which uses different lasers to remove prostate tissue and improve symptoms. Uro Lift is a minimally invasive procedure that uses implants to lift and hold the enlarged prostate tissue out of the way, opening the urethra and reducing symptoms (Franco et al., 2023). In some cases, a combination of drug therapies can improve symptoms and prevent disease progression, for example, an alpha-blocker combined with a 5-alpha-reductase inhibitor (Devlin et al., 2021). Ongoing research is exploring new therapeutic options such as phospho diesterase inhibitors (Miernik and Gratzke, 2020).

18.5.4 Prostate Cancer

Prostate cancer, a prevalent malignancy among men, has witnessed an evolution in treatment options over the years. Current therapies for prostate cancer, as outlined by Wasim et al. (2022), encompass a range of strategies tailored to meet individual patient needs, aiming to enhance outcomes and improve the quality of life for those affected.

For instance, active surveillance is often employed for early-stage or slow-growing tumors, while surgery such as radical prostatectomy is a viable option for localized cases. Radiation therapy, including external beam radiation and brachytherapy, targets cancer cells and is frequently used for localized prostate cancer. Hormonal therapy, or androgen deprivation, is a common treatment involving reducing testosterone levels to impede prostate cancer growth. Chemotherapy featuring docetaxel and cabazitaxel is effective for extending life expectancy. Targeted therapy, exemplified by PARP inhibitors, is combined with hormone therapy to treat advanced prostate cancer. Immunotherapy, typified by sipuleucel-T, serves as a personalized vaccine for advanced cases. Ongoing clinical trials explore novel treatment options like proton beam radiation therapy, and photodynamic therapy, with the goal of enhancing outcomes and elevating the quality of life for prostate cancer patients (Carlsson and Vickers, 2020; Williams et al., 2022).

18.5.5 Varicoceles

Varicoceles are common and are characterized by enlarged scrotum veins that impair blood flow and increase temperatures in the testicles. While often asymptomatic, varicoceles can lead to fertility issues and discomfort. Common causes include faulty valves in the veins and increased pressure within the blood vessels, and risk factors include family history, age, and conditions causing increased abdominal pressure. Diagnosing varicoceles typically involves a thorough physical examination, but scrotal ultrasound can be used to visualize and confirm the enlargement of the veins (Fang et al., 2021). How to manage varicoceles depends on their severity; asymptomatic cases that are not affecting fertility only call for a conservative approach involving observation. Wearing supportive underwear and avoiding excessive heat exposure to the scrotum helps alleviate discomfort and minimize the impact on sperm quality.

Surgical procedures such as varicocelectomy involve ligating or blocking the affected veins to redirect blood flow via open surgery or minimally invasive techniques like laparoscopy or microsurgery. When the varicoceles are contributing to male infertility, surgery corrects the sperm parameters before the man undergoes assisted reproductive techniques such as IVF or IUI. Finally, varicocele embolization is a nonsurgical approach that involves blocking the affected veins using a catheter and is called for especially for recurrent cases or in individuals with contraindications for surgery, and research is ongoing exploring novel approaches, including using antioxidants and anti-inflammatory agents to manage varicoceles and improve sperm quality. These approaches aim to address the underlying factors contributing to varicocele-related infertility (Barati et al., 2020; Minhas et al., 2021; Huyghe et al., 2023).

18.5.6 Hypogonadism

Hypogonadism is a condition characterized by diminished gonadal function that can profoundly impact reproductive and overall health in males. It stems from disruptions in the hypothalamic–pituitary–gonadal axis, leading to insufficient production of sex hormones. It is categorized as primary (hypergonadotropism) or secondary (hypogonadotropism), each with distinct etiologies. Common causes include aging, genetic disorders, testicular trauma, and systemic illnesses. Accurate diagnosis of hypogonadism involves a comprehensive evaluation encompassing clinical assessments, hormonal assays, and imaging studies. Symptoms such as fatigue, decreased libido, and infertility, coupled with laboratory measurements of testosterone levels and additional hormonal markers, aid in confirming the diagnosis. Contemporary management approaches for hypogonadism begin with testosterone replacement therapy, a cornerstone in managing hypogonadism that aims to restore testosterone levels to the normal physiological range. Modalities include intramuscular injections, transdermal patches, gels, and buccal formulations.

For other medical options, clomiphene citrate acts on the hypothalamus and pituitary, increasing gonadotropin release and subsequently enhancing testosterone production. It is an option for those desiring fertility preservation. Humanchorionic gonadotropin, a hormone that mimics luteinizing hormone, stimulates testicular Leydig cells to produce testosterone. It is particularly considered in hypogonadal men with preserved testicular responsiveness. Nonmedical lifestyle interventions including weight loss and regular exercise can positively impact testosterone levels, especially in cases associated with obesity or sedentary habits, and in terms of emerging therapies, selective estrogen receptor modulators are being explored for their potential in modulating estrogenic effects and enhancing testosterone secretion (Lundy and Sabanegh, 2017; Swee and Quinton, 2019).

18.6 CONCLUSION

In conclusion, the MRT is a finely tuned system that is crucial for human reproduction, and disorders affecting this system have profound implications for fertility, sexual health, and overall wellbeing. Understanding the anatomy, functions, and potential disorders of the MRT is essential for healthcare professionals and individuals alike. Advances in research and therapeutic interventions continue to shape the landscape of male reproductive health, offering hope for improved outcomes and increased understanding in the future.

REFERENCES

Al-Obaidy, K. I., Magers, M. J., & Idrees, M. T. (2022). Testicular cancer: Contemporary updates in staging. *Surgical Pathology Clinics*, *15*(4), 745–757. https://doi.org/10.1016/j.path.2022.07.010.

American Urological Association. (2023). *Management of lower urinary tract symptoms attributed to benign prostatic hyperplasia: AUA guideline*. Available from: https://www.auanet.org/guidelines-and-quality/guidelines/benign-prostatic-hyperplasia-(bph)-guideline.

Babakhanzadeh, E., Nazari, M., Ghasemifar, S., & Khodadadian, A. (2020). Some of the factors involved in male infertility: A prospective review. *International Journal of General Medicine*, *13*, 29–41. https://doi.org/10.2147/IJGM.S241099.

Barati, E., Nikzad, H., & Karimian, M. (2020). Oxidative stress and male infertility: Current knowledge of pathophysiology and role of antioxidant therapy in disease management. *Cellular and Molecular Life Sciences: CMLS*, *77*(1), 93–113. https://doi.org/10.1007/s00018-019-03253-8.

Carlsson, S. V., & Vickers, A. J. (2020). Screening for prostate cancer. *The Medical Clinics of North America*, *104*(6), 1051–1062. https://doi.org/10.1016/j.mcna.2020.08.007.

Chovanec, M., & Cheng, L. (2022). Advances in diagnosis and treatment of testicular cancer. *BMJ (Clinical Research Ed.)*, *379*, e070499. https://doi.org/10.1136/bmj-2022-070499.

Choy, J. T., & Eisenberg, M. L. (2018). Male infertility as a window to health. *Fertility and Sterility*, *110*(5), 810–814. https://doi.org/10.1016/j.fertnstert.2018.08.015.

Dagur, G., Warren, K., Suh, Y., Singh, N., & Khan, S. A. (2016). Detecting diseases of neglected seminal vesicles using imaging modalities: A review of current literature. *International Journal of Reproductive Biomedicine*, *14*(5), 293–302.

de Souza, D. A. S., Faucz, F. R., Pereira-Ferrari, L., Sotomaior, V. S., & Raskin, S. (2018). Congenital bilateral absence of the vas deferens as an atypical form of cystic fibrosis: Reproductive implications and genetic counseling. *Andrology*, *6*(1), 127–135. https://doi.org/10.1111/andr.12450.

Devlin, C. M., Simms, M. S., & Maitland, N. J. (2021). Benign prostatic hyperplasia – what do we know? *BJU International*, *127*(4), 389–399. https://doi.org/10.1111/bju.15229.

Dimitriadis, F., Borgmann, H., Struck, J. P., Salem, J., & Kuru, T. H. (2023). Antioxidant supplementation on male fertility-a systematic review. *Antioxidants (Basel, Switzerland)*, *12*(4), 836. https://doi.org/10.3390/antiox12040836.

Dupesh, S., Pandiyan, N., Pandiyan, R., Kartheeswaran, J., & Prakash, B. (2020). Ejaculatory abstinence in semen analysis: Does it make any sense? *Therapeutic Advances in Reproductive Health*, *14*. https://journals.sagepub.com/doi/10.1177/2633494120906882.

Fang, Y., Su, Y., Xu, J., Hu, Z., Zhao, K., Liu, C., & Zhang, H. (2021). Varicocele-mediated male infertility: From the perspective of testicular immunity and inflammation. *Frontiers in Immunology*, *12*, 729539. https://doi.org/10.3389/fimmu.2021.729539.

Franco, J. V. A., Tesolin, P., & Jung, J. H. (2023). Update on the management of benign prostatic hyperplasia and the role of minimally invasive procedures. *Prostate International*, *11*(1), 1–7. https://doi.org/10.1016/j.prnil.2023.01.002.

Gonzales, G. F. (2001). Function of seminal vesicles and their role on male fertility. *Asian Journal of Andrology*, *3*(4), 251–258.

Gurung, P., Yetiskul, E., & Jialal, I. (2023). Physiology, male reproductive system. [Updated 2023 May 1]. In: *StatPearls* [Internet]. Treasure Island, FL: StatPearls Publishing. Available from: https://www.ncbi.nlm.nih.gov/books/NBK538429/.

Huyghe, E., Methorst, C., & Faix, A. (2023). Varicocèle et infertilité masculine [Varicocele and male infertility]. *Progres en urologie: journal de l'Association francaise d'urologie et de la Societe francaise d'urologie*, *33*(13), 624–635. https://doi.org/10.1016/j.purol.2023.09.003.

InformedHealth.org [Internet]. (2011). How does the prostate work? In: *Cologne, Germany: Institute for Quality and Efficiency in Healthcare (IQWiG); 2006-*. [Updated 2016 August 23]. Available from: https://www.ncbi.nlm.nih.gov/books/NBK279291/.

Irwin, G. M. (2019). Erectile dysfunction. *Primary Care*, *46*(2), 249–255. https://doi.org/10.1016/j.pop.2019.02.006.

James, E. R., Carrell, D. T., Aston, K. I., Jenkins, T. G., Yeste, M., & Salas-Huetos, A. (2020). The role of the epididymis and the contribution of epididymosomes to mammalian reproduction. *International Journal of Molecular Sciences*, *21*(15), 5377. https://doi.org/10.3390/ijms21155377.

Kaltsas, A., Zachariou, A., Markou, E., Dimitriadis, F., Sofikitis, N., & Pournaras, S. (2023). Microbial dysbiosis and male infertility: Understanding the impact and exploring therapeutic interventions. *Journal of Personalized Medicine*, *13*(10), 1491. https://doi.org/10.3390/jpm13101491.

Kim, S., Cho, M. C., Cho, S. Y., Chung, H., & Rajasekaran, M. R. (2021). Novel emerging therapies for erectile dysfunction. *World Journal of Men's Health*, *39*(1), 48–64. https://doi.org/10.5534/wjmh.200007.

Krassioukov, A., & Elliott, S. (2017). Neural control and physiology of sexual function: Effect of spinal cord injury. *Topics in Spinal Cord Injury Rehabilitation*, *23*(1), 1–10. https://doi.org/10.1310/sci2301-1.

Langan, R. C. (2019). Benign prostatic hyperplasia. *Primary Care*, *46*(2), 223–232. https://doi.org/10.1016/j.pop.2019.02.003.

Leslie, S. W., Soon-Sutton, T. L., & Khan, M. A. B. (2024, February 25). Male infertility. In: *StatPearls*. StatPearls Publishing. https://www.ncbi.nlm.nih.gov/books/NBK562258/.

Lundy, S. D., & Sabanegh, E. S., Jr. (2017). Varicocele management for infertility and pain: A systematic review. *Arab Journal of Urology*, *16*(1), 157–170. https://doi.org/10.1016/j.aju.2017.11.003.

Miernik, A., & Gratzke, C. (2020). Current treatment for benign prostatic hyperplasia. *Deutsches Arzteblatt International*, *117*(49), 843–854. https://doi.org/10.3238/arztebl.2020.0843.

Minhas, S., Bettocchi, C., Boeri, L., Capogrosso, P., Carvalho, J., Cilesiz, N. C., Cocci, A., Corona, G., Dimitropoulos, K., Gül, M., Hatzichristodoulou, G., Jones, T. H., Kadioglu, A., Martínez Salamanca, J. I., Milenkovic, U., Modgil, V., Russo, G. I., Serefoglu, E. C., Tharakan, T., Verze, P., . . . EAU Working Group on Male Sexual and Reproductive Health. (2021). European Association of Urology Guidelines on Male Sexual and Reproductive Health: 2021 update on male infertility. *European Urology*, *80*(5), 603–620. https://doi.org/10.1016/j.eururo.2021.08.014.

Moore, S. L., Chebbout, R., Cumberbatch, M., Bondad, J., Forster, L., Hendry, J., Lamb, B., MacLennan, S., Nambiar, A., Shah, T. T., Stavrinides, V., Thurtle, D., Pearce, I., & Kasivisvanathan, V. (2021). Orchidopexy for testicular torsion: A systematic review of surgical technique. *European Urology Focus*, *7*(6), 1493–1503. https://doi.org/10.1016/j.euf.2020.07.006.

Nguyen, V., Ngo, L., & Jaqua, E. E. (2023). Cryptorchidism (undescended testicle). *American Family Physician*, *108*(4), 378–385.

Nickel, J. C. (2003). Chronic epididymitis: A practical approach to understanding and managing a difficult urologic enigma. *Reviews in Urology*, *5*(4), 209–215.

O'Donnell, L., Stanton, P., & de Kretser, D. M. (2000). Endocrinology of the male reproductive system and spermatogenesis. [Updated 2017 January 11]. In: Feingold, K. R., Anawalt, B., Blackman, M. R., et al., editors. *Endotext* [Internet]. South Dartmouth, MA: MDText.com, Inc. Available from: https://www.ncbi.nlm.nih.gov/books/NBK279031/.

Pandiyan, N., & Khan, S. D. (2017). A clinical approach to male infertility. *Male Infertility: A Clinical Approach*, 41–54. https://doi.org/10.1007/978-81-322-3604-7_4.

Patel, D. V., Halls, J., & Patel, U. (2012). Investigation of erectile dysfunction. *British Journal of Radiology*, *85*(Spec Iss. 1), S69–S78. https://doi.org/10.1259/bjr/20361140.

Revenig, L., Leung, A., & Hsiao, W. (2014). Ejaculatory physiology and pathophysiology: Assessment and treatment in male infertility. *Translational Andrology and Urology*, *3*(1), 41–49. https://doi.org/10.3978/j.issn.2223-4683.2014.02.02.

Sciacqua, L. V., Vanzulli, A., Di Meo, R., Pellegrino, G., Lavorato, R., Vitale, G., & Carrafiello, G. (2023). Minimally invasive treatment in Benign Prostatic Hyperplasia (BPH). *Technology in Cancer Research & Treatment*, *22*. https://doi.org/10.1177/15330338231155000.

Sengupta, P., Dutta, S., Jegasothy, R., Slama, P., Cho, C. L., & Roychoudhury, S. (2024). 'Intracytoplasmic sperm injection (ICSI) paradox' and 'andrological ignorance': AI in the era of fourth industrial revolution to navigate the blind spots. *Reproductive Biology and Endocrinology: RB&E*, *22*(1), 22. https://doi.org/10.1186/s12958-024-01193-y.

Smith, L. B., & Walker, W. H. (2014). The regulation of spermatogenesis by androgens. *Seminars in Cell & Developmental Biology*, *30*, 2–13. https://doi.org/10.1016/j.semcdb.2014.02.012.

Stewart, J. L., & Shipley, C. F. (2021). Management of reproductive diseases in male small ruminants. *The Veterinary Clinics of North America. Food Animal Practice*, *37*(1), 105–123. https://doi.org/10.1016/j.cvfa.2020.10.005.

Strauss, A. C., & Dimitrakov, J. D. (2010). New treatments for chronic prostatitis/chronic pelvic pain syndrome. *Nature Reviews. Urology*, *7*(3), 127–135. https://doi.org/10.1038/nrurol.2010.4.

Suede, S. H., Malik, A., & Sapra, A. (2023). Histology, spermatogenesis. [Updated 2023 Mar 6]. In: *StatPearls*[Internet]. Treasure Island, FL: StatPearls Publishing. Available from: https://www.ncbi.nlm.nih.gov/books/NBK553142/.

Swee, D. S., & Quinton, R. (2019). Managing congenital hypogonadotrophic hypogonadism: A contemporary approach directed at optimizing fertility and long-term outcomes in males. *Therapeutic Advances in Endocrinology and Metabolism*, *10*. https://doi.org/10.1177/2042018819826889.

Taitson, P. F., Mourthé Filho, A., & Rodrigues, L. M. F. (2022). Treating male infertility. *JBRA Assisted Reproduction*, *17*(6), 351–352.

Towe, M., Peta, A., Saltzman, R. G., Balaji, N., Chu, K., & Ramasamy, R. (2022). The use of combination regenerative therapies for erectile dysfunction: Rationale and current status. *International Journal of Impotence Research*, *34*(8), 735–738. https://doi.org/10.1038/s41443-021-00456-1.

Verze, P., Cai, T., & Lorenzetti, S. (2016). The role of the prostate in male fertility, health and disease. *Nature Reviews Urology*, *13*(7), 379–386. https://doi.org/10.1038/nrurol.2016.89.

Wasim, S., Lee, S. Y., & Kim, J. (2022). Complexities of prostate cancer. *International Journal of Molecular Sciences*, *23*(22), 14257. https://doi.org/10.3390/ijms232214257.

Williams, I. S., McVey, A., Perera, S., O'Brien, J. S., Kostos, L., Chen, K., Siva, S., Azad, A. A., Murphy, D. G., Kasivisvanathan, V., Lawrentschuk, N., & Frydenberg, M. (2022). Modern paradigms for prostate cancer detection and management. *Medical Journal of Australia*, *217*(8), 424–433. https://doi.org/10.5694/mja2.51722.

Section V

Emerging Technologies in Drug Development, Design, and Delivery Systems Including Drug Repurposing

19 Biotechnological Advances in Geriatric Medicine

Brij Bharti*
Department of Life Science, JC Bose University of Science and Technology, Faridabad, Haryana, India

19.1 INTRODUCTION

Biotechnological advancement encourages the scientists and researchers to identify, characterize, and to find a solution or treatment of a disease. In the area of geriatric, the use of biotechnologies is limited due to specialized prognosis, care and subjective treatment of age associated complexity and biology of diseases.

The medical advancement of geriatric medicine showed that statins like pravastatin decreased the risk of death caused by stroke, coronary events (Lemaitre et al., 1998) and cardiovascular risk. Likewise, diuretics and beta-blockers reduce the risk of strokes and cardiovascular mortality in geriatric people, and calcium and vitamin D decrease the risk of non-vertebral fractures in healthy people over 65 (Sharon, 2001). However geriatric medicine helps geriatric people to overcome the diseased condition but there are so many challenges faced them. The main limitation of geriatric medicine is that critically ill geriatric patients is at an increased risk of adverse drug reaction due to changes in physiology, metabolism, and polypharmacy.

19.2 GERIATRIC DISEASES AND MEDICINES

Geriatric explores the aging process, the impact of aging on illness patterns, drug therapy in seniors, health maintenance, and rehabilitation. Geriatrics help senior person manage their overall health and wellbeing and maintain their independence for as long as possible. It deals with the unique health needs of the elderly, including chronic diseases, nutritional problems, skin ailments, cognitive loss, memory impairment, adverse effects of medications, immobility, and balances issue created by natural aging process. Aging poses profound health-related challenges that need to be tackled to reduce the social and economic burden on our aging society. Multidisciplinary perspectives require an interrelation between physical mental and social factor will be of tremendous importance to understand the underlying molecular processes of aging and to accelerate the discovery and development of effective aging interventions. It is therefore indispensable that industry and academia develop deeper cooperation and greater interchange of knowledge and technology (Bakula et al., 2019). This would be a great approach to take care of elderly people in all respect.

Geriatric diseases represent a group of changes which can be divide in to sub domains. These sub-domains include normal aging common diseases, and functional, cognitive, psychiatric, and social changes (Jaul and Barron, 2017). However, these changes can be observed in every organ system but for the sake of convenience it can be describes as sensory changes (Davis et al., 2016), muscle health and fat changes (Xue Ql et al., 2010), immunosenescence (Bandaranayake, 2016) and urological changes (Ariathianto, 2011) by the Jaul and Barron in 2017. On the other hand, geriatric diseases also include somatic and multiple chronic condition which take in cardiovascular diseases, hypertension, osteoarthritis, diabetes, osteoporosis, and cancer, respectively (Wang, 2012; Odden et al., 2014; Gerteis et al., 2014). These reports propose monitoring of person's condition before time prevent from major consequence of disease.

The goal of GM is to integrate the management of these conditions under the aim of preserving function, maintaining autonomy as long as possible, promoting wellbeing and social engagement, and, at the end of one's trajectory, humane end-of-life care. Life prolongation is not an absolute goal in GM and aspects such as quality of life; personal values; individualized health priorities; and rational, often modest, objectives frequently guide clinical decisions.

19.3 CHALLENGES IN GERIATRIC MEDICINE

In developing countries, many geriatric patients face socioeconomic challenges such as lack of infrastructure; changing family structure, lack of social support; social inequality; availability, accessibility, and affordability of healthcare; and economic dependency (WHO, 1948). Many of these problems cannot be resolved with medicine, but increasing awareness of the needs of geriatric patients can extend life expectancy. Geriatric medicine focused on some basic principles like fighting against immobility, the fight against ageism, description of geriatric syndrome and introduction of comprehensive geriatric assessment (Ribera Casado, 2012).

* Corresponding author: brijbharti@jcboseust.ac.in

DOI: 10.1201/9781032660592-24

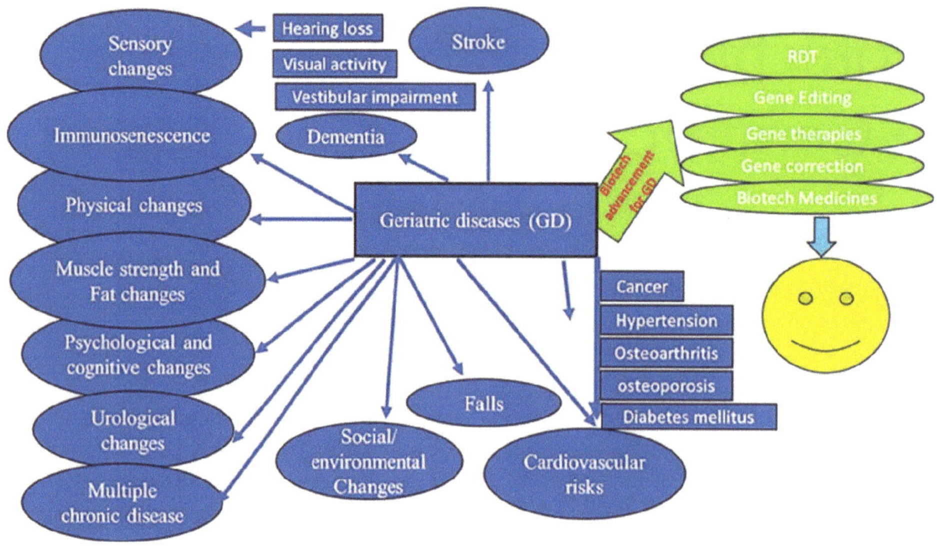

FIGURE 19.1 Geriatric diseases and potential biotechnological treatment approaches.

19.4 GENETIC AND MOLECULAR ADVANCEMENTS IN GERIATRIC MEDICINE

Biomedical and genetic techniques are making it possible to find genes on chromosomes and sequence of nucleotides in gene and genetic polymorphism related with disease but it does not answer clinical question about the disease. Detection and treatment of disease at any stage require morphological, physiological, molecular, and genetic aspect about the disease. For example, telomere i.e. the sequence of DNA and responsible for the cell division, apoptosis. It also play an important role in aging, so by taking this in account we can slow down the process of aging.

Understanding aging mechanisms at the genetic and molecular levels has indicated that chronic sterile inflammation, macromolecular and organelle dysfunction, stem and progenitor cell dysfunction, and cellular senescence are the four basic processes leading to geriatric syndromes (Chang et al., 2016). The proteomic and genomic studies of age-related disorders provide ideas for treating diseases. In recent studies, the age proteome is highly tissue specific (Long, 2020), but few proteins change systematically with aging across selected tissues, such as the heart, lung and whole blood, suggesting a tighter co-aging pattern than in other tissues such as muscle (Yang et al., 2015).

Omics technologies are leading to the development of new biomarkers that target the aging process and overcome the multiple challenges in the advanced treatment of geriatrics can be overcome. The molecular and genetic differences associated with geriatric disease can be elucidated by using DNA microarray techniques which scan entire genome for gene that changes expression with age. Changes in transcriptional and translational profile with the age provide vision to development of novel biomarkers. The term "aging clock" was developed to represent a set of molecules capable of predicting a patient's chronological age (Lehallier et al., 2020).

The molecular signatures of various geriatric diseases have been identified; for example, Alzheimer's disease involves amyloid beta protein and hyperphosphorylation of tau protein (Knopman et al., 2021). Recently, Nir Barzilai's targeting of aging with metformin became a proof of concept that could pave the way to clinical trials leading to healthy. Since that time, more than 250 compounds have been shown to increase lifespan in aging model organisms by targeting cellular processes such as autophagy, cellular senescence, or DNA repair (Castillo-Quan et al., 2019). Advances in big data analysis allowed for new aging clocks to be developed. The epigenetic clock is a promising biomarker of aging and can accurately predict human chronological age (Horvath, 2015).

19.5 BIOTECHNOLOGY APPLICATIONS IN GERIATRIC MEDICINE

In the field of geriatric medicine (GM), biotechnology techniques, therapies, and medicines are now available for the effective treatment of diseases. These techniques like single gene analysis through multiplex, nested/semi-nested, broad range, reverse transcriptase and real-time PCR; cell/gene therapy; molecular immune response; biospecific antibodies; and clinical bioinformatics allow us to focus on the root causes of disorders rather than the symptoms (Dwivedi et al., 2017). Drugs and vaccines generated through gene editing have had significant impacts on GM. Indeed, in older age, the prevalence of both therapeutic failures and adverse drug reactions is high, and diseases are not always responsive to treatment, and biotech tools such as gene therapy and gene editing are increasing the promise of successful geriatric disease treatment.

Gene editing has proven to be effective in treating geriatric diseases by allowing the direct editing of genes *in vivo* and *in vitro* (Cong et al., 2013; Jinek et al., 2012). Gene therapies have also been applied in preclinical trials to treat many diseases including geriatric diseases (Song et al., 2022; Maeder and Gersbach, 2016). However, aging is characterized by a comprehensive decline in the functions of multiple organs, which can hardly be attributed to and therefore not fixed by a single gene. Instead of fixing a specific gene, another way to combat aging and its related diseases is cell transplantation (Turinetto et al., 2016). Functional loss of stem cells is a hallmark of aging, and supplementation or rejuvenation of these cells is a potential treatment to reverse aging (Schultz and Sinclair, 2016). Mesenchymal stem cells, induced pluripotent stem cells, and embryonic stem cells are used to treat various chronic diseases including geriatric syndrome (Shi et al., 2017).

The development of new techniques, especially genetic enhancement, creates an environment for effectively treating disease, but there are some technical limitations such as off-target modification and ethical issues related to genome modification, and so on. With better understanding longevity pathways, healthy aging, stress resistance, and anticancer mechanisms, researchers will develop more strategies to revolutionize cell therapy.

19.6 CONCLUSION

Biotechnological advancements in geriatric medicine create an interdisciplinary approach to studying, exploring, and treating geriatric disease, and researchers are studying different types of molecular damage and the mechanisms of selective changes in gene activity. The powerful tools of recombinant DNA genetics are being applied to GM including gene therapies, gene editing, and biotech medicine. In the future, genetically enhanced cells will potentially serve as resources in cell therapy providing powerful, standardized, and commercialized products to treat multiple age-associated disorder and disease.

REFERENCES

Ariathianto Y. Asymptomatic bacteriuria – prevalence in the elderly population. Aust Fam Physician (2011) 40:805–9.

Bakula D, Ablasser A, Aguzzi A, Antebi A, Barzilai N, Bittner MI, Jensen MB, Calkhoven CF, Chen D, Grey ADNJ, Feige JN, Georgievskaya A, Gladyshev VN, Golato T, Gudkov AV, Hoppe T, Kaeberlein M, Katajisto P, Kennedy BK, Lal U, Martin-Villalba A, Moskalev AA, Ozerov I, Petr MA, Reason, Rubinsztein DC, Tyshkovskiy A, Vanhaelen Q, Zhavoronkov A, Scheibye-Knudsen M. Latest advances in aging research and drug discovery. Aging (Albany NY). 2019 Nov 21;11(22):9971–9981. doi: 10.18632/aging.102487. Epub 2019 Nov 21. PMID: 31770722; PMCID: PMC6914421..

Bandaranayake T, Shaw AC. Host resistance and immune aging. Clin Geriatr Med (2016) 32:415–32.

Castillo-Quan JI, Tain LS, Kinghorn KJ, et al. A triple drug combination targeting components of the nutrient-sensing network maximizes longevity. Proc Natl Acad Sci USA (2019) 116:20817–19.

Chang J, Wang Y, Shao L, et al. Clearance of senescent cells by ABT263 rejuvenates aged hematopoietic stem cells in mice. Nat Med (2016) 22:78–83.

Cong L, Ran FA, Cox D, et al. Multiplex genome engineering using CRISPR/Cas systems. Science (2013) 339:819–23.

Davis A, McMahon CM, Pichora-Fuller KM, et al. Aging and hearing health: The life-course approach. Gerontologist (2016) 56(Suppl 2):S256–67.

Dwivedi S, Purohit P, Mishra R, et al. Diseases and molecular diagnostics: A step closer to precise medicine. Ind J Clin Biochem (2017) 32(4):374–98.

Gerteis J, Izrael D, Deitz D, et al. *Multiple Chronic Conditions Chartbook*. Rockville, MD: Agency for Healthcare Research and Quality (2014). AHRQ Publications No. Q14–0038.

Horvath S. DNA methylation age of human tissues and cell types. Genome Biol (2015) 16:96.

Jaul E, Barron J. Age-related diseases and clinical and public health implications for the 85 years old and over population. Front Pub Health (2017) 5:335.

Jinek M, Chylinski K, Fonfara I, et al. A programmable dual-RNA-guided DNA endonuclease in adaptive bacterial immunity. Science (2012) 337:816–21.

Knopman DS, et al. Alzheimer disease. Nat Rev Dis Prim (2021) 7:33.

Lehallier B, Shokhirev MN, Wyss-Coray T, Johnson AA. Data mining of human plasma proteins generates a multitude of highly predictive aging clocks that reflect different aspects of aging. Aging Cell (2020) 19 (11).

Lemaitre RN, Furberg CD, Newman AB, et al. Time trends in the use of cholesterol-lowering agents in older adults: The cardiovascular health study. Arch Intern Med (1998) 158:17618.

Long JZ. Proteomics illuminates fat as key tissue in aging. Proc Natl Acad Sci USA (2020) 117:10111–12.

Maeder ML, Gersbach CA. Genome-editing technologies for gene and cell therapy. Mol Ther (2016) 24:430–46.

Odden MC, Shlipak MG, Whitson HE, et al. Risk factors for cardiovascular disease across the spectrum of older age: The cardiovascular health study. Atherosclerosis (2014) 237:336–42.

Ribera Casado MJ, The history of geriatric medicine. The present: Problems and opportunities. Eur Geriatric Med (2012) 3 (4):228–232.

Schultz MB, Sinclair DA. When stem cells grow old: Phenotypes and mechanisms of stem cell aging. Development (2016) 143:3–14.

Sharon E. Straus recent advances geriatric medicine. BMJ (2001) 322:86–9.

Shi Y, Inoue H, Wu JC, et al. Induced pluripotent stem cell technology: A decade of progress. Nat Rev Drug Discov (2017) 16:115–30.

Song G, Ma Y, Gao X, et al. CRISPR/Cas9-mediated genetic correction reverses spinocerebellar ataxia 3 disease-associated phenotypes in differentiated cerebellar neurons. Life Med (2022) 1(1):27–44.

Turinetto V, Vitale E, Giachino C, et al. Senescence in human mesenchymal stem cells: Functional changes and implications in stem cell-based therapy. Int J Mol Sci (2016) 17:1164.

Wang JC, Bennett M. Aging and atherosclerosis mechanisms, functional consequences, and potential therapeutics for cellular senescence. Circ Res (2012) 111:245–59.

World Health Organization. WHO definition of health. Preamble to the constitution of the World Health Organization as adopted by the International Health Conference, New York. (1948) [online]. Available from www.who.int/about/definition/en/print.html [Accessed 4 September 2019].

Xue QL, Beamer BA, Chaves PH, Guralnik JM, Fried LP. Heterogeneity in rate of decline in grip, hip, and knee strength and the risk of all-cause mortality: The women's health and aging study II. J Am Geriatr Soc (2010) 58:2076–84.

Yang J, Huang T, Petralia F, et al. Synchronized age-related gene expression changes across multiple tissues in human and the link to complex diseases. Sci Rep (2015) 5:15145. doi: 10.1038/srep15145

20 Emerging Technologies and Trends in Drug Discovery

Neha Sylvia Walter,[1] *Varun Gorki,*[2]* *Shalmoli Bhattacharyya,*[1] *Bikash Medhi,*[2] *and RC Sobti*[3]

[1] Department of Biophysics, PGIMER, Chandigarh, India
[2] Department of Pharmacology, PGIMER, Chandigarh, India
[3] Department of Biotechnology, Panjab University, Chandigarh, India

20.1 INTRODUCTION

Despite the development of novel technologies and an understanding of key biological processes, the task of developing a new drug molecule to treat an ailment is long, costly and full of uncertainties and with a high attrition rate. This process includes three stages *viz.* drug discovery, preclinical studies and clinical trials. The initial stage begins with identification of a 'hit' or a lead compound that illustrates considerable efficacy during primary screening. This is followed by optimisation of the molecule for improving affinity, selectivity, solubility, absorption, distribution, metabolism, and excretion properties that will convert the hit compound to a lead molecule.

The second stage includes preclinical development primarily focusing on understanding the mechanism of drug action, efficacy on animals, pharmacokinetic studies, bioavailability, and route of excretion. The last stage includes the clinical trials which are the longest and the most costly stage, consisting of three different phases: Phase I, Phase II, and Phase III. In the Phase I clinical trials the drug is tested in a small group of healthy subjects in order to ascertain drug safety and its pharmacokinetic characteristics. In Phase II the drug is tested in a larger number of patients to assess its efficacy and to determine optimum dose. In case of Phase III clinical trials the drug is tested in a large number of patients in order to confirm the observed efficacy, safety and for obtaining approval from regulatory authorities. After the successful completion of all the phases, the drug is registered and marketed, however, the surveillance continues after this until the drug is on the market. Furthermore, the marketed drugs continue to be improved and optimized by the development of novel formulations and also through post-marketing studies.

The drug discovery and development process involves a close co-operation between different scientific fields concomitant with the employment of several teams by biotechnology and pharmaceutical companies to convert the discovered lead compound into a therapeutic product that passes through different drug development stages [1]. Drug design is also a critical process in developing new drug molecules that plays a key role in improving outcomes of drug treatment in patients concomitant with improving our understanding of different diseases.

20.2 DIFFERENT APPROACHES TO DRUG DISCOVERY

The oldest approach to drug discovery has been serendipity which implies chance discovery. There are several instances wherein the drugs were discovered by serendipity notable among which was penicillin, Chlordiazepoxide etc. There are also examples where drug developed for one condition showed promise for treatment of another therapeutic condition. This approach to identify new drugs is referred to as drug repurposing, repositioning, re-tasking or re-profiling. This strategy involves the novel use of an existing drug to treat a different therapeutic condition. This approach is based on the fact that different diseases may have common molecular pathways. The main advantage of this is low development cost and time saving as the compound has been researched extensively and all data pertaining to its safety is already available negating the requirement of phase I clinical trials. There are several successful instances of repositioning such as cyclosporine, eflornithine, sildenafil, thalidomide, and so on.

Another strategy to discover new drugs is by inducing chemical changes in known compounds or natural products. Introduction of small modifications in parent structure leads to improved therapeutic profiles *viz.* enhanced half-life, increased stability, bioavailability, higher potency and reduced side effect(s) of the drugs.

Another effective approach for drug discovery involves virtual screening of databases or by HTS (high-throughput) assays [2]. High-throughput screening enables testing of large number (thousands to millions) of samples for their activity using automated equipment. The pharmaceutical and biotechnology companies employ this technique to identify hit molecules possessing considerable biological or pharmacological activity which can then be optimised using medicinal chemistry for discovery and development of new drugs. The drug Prontosil and Paclitaxel (anti-tumour agent) were identified by HTS.

Combinatorial chemistry is another approach used by the pharmaceutical and biotechnological companies to discover

* Corresponding author: ghai2910@gmail.com

DOI: 10.1201/9781032660592-25

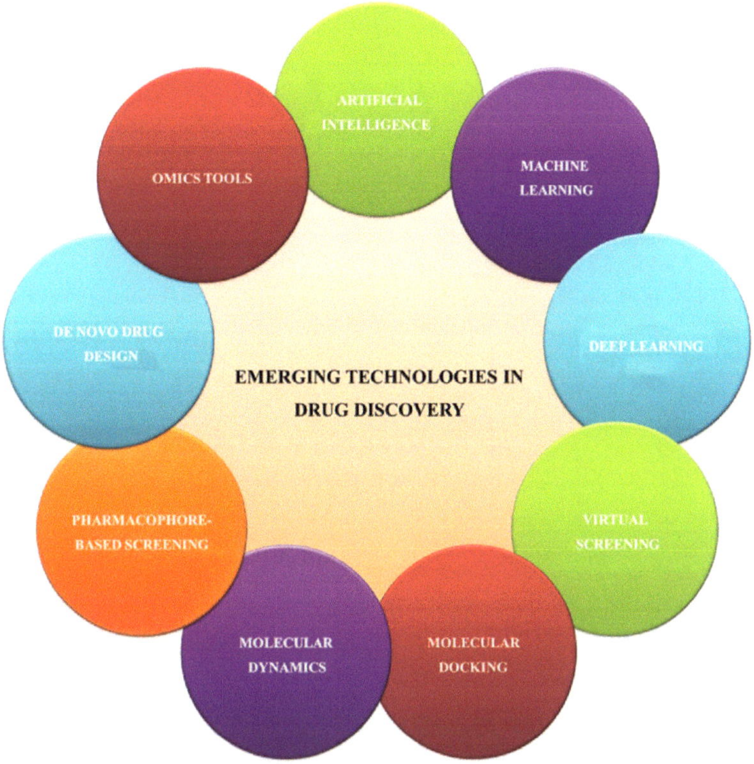

FIGURE 20.1 Emerging technologies in drug discovery.

new drug molecules. It is a high-throughput synthesis technique to synthesize a large number of small organic compounds which are then screened for their biological activity. There are three steps in combinatorial chemistry *viz.* synthesis, candidate screening and library product purification. This technique has altered the drug development process for developing hit-to-lead libraries.

Rational based drug design is an advanced method for discovering new drug. It is also referred to as reverse pharmacology. The initial step in this method is to identify the biological target; for example, any enzyme or molecule involved in the disease development or in the signalling pathway. After this the key step involves identification of the hit molecule, i.e., the ligand interacting with this target molecule to inhibit the process or the pathway. This is followed by optimization of the structure to develop a compound (lead molecule) with favourable properties (potency, selectivity, affinity, no toxicity, permeability, bioavailability, solubility.) which can become a drug [3]. Two main approaches are involved in the rational based drug design: ligand-based and structure-based. The ligand-based drug design approach is preferred for drug discovery and lead optimization when the 3D (three-dimensional) structure of potential drug target is unknown. It is an effective method that facilitates understanding of the ligand features that are crucial for their biological efficacy when the receptor structure is absent. In the case of SBDD (structure based drug design), the target's 3D structure is known and the ligand is so designed to be complementary to the binding site on the target molecule.

All the aforementioned strategies have been employed to identify new hit molecules during the drug discovery process. However, the advent of bioinformatics tools as well as the application of "omics" techniques has provided a new impetus to the entire process. The following section outlines the recent trends or strategies in the drug discovery process.

20.3 RECENT TRENDS AND EMERGING TECHNOLOGIES IN DRUG DISCOVERY

The advancement in technology and the recent developments in science *viz.* advances in microbial culturing, genome mining assignments, improvement in analytical tools, employment of genomics and proteomic techniques in drug discovery are overcoming the current challenges and enabling development of novel tools to reshape the drug discovery process. In recent years several technologies have emerged to develop new drugs for the treatment of various diseases (Figure 20.1).

20.3.1 Artificial Intelligence

Artificial intelligence (AI), a technology-based system, can imitate the intelligence of humans using a range of innovative tools and networks. This tool is now widely being used in various sectors of society. It is also being employed in all aspects of drug discovery as it involves high-performance computers and efficient algorithms which are capable of

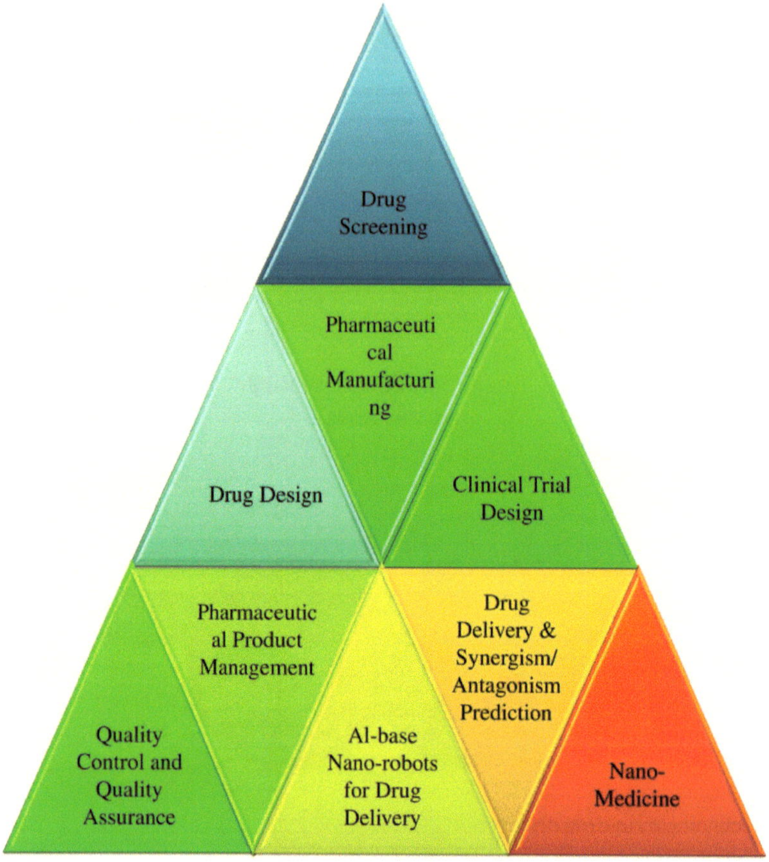

FIGURE 20.2 Applications of AI in drug discovery and development.

integrating enormous volumes of data to train expert systems to execute a challenging task [4].

AI can be used effectively to analyse enormous amounts of data and relevant patterns that cannot be easily detected by humans. This technology could help to design small molecules with desired properties to overcome the main bottleneck for advancing new drug candidates to the clinic. Moreover, AI would accelerate the drug discovery process to develop more targeted, potent, safer and affordable drugs. AI takes computational screening of the drugs to the higher level and also provides hope to provide novel drugs, particularly personalised medicines quickly to the market at low cost. AI can help in the identification of hit and lead molecules, concomitant with expediting the validation of therapeutic target and structural design optimisation [5].

AI techniques can be used in all aspects of drug discovery and development ranging from drug screening, drug design, and drug delivery as well as in designing clinical trials, pharmaceutical manufacturing, quality control and assurance as well as management of pharmaceutical products (Figure 20.2).

In the case of drug screening AI has been used for the prediction of physicochemical properties, bioactivity and toxicity. As far as drug designing is concerned AI has application in speculating the structure of target protein and drug-target binding affinity prediction. In addition, its advanced applications include drug delivery by AI-based nanorobots, combination drug delivery using AI for predicting synergism/antagonism, nanomedicine and oligonucleotide design (Figure 20.2). In addition, AI is also playing an important role in drug discovery inspired by natural products, *de novo* drug design, drug-target interaction (DTI) prediction, and also in drug repurposing.

Hence the application of AI in discovering and developing new drugs would not only expedite the duration for the new molecule/drug to arrive in the market, but will also escalate the product quality, reduce the cost of production and ensure production process safety in addition to providing a better use of resources. AI can also be employed to formulate the developed drugs as per their correct dosage form and also their optimization. It will enable the manufacture of products quickly with better-quality whilst assuring consistency among the batches. Moreover, this technology can also establish the efficacy and safety of the drugs in clinical trials, concomitant with ensuring their proper market position and cost by comprehensive market prediction and analysis [6]. Though there are several challenges in the application of this technology; however, it will develop into an indispensable tool for the pharmaceutical industry in the near future.

20.3.2 Machine Learning

Machine learning is a branch of AI that can be employed for developing computer algorithms which can process raw data to perform a given query or task. A number of ML methods are in the drug discovery process predicting biological activity, drug side effects, molecular characteristics, and drug–drug interactions. AI algorithms can be used to perform several tasks such as clustering or pattern recognition, regression and classification in the given data set.

Historically, the Hansch equation was the first linear regression model to illustrate the 2D structure-activity relationship utilizing physicochemical descriptors (i.e., hydrophobicity, steric parameters) [7]. It was when the Quantitative structure-activity relationship (QSAR) study field initiated employing predictive algorithms *viz.* molecular descriptors of the sequences and linear regression. The QSAR study was followed by the idea of Drug Likeness (whether a molecule has drug-like properties or not) employing 1D and 2D molecular descriptors which pioneered the utilization of ML algorithms in drug discovery.

Moreover, with the advancement of biotechnology and computational techniques a large amount of data pertaining to the molecules was generated that ultimately led to development of large number of public databases for instance PubChem, ZINC, DrugBank and ChEMBL. Due to the availability of these databases new Machine Learning models were developed that aided screening of new drug molecules.

Some of the methods used by the pharmaceutical industry in drug discovery include Naive Bayes (to predict possible drug targets), Support Vector Machines (to segregate drugs on the basis of their KEGG classification), Random Forest (best in terms of performance, speed and generalizability) and Artificial Neural Networks (for drug prediction) [8].

The methodology of ML in drug discovery involves the following steps *viz.* collection of data, mathematical descriptors generation, searching for best variables subset, model training and validation.

The initial step involves data collection which must possess certain attributes (physical–chemical attributes such as low toxicity, absorption and specificity). This is followed by mathematical descriptors generation to form a dataset that can be processed by the ML model. The data set is classified into two subsets *viz.* model training (having higher data percentage) and model testing (having smaller data). The next step involves the best subset variables search within the training set in order to reduce the redundant and useless variables.

After obtaining the ideal variables subset, the training of the model is carried out. This includes selection of algorithms and their parameters followed by experimentation with the training data. During this process again the initial dataset is classified into two subsets *viz.* training set and validation set. Cross-validation (CV) techniques are then run for selection of the best parameters combination for each technique which are then used to measure the performance of each model. The model that exhibits high performance at low cost is defined as the best model. Finally, the best model selected using CV techniques is subjected to final validation and if the result obtained is statistically significant, creation of the new predictive drug model can be concluded.

20.3.3 Deep Learning

As already discussed, various computational algorithms are recently being used to accelerate the process of drug discovery, as it is a highly laborious and costly process. Recently, novel deep learning (DL)-based techniques are being utilized at different drug development stages [9]. Deep learning can be employed in various aspects of drug discovery which includes drug–drug similarity, drug-target interactions, side effects of drug combinations, and prediction of drug sensitivity and response.

A far as drug-target interactions are concerned deep learning can help overcome the limitations of molecular docking and drug-based techniques by employing unstructured-based strategies that do not require docking or 3D structural data for Drug-Target Interaction (DTI) prediction. The use of deep neural network models can slowly decrease the disparity between empirical study and *in silico* predictions, while deep learning technology can lessen the time otherwise required for manual operations and other time-consuming protocols. Deep neural networks can also be used for the analysis of medicines and for prediction of properties of drugs i.e. efficacy as well as their physicochemical properties due to the availability of MoleculeNet and DeepChem packages.

Deep learning can also be successfully employed for prediction of drug sensitivity and response. It receives the protein and drug interactions heterogenous network as input based on which response scores are predicted. Deep neural network (DNN), pre-trained CNN, Graph Neural Network (GNN) based techniques are also being employed for drug response prediction [10].

Moreover, Deep learning technique can also be used for the prediction of drug–drug interactions (DDIs) side-effects. Recently, Han *et al.* reported this technique is to be used for the prediction DDIs by employing graph theory and drug structural properties [11]. This will be of immense benefit for making informed decisions prior to prescribing a medication to patients having complicated conditions. A DL system can be accurately used for predicting DDIs results employing a deep feed-forward neural network and auto-encoders. In addition, neural graph networks were also used to predict DDI event by evaluating chemical structural similarity [12].

Drug–drug similarity prediction can also be carried out employing DL on the basis of similarity in chemical structure, drug-induced pathways, target protein sequence and target protein function. Several studies have employed different strategies such as Convolutional Neural Network (CNN), Semi-Supervised Learning Technique (DDI-IS-SL), Anatomical Therapeutic Chemical (ATC), Similarity-based ML for Drug Interaction Prediction (SMDIP) and Ranking-based k-Nearest Neighbour (Re-KNN), for predicting drug–drug similarity. Nonetheless, the deep learning technique can also be used for the optimization of drug dose.

On account of the success of DL strategies and their implication in the pharmaceutical industry for identification of new drugs, this technique can be successfully used in the future for extensive data analysis and assessment for drug discovery and development. Thus, DL represents a promising open research area for addressing challenges in drug discovery.

20.3.4 Virtual Screening

Virtual screening (VS) is a computational technique for the identification of hit molecules. It is a faster, cost-effective, and less resource-intensive approach for developing lead compounds than the experimental strategies. The approach involves the computational screening of libraries of different types of drug-like or lead compounds against the 3D structures of target proteins. Virtual screening is either ligand-based or structure-based. In ligand-based virtual screening, structure-activity data from a known set of active compounds/molecules is utilized to identify new drug candidates that can be further evaluated employing experimental approaches. QSAR, 3D shape matching, pharmacophore-based search, similarity and substructure-based search are also included in ligand-based virtual screening [13].

In contrast, structure-based virtual screening utilizes molecular docking wherein, the target protein receptor's 3D structure is used to dock the drug candidates, which are then ranked on the basis of their docking score (predicted binding affinity). Structure-based virtual screening can also employ the protein-ligand fingerprints and ePharmacophore techniques. Both ligand-based and structure-based screening can be used singly, concurrently or as a layered screening technique based on the research under question.

The first step in virtual screening is selecting the requisite chemical library from the available commercial and public databases; next is preprocessing the chemical library, that is, removing tautomers, duplicate structures, and counter ions. Filtering drug-like molecules involves using filtering tools such as Lipinski's rule of five and the ADMET filter (absorption, distribution, metabolism, excretion, and toxicity). The final step is screening the drug-like molecules using ligand or structure to identify the hit molecules, which can be synthesized or purchased and further screened for their pharmacological activity employing *in vitro* and *in vivo* assays. Lead compounds illustrating promising biological and pharmacological activity are further evaluated in preclinical and clinical trials to ultimately identify them as efficacious drug molecules.

20.3.5 Molecular Docking

Docking is a tool that can be employed for the prediction of the protein (enzyme) ligand (small molecules) interaction. This molecular modeling technique is utilized for the prediction of mutual orientation, mode of binding, binding energy and conformation of each molecule. The lower the binding energy of the complex, the more is the stability. This is employed for filtering the libraries of molecules based on their binding affinities. This technique enables identification of hit molecules from huge libraries or datasets that exhibit affinity to a specific binding site on the molecule [14].

Docking is carried out by placing the given fragments or molecules into one of the protein's active site utilizing different strategies for instance clique-searching, pose clustering or geometric hashing. The initial step of molecular docking involves the generation of all possible orientations and conformations of the protein-ligand complex followed by the scoring function that returns a number based on the input which is indicative of favorable interaction as the second step.

The basic steps of docking include ligand preparation, protein preparation and ligand protein docking. Docking can be employed at various stages in drug design and discovery, for instance to identify new ligands by virtual screening and predict the binding affinities of active compounds. In addition, molecular docking can also be used to identify a potential target for different ailments [15].

20.3.6 Molecular Dynamics

Molecular dynamics (MD) is a technique used for simulation of molecular movement as well as their interactions in various mediums on the basis of their force field [16]. The major advantage of MD is that information pertaining to the 3D structure resolution obtained using this technique cannot be provided by any other methods such as spectroscopy, microscopy and crystallography.

The fundamental idea behind this technique is that it enables calculation of the force applied on each atom by every other atom in a biomolecular system. Moreover, Newton's laws of motion can be then used for estimating each atom's spatial position as a function of time ultimately leading to information about the configuration at the atomic level of a biomolecular system at every time point all through the simulation interval [17].

It is possible to pinpoint the effect of a vast range of molecular disturbances by providing a comparison of simulations that were performed under different conditions. The time steps during MD simulation should be very short (only for 100–200ns) to ensure numerical stability. Moreover, use of highly specialized hardware enables enhanced speed, which allows certain simulations to reach a timescale of millisecond.

MD simulation has numerous applications in drug discovery, the most basic being modifying the ligand to improve its biological activity and other properties. This technique is helpful in determining the major interactions of the ligand with the binding pocket, prediction of binding pocket rearrangements upon ligand binding, testing and refining of ligand poses [18] and revelation of the ligand pose and its binding site [19] as well as accurate estimation of ligand binding affinities (free energies) [20].

MD simulation technique is also beneficial for virtual screening enabling selection of a ligand that binds to a given target. As the binding pocket might be highly flexible, docking against a particular receptor will help to identify only a few binder subsets. Identification of various plausible structures by simulation will enhance the heterogeneity of ligands binding [21].

Simulation technique is also employed for designing allosteric drugs that bind a target different from that of the ligand. These drugs are of immense importance as they illustrate potential to modulate the natural signalling patterns of the host, enhance selectivity among receptor subtypes that are closely related and exhibit activity against targets that seemed undruggable [22].

This technique is also used to design drugs exhibiting desired properties such as binding and dissociation kinetics that are important for drug safety and efficacy. Various elements governing binding and dissociation kinetics at different targets have been illustrated by several simulation studies [17]. These studies can be helpful in the rational design of ligands with desirable kinetic properties. MD-based methods are also being used to rank the related ligands on the basis of their dissociation rates.

Thus, MD simulations can be employed for drug discovery along with other experimental approaches for enhancing the successful identification of a new drug lead and this opportunity will further increase as this technique becomes more accessible, accurate, cheaper, and faster.

20.3.7 Pharmacophore-Based Screening

Pharmacophore-based screening is another type of virtual screening that can speed up the search for a desired molecule from compound libraries that is active against a specific target. Pharmacophore is an abstract concept used to describe electrostatic and steric complementarities of active molecules with defined target and contrary to popular misconception. It is not a depiction of an actual molecule or a true arrangement of functional groups [23].

Pharmacophore modeling techniques are emerging as a popular tool for discovering new drug molecules. The 3D pharmacophore models should provide a description about the location and nature of functional groups involved in protein-ligand interaction, various types of non-covalent bonds and their attributes so as to be helpful in drug designing.

The initial step in this technique is creation of a pharmacophore model query that describes the kind of chemical properties and their geometric constraints which the screened molecules must match. Two different approaches i.e. structure-based and ligand-based are used in 3D pharmacophore-based virtual screening. In the structure-based method the compatibilities between a ligand and its binding-site are taken into consideration. This approach requires information pertaining to the active configuration of the binding ligand and the macromolecule. The major advantage of structure-based approach is that it takes into account information regarding the directional nature of binding-site associations in pharmacophore models, which can result in models that are exceedingly restricted and with attributes that are orientation constrained [24].

On the contrary if the macromolecule 3D structure is not available then the ligand-based approach is employed. The perception of chemical characteristics shared by a group of ligands (training set) that are well known to exhibit the required biological efficacy for the target can be used to create 3D pharmacophore models in a ligand-based manner. If sufficient data pertaining to the ligand is obtainable and the training set compounds are known to bind to the protein at a particular site, the latter method yields satisfactory results [25]. Moreover, 3D pharmacophore models can also be successfully used as filters for virtually screening of large compound libraries.

20.3.8 *De Novo* Drug Design

The term *de novo* means "from the beginning," and this approach enables generating new compounds without using a template [26]. DNDD is the process of creating new chemical compounds which adhere to a certain set of parameters employing computational growth techniques. DNDD results in the development of new compounds only on the basis of the data pertaining to the known active ligand or its biological target or receptor. The creation of the molecules (sampling), receptor active site description or pharmacophore modeling of ligand, and assessment of the created compounds are the three main steps in DNDD.

Traditional approaches to DNDD include structure-based and ligand-based design employing evolutionary algorithms. In the structure based approach the structure of the receptor is known by NMR, X-ray crystallography or electron microscopy. Homology modeling can be used to obtain a relevant structure for DNDD if receptor structure is unknown [27]. Structure based DNDD initiates after describing the receptors active site, which is analyzed for determination of the shape constraints and the non-covalent interactions (hydrogen-bonds, hydrophobic and electrostatic interactions) for a ligand [28]. The active sites are reported to play a crucial part in increasing the selectivity by decreasing the large amount of generated structures.

On the other hand, the ligand-based approach is employed when three-dimensional structure data of the given target are not available therefore, only a few known binders serve as an alternative approach for DNDD [29]. Active binders can be searched by screening literature, databases such as ChEMBL and structure–activity relationship studies. SYNOPSIS, TOPAS, and DOGS are some examples of ligand-based DNDD tools.

Recently, AI including deep reinforcement learning (DRL) is also being used to develop new DNDD approaches by employing artificial neural networks such as autoencoders, convolutional neural networks, recurrent neural networks, and generative adversarial networks. This approach has the potential to accelerate the drug discovery and development in addition to overcoming the major challenges (cost and time constraints) encountered especially during the early stages of drug discovery. This can be achieved by employing *in silico* studies for DNDD, predicting synthesis and bioactivity. DNDD approaches can overcome the drawbacks of conventional strategies as they are versatile, flexible, data-driven and can use enormous volumes of data available from the databases and scientific literature [30].

20.3.9 Omics Techniques in Drug Discovery

In recent years, the use of computational tools in drug discovery has enabled exploration, organization, and analysis of tremendous amounts of biological data to identify molecular drug targets. The data for this drug target-based drug discovery are obtained from a wide variety of experimental, pharmacological mechanistic and more recently, omics-based data sources [31]. Omics tools have enabled screening at gene, transcript, protein, and metabolite as well as their interaction network level to identify novel targets for drug discovery.

Omics tools help with identifying drug target-disease associations and better understanding of molecular mechanisms underlying complex diseases, advanced detection of drug-induced side effects, mechanism of drug action and information on drug treatments. This information can be combined with already available published scientific literature so as to enhance the chances to identify the correct drug candidates. Next we discuss various omics tools and their relevance in drug discovery.

20.3.9.1 Genomics

Genomics is defined as the study of a complete or a part of the genetic or epigenetic sequence of an organism to expedite drug development. Various conceptual frameworks and methods support target prioritization, identification, tractability, and result prediction following pharmaceutical modifications [32]. Next-generation sequencing (NGS) and DNA microarray technology has opened a new era of genetics-informed drug discovery. The use of these techniques has helped to perform high-throughput genotype–phenotype relationships analysis on human populations,

Genome-wide association studies (GWAS) and whole genome sequencing provide essential tools to identify and validate novel drug targets, evaluation of their therapeutic activity and toxicity profile. DNA microarrays have been used to conduct GWASs to locate single nucleotide polymorphisms that have been associated with diseases risk and traits. GWAS has a wide range of application in drug discovery for instance, discovery of genes contributing to complex genetic traits, identification of biological pathways conferring disease susceptibility, developing *in silico* methods to identify drug target and disease genes [33]. GWAS analysis can also be merged with transcriptomics data, to locate disease related genes. This information is obtained by correlating gene expression profiles of disease and control groups.

GWASs have also been reported to change the pharmacogenomic research, which is the study of the influence of individual genetic variations on drug response/metabolism. Such studies therefore, enable better understanding patients' specific genomic characteristics such as drug efficacy and toxicity that can impact their response to treatment [34]. Some of the genomics databases include PharmGKB, GWAS Catalog, GWAS Central, NCBI, and dbGaP. In addition, CRISPR screen can be used in combination with drug treatment to discover genes that are involved in enhancing or suppressing drug treatment [35] and also to validate target specificity.

20.3.9.2 Transcriptomics

Transcriptomics refers to the study of complete set of the RNA transcribed by cells and tissues. It covers all aspects of the RNA molecules *viz.* transcription, functions, locations, trafficking, expression levels, and degradation. Transcriptomics can also play a significant role in discovering and developing new drugs. The molecular basis of particular biological processes in diseases can be discovered through transcriptome research, which can show gene activities and structures at the whole-genome level. In biological, medical, clinical, and therapeutic research, gene expression microarray technology and high-throughput RNA-sequencing have been extensively used. Gene expression microarray can be employed to understand the positive and negative impacts of drug treatment and screening of drugs at the cellular or tissue level. This can drastically reduce the use of animals as well as other experiment costs. This technique is widely used in drug screening and drugs biomarker detection on account of its, speed, simple analysis, high throughput, faster detection, and low cost.

High-throughput RNA-sequencing on the other hand is an effective technique for identifying transcripts that are differently expressed in particular biological activities and can also help to identify molecular mechanisms underlying the disease. The research of disease mechanisms, clinical diagnostics, and drug development are greatly being benefited from the widespread use of RNA-seq. This technique also helps to locate the genes involved in drug resistance and sensitivity [36]. DrugMatrix, TG-GATEs, LINCS 1000, Expression Atlas, and GEO repository are some of the transcriptomic databases for targeted drug discovery.

20.3.9.3 Proteomics

Proteomics is the study of the structure and the function of entire protein content of a cell, an organism. A protein's expression at various levels can be analyzed using proteomics, enabling the assessment of certain cellular responses – both quantitative and qualitative – related to that protein. Qualitative and quantitative study of proteomes is performed at genomic, transcriptomic, post-transcriptional, and post-translational levels.

A number of methods have been developed to study the proteome of an organism, but mass spectrometry (MS) has proven to be the most useful method for profiling proteins. MS can be used for understanding protein–protein interactions, expression, and to locate the sites of protein modification that enables it to be useful in targeted drug discovery [37]. As far as drug development is concerned, proteomics can be successfully used for assay development for drug screening, identifying and validating drug targets and generating biomarkers for disease stratification, toxicity, and effectiveness. Additionally, quantitative proteomics can offer in-depth understandings of cellular processes [38], disease mechanisms, and biomarker discoveries. Proteomic databases for targeted drug discovery include PRIDE Archive, ProteomicsDB, Human Proteome Map, and Human Proteome Atlas.

Chemoproteomics is a new approach that enables extensive protein research by integrating chemical techniques with MS proteomics. Chemoproteomics have played a significant role in early stage drug discovery by offering platforms for testing small-molecule libraries against members of big protein classes which are poorly characterised, locating the proteins that are the targets for bioactive small compounds found in phenotypic screens and enhancing the selectivity of lead probes through parallel screening against several proteins in native proteomes and occasionally *in vivo*. Chemoproteomics is becoming the go-to technique for solving the dreaded "target identification problem" in phenotypic screening programmes and can even be used to more thoroughly reevaluate the proteins that interact with medications with well-known mechanisms of action [39].

20.3.9.4 Metabolomics

The detailed study of metabolites (small compounds) found in cells, biofluids, tissues, or living things is called metabolomics. It presents a summary of the metabolic state and major biochemical processes in relation to a given cellular or biological system.

Clinical testing and applications based on metabolomics are rapidly developing and can provide a fresh perspective on disease mechanisms, identification of biomarkers and metabolic causes for chronic diseases, utilize efficient precision medicine strategies like individualized drug-response monitoring and personalized phenotyping and evaluation of toxicity of drug targets. In addition, metabolomic analysis can help identify off-targets of a drug candidate and hence illustrate a mechanistic explanation of its toxicity [40].

This technique offers several advantages such as reducing the costs of toxicity screening, generating a new generation of biomarkers, quantification of drug efficacy and safety, improvement in designing the clinical trial, better selection of patients, understanding mechanisms of action of toxicity of drugs and monitoring, and decreasing the time for drug development [31]. Examples of metabolomic databases include the Human Metabolome Database, Golm Metabolome Database, MetaboLights, and MetabolomeExpress.

Thus, omics technologies are emerging as an integral part of drug discovery and development. Data from genomic, transcriptomic, and proteomic analyses could be leveraged more effectively to develop novel approaches for validating and matching targets to diseases. Systematic integration of the omics-driven molecular disease profiles and drug treatments can considerably accelerate drug discovery and development process.

20.4 CONCLUSION

As discussed in the aforementioned sections various different technologies are emerging and showing promise in drug discovery and development process. These strategies are important tools for overcoming the drawbacks of the conventional approaches as drug discovery is a long, laborious, exorbitant and time consuming process. AI is providing novel, efficacious and cost-effective methods for discovering new drugs as it can analyze enormous amounts of data in a short duration of time. Use of the computational tools such a molecular docking, MD simulation, pharmacophore based screening along with the availability of machine learning and deep learning will decrease the likelihood of failure of a drug in the late stages of drug development particularly during clinical trials. All these approaches can be employed singly or in combination to accelerate drug discovery and development ultimately to be able to design and develop efficacious, target oriented, non-toxic, patient- customized, specific drugs in a short duration of time in the future.

REFERENCES

[1] William J. Emerging trends in drug discovery and innovations in drug design and development. Drug Des. 2023; 12: 226.

[2] Butkiewicz M, Wang Y, Bryant SH, Lowe EW Jr, Weaver DC, Meiler J. High-throughput screening assay datasets from the PubChem database. Chem Inform. 2017; 3(1): 1.

[3] Young DC (ed.). Computational drug design. In Properties that make a molecule a good drug. John Wiley & Sons: Hoboken, NJ, 2009, pp. 9–39.

[4] Sadybekov AV, Katritch V. Computational approaches streamlining drug discovery. Nature. 2023; 616(7958): 673–685.

[5] Mak KK, Pichika MR. Artificial intelligence in drug development: Present status and future prospects. Drug Discov Today. 2019; 24(3): 773–780.

[6] Paul D, Sanap G, Shenoy S, Kalyane D, Kalia K, Tekade RK. Artificial intelligence in drug discovery and development. Drug Discov Today. 2021; 26(1): 80–93.

[7] Menden MP, Wang D, Mason MJ, Szalai B, Bulusu KC, Guan Y, Yu T, Kang J, Jeon M, Wolfinger R, Nguyen T, Zaslavskiy M; AstraZeneca-Sanger Drug Combination DREAM Consortium; Jang IS, Ghazoui Z, Ahsen ME, Vogel R, Neto EC, Norman T, Tang EKY, Garnett MJ,

Veroli GYD, Fawell S, Stolovitzky G, Guinney J, Dry JR, Saez-Rodriguez J. Community assessment to advance computational prediction of cancer drug combinations in a pharmacogenomic screen. Nat Commun. 2019; 10(1): 2674.

[8] Carracedo-Reboredo P, Liñares-Blanco J, Rodríguez-Fernández N, Cedrón F, Novoa FJ, Carballal A, Maojo V, Pazos A, Fernandez-Lozano C. A review on machine learning approaches and trends in drug discovery. Comput Struct Biotechnol J. 2021; 19: 4538–4558.

[9] Kim J, Park S, Min D, Kim W. Comprehensive survey of recent drug discovery using deep learning. Int J Mol Sci. 2021; 22(18): 9983.

[10] Askr H, Elgeldawi E, Aboul Ella H, Elshaier YAMM, Gomaa MM, Hassanien AE. Deep learning in drug discovery: An integrative review and future challenges. Artif Intell Rev. 2023; 56(7): 5975–6037.

[11] Han X, Xie R, Li X, Li J. SmileGNN: Drug-drug interaction prediction based on the SMILES and graph neural network. Life (Basel). 2022; 12(2): 319.

[12] Huang K, Xiao C, Hoang TN, Glass LM, Sun J. CASTER: Predicting drug interactions with chemical substructure representation. Proc AAAI Conf Artif Intell. 2020; 34(1): 702–709.

[13] Jahn A, Hinselmann G, Fechner N, Zell A. Optimal assignment methods for ligand-based virtual screening. J Cheminform. 2009; 1: 14.

[14] Shoichet BK. Virtual screening of chemical libraries. Nature. 2004; 432(7019): 862–865.

[15] Kroemer RT. Structure-based drug design: Docking and scoring. Curr Protein Pept Sci. 2007; 8(4): 312–328.

[16] Ganesan A, Coote ML, Barakat K. Molecular dynamics-driven drug discovery: Leaping forward with confidence. Drug Discov Today. 2017; 22(2): 249–269.

[17] Hollingsworth SA, Dror RO. Molecular dynamics simulation for all. Neuron. 2018; 99(6): 1129–1143.

[18] Spahn V, Del Vecchio G, Labuz D, Rodriguez-Gaztelumendi A, Massaly N, Temp J, Durmaz V, Sabri P, Reidelbach M, Machelska H, Weber M, Stein C. A nontoxic pain killer designed by modeling of pathological receptor conformations. Science. 2017; 355(6328): 966–969.

[19] Shan Y, Kim ET, Eastwood MP, Dror RO, Seeliger MA, Shaw DE. How does a drug molecule find its target binding site? J Am Chem Soc. 2011; 133(24): 9181–9183.

[20] Perez A, Morrone JA, Simmerling C, Dill KA. Advances in free-energy-based simulations of protein folding and ligand binding. Curr Opin Struct Biol. 2016; 36: 25–31.

[21] Amaro RE, Baron R, McCammon JA. An improved relaxed complex scheme for receptor flexibility in computer-aided drug design. J Comput Aided Mol Des. 2008; 22(9): 693–705.

[22] Conn PJ, Christopoulos A, Lindsley CW. Allosteric modulators of GPCRs: A novel approach for the treatment of CNS disorders. Nat Rev Drug Discov. 2009; 8(1): 41–54.

[23] Seidel T, Ibis G, Bendix F, Wolber G. Strategies for 3D pharmacophore-based virtual screening. Drug Discov Today Technol. 2010; 7(4): e203–e270.

[24] Leach AR, Gillet VJ, Lewis RA, Taylor R. Three-dimensional pharmacophore methods in drug discovery. J Med Chem. 2010; 53(2): 539–558.

[25] Evers A, Hessler G, Matter H, Klabunde T. Virtual screening of biogenic amine-binding G-protein coupled receptors: Comparative evaluation of protein- and ligand-based virtual screening protocols. J Med Chem. 2005; 48(17): 5448–5465.

[26] Devi RV, Sathya SS, Coumar MS. Evolutionary algorithms for de novo drug design – A survey. Appl Soft Comput. 2015; 27: 543–552.

[27] Muhammed MT, Aki-Yalcin E. Homology modeling in drug discovery: Overview, current applications, and future perspectives. Chem Biol Drug Des. 2019; 93(1): 12–20.

[28] Schneider G, Fechner U. Computer-based de novo design of drug-like molecules. Nat Rev Drug Discov. 2005; 4(8): 649–663.

[29] Fischer T, Gazzola S, Riedl R. Approaching target selectivity by de novo drug design. Expert Opin Drug Discov. 2019; 14(8): 791–803.

[30] Mouchlis VD, Afantitis A, Serra A, Fratello M, Papadiamantis AG, Aidinis V, Lynch I, Greco D, Melagraki G. Advances in de novo drug design: From conventional to machine learning methods. Int J Mol Sci. 2021; 22(4): 1676.

[31] Paananen J, Fortino V. An omics perspective on drug target discovery platforms. Brief Bioinform. 2020; 21(6): 1937–1953.

[32] Spreafico R, Soriaga LB, Grosse J, Virgin HW, Telenti A. Advances in genomics for drug development. Genes (Basel). 2020; 11(8): 942.

[33] Yin W, Gao C, Xu Y, Li B, Ruderfer DM, Chen Y. Learning opportunities for drug repositioning via GWAS and PheWAS findings. AMIA Jt Summits Transl Sci Proc. 2018; 2017: 237–246.

[34] Weinshilboum RM, Wang L. Pharmacogenomics: Precision medicine and drug response. Mayo Clin Proc. 2017; 92(11): 1711–1722.

[35] Colic M, Wang G, Zimmermann M, Mascall K, McLaughlin M, Bertolet L, Lenoir WF, Moffat J, Angers S, Durocher D, Hart T. Identifying chemogenetic interactions from CRISPR screens with drugZ. Genome Med. 2019; 11: 52.

[36] Yang X, Kui L, Tang M, Li D, Wei K, Chen W, Miao J, Dong Y. High-throughput transcriptome profiling in drug and biomarker discovery. Front Genet. 2020; 11: 19.

[37] Schirle M, Bantscheff M, Kuster B. Mass spectrometry-based proteomics in preclinical drug discovery. Chem Biol. 2012; 19(1): 72–84.

[38] Agnetti G, Kane LA, Guarnieri C, Caldarera CM, VanEyk JE. Proteomic technologies in the study of kinases: Novel tools for the investigation of PKC in the heart. Pharmacol Res. 2007; 55(6): 511–522.

[39] Moellering RE, Cravatt BF. How chemoproteomics can enable drug discovery and development. Chem Biol. 2012; 19(1): 11–22.

[40] Wang P, Shehu AI, Ma X. The opportunities of metabolomics in drug safety evaluation. Curr Pharmacol Rep. 2017; 3(1): 10–15.

21 A Novel Approach to Anticancer Research

Nonmammalian Models for Drug Development

*Iyshwarya BK and Ramakrishnan Veerabathiran**
Human Cytogenetics and Genomics Laboratory, Faculty of Allied Health Sciences, Chettinad Hospital and Research Institute, Chettinad Academy of Research and Education, Kelambakkam, Tamil Nadu, India

21.1 INTRODUCTION

Cancer may be characterized by, among other factors, changes in gene expression patterns altering the balance in the homeostatic processes governing cell proliferation and death. Among other mechanisms, genetic and epigenetic mechanisms may play a pivotal role in the cells bypassing cell cycle checkpoints, which generally cause the cell to die. As they multiply uncontrollably, they turn into cancer cells. If left untreated, these cells can spread to distant locations and infect nearby tissues, causing severe morbidity and death. A molecular and cell biological perspective may suggest that cancer is a complex family of diseases. Still, it is a disease resulting from DNA mutations caused by similar molecular flaws. Epigenetic mechanisms play a pivotal role in cancer development and progression, influencing gene expression through modifications to DNA or associated proteins. DNA methylation alterations, such as hypermethylation of tumor suppressor genes and targeted hypomethylation for immune response enhancement, contribute to uncontrolled cell growth and genomic instability in cancer. Histone modifications, including acetylation associated with active gene transcription and methylation influencing gene expression, are targeted by inhibitors like vorinostat and romidepsin to induce cancer cell death. Non-coding RNAs, such as microRNAs acting as tumor suppressors or oncogenes and long non-coding RNAs influencing gene expression, offer potential therapeutic avenues. Chromatin remodeling through ATP-dependent complexes, impacting nucleosome positioning, is implicated in cancer, with emerging strategies targeting chromatin remodelers. RNA methylation, particularly N6-methyladenosine (m6A), affecting mRNA stability and translation, is explored for cancer treatment through inhibitors of m6A writers and erasers. The understanding of these epigenetic mechanisms has driven the development of targeted therapies, providing promising avenues for personalized cancer treatments (Cheng et al., 2019).

Due to mutations in certain genes, cancer is considered a genetic illness, but it is not necessarily inherited (Dhanya et al., 2019). Worldwide, cancer is considered to be the sixth leading cause of death. According to estimates, 9.6 million people worldwide died from cancer in 2018 (Bray et al., 2020). Moreover, according to 2008 research, the number of new cancer cases worldwide in 2008 was 12.7 million; by 2020, that number is expected to reach 15–17 million. It is estimated that by 2040, 27.5 million new cases will be annually (López-Gómez et al., 2013). Malignant cells multiply rapidly once they have acquired enough genetic and epigenetic abnormalities in a tumor microenvironment. This microenvironment is essential for the growth, migration, and proliferation of cancer cells (Gheybi and Adeli, 2015; Diao et al., 2011; Oerlemans et al., 2010). The primary method of treating cancer is thought to be surgical removal of the afflicted tumor, although this has various unpleasant side effects include swelling, infection, Pain, Fatigue, appetite loss, organ dysfunction. Other available include radiation, immunotherapy, chemotherapy, photodynamic therapy, hyperthermia, unconventional therapy using natural bioactive materials, and tumor vaccination (Feng et al., 2003; Zeng et al., 2014). Immune proteins are used as radioactivity carriers for targeted radioimmunotherapy (Larson et al., 2015; Milenic et al., 2004). Multiple forms of chemotherapeutics are available, but may have undesirable side effects, due to the lack of selectivity of drug action (Clementi et al., 2011).

21.2 DRUG DISCOVERY MECHANISMS

The most significant advances in the design of anticancer medicine have been summarized to provide a general overview of the development of anticancer therapy. A number of recent approaches have been discussed, including next-generation sequencing and RNA interference, in light of the complexity of tumor-specific epigenetic mutation-induced effects and the identification of multiple targets. There are

* Corresponding author: rkgenes@gmail.com

DOI: 10.1201/9781032660592-26

numerous genes implicated in cancer, which makes the disease more challenging to treat. Oncogene activation and tumor suppressor gene silencing are determined by abnormal genetic constitutions that result from mutations and non-mutations (epigenetic factors) (Jayashree et al., 2015).

Studies involving alterations in the signaling pathways and networks in cancer can be better understood using a combination of experimental and computational strategies. The inherent complexities in terms of the regulatory aspects, variations in the kinetics as well as cross-talk among the pathways warrants a systems approach for a better comprehension and recapitulation of the hallmark features of neoplasia as well as to better discriminate the early and late stages. One approach can be to identify key proteins/RNA molecules or segments of DNA that can be represented as nodes in a computational-based signaling pathway study (Hornberg et al., 2006). This approach aims to analyze the drugs used specifically, and refine the current treatment methods, using a disease-focused targeted approach, Data sets that are analyzed include those providing a spectrum of mutations; the signalosome, or the metabolome of molecular networks (Csermely et al., 2013).

Suppressing a specific cancer characteristic will not always result in a permanent cure/remission of the cancer; contrary to popular belief, it can increase the likelihood of acquiring neoplastic properties. For instance, specific angiogenesis inhibitors are associated with an increased rate of metastasis (Wilson et al., 2013). Creating new and effective, target-based anticancer medications involves disabling the genetic and epigenetic characteristics and processes controlling signaling pathways in neoplastic cells. Drug development for anticancer agents should aim to inhibit, among other pathologic features, resistance to apoptosis, increased angiogenesis, and vascular disruption (Ma et al., 2009; Lal et al., 2013).

A variety of signaling pathways and kinases are considered targets for the development of small-molecule inhibitors. The most significant of these are insulin-like growth receptors, hepatocyte growth factor, c-Met, and human epidermal growth receptors (HER), as well as inhibitors of PI3K, Akt, MTOR, and hepatocyte growth factor. There have been several anticancer drugs approved that target these kinases, including imatinib, gefitinib, erlotinib, sorafenib, cetuximab, and bevacizumab (Akhdar et al., 2010).

Apart from the aforesaid mechanisms, other targets include those contributing to drug efflux, DNA damage repair, drug target modification, cell death inhibition, and epithelial–mesenchymal transition (Rochat, 2009). When normal cells transform into cancerous cells, they also undergo metabolic changes that make them resistant to chemotherapies (Housman et al., 2014). The effectiveness of different drug-like or FDA-approved compounds has been tested *in vitro* on a variety of tumor models without success. Several studies have demonstrated the effectiveness of tricyclic antidepressants in treating non-small cell lung cancer (Jahchan et al., 2013), while two compounds have been identified from a screening of 796 FDA-approved drugs as

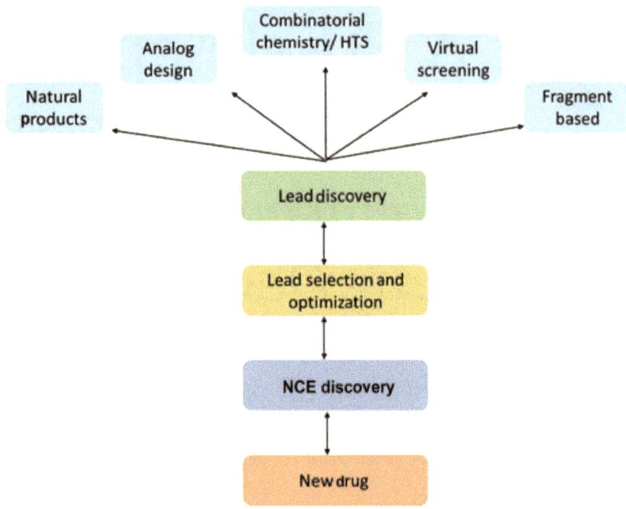

FIGURE 21.1 The drug discovery process.

inhibiting the proliferation of metastatic gastrointestinal stromal tumors (Pessetto et al., 2013).

There are several compounds from an FDA clinical library that are sensitizing to tumor necrosis factor-related apoptosis-inducing ligand, suggesting they could be effective as chemotherapy agents (Taylor et al., 2011). The relationship between drug resistance and cancer cell hypoxia is inextricably linked. As a result of hypoxia, several enzymes that are directly involved in the metabolism and activation of medicines are affected. A unique technique involving cells' induction hypoxia is required to develop novel therapeutic drugs. Drug resistance has been linked to several factors, including nuclear transcription factors such as apoptosis, DNA repair, cellular damage, and the presence of certain receptors such as pregnane X receptor, constitutive androstane receptor, nuclear factor-erythroid 2p45-related factor 2 (NRF2), and hypoxia-inducible factor-1 (HIF-1) (Akhdar et al., 2012).

Besides regulating tumor-specific genes, HIF-1 is also essential for anaerobic energy metabolism, angiogenesis, cell survival, treatment resistance and cell invasion. There is an increase in HIF-1a overexpression in solid tumors under hypoxic conditions. Therefore, disorders induced by hypoxia or ischemia might benefit from HIF-1 activity modulation or suppression. During hypoxic conditions, some prodrugs are converted into 6-MP and 6-TG, since some pharmaceutical compounds are transformed into toxic compounds at a low oxygen level. There are several types of bioreductive agents, including quinones, nitroaromatics, nitro-heterocyclics, aliphatic oxides, and heteroaromatic N-oxides (Yeo et al., 2004). See Figure 21.1 for an illustration of the process of drug discovery.

21.3 *IN VITRO* SCREENING METHODS

Screening using animal models, as was traditionally done, is extremely immoral and has been outlawed in certain countries, such as Europe and India. Animal experimentation in

India is regulated by the Indian Committee for Control and Supervision of Experiments on Animals, which adheres to four principles: replacement, reduction, refinement, and rehabilitation (Pereira et al., 2005). A high-throughput test is frequently conducted on cells or targets prior to the *in vivo* testing of anticancer medications.

21.3.1 Cell Line Prescreens

As they are the most controllable test systems due to their immortality and reproducible growth features, permanent human tumor cell lines are utilized as the main component of cellular screens in cancer research. As part of its *in vitro* primary screening program, the NCI features a panel of 60 different human tumor cell types from nine different types of cancer (leukemia, non-small cell lung cancer, colon cancer, brain cancer, melanoma, ovarian cancer, renal cancer, prostate cancer, and breast cancer). These cell lines are used for testing against compounds over a predetermined concentration range to determine how much each cell line is affected by cytotoxicity or growth inhibition (Boyd and Paull, 1995).

The dose–response curve of sixty cell lines screen is constructed and operated so that the characteristic "fingerprint" of cellular responses resulting from the screen can be used for pattern recognition algorithms for each drug tested. A test substance's mechanism of action can be derived using one of these algorithms. Depending on the response pattern, they can also determine if it is unique or similar to any standard compounds in the database (Boyd, 1997).

21.3.2 Determining Cell Growth

Various approved techniques can be used to measure cell growth by excluding certain dyes from living membranes. The selection of a technique is a variety of influences of factors and a minimal number of cells must be included, including sensitivity. A variety of assays are used to determine cytotoxicities, such as sulforhodamine B (SRB), propidium iodide, luciferase, and 3-(4,5-dimethylthiazol-2-yl)-2,5-diphenyltetrazolium (MTT).

21.3.2.1 SRB

The SRB test involves binding a bright pink anionic dye to the essential amino acids of trichloroacetic acid-fixed cells, based on electrostatic interactions (Mngadi et al., 2020). Among the benefits of this method are its speed, accuracy, and cost-effective nature. In order to measure the protein concentration indirectly, Tris base (tris(hydroxymethyl) aminomethane) can be used to recover the acid-bound dye after the unbound dye has been washed away. Further, this method may be used both for routine laboratory tests and for screening anticancer agents on a large scale (Kawashima et al., 2014). Tests with this endpoint can be performed nondestructively, are time-independent (stable), and can be compared with other fluorescence tests (Hassoun et al., 2020). Despite the labor-intensive nature of the SRB assay (multiple washing operations), the results obtained from this assay are comparable with those obtained from MTT (Kumar et al., 2016).

21.3.2.2 Propidium Iodide

There has been evidence that ethidium bromide and PI intercalate with DNA and pass exclusively through the membrane of dying or dead cells (Krishan, 1975). Due to the binding of these dyes to DNA, they are fluorescent (more intensely with PI). Nuclear localization of the fluorescent signal is indicative of cell death and can be an endpoint for the assessment of chemotherapeutic potential of the anticancer agent. Due to their stability and viability, these dyes are more accurate and trustworthy than other dyes. Fluorescence detection (first measurement) determines the number of nonviable cells after the cells have been incubated with PI. After the cells have been frozen at 20 °C for 24 hours, the second measurement will be performed. A difference between the two readings will reveal the number of living cells due to the intercalation of PI into DNA during freezing. There is no need to wash the cells before the test, and only 150–500 cells per well are required for drug detection. As PI binds to double-stranded RNA in the cytoplasm, this experiment may have some drawbacks. However, this may be avoided by using eliminating the RNA using RNAase (Dengler et al., 1995).

21.3.2.3 Luciferase

The trypan blue, MTT, and SRB all have limitations that affect their accuracy and reliability. The trypan blue assay can be influenced by the expertise of the analyst and the dye can be incorporated by live cells, leading to inaccurate results (Avelar-Freitas et al., 2014). The MTT can be interfered with by certain compounds, such as glycolysis inhibitors, and has a higher variability in detecting changes in cell number compared to other assays. SRB, while more reproducible, can also be influenced by the presence of certain compounds (Van Tonder et al., 2015). The LDH assay, while not specifically mentioned in the provided papers, can also be affected by various factors such as the presence of certain compounds and the type of cells being tested. Efforts to resolve the fluorescence quenching issue in the trypan blue assay have been made, with studies showing that trypan blue can be used to quench the autofluorescence of cells, improving the accuracy of protein expression analysis (Srivastava et al., 2011).

The primary free energy source, also known as the "energy currency," is the nucleotide adenosine triphosphate (ATP), which all cells need to survive and perform their unique functions. Damage or hypoxia causes cytoplasmic ATP levels to decrease, thereby providing an quantifiable endpoint for assessment of the vitality. High-quantum chemiluminescence is produced due to the reaction between cellular ATP, luciferin, and luciferase after cell lysis. Under ideal circumstances, the ATP concentration is inversely correlated with the light intensity (Crouch et al., 1993).

For the MTT test to yield readings above the background, at least 25,000 cells were needed per well; in contrast, the

ATP assay detected cell viability at a lower density of 2000 cells/well and was more reproducible. The downside of the ATP assay is that sample quenching can impair the luminescence readout (Petty et al., 1995, Van Meerloo et al., 2011). Cell viability can also be determined using other methods, but several issues constrain their utility. For example, cells must be counted within three to five minutes using the trypan blue dye exclusion assay. Dead cells accumulate with time, which is why the number increases. When the agent under investigation only affects intracellular processes, the results of the lactate dehydrogenase assay may be misleading (Weyermann et al., 2005).

Monolayer cellular screening is the most common and practical method for cytotoxicity studies, but it cannot replicate 3D *in vivo* development. Several drugs are not appropriate for screening monolayer cells, such as signal transduction inhibitors, antibodies, bioreductive treatments, antiangiogenic peptides, and antitelomerases. Furthermore, growth in adherent conditions acquires mutations and genetic/epigenetic changes that are not present in the original tumor from which the cells were derived (Matossian et al., 2019).

21.3.2.4 MTT

MTT was developed by Mosmann in 1983 and involves converting a colorless tetrazolium salt into a bright, insoluble formazan product the formation of which is inversely proportional to the number of living cells. During the reduction process, NADH, NADPH, and other oxidized substrates or coenzymes are used to obtain electrons from the tetrazolium salt. Mitochondria are one of the many places in the cell where MTT levels are lower. Although MTT tests are rapid, useful, and easy to perform, many variables can affect their results. Additionally, MTT formazan production differs among cell types, and the length of MTT exposure influences the drug cytotoxicity. Each cell line should be subjected to standardized assay conditions in order to minimize the effects they may have on the results. The MTT concentration and exposure time should be standardized, as well as the cell inoculation density and test duration, so as to prevent depletion of the medium's nutrients (Vistica et al., 1991).

21.4 SCREENING ANTICANCER ACTIVITY

Before the discovery of antiproliferative drugs, all attempts to treat cancer focused on preventing uncontrolled cell division, a root cause of cancer that was previously believed to exist. Preclinical and clinical research focused on reducing tumor size; numerous therapeutically relevant anticancer medicines were identified when new compounds were examined in drug discovery programs employing rapidly expanding animal cancer models.

Cancers that develop quickly, such as lymphomas, juvenile leukemia, and germline tumors, receive the most effective treatments; lung, breast, and colorectal cancer have relatively few effective treatments (Suggitt and Bibby, 2005). The conventional method of screening for anticancer drugs involves identifying drugs that have cytostatic or cytotoxic effects on tumor cells and that decrease tumor growth in mouse tumor models. The chemicals were discovered either by accident or as blockers of metabolic pathways (Zhang and Tang, 2018). This strategy was highly successful, but the identified drugs typically had harmful side effects.

Recent advances in the molecular biology of cancer have led to the development of predesigned chemicals that target specific molecular markers that are essential for cancer development. Traditional anticancer agent screening entails investigating tumor regression and survival in animal screening models to uncover cytotoxic medicines, and human clinical trials are used to evaluate doses and dosage regimens based on preclinical data. The adverse effects of cytotoxic drugs, as well as their capacity to promote tumor regression, are caused by similar processes. The medications that target these molecules include receptor tyrosine kinase inhibitors, matrix metalloproteinase inhibitors, and anti-inflammatory agents and farnesyl transferase inhibitors. These medications affect intracellular nuclear, intracellular nonnuclear, transmembrane, and extracellular processes (McKeage, 2008). ["Recent advances in the molecular biology of cancer have led to the development of various cancer targets. These targets are addressed using predesigned chemicals that focus on specific molecular markers essential for cancer development, inhibiting or altering them. Traditional anticancer agent screening involves investigating tumor regression and survival in animal models to uncover cytotoxic medicines, followed by human clinical trials to evaluate doses and dosage regimens based on preclinical data. The adverse effects of cytotoxic drugs, as well as their ability to promote tumor regression, arise from similar processes. Medications that target these molecules include receptor tyrosine kinase inhibitors, matrix metalloproteinase inhibitors, anti-inflammatory agents, and farnesyl transferase inhibitors. These medications influence intracellular nuclear, intracellular nonnuclear, transmembrane, and extracellular processes."] During animal experiments, these medications often inhibit tumor growth rather than reverse it.

There are several key components of preclinical and clinical screening, including physicochemical variables. For example, the solubility and stability of molecularly targeted drugs in the solid state and in solution are important determinants of the pharmacokinetic behavior of candidate drugs. In addition, researchers consider pH profiling, identifying degradation pathways, and evaluating absorption in cell culture and animal models, among other factors. Researchers use cell culture models to conduct cell growth inhibitory and clonogenic studies *in vitro* to assess anticancer effects as well as using animal models containing human xenografts or murine allografts (Amundson et al., 2008).

Better anticancer medications are required for cancer patients, and trustworthy screening techniques are essential for a productive drug development process. These techniques should be able to identify substances with tremendous therapeutic potential while also determining if that potential is sufficient to warrant further study. The use of animal testing on a large scale is unethical and is severely prohibited nowadays, especially in Europe (Kiani et al., 2022). The majority of anticancer medicines will be tested *in vitro* before they are tested *in vivo* via high-throughput cellular or target-based assays. During the pre-clinical phase, screening plays a crucial role in determining which drugs require additional development. The purpose of screening procedures is to determine whether molecular cancer therapies are working as intended to suppress tumor growth. Tumor xenograft and cell line models that closely resemble the characteristics and responses of patients are used in drug development and screening procedures as part of an integrated strategy to provide corroborative evidence (Fiebig et al., 2007).

Several functional antibodies have been developed based on target-driven drug discovery. These antibodies block growth factor receptors and cell signals. Before being tested on people, each potential anticancer drug must complete a qualification study. At first, *in vitro* screening is conducted using the NCI60 human tumor cell line anticancer drug screening approach in order to determine whether a potential candidate can halt the growth of tumor cells in culture. The National Cancer Institute (NCI) developed this modern *in vitro* pharmacological screening, which includes 60 human tumor cell lines. Currently, the NCI60 is one of the most frequently used methods for evaluating potential anticancer medications at an early stage of development (Shoemaker, 2006). *In vivo*, animal testing is carried out after *in vitro* research reveals possible anticancer activity. This step in the evaluation, which is crucial for comprehending *in vivo* anticancer activity, often employs inbred laboratory mice, dogs, and primates (Workman et al., 2010). Human and mouse genomes are similar, as well as have low maintenance costs, short gestation periods, high reproduction rates, and the ease of producing implanted tumors. As a result, mouse tumor models are used in cancer research and for assessing potential anticancer drugs.

21.5 SCREENING MODELS FOR ANTICANCER STUDIES

Cancer research relies heavily on animal models, both *in vitro* and *in vivo*. This is because they provide the opportunity to identify carcinogens, develop cancer therapies, screen drugs, and learn more about the molecular processes that are responsible for tumor growth and metastasis. A wide range of microenvironments are affected by the metastatic process, including stroma, blood vessel endothelium, the circulation system, and secondary tissues (Chambers et al., 2002; Fidler, 2003; Steeg, 2006). The interaction between a cancer cell and its local microenvironment plays an imperative role in determining whether these processes will be effectively addressed. Although metastases account for more than 90% of cancer-related deaths, many aspects of the metastatic cascade remain unknown (Wirtz et al., 2011; Weigelt et al., 2005; Mehlen and Puisieux, 2006). The development of other mouse models of primary and metastatic cancers has also been carried out. These include genetically modified mice, immunocompromised mice with human tumor xenografts, and environmental models (Cekanova and Rathore, 2014; Denayer et al., 2014). Despite *in vivo* models demonstrating the complexity of metastatic processes, distinguishing between the different phases and identifying quantitative mechanistic information can be challenging.

However, despite having limited physiological relevance and capturing only a small portion of the tumor microenvironment, *in vitro* models allow for a more precise control and quantitative analysis of the most experimental variables. Models of solid tumors *in vitro* are available at various levels of complexity, including cell lines derived from tumors and 3D models of the tumor microenvironment (Hulkower and Herber, 2011; Infanger et al., 2013; Vidi et al., 2013). The development of several models has led to mechanistic knowledge regarding tumor development and proliferation, migration and invasion, matrix remodeling, dormancy, intravasation, and extravasation, that has translational ramifications into terms of the development of drugs in a site-specific/selective manner. Cancer cells must be available for use in any *in vitro* tumor model. This is because they are easy to cultivate, enable direct comparisons with experimental data, and are inexpensive to maintain (Greshock et al., 2007; Holliday and Speirs, 2011). *In vitro* tumor models include transwell, spheroid-based, hybrid, and microvessel-based tumor models. A discussion of the pros and cons of these models, as well as the variables that may affect the tumor microenvironment, is provided.

21.6 *IN VITRO* SCREENING MODELS: LIMITATIONS AND BENEFITS

21.6.1 Imaging Methods

A range of *in vitro* and *in vivo* models have been developed for screening the imaging methods of anticancer drugs. *In vivo* models, such as subcutaneous xenotransplant and orthotopic tumor models, have been widely used for preclinical screening (Blatt et al., 2013). *In vitro* models, including 3D cell culture, microfluidic systems, and 3D bioprinting, have also been developed to mimic *in vivo* cancer and study genetic, epigenetic, and biochemical changes (Kitaeva et al., 2020). These models have been instrumental in the development and testing of new anticancer agents, with a focus on identifying drugs with novel antitumor activity (Siddiqui et al., 2017).

21.6.1.1 A Transwell-Based Model

To measure cancer cell migration and invasion, transwell-based assays are often utilized (Figure 21.2) (Marshall, 2011; Kramer et al., 2013). The metastatic cascade is centered on migration, which is the movement of cells from one site to another in humans and animal model systems (Madsen et al., 2015). Gradients in soluble substances,

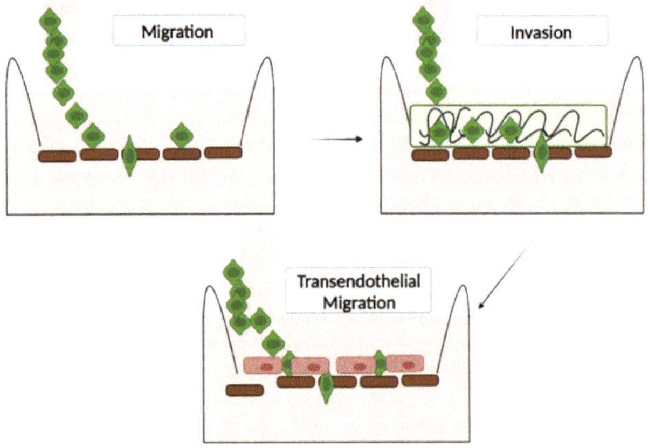

FIGURE 21.2 The transwell-based model.

electric fields, or matrix stiffness can guide or cause random cell migration (Rorth, 2009). Cell movement in a 3D ECM is called invasion (Friedl and Alexander, 2011; Friedl et al., 2012). Transwell assays can be divided into three categories: migration, invasion, and transendothelial migration. When cancer cells are seeded in invasion studies, they are placed on top of a permeable membrane. In migration studies, the chemotactic capability of cells toward a chemo-attractant. Confluent endothelial cells are layered on the membrane in transendothelial migration tests (Smith et al., 1991).

21.6.1.1.1 Migration Assays

Plating cancer cells on top of a porous membrane is the purpose of this experiment. An analysis is carried out to determine how many cells will cross a membrane and migrate toward a chemoattractant. After harvesting, cells are routinely fixed for imaging 24 hours later. Cell migration is usually determined by staining or labeling with crystal violet. During the process of imaging the membrane, the bottom surface is scanned and quantified as the cells travel along the field of view. Despite the appearance of simplicity, this experiment provides a high-throughput *in vitro* model of tumor invasion through leaky vessels. The serum is used as a surrogate for and is representative of the components of the blood and the holes in the membrane are considered to mimic the gaps in the endothelial cells in this model (Pouliot et al., 2000).

21.6.1.1.2 Assays for Invasion

Invasion experiments significantly increase the complexity of this model. The poros membrane is covered with an ECM layer that is intended to resemble the basement membrane of the vasculature. Collagen and laminin are also utilized in this layer, but Matrigel is employed more frequently (Rath et al., 2013; Wang et al., 2009). A layer of ECM can be up to 1 mm thick. Although the tests used to assess cell invasion and migration are generally comparable, it is essential to remember that some medications and genetic modifications have a higher impact on the suppression of invasion than on the former.

21.6.2 Migration Assays for Transendothelial Cells

The importance of isolating and characterizing circulating cancer stem cells, which play a crucial role in driving metastasis, is highlighted by (Yang *et al.*, 2015). Advanced imaging techniques are crucial for *in vivo* detection of these elusive cells, as emphasized by (Hart and El-Deiry, 2008). The concept of migrating cancer stem cells, which are responsible for the dissemination of tumor cells and the formation of metastases, is proposed by (Brabletz et al., 2005). The need for functional imaging methodologies to assess the efficacy of therapeutic agents in the clinic, particularly in the context of metastasis, is underscored by (Messiou and deSouza, 2010). These studies collectively underscore the significance of advanced imaging tools in understanding and targeting metastasizing cancer cells and cancer stem cells. An endothelial monolayer is placed on a porous support when transendothelial migration tests are conducted. This monolayer is expected to recapitulate the hurdles imposed by cell-cell interactions and the ECM elaborated by the endothelial cells. A transendothelial test provides a highly efficient extravasation model considering that cancer cells must first pass through the endothelium before they can cross through the basement membrane (Rahn et al., 2005). Endothelial cells can be grown to confluence on the bottom of the chamber. Cancer cells may be inserted into the chamber on top in order to simulate intravasation (Pignatelli et al., 2014). Transendothelial methods are also commonly employed to examine the tight cell-cell endothelium in brain capillaries (Lin et al., 2015).

21.6.2.1 Spheroids

The spheroids consist of several cells growing in suspension or in a three-dimensional matrix. Multicellular tumor spheroids, also known as cancer cell spheroids, are non-vascular nodules or micrometastases from tumors (Friedrich et al., 2009). Although 3D spheroids are the process is more time consuming and expensive than 2D cell culture, they are often utilized to screen drugs, evaluate tumor growth and proliferation (Gottfried et al., 2006; Hirschhaeuser et al., 2010; LaBarbera et al., 2012; Fennema et al., 2013). Through the use of 3D spheroids, transportation features and interactions between tumor cells and their surroundings can be reproduced. Compared to two-dimensional culture, tumor cell protein and gene expression patterns in spheroids are far more similar to those observed in clinical and *in vivo* studies (Kapałczyńska et al., 2018).

21.6.2.1.1 Spheroid Formation

The spheroid formation, a crucial aspect of 3D cell culture, relies on key factors for reproducibility. Consistent cell density is paramount, ensuring uniformity in spheroid size and structure. The choice of matrix or scaffold material, along with culture medium composition, significantly influences reproducibility. Standardized incubation conditions, including temperature, humidity, and CO2 levels, play a pivotal role in promoting consistent spheroid development. Rigorous adherence to established protocols and regular quality control checks further enhance reproducibility in spheroid formation experiments (Białkowska et al., 2020). It may be accomplished using four broad techniques: Techniques for suspension culture, non-adherent surfaces, hanging drop techniques, and microfluidics have been proposed (Mehta et al., 2012). Each technique has specific benefits and drawbacks. By agitating the cells to keep them suspended or by making the fluid more viscous, suspension culture techniques encourage the creation of spheroid structures. High throughput is possible with suspension culture, but size and homogeneity cannot be controlled (Lin and Chang, 2008; Metzger et al., 2011).

Spheroid development is aided by cell culture on non-adherent surfaces because it hinders substrate adhesion. Agar, 1%–1.5% agarose, or another non-adherent character is used to cultivate suspended cells using the liquid overlay method (LOT) (Ivascu and Kubbies, 2006). A non-adhesive surface can be applied in a straightforward fashion. However, the spheroid's size and homogeneity cannot be controlled. Spheroid growth on microarrays increases throughput and enables the control of spheroid size (Hsiao et al., 2012). To regulate the growth of spheroid surfaces, stamped agarose microwells or non-adherent 96-well plates with a round bottom may be used. The hanging drop method and microfluidic device-based procedure provide better control over the size and composition of the spheroid, despite the fact that they are more challenging (Aijian and Garrell, 2015).

An adhesive is used to suspend cells from the underside of a tissue culture cover in the hanging drop technique. Due to their ability to precisely regulate spheroid formation, microfluidic devices are becoming more and more prevalent (Fu et al., 2014). The quicker development and improved size homogeneity of a spheroid is made possible by continuous perfusion under physiological circumstances. Spheroids may also be created, maintained, and tested using an all-in one microfluidic systems, one microfluidic systems offer a comprehensive approach to create, maintain, and test cellular spheroids. These systems integrate various functions, including cell loading, culture maintenance, and analytical assessments within a single microfluidic platform. The design allows precise control over spheroid size and composition while providing a user-friendly, streamlined process. Such systems enhance experimental reproducibility and enable efficient high-throughput screening. The all-in-one microfluidic approach minimizes manual interventions, reducing the risk of contamination and optimizing the workflow for spheroid-based assays in biomedical research and drug development (Wu et al., 2008).

21.6.2.2 Hybrid Models

Transwell- or spheroid *in vitro* tumor models are only one of the numerous subcategories. Models of 3D invasion, ex vivo tumor sections implanted in the human body, and microfluidic avascular models are some of them. A combination of the complex tumour microenvironment and the relative simplicity of the *in vitro* model is utilized in these studies. A tumor slice implanted from a biopsied patient can be utilized for both basic research on tumor development and invasion as well as tailored treatment regimen selection. With 3D invasion models, detailed imaging of cellular interactions can be performed. This allows investigations of the metastatic cascade, which includes a variety of events. This eliminates some of the challenges associated with invasive biopsy methods. Among the various types of microfluidic models, avascular microfluidic models represent the simplest. In spite of this, the methods such as atomic force microscopy (AFM), micropipette aspiration, magnetic bead microrheology, employ cover a wide variety of approaches for examining tumor cell migration across heterogeneous geometries (Zuela-Sopilniak and Lammerding, 2019). Heterogeneous geometries in tumor cell migration encompass diverse microenvironments, such as substrates with varying topography and stiffness. Atomic force microscopy (AFM) enables the study of cell responses to intricate surface features, while micropipette aspiration explores cell deformation in confined spaces. Magnetic bead microrheology delves into the viscoelastic properties of the microenvironment, creating gradients in stiffness for understanding how tumor cells adapt to mechanical cues. These techniques collectively provide valuable insights into the complex interplay between tumor cells and heterogeneous geometric landscapes during migration (Yabushita et al., 2004).

21.6.2.3 Models of Tumor Microvessels and Microvessel Self-Assembly

As a crucial part of the tumor microenvironment, the tumor vasculature supplies nutrients that are vital to the development of the tumor (Figure 21.3). It is well known that the endothelial cells that line blood arteries release substances that both encourage and inhibit the formation of tumors (Butler et al., 2010; Ghajar et al., 2013). Furthermore, the tumor vasculature is responsible for influencing the invasion, intravasation, and extravasation phases of the

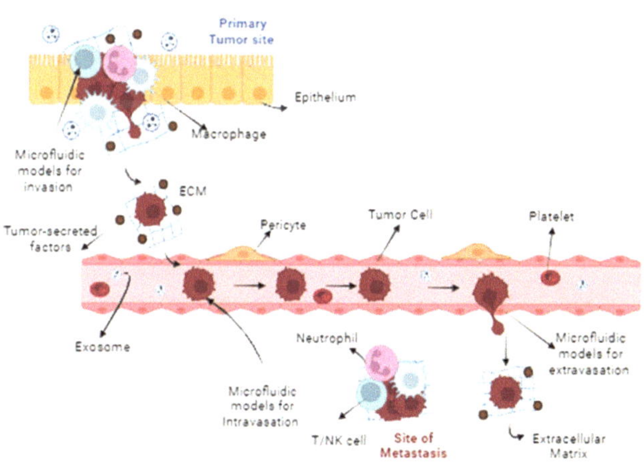

FIGURE 21.3 The tumor microenvironment.

metastatic cascade. An *in vitro* tumour model recapitulating the salient pathologic features of the microvessels has been incorporated in tumor model systems (references -most recent). This type of model enables the investigation of the relationship between tumor vasculature and cancer cells. Most of the time, endothelial cells are seeded onto predetermined ECM scaffolds to create microvessels. The successful conversion of seeded or dispersed endothelial cells into circular microvessels involves several pivotal factors, which collectively contribute to the formation of functional and organized vascular structures such as ECM scaffolds, angiogenic signals, and cell–cell interactions (Lovett et al., 2009). However, endothelial cells can also self-assemble through matrix remodeling after being dispersed randomly within an ECM (Bogorad et al., 2015).

Using endothelial cells deposited on certain surfaces of ECM, the devices generate vascular endothelium. Despite this, they can only form microvessels that are greater than 50 μm in diameter due to constraints related to homogenous cell seeding. Upon injection into the matrix, endothelial cells, the initial formation of a random network then self-assembles with the eventual formation of vessels of an architecture that resembles microvessels formed/present *in vivo* with capillary diameters of 10–20 mm. A network can be anastomosed or joined with the surrounding media ports or channels before blood can be perfused through the vessels. Activating or suppressing tumor cells concealed within the surrounding ECM interact with the vascular endothelium at the edges of self-assembled network vessels (Moya et al., 2013) (see Table 21.1).

TABLE 21.1
Advantages and Disadvantages of In Vitro Tumor Models

Models	Advantages	Disadvantages
	Transwell model	
Migration	Low cost Easily implemented	Low physiological relevance
Invasion	High throughput	Only assay single cell motility
Transendothelial migration	Used to compare metastatic potential of cells	Migration and invasion assays can result in conflicting data
	Spheroids	
Cell suspension culture	Simple Mass production Long term culture	No individual compartments for spheroids Cannot control uniformity
Non-adherent surfaces	Simple Better efficiency (MW) High throughput (MW) Uniform spheroid size	Low throughput Long-term culture difficult
Hanging drop	Control of spheroid size Uniform spheroid size Allows coculture with defined cell types	Low throughput Long-term culture difficult Inefficient
Microfluidic devices	Control of spheroid size, Control of spheroid growth parameters Continuous perfusion Faster spheroid formation	Difficulty collecting cells for analysis
	Hybrid Models	
Embedded ex vivo Tumor sections	Maintains tumor heterogeneity Patient-specific assay Mimics outgrowth into surrounding tissues	Lacks flow through vasculature
3D invasion models	3D microenvironment allows real-time tracking of cells Balance of complexity and experimental control	Lacks vasculature Lacks tumor complexity
Avascular microfluidic	Simple migration assay Easy to isolate effects of variables Allows real-time tracking of cells	Lacks vasculature, typically lacks 3D environment
	Tumor microvessels	
Predefined ECM scaffold	Well-defined vessel endothelium and shear stress	Limited vessels diameter ranges (>50 μm) and simple liner geometries
Microvessel self- assembly	Generates vessel sizes from capillaries to small microvessels and complex networks	Random vessel network with unpredictable flow

21.6.2.3.1 ECM Scaffold Predefined

The subtractive template technique can be utilized to create cylindrical microvessels with diameters as small as 50 mm. An ECM, most commonly collagen type I or fibrin, is used to form microvessels by seeding endothelial cells within the channel. These ECM surfaces are used to seed cells onto the surfaces and allow them to self-assemble into continuous monolayers for the determination of functional characteristics, including vessel permeability, the expression of junctional proteins, and the responses to vascular mediators and inflammatory cytokines (Chrobak et al., 2006).

A wide variety of cell types can be produced by these three-dimensional cylinders, which are capable of coping with shear stress, possess a physiologically relevant form, and produce a wide range of shear stress-resistant cell types. A tumor model may contain smooth muscle cells, pericytes, and lymphatic drainage, but their absence indicates abnormal development of the tumor's vasculature (Wong et al., 2013). Developing endothelial-lined microvessels with tumor cells allows the examination of a variety of tumor-endothelial interactions, such as paracrine signaling from the endothelium, tumor-driven angiogenesis, and intravasation and extravasation (Wang et al., 2015).

Recent studies have highlighted the potential of using decellularized ECM from normal and tumor tissues as scaffolds for cancer organoid cultures (Varinelli et al., 2021). The ECM plays a crucial role in tumor progression and metastasis, with its biochemical and biophysical properties influencing cancer cell behavior. Tailoring the ECM to tissue-specific requirements can be achieved through electrospun scaffolds, which have shown promise in tissue engineering applications. Furthermore, the use of matrix mimetic culture systems, including naturally derived and synthetic scaffolds, has facilitated the study of tumor–ECM interactions in 3D environments (Gill et al., 2014). These studies collectively underscore the importance of the ECM in cancer biology and the potential of ECM-based scaffolds in cancer organoid cultures.

The potential of microfluidic-based 3D microvascular network models in studying tumor microvessels. Wang et al. (2018) discussed the integration of tumor cells into these models, highlighting their potential in cancer metastasis and drug screening. Dávila et al. (2021) further advanced this work by developing a microvessel-on-chip device to study nanomedicine transport dynamics, providing a platform for testing and developing onco-nanomedicine. Gerigk et al. (2019) and Menna et al. (2019) both focused on specific tumor types, with Gerigk developing a microvessel-on-a-chip model to study glioma-vascular interactions and Menna using a microvasculature-on-chip device to assess the metastatic potential of prostate cancer cells. These studies collectively demonstrate the potential of microfluidic-based models in studying tumor microvessels and their applications in cancer research.

21.7 CONCLUSION

Since precision medicine is becoming more widespread, *in vitro* tumor models are becoming increasingly relevant for investigating patient-specific therapy and determining metastatic potential. Testing an anticancer agent's antitumor efficacy against a panel of malignant cell lines is the first step in its development. Using these assays, it is possible to identify substances that need further testing in animal models. Several technologies will be incorporated into next-generation *in vitro* tumor models to address specific applications. Developing new technologies models of cancer screening for therapy is essential for the advancement of cancer treatment as well as for the introduction of more effective chemotherapeutic options. Despite significant progress in understanding and targeting epigenetic mechanisms in cancer, current models face several deficiencies. Challenges include the complexity of epigenetic interactions, limited knowledge of context-specific epigenetic alterations, and the need for more comprehensive biomarkers for patient stratification.

Future directions should prioritize the development of integrated, multi-omics approaches, advancements in single-cell epigenomics, and a deeper understanding of the dynamic nature of epigenetic changes over time. Overcoming these challenges will be crucial for advancing personalized cancer treatments and maximizing the therapeutic potential of targeting epigenetic dysregulation. Many anticancer medications are developed using screening models to reduce the costs and resources associated with their development. In recent years, researchers have found that distinct genetic modifications are responsible for the formation of various tumor types, allowing them to test certain novel drugs on tumor cell lines with these abnormal genetic characteristics to evaluate their antitumor activity. The evolving comprehension of epigenetic intricacies has led to the development of targeted therapies, marking a paradigm shift toward personalized cancer treatments.

The ability to modulate specific epigenetic alterations offers the potential for more effective and precise interventions, bringing us closer to a future where cancer treatment is tailored to the unique molecular profile of each patient. As research in this field progresses, the prospects for innovative and tailored therapeutic approaches continue to grow, offering hope for improved outcomes in the fight against cancer. In particular, the expansion of new tumor models can benefit from advances in organogenesis/embryonic development, a process in which progenitor cells self-organize into specific tissues or organs. The progress of *in vitro* tumor models has become increasingly significant with the push toward precision medicine by providing patient-specific treatments, clinical care, and metastatic assessment.

21.8 ACKNOWLEDGMENTS

The authors are grateful to the Chettinad Academy of Research Education for its continued support and encouragement.

REFERENCES

Aijian AP, Garrell RL. Digital microfluidics for automated hanging drop cell spheroid culture. Journal of Laboratory Automation. 2015 Jun;20(3):283–95.

Akhdar H, Legendre C, Aninat C, Morel F. Anticancer drug metabolism: Chemotherapy resistance and new therapeutic approaches. Topics on Drug Metabolism. 2012 Feb 22:138–70.

Akhdar H, Legendre C, Aninat C, Morel F. BCR ABL kinase inhibitors for cancer therapy. International Journal of Pharmaceutics. 2010:2(2):80–90.

Amundson SA, Do KT, Vinikoor LC, Lee RA, Koch-Paiz CA, Ahn J, Reimers M, Chen Y, Scudiero DA, Weinstein JN, Trent JM. Integrating global gene expression and radiation survival parameters across the 60 cell lines of the National Cancer Institute Anticancer Drug Screen. Cancer Research. 2008 Jan 15;68(2):415–24.

Avelar-Freitas BA, Almeida VG, Pinto MC, Mourão FA, Massensini AR, Martins-Filho OA, Rocha-Vieira E, Brito-Melo GE. Trypan blue exclusion assay by flow cytometry. Brazilian Journal of Medical and Biological Research. 2014 Mar 18;47:307–15.

Białkowska K, Komorowski P, Bryszewska M, Miłowska K. Spheroids as a type of three-dimensional cell cultures – examples of methods of preparation and the most important application. International Journal of Molecular Sciences. 2020 Aug 28;21(17):6225.

Blatt NL, Mingaleeva RN, Khaiboullina SF, Kotlyar A, Lombardi VC, Rizvanov AA. In vivo screening models of anticancer drugs. Life Science Journal. 2013;10(4):1892–900.

Bogorad MI, DeStefano J, Karlsson J, Wong AD, Gerecht S, Searson PC. In vitro microvessel models. Lab on a Chip. 2015;15(22):4242–55.

Boyd MR. The NCI in vitro anticancer drug discovery screen: Concept, implementation, and operation, 1985–1995. Anticancer drug development guide: Preclinical screening, clinical trials, and approval. 1997:23–42.

Boyd MR, Paull KD. Some practical considerations and applications of the National Cancer Institute in vitro anticancer drug discovery screen. Drug Development Research. 1995 Feb;34(2):91–109.

Brabletz T, Jung A, Spaderna S, Hlubek F, Kirchner T. Migrating cancer stem cells – an integrated concept of malignant tumour progression. Nature Reviews Cancer. 2005 Sep;5(9):744–9.

Bray FF, Ferlay J, Soerjomataram I, Siegel RL, Torre LA, Jemal AJ. Erratum: Global cancer statistics 2018: GLOBOCAN estimates of incidence and mortality worldwide for 36 cancers in 185 countries. CA: A Cancer Journal for Clinicians. 2020;70(4):313.

Butler JM, Kobayashi H, Rafii S. Instructive role of the vascular niche in promoting tumour growth and tissue repair by angiocrine factors. Nature Reviews Cancer. 2010 Feb;10(2):138–46.

Cekanova M, Rathore K. Animal models and therapeutic molecular targets of cancer: Utility and limitations. Drug Design, Development and Therapy. 2014 Oct 14:1911–22.

Chambers AF, Groom AC, MacDonald IC. Dissemination and growth of cancer cells in metastatic sites. Nature Reviews Cancer. 2002;2:563–72.

Cheng Y, He C, Wang M, Ma X, Mo F, Yang S, Han J, Wei X. Targeting epigenetic regulators for cancer therapy: Mechanisms and advances in clinical trials. Signal Transduction and Targeted Therapy. 2019 Dec 17;4(1):62.

Chrobak KM, Potter DR, Tien J. Formation of perfused, functional microvascular tubes in vitro. Microvascular Research. 2006;71:185–96. doi:10.1016/j.mvr.2006.02.005

Clementi C, Miller K, Mero A, Satchi-Fainaro R, Pasut G. Dendritic poly (ethylene glycol) bearing paclitaxel and alendronate for targeting bone neoplasms. Molecular Pharmaceutics. 2011;8:1063–72.

Crouch SP, Kozlowski R, Slater KJ, Fletcher J. The use of ATP bioluminescence as a measure of cell proliferation and cytotoxicity. Journal of Immunological Methods. 1993;160:81–8.

Csermely P, Korcsmáros T, Kiss HJM, London G, Nussinov R. Structure and dynamics of molecular networks: A novel paradigm of drug discovery: A comprehensive review. Pharmacology & Therapeutics. 2013:138(3):333–408.

Dávila S, Cacheux J, Rodríguez I. Microvessel-on-chip fabrication for the in vitro modeling of nanomedicine transport. ACS Omega. 2021 Sep 23;6(39):25109–15.

Denayer T, Stöhr T, Vanroy M. Animal models in translational medicine: Validation and prediction. New Horizons in Translational Medicine. 2014;2:5–11. doi:10.1016/j. nhtm. 2014.08.001

Dengler WA, Schulte J, Berger DP, Mertelsmann R, Fiebig HH. Development of a propidium iodide fluorescence assay for proliferation and cytotoxicity assays. Anticancer Drugs. 1995;6:522–32.

Dhanya KC, Menon A, Rai LS. In-vitro models in anticancer screening. In Phytochemistry: An in-silico and in-vitro update (pp. 251–65). Singapore: Springer, 2019.

Diao YY, Li HY, Fu YH, Han M, Hu YL, Jiang HL, Tsutsumi Y, Wei QC, Chen DW, Gao JQ. Doxorubicin-loaded PEG-PCL copolymer micelles enhance cytotoxicity and intracellular accumulation of doxorubicin in adriamycin-resistant tumor cells. International Journal of Nanomedicine. 2011;6:1955–62.

Feng SS, Chien S. Chemotherapeutic engineering: Application and further development of chemical engineering principles for chemotherapy of cancer and other diseases. Chemical Engineering Science. 2003;58:4087–114.

Fennema E, Rivron N, Rouwkema J, Van Blitterswijk C, De Boer J. Spheroid culture as a tool for creating 3D complex tissues. Trends in Biotechnology. 2013;31:108–15. doi:10.1016/j. tibtech.2012.12.003

Fidler IJ. The pathogenesis of cancer metastasis: The 'seed and soil' hypothesis revisited. Nature Reviews Cancer. 2003; 3:453–58. doi:10.1038/nrc1098

Fiebig HH, Schüler J, Bausch N, Hofmann M, Metz T, Korrat A. Gene signatures developed from patient tumor explants grown in nude mice to predict tumor response to 11 cytotoxic drugs. Cancer Genomics Proteomics. 2007;4:197–209.

Friedl P, Alexander S. Cancer invasion and the microenvironment: Plasticity and reciprocity. Cell. 2011;147:992–1009. doi:10.1016/j.cell.2011.11.016

Friedl P, Locker J, Sahai E, Segall JE. Classifying collective cancer cell invasion. Nature Cell Biology. 2012;14:777–83. doi:10.1038/ncb2548

Friedrich J, Seidel C, Ebner R, Kunz-Schughart LA. Spheroid-based drug screen: Considerations and practical approach. Nature Protocols. 2009;4:309–24. doi:10.1038/nprot.2008.226

Fu CY, Tseng SY, Yang SM, Hsu L, Liu CH, Chang HY. A microfluidic chip with a U-shaped microstructure array for multicellular spheroid formation, culturing and analysis. Biofabrication. 2014;6:015009. doi:10.1088/1758-5082/6/1/015009

Gerigk M, Bulstrode HJ, Huang YY. Microvessel-on-a-chip for investigating glioma-vascular interactions. Cancer Research. 2019 Jul 1;79(13_Supplement):39.

Ghajar CM, Peinado H, Mori H, Matei IR, Evason KJ, Brazier H, et al. The perivascular niche regulates breast tumour dormancy. Nature Cell Biology. 2013;15:807–17. doi:10.1038/ncb2767

Gheybi H, Adeli M. Supramolecular anticancer drug delivery systems based on linear dendritic copolymers. Polymer Chemistry. 2015;6:2580–615.

Gill BJ, West JL. Modeling the tumor extracellular matrix: Tissue engineering tools repurposed towards new frontiers in cancer biology. Journal of Biomechanics. 2014 Jun 27;47(9):1969–78.

Gottfried E, Kunz-Schughart LA, Andreesen R, Kreutz M. Brave little world: Spheroids as an in vitro model to study tumor-immune-cell interactions. Cell Cycle. 2006;5:691–95. doi:10.4161/cc.5.7.2624

Greshock J, Nathanson K, Martin AM, Zhang L, Coukos G, Weber BL, et al. Cancer cell lines as genetic models of their parent histology: Analyses based on array comparative genomic hybridization. Cancer Research. 2007;67:3594–600. doi:10.1158/0008-5472.CAN-06-3674

Hart LS, El-Deiry WS. Invincible, but not invisible: Imaging approaches toward in vivo detection of cancer stem cells. Journal of Clinical Oncology. 2008 Jun 10;26(17):2901–10.

Hassoun A, Cropotova J, Rustad T, Heia K, Lindberg SK, Nilsen H. Use of spectroscopic techniques for a rapid and non-destructive monitoring of thermal treatments and storage time of sous-vide cooked cod fillets. Sensors. 2020 Apr 23;20(8):2410.

Hirschhaeuser F, Menne H, Dittfeld C, West J, Mueller-Klieser W, KunzSchughart LA. Multicellular tumor spheroids: An underestimated tool is catching up again. Journal of Biotechnology. 2010;148:3–15. doi:10.1016/j.jbiotec.2010.01.012

Holliday DL, Speirs V. Choosing the right cell line for breast cancer research. Breast Cancer Research. 2011;13:215. doi:10.1186/bcr2889

Hornberg JJ, Bruggeman FJ, Westerhoff HV, Lankelma J. Cancer: A systems biology disease. Biosystems. 2006;83(2–3):81–90.

Housman G, Byler S, Heerboth S, Lapinska K, Longacre M, Snyder N, Sarkar S. Drug resistance in cancer: An overview. Cancers (Basel). 2014;6(3):1769–92.

Hsiao AY, Tung YC, Kuo CH, Mosadegh B, Bedenis R, Pienta KJ, Takayama S. Micro-ring structures stabilize microdroplets to enable long term spheroid culture in 384 hanging drop array plates. Biomedical Microdevices. 2012;14:313–23. doi:10.1007/s10544-011-9608-5

Hulkower KI, Herber RL. Cell migration and invasion assays as tools for drug discovery. Pharmaceutics. 2011;3:107–24. doi:10.3390/pharmaceutics3010107

Infanger DW, Cho Y, Lopez BS, Mohanan S, Liu SC, Gursel D, Boockvar JA, Fischbach C. Glioblastoma stem cells are regulated by interleukin-8 signaling in a tumoral perivascular niche. Cancer Research. 2013;73:7079–89. doi:10.1158/0008-5472.CAN-13-1355

Ivascu A, Kubbies M. Rapid generation of single-tumor spheroids for high-throughput cell function and toxicity analysis. Journal of Biomolecular Screening. 2006;11:922–32. doi:10.1177/1087057106292763

Jahchan NS, Dudley JT, Mazur PK, Flores N, Yang D, Palmerton A, Zmoos A-F, Vaka D, Tran KQT, Zhou M, Krasinska K, Riess JW, Neal JW, Khatri P, Park KS, Butte AJ, Sage J. A drug repositioning approach identifies tricyclic antidepressants as inhibitors of small cell lung cancer and other neuroendocrine tumors. Cancer Discovery. 2013;3:1364–77.

Jayashree BS, Nigam S, Pai A, Patel HK, Reddy ND, Kumar N, Rao CM. Targets in anticancer research – A review. Indian Journal of Experimental Biology. 2015;53(8):489–507.

Kapałczyńska M, Kolenda T, Przybyła W, Zajączkowska M, Teresiak A, Filas V, Ibbs M, Bliźniak R, Łuczewski Ł, Lamperska K. 2D and 3D cell cultures–a comparison of different types of cancer cell cultures. Archives of Medical Science. 2018 Jun 19;14(4):910–9.

Kawashima Y, Kodera Y, Singh A, Matsumoto M, Matsumoto H. Efficient extraction of proteins from formalin-fixed paraffin-embedded tissues requires higher concentration of tris (hydroxymethyl) aminomethane. Clinical Proteomics. 2014 Dec;11(1):1–6.

Kiani AK, Pheby D, Henehan G, Brown R, Sieving P, Sykora P, Marks R, Falsini B, Capodicasa N, Miertus S, Lorusso L. Ethical considerations regarding animal experimentation. Journal of Preventive Medicine and Hygiene. 2022 Jun;63(2 Suppl 3):E255.

Kitaeva KV, Rutland CS, Rizvanov AA, Solovyeva VV. Cell culture based in vitro test systems for anticancer drug screening. Frontiers in Bioengineering and Biotechnology. 2020 Apr 9;8:322.

Kramer N, Walzl A, Unger C, Rosner M, Krupitza G, Hengstschläger M, Dolznig H. In vitro cell migration and invasion assays. Mutat. Res. 2013;752:10–24. doi:10.1016/j.mrrev.2012.08.001

Krishan A. Rapid flow cytofluorometric analysis of mammalian cell cycle by propidium iodide staining. Journal of Cell Biology. 1975;66:188–93.

Kumar S, Bajaj S, Bodla RB. Preclinical screening methods in cancer. Indian Journal of Pharmacology. 2016 Sep;48(5):481.

LaBarbera DV, Reid BG, Yoo BH. The multicellular tumor spheroid model for high-throughput cancer drug discovery. Expert Opinion on Drug Discovery. 2012;7:819–30. doi:10.1517/17460441.2012.708334

Lal H, Kolaja KL, Force T. Cancer genetics and the cardiotoxicity of the therapeutics. Journal of the American College of Cardiology. 2013;61(3):267–74.

Larson SM, Carrasquillo JA, Cheung NK, Press OW. Radioimmunotherapy of human tumours. Nature Reviews Cancer. 2015;15:347–60.

Lin HH, Chiang MT, Chang PC, Chau LY. Myeloid heme oxygenase-1 promotes metastatic tumor colonization in mice. Cancer Science. 2015;106:299–306. doi:10.1111/cas.12604

Lin RZ, Chang HY. Recent advances in three-dimensional multicellular spheroid culture for biomedical research. Biotechnology Journal. 2008;3:1172–84. doi:10.1002/biot.200700228

López-Gómez M, Malmierca E, de Górgolas M, Casado E. Cancer in developing countries: The next most preventable pandemic. The global problem of cancer. Critical Reviews in Oncology/Hematology. 2013;88:117–22.

Lovett M, Lee K, Edwards A, Kaplan DL. Vascularization strategies for tissue engineering. Tissue Engineering Part B: Reviews. 2009 Sep 1;15(3):353–70.

Ma WW, Adjei AA, Adjei AA. Novel agents on the horizon for cancer therapy. CA: A Cancer Journal for Clinicians. 2009;59(2):111–37.

Madsen CD, Hooper S, Tozluoglu M, Bruckbauer A, Fletcher G, Erler JT, Bates PA, Thompson B, Sahai E. STRIPAK components determine mode of cancer cell migration and metastasis. Nature Cell Biology. 2015;17:68–80. doi:10.1038/ncb3083

Marshall J. Transwell® invasion assays. In Cell migration, eds CM Wells, M Parsons (pp. 97–110). London: Springer, 2011.

Matossian MD, Burks HE, Elliott S, Hoang VT, Bowles AC, Sabol RA, Wahba B, Anbalagan M, Rowan B, Abazeed ME, Bunnell BA. Drug resistance profiling of a new triple negative breast cancer patient-derived xenograft model. BMC cancer. 2019 Dec;19(1):1–7.

McKeage MJ. The potential of DMXAA (ASA404) in combination with docetaxel in advanced prostate cancer. Expert Opinion on Investigational Drugs. 2008;17:23–9.

Mehlen P, Puisieux A. Metastasis: A question of life or death. Nature Reviews Cancer. 2006;6:449–58. doi:10.1038/nrc1886

Mehta G, Hsiao AY, Ingram M, Luker GD, Takayama S. Opportunities and challenges for use of tumor spheroids as models to test drug delivery and efficacy. Journal of Controlled Release. 2012;164:192–204. doi:10.1016/j. jconrel.2012.04.045

Menna MD, Sousa ER, La Manna F, Zeinali S, Guenat O, Kruithof-de Julio M, Thalmann GN. MP81-01 microvasculature on chip: Cancer cells extravasation in vitro modeling. Journal of Urology. 2019 Apr;201(Supplement 4):e1170–1.

Messiou C, deSouza NM. Metastasis imaging: Current concepts and future challenges. Cancer Biomarkers. 2010 Jan 1;7(4–5):171–2.

Metzger W, Sossong D, Bachle A, Putz N, Wennemuth G, Pohlemann T, Oberringer M. The liquid overlay technique is the key to formation of co-culture spheroids consisting of primary osteoblasts, fibroblasts and endothelial cells. Cytotherapy 2011;13:1000–12. doi:10.3109/14653249.2011.583233

Milenic D, Garmestani K, Dadachova E, Chappell L, Albert P, Hill D, Schlom J, Brechbiel M. Radioimmunotherapy of human colon carcinoma xenografts using a 213Bilabeled domain-deleted humanized monoclonal antibody. Cancer Biotherapy and Radiopharmaceuticals. 2004;19:135–47.

Mngadi S, Mokhosi S, Singh M, Mdlalose W. Chitosan-functionalized Mg0.5Co0.5Fe2O4 magnetic nanoparticles enhance delivery of 5-fluorouracil in vitro. Coatings. 2020 May 2;10(5):446.

Moya M, Tran D, George SC. An integrated in vitro model of perfused tumor and cardiac tissue. Stem Cell Research & Therapy. 2013;4(Suppl. 1):S15. doi:10.1186/scrt376

Oerlemans C, Bult W, Bos M, Storm G, Nijsen JF, Hennink WE. Polymeric micelles in anticancer therapy: Targeting, imaging and triggered release. Pharmaceutical Research. 2010;27:2569–89.

Pereira S, Tettamanti M. Ahimsa and alternatives – The concept of the 4th R. The CPCSEA in India. ALTEX. 2005;22:3–6.

Pessetto ZY, Weir SJ, Sethi G, Broward MA, Godwin AK. Drug repurposing for gastrointestinal stromal tumor. Molecular Cancer Therapeutics. 2013;12:1299–309.

Petty RD, Sutherland LA, Hunter EM, Cree IA. Comparison of MTT and ATP-based assays for the measurement of viable cell number. Journal of Bioluminescence and Chemiluminescence. 1995;10:29–34.

Pignatelli J, Goswami S, Jones JG, Rohan TE, Pieri E, Chen X, Adler E, Cox D, Maleki S, Bresnick A, Gertler FB, Condeelis JS, Oktay MH. Invasive breast carcinoma cells from patients exhibit MenaINV- and macrophage-dependent transendothelial migration. Science Signaling. 2014;7:ra112. doi:10.1126/scisignal.2005329

Pouliot N, Pearson HB, Burrows A. Investigating metastasis using in vitro platforms. Austin, TX: Landes Bioscience, 2000.

Rahn JJ, Chow JW, Horne GJ, Mah BK, Emerman JT, Hoffman P, Hugh JC. MUC1 mediates transendothelial migration in vitro by ligating endothelial cell ICAM-1. Clinical & Experimental Metastasis. 2005;22:475–83. doi:10.1007/s10585-005-3098-x

Rath BH, Fair JM, Jamal M, Camphausen K, Tofilon PJ. Astrocytes enhance the invasion potential of glioblastoma stem-like cells. PLoS ONE. 2013;8:e54752. doi:10.1371/journal.pone.0054752

Rochat B. Importance of influx and efflux systems and xenobiotic metabolizing enzymes in intratumoral disposition of anticancer agents. Current Cancer Drug Targets. 2009;9(5):652–74.

Rorth P. Collective cell migration. Annual Review of Cell and Developmental Biology. 2009;25:407–29. doi:10.1146/annurev.cellbio.042308.113231

Shoemaker RH. The NCI60 human tumour cell line anticancer drug screen. Nature Reviews Cancer. 2006;6:813.

Siddiqui S, Khanvilkar VV, Jagdale D. In-vitro methods of screening of anticancer agent. World Journal of Pharmacy and Pharmaceutical Sciences. 2017 Jul 10;6(9):409–24.

Smith W, Gamble J, Clark-Lewis I, Vadas M. Interleukin-8 induces neutrophil transendothelial migration. Immunology. 1991;72:65.

Srivastava GK, Reinoso R, Singh AK, Fernandez-Bueno I, Hileeto D, Martino M, Garcia-Gutierrez MT, Merino JM, Alonso NF, Corell A, Pastor JC. Trypan Blue staining method for quenching the autofluorescence of RPE cells for improving protein expression analysis. Experimental Eye Research. 2011 Dec 1;93(6):956–62.

Steeg PS. Tumor metastasis: Mechanistic insights and clinical challenges. Nature Medicine. 2006;12:895–904. doi:10.1038/nm1469

Suggitt M, Bibby MC. 50 years of preclinical anticancer drug screening: Empirical to target-driven approaches. Clinical Cancer Research. 2005;11:971–81.

Taylor DJ. Parallel screening of FDA-approved antineoplastic drugs for identifying sensitizers of TRAIL-induced apoptosis in cancer cells. BMC Cancer. 2011;11:470.

Van Meerloo J, Kaspers GJ, Cloos J. Cell sensitivity assays: The MTT assay. In Cancer cell culture: Methods and protocols (pp. 237–45), 2011.

Van Tonder A, Joubert AM, Cromarty AD. Limitations of the 3–(4, 5-dimethylthiazol-2-yl)-2, 5-diphenyl-2H-tetrazolium bromide (MTT) assay when compared to three commonly used cell enumeration assays. BMC Research Notes. 2015 Dec;8:47.

Varinelli L, Guaglio M, Brich S, Zanutto S, Belfiore A, Zanardi F, Iannelli F, Oldani A, Costa E, Chighizola M, Lorenc E. Decellularized normal and tumor extracellular matrix as scaffold for cancer organoid cultures of colorectal peritoneal metastases. bioRxiv. 2021 Jul; 14(11).

Vidi PA, Bissell MJ, Lelievre SA. Three-dimensional culture of human breast epithelial cells: The how and the why. Methods in Molecular Biology. 2013;945:193–219. doi:10.1007/978-1-62703-125-7_13

Vistica DT, Skehan P, Scudiero D, Monks A, Pittman A, Boyd MR. Tetrazolium-based assays for cellular viability: A critical examination of selected parameters affecting formazan production. Cancer Research. 1991;51:2515–20.

Wang F-Q, Ariztia EV, Boyd LR, Horton FR, Smicun Y, Hetherington JA, Smith PJ, Fishman DA. Lysophosphatidic acid (LPA) effects on endometrial carcinoma in vitro proliferation, invasion, and matrix metalloproteinase activity. Gynecol. Oncol. 2010;117:88–95. doi:10.1016/j.ygyno.2009.12.012

Wang X, Sun Q, Pei J. Microfluidic-based 3D engineered microvascular networks and their applications in vascularized microtumor models. Micromachines. 2018 Sep 27;9(10):493.

Wang XY, Pei Y, Xie M, Jin ZH, Xiao YS, Wang Y, Zhang L-N, Li Y, Huang W-H. An artificial blood vessel implanted three-dimensional microsystem for modeling transvascular migration of tumor cells. Lab on a Chip. 2015;15:1178–87. doi:10.1039/c4lc00973h

Weigelt B, Peterse JL, van't Veer LJ. Breast cancer metastasis: Markers and models. Nature Reviews Cancer. 2005;5:591–602. doi:10.1038/nrc1670

Weyermann J, Lochmann D, Zimmer A. A practical note on the use of cytotoxicity assays. International Journal of Pharmaceutics. 2005;288:369–76.

Wilson JL, Hemann MT, Fraenkel E, Lauffenburger DA. Integrated network analyses for functional genomic studies in cancer. Seminars in Cancer Biology. 2013;23(4):213–18.

Wirtz D, Konstantopoulos K, Searson PC. The physics of cancer: The role of physical interactions and mechanical forces in metastasis. Nature Reviews Cancer. 2011;11:512–22. doi:10.1038/nrc3080

Wong KH, Truslow JG, Khankhel AH, Chan KL, Tien J. Artificial lymphatic drainage systems for vascularized microfluidic scaffolds. Journal of Biomedical Materials Research Part A. 2013;101:2181–190. doi:10.1002/jbm.a.34524

Workman P, Aboagye EO, Balkwill F, Balmain A, Bruder G, Chaplin DJ, Double JA, Everitt J, Farningham DA, Glennie MJ, Kelland LR. Guidelines for the welfare and use of animals in cancer research. British Journal of Cancer. 2010;102: 1555.

Wu LY, Di Carlo D, Lee LP. Microfluidic self-assembly of tumor spheroids for anticancer drug discovery. Biomedical Microdevices. 2008;10:197–202. doi:10.1007/s10544-007-9125-8

Yabushita H, Ohnishi M, Komiyama M, Mori T, Noguchi M, Kishida T, Noguchi Y, Sawaguchi K, Noguchi M. Usefulness of collagen gel droplet embedded culture drug sensitivity testing in ovarian cancer. Oncology Reports. 2004;12:307–11. doi:10.3892/or.12.2.307

Yang MH, Imrali A, Heeschen C. Circulating cancer stem cells: The importance to select. Chinese Journal of Cancer Research. 2015 Oct;27(5):437.

Yeo EJ, Chun YS, Park JW. New anticancer strategies targeting HIF-1. Biochemical Pharmacology. 2004;68(6):1061–69.

Zeng X, Morgenstern R, Nyström AM. Nanoparticle-directed subcellular localization of doxorubicin and the sensitization breast cancer cells by circumventing GST-mediated drug resistance. Biomaterials. 2014;35:1227–39.

Zhang Z, Tang W. Drug metabolism in drug discovery and development. Acta Pharmaceutica Sinica B. 2018 Sep 1;8(5):721–32.

Zuela-Sopilniak N, Lammerding J. Engineering approaches to studying cancer cell migration in three-dimensional environments. Philosophical Transactions of the Royal Society B. 2019 Aug 19;374(1779):20180219.

22 Crystal Engineering
A Versatile Platform for Oral Drug Delivery

Mansi Prashar,[1] *Satyam Raj,*[1] *Poonam Arora,*[2] *and Neelima Dhingra*[1]*
[1] University Institute of Pharmaceutical Sciences, Panjab University, Chandigarh, India
[2] Chitkara College of Pharmacy, Chitkara University, Rajpura, Patiala, Punjab, India

22.1 BACKGROUND

Development of a new drug molecule is costly and lengthy; bringing a new drug from initial discovery to final regulatory approval currently takes about 10 years and costs more than $130 million (Rennane et al. 2021). Ongoing challenges such as increasing cost of research and development, reduced effective patent duration, fewer companies engaging in drug discovery, and alternative investment opportunities have contributed to a decrease in the launch of new chemical entities since the late 1950s. The safety of "old" drugs has been improved using different methods such as dose titration, individualizing drug therapy and therapeutic drug monitoring. Delivering drug at the controlled rate, slow delivery, targeted delivery are some of the other attractive methods and have been pursued vigorously by drug companies (Tiwari et al. 2012).

Drug delivery systems (DDSs) are technologies that enable the administration of therapeutic agents in the body. The goal of DDSs is to enhance the effectiveness and safety of drugs by controlling the release rate, timing, location of drugs in the body and by reducing unwanted accumulation of the drug (Jain 2020). To improve the outcomes of drug therapy, several strategies have been applied to modify the drug properties and behavior in the body. These include lowering the drug toxicity, adjusting the drug pharmacokinetics, enhancing the drug delivery to the target site and increasing patient adherence and satisfaction (Henry 2023; Vargason et al. 2021). DDSs demonstrate noteworthy advantages for health and disease management due to their ability to prolong drug circulation (Ezike et al. 2023). Despite the advances in DDSs, some obstacles remain like biocompatibility and acceptability; synthetic materials interact differently with human cells than biological ones (Kohane and Langer 2010). Moreover, the harsh conditions in the GI tract limit drug bioavailability via the oral route.

Therefore, in order to improve biopharmaceutical parameters (solubility and dissolution rate), researchers have developed alternative formulation such as salts, solid dispersions, amorphous and polymorphic forms, inclusion complexes and co-crystals. Co-crystallization alters the molecular interactions and composition of pharmaceutical materials (Savjani et al. 2012), and is considered to be a better alternative to optimize drug properties. Against salts and polymorphic and amorphous forms, which all have their own limitations in utility, co-crystals offer a viable alternative (Yadav et al. 2009). In the present chapter, we summarize oral dosing as a convenient and patient-friendly method for administering drugs and discuss how innovative co-crystals complement this approach by enabling the modification of drug properties for improved solubility and bioavailability and controlled drug release.

22.2 ORAL DRUG DELIVERY AND BARRIERS

The route of drug administration refers to the way a drug enters the body to achieve its intended therapeutic goal (Kim and De Jesus 2025). The drug must reach the target tissues and cells through the bloodstream or other means, which requires choosing an appropriate method of administration. Different routes of administration have their respective advantages and disadvantages, and the route is selected based on factors such as desired onset of action, patient's co-operation, convenience, effect of gastric pH, first-pass metabolism, drug availability and cost, and potential adverse effects (Alqahtani et al. 2021). Intravenous, intramuscular, intranasal, intradermal/transdermal and oral are some of the different ways to deliver drugs into the body (Figure 22.1). Parenteral routes of administration avoid the digestive system and exhibit higher bioavailability than other routes and thereby offer the greatest control over the real drug amount in systemic circulation. However, fear, pain, tissue damage and infections are some drawbacks associated with parenteral administration along with irreversible drug administration (Homayun et al. 2019).

The oral route is the most common drug delivery method because of benefits such as simple and convenient administration, low cost, patient satisfaction, and ease of production. The oral route is preferred for most small-molecule drugs, holding 90% of the global market, with 84% of the top-selling drugs being orally delivered (Prasad et al. 2017). The oral formulations sector generated revenue worth US$35 billion in 2023 and was expected to grow annually by 3.37% CAGR from 2023 to 2028 (Healthcare 2022). Conditions in the gastrointestinal (GI) tract can limit the bioavailability of medicines, but dosage forms such as solutions, capsules, tablets, granules, powders, suspensions, and emulsions not only address the issue but offer more flexibility than other routes of administration (Bhutani et al. 2021).

* Corresponding author: neelimad08@gmail.com

DOI: 10.1201/9781032660592-27

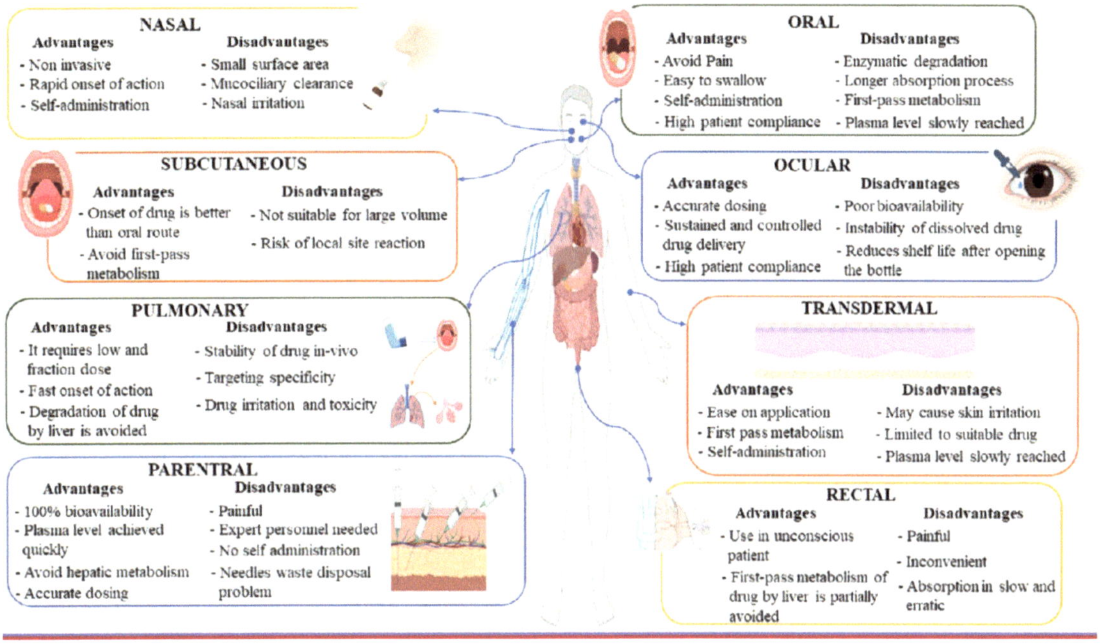

FIGURE 22.1 Drug delivery routes. (From Benoit et al. 2014.)

Further, the bioavailability of drugs can be improved by designing novel drug delivery systems that can transport the drug to the site of action with reduced impact on vital tissues and fewer side effects (Upadhyay 2014). The rationale is to increase the accumulation of therapeutic compounds in the target site at a controlled rate by providing slow, targeted delivery. Oral control-released drug delivery addresses the second obstacle to oral delivery by decreasing the frequency of dosing and offers benefits over traditional dosage forms like enhanced patient adherence and lower steady-state drug levels, resulting in the better control of disease and reduced local or systemic side effects. It also improves the safety margin of highly potent drugs by controlling plasma levels, increasing bioavailability and ultimately reducing total administered doses (Srikanth et al. 2013).

22.2.1 Barriers to Oral Drug Delivery

Easy painless application, affordable prices, wide drug absorption/distribution, and high patient compliance has made the oral route as one of the most common route for patients. However, the efficiency of many oral medicines is still limited due to anatomical, biochemistry factors, and physiology barrier in GIT, resulting in low permeability, and drug degradation (Lou et al. 2023).

The different anatomical characteristics associated with various parts of GIT like oral cavity, esophagus, stomach, small intestine and colon have varying effects on drug absorption and drug delivery (Greenwood-Van Meerveld et al. 2017). Undoubtedly the oral mucosa present in the oral cavity, exerts a mild microenvironment for easy accessibility, good permeability, facile access to circulation, and absorption of drugs (Targhotra and Chauhan 2020). However, its limited surface, presence of saliva and enzyme with different composition in oral cavity extend barriers in delivery of drugs in mouth (Fox et al. 2015). The strong acidic environment in stomach, with a pH range of 1.0–2.5 that can break down food, ectogenic pathogens (Yoshida et al. 2013) and acid-labile drugs (Ensign et al. 2012), poses the most significant barrier to drug absorption. Further, the stomach possesses extrinsic epithelial cells and a mucin–bicarbonate barrier (Drucker 2020).

The small intestine with a large surface area due to the villi and microvilli in the intestinal lumen is considered as a prime site for oral drug delivery. Also, the intestinal mucosa helps in recognizing and transferring the ectogenic antigens to the immune system, but still there are some challenges arising in drug delivery because of the special physiology of small intestine. The harsh stomach chemical microenvironment, bile salts, pancreatic enzymes, and the mucosal layer have been found to decrease the drug bioavailability. It is known that drug delivery systems that can increase the retention time of drug at villi and microvilli, or improve lipid solubility, also interact with specific receptor or carrier got the ability to increase the overall bioavailability (Coffey et al. 2021).

The colon exhibits a higher-pH environment, longer residence time than in the upper GI tract, and relatively low enzyme activity (Hua et al. 2015). Different pH environments and digestive enzymes pose biochemical barriers in oral drug delivery. The variation in the pH in different parts of the GI tract–the gradual increase from 1 to 8 from the stomach to the colon—affects drug delivery and is therefore a deciding factor for route selection in designing targeted oral drugs (Barbari et al. 2017). The main anatomical and biochemical barriers of oral administration are summarized in Figure 22.2.

Challenges in Route Drug Delivery System

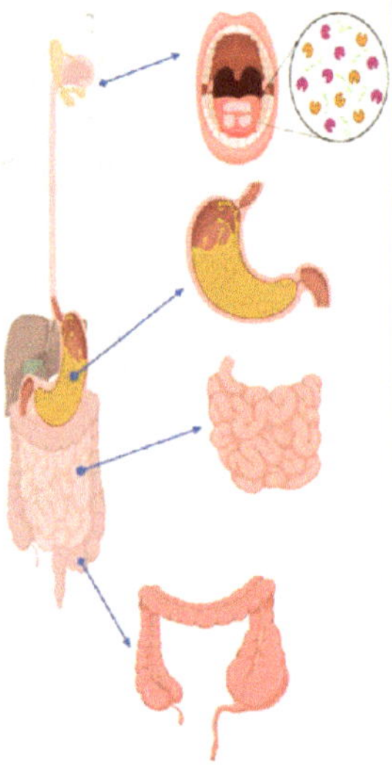

Biological Barrier	Anatomical Factors	pH Values	Enzyme	Microbiome
Mouth Cavity	Limited Surface, Saliva, Enzymatic	6.6–7.1	Salivary Amylase	Gemella, Streptococcus, Veillonella, Haemophilus, Neisseria, Porphyromonas, Actinomyces, Fusobacterium, Prevotella
Stomach	Gastric Acid, Mucin-bicarbonate Barrier, Pepsin, Limited Absorption Area	1.5–2.2	Pepsin Amylase	Prevotella, Streptococcus, Veillonella, Rothia, Haemophilus
Small Intestine	Pancreatic Enzymes, Bile Salts, Mucosal Layer	5.9–7.8	Pancreatic Amylase, Maltase, Trypsin, Peptidases, Aminopeptidase, Nuclease, Nucleosidases, Lipase	Clostridiales, Streptococcus, Bacteroides, Actinomyxinae, Lactobacillus, Corynebacteria, Enterococcus
Colon	Gut Microflora, Enzymes Produced By Microflora	6–6.7	Aminopeptidase, Azoreductase, Polysaccharase, Glycosidase, Nitroreductase	Porphyromonas, Eubacterium, Bacteroides, Bifidobacterium, Streptococcus, Enterobacteriaceae, Enterococcus, Clostridium, Lactobacillus, Ruminococcus, Escherichia Coli

FIGURE 22.2 Challenges in drug delivery. (From Lou et al. 2023.)

22.3 CRYSTAL ENGINEERING: A RATIONAL APPROACH TO IMPROVING BIOAVAILABILITY

The Biopharmaceutics Classification System (BCS) labels drugs into four classes based on solubility and permeability (Kumar et al. 2018). The solubility criterion is whether the maximum dose of a drug can dissolve in 250 mL or less of aqueous media with pH ranging from 2 to 7.5. The permeability criterion is whether at least 90% of the administered dose is absorbed (Papich and Martinez 2015). BCS Class I drugs with high solubility and high permeability are suitable for oral delivery, but the others are not: Class II, low solubility; Class III, low permeability; Class IV, both (Tsume et al. 2014).

Various strategies have been used to enhance the solubility of drugs in water. A common method of improving oral drug absorption is using a mixture of surfactants. Surfactants, which have a hydrophilic head and a hydrophobic tail, assist drug molecules in locating at the interface – reducing interfacial tension – and temporarily open tight junctions between cells (Sareen et al. 2012). However, high concentrations of surfactants can pose a safety risk and require caution. Substantially reducing drugs' particle sizes, into micro-/nanoparticles, co-crystals, and salts, for instance, also significantly enhances drug bioavailability of drugs by increasing their surface area and consequently their dissolution rate (Renukuntla et al. 2013).

Crystal engineering involves modifying the intermolecular interactions that regulate the breaking and formation of noncovalent bonds, such as hydrogen bonding, van der Waals force, π-stacking, electrostatic interactions, and halogen bonding; it relies on the directionality, specificity and predictability of intermolecular hydrogen bonds to construct supramolecular structures with well-controlled dimensionality. Carboxylic acids, amides, carbohydrates, alcohols, and amino acids are able to co-crystalize with active pharmaceutical ingredients (Nangia and Desiraju 2019; Nugrahani and Jessica 2021).

A growing number of researchers have acknowledged the importance of intermolecular interactions or supramolecular synthons (Figure 22.3), a term coined by Desiraju in 1995, in the design of crystals with specific functional groups that can complement each other; a synthon is a part of a molecule that can be used to build larger molecules and is the essence of crystal engineering, which exploits the principles of supramolecular chemistry to create novel materials and structures (Shishkin et al. 2014).

The aim of crystal engineering is to design and use synthons that can create different patterns and networks in one, two or three dimensions. The Cambridge Structural Database can help to find common hydrogen bonding patterns that can be used as synthons in different crystal structures. The concept of supermolecules, introduced by Lehn, compares the intermolecular bond that connects molecules in solid-state structures to the covalent bond that connects atoms in molecules. Crystal engineering, which originated from the term 'molecular engineering' was coined by von Hippel in 1962, is based on this analogy and aims to design and construct solid-state supermolecules using intermolecular interactions as the building tools (Pidcock et al. 2012).

The field emerged from the study of topochemistry, which investigates how the crystal packing of molecules affects their solid-state reactions. With the advancement of crystallographic techniques such as four circles' diffractometers and area detector technology, crystal engineering has expanded to cover various aspects of structure prediction, intermolecular interactions, control and rationalization, as

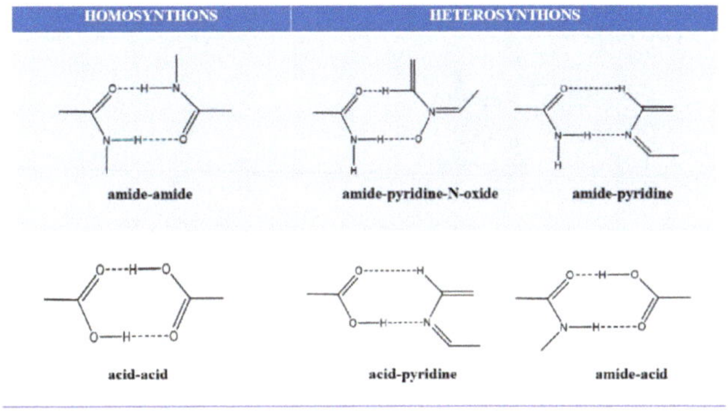

FIGURE 22.3 Homo- and hetrosynthons in crystal engineering. (From Saha and Desiraju 2018.)

well as the preparation of new molecular components and crystalline materials. The key ideas underlying crystal engineering are the molecular mechanism of crystallization, the packing of molecules in crystals, the intermolecular forces, and the principles of molecular recognition that guide the assembly of supermolecules (Biradha and Santra 2013).

Crystallization involves forming crystals from a melt or a solution, which requires a minimum number of molecules to cluster together. Key issues include the formation of crystal nuclei, the influence of crystallization conditions, and the development of crystal shape over time. These factors involve understanding how crystals form, how the conditions affect their formation, and how their shape and structure develop over time (Jones and Bimbo 2020).

Co-crystals are crystalline materials that consist of two or more molecules, usually an active pharmaceutical ingredient (API) and a co-former, in the same crystal lattice. They represent a promising way to enhance and modify the physicochemical properties of APIs without altering their molecular structure. For example, co-crystals can improve the solubility, bioavailability, stability, hygroscopicity, mechanical, and flow properties of drugs. Co-crystals also have potential applications in purification and enantiomeric separation (Guo et al. 2021). In the past 20 years, there has been much research on cocrystals, both in academia and industry (Stahly 2009). Some co-crystals have already reached the market or are in clinical trial phases, such as Entresto®, Lexapro® (Erxleben 2020), ertuglifozin-L-pyroglutamic acid (Duggirala et al. 2020) and tramadol-celecoxib (Almansa et al. 2019). The patentability of co-crystals prepared with different co-formers is still ambiguous, despite the regulations of USFDA and EMA that allow separate patents for each co-crystal. According to the WIPO patent database, there has been a significant rise in the patent applications of co-crystals in recent years (Trask 2007).

Morissette et al. define co-crystals (of drugs and drug candidates) as a subset of multi-component crystal systems that also comprise salts, solvates, clathrates, inclusion crystals and hydrates. They distinguish solvates from co-crystals based on the physical state of the individual components at room temperature. Solvates are crystals that contain a liquid component, while co-crystals consist of two solid components (Aitipamula et al. 2012).

Solvates are often formed accidentally during crystallization from solution and improve drug dissolution rate, as demonstrated for the solvated forms of spironolactone. However, solvated crystals are prone to instability, resulting in desolvation during storage and possible crystallization of the amorphous phase into less soluble forms. Moreover, solvated crystals can contain solvent levels that are unacceptable to regulatory authorities or toxicologically problematic. Co-crystals, on the other hand, are usually designed deliberately and are more stable, especially as the co-crystallizing agents are solids at room temperature (Aitipamula et al. 2012).

Co-crystal screening is a technique that can be applied to high-throughput technologies and is analogous to salt screening. The choice of a co-crystallizing agent for an API should be based on its pharmaceutical acceptability and non-toxicity, so that the final product is suitable for human use. This narrows the options to those that have been approved as human consumables, such as pharmaceutical excipients and GRAS (generally recognized as safe for human consumption as classified by the US Food and Drug Administration (FDA). However, GRAS status does not necessarily imply that a substance can be used as a co-crystal forming agent, especially if there is no previous record of its use in pharmaceuticals or if it lacks a pharmacopoeia monograph. Even if there are precedents, the amount and use of additives are limited by their safety levels in existing pharmaceutical products (Gokhale and Mantri 2017).

Forming co-crystals requires considering the hydrogen bond donors and acceptors of the materials that participate in co-crystallization and how they can interact with each other. Etter and coworkers suggested some rules to help design hydrogen bonded solids deliberately based on their detailed study of packing preferences and hydrogen-bond patterns in various organic crystals: (a) The best proton donor and acceptor left after intramolecular hydrogen-bond formation will form intermolecular hydrogen bonds with each other; (b) all good proton donors and acceptors participate in hydrogen bonding; and (c) intramolecular hydrogen bonds in six-membered rings are preferred over intermolecular hydrogen bonds (Ghadi 2014).

22.3.1 Preparing Co-Crystals

Co-crystals are solid forms of existing or available active pharmaceutical ingredients with superior physicochemical properties prepared using various solution-based or solid-based methods (Figure 22.4). Some of these methods are conventional and well-established in the laboratory, while others are innovative and novel and enable continuous and scalable co-crystal formation. All methods aim to enhance the solubility, bioavailability, flowability, and tablet manufacturing ability of the co-crystals (Guo et al. 2021).

Solid-based co-crystallization involves the spontaneous formation of co-crystals by mixing pure API and co-former

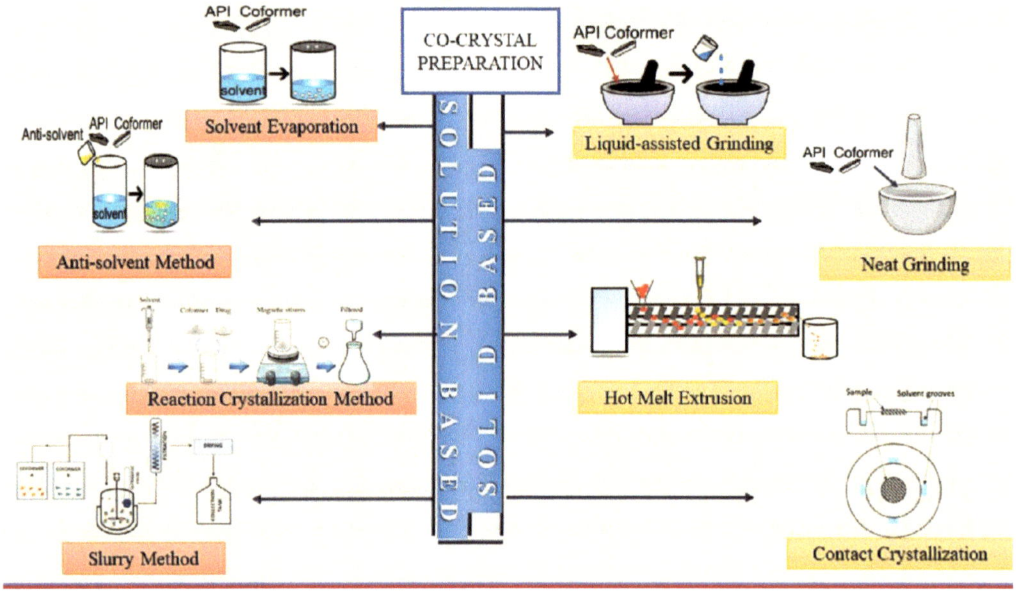

FIGURE 22.4 Preparing co-crystals. (From Karimi-Jafari et al. 2018.)

in a controlled atmospheric environment, without any mechanical agitation or grinding. However, some cases may involve brief grinding of pure components separately before mixing them. Rodriguez-Hornedo et al. studied the effect of pre-milling starting materials on the co-crystallization time of carbamazepine and nicotinamide and found that co-crystallization time decreased from 80 days for unmilled reactants to 12 for premilled reactants (O'Sullivan et al. 2022).

Moreover, higher co-crystallization rates have been reported for the same system at higher temperature and relative humidity, regardless of the mechanical activation. Co-crystallization under deliquescent conditions involves three steps: absorption of moisture from the environment, the dissolution of the reactant molecules in the liquid phase; and finally, crystallization from solution, the most common employed method for producing single-component crystals (Jayasankar et al. 2007). However, the lack of control over the nucleation, crystallization and phase evolution of the co-crystals is still a significant challenge (Garg and Azim 2021).

Understanding the basic principles of solution crystallization is helpful before attempting to grow co-crystals, and the factor that drives crystallization is supersaturation, the difference between the actual concentration of the solute in the solution and the concentration at which the solute is in equilibrium with the solid phase at a given temperature in that solvent. The degree of supersaturation for co-crystallization depends on the concentrations of both solutes (API and co-former) relative to the solubility of the co-crystal in the same solvent (Garg and Azim 2021). Researchers have begun attempting to optimize co-crystals for desired properties by combining solution co-crystallization and grinding methods. As an example, the potential of co-crystal formation of caffeine with dicarboxylic acids was investigated by the solution co-crystallization of caffeine/L-malic acid (Leyssens et al. 2014).

22.4 POTENTIAL CO-CRYSTALS IN PRECLINICAL DEVELOPMENT

The pharmaceutical industry's screening programs indicate that around 40% of new chemical entities face several difficulties at the formulation and developmental stages, especially BCS class II and IV drugs, with limited therapeutic efficacy because of low solubility and bioavailability (Bhalani et al. 2022). Co-crystal formation has been applied to APIs regardless of their acid–base or ionization properties while maintaining thermodynamic stability (Joshi et al. 2023). In the past two decades, the research and development on pharmaceutical co-crystals has gained momentum and led to enhanced scientific knowledge, evolved regulatory guidelines and an expanded in intellectual property landscape. Further, according to Kale et al. selecting appropriate formulation and process design along with understanding material properties during pre-formulation stage can mitigate the challenges in the translation development of pharmaceutical co-crystals. Some of these co-crystals-based products have been granted patents and more in preclinical development stage (Kale et al. 2017).

Ahirrao et al. (2022) developed a co-crystal form of Etodolac an NSAID (nonsteroidal anti-inflammatory drug) used to treat rheumatoid arthritis using glutaric acid as a co-former. They found that the co-crystal form had much higher solubility and dissolved two to three times more quickly than pure etodolac. Another NSAID, piroxicam, is an enolic acid that relieves the symptoms of rheumatoid arthritis and osteoarthritis, but it has slow oral absorption and low bioavailability, which limit its analgesic efficacy. Researchers studied co-crystals of piroxicam with different carboxylic acids, which can modify its physicochemical properties and molecular interactions in co-crystals. They discovered 50 co-crystals of piroxicam with 23 carboxylic acids and determined the crystal structures of eight of them, include the dioxane solvate of piroxicam. The results showed that piroxicam has a high propensity to form multicomponent solid forms with various guests (Childs and Hardcastle 2007).

A promising area of crystal engineering is drug–drug co-crystals, which can modify the physicochemical properties of APIs and offer intellectual property protection due to their novelty, utility, and non-obviousness. Drug–drug co-crystals can alter fundamental properties and enable the development of combined drug therapies with potential benefits such as synergistic and/or additive effects, dose reduction, cost-effectiveness and improved patient compliance (Wang et al. 2021). To address the drawbacks of single drug therapy, Eesam et al. prepared Carvedilol-hydrochlorothiazide (CAR-HCT) co-crystals (molar ratio 2:0.5) using drug–drug co-crystal technology; further evaluated by thermal, diffractometric and spectral (FT-IR and Raman) analyses. SEM analysis revealed that the CAR-HCT co-crystals had distinct morphological behavior from that of either CAR or HCT alone. The co-crystals demonstrated significantly greater solubility, 7.3-fold higher than pure CAR in 0.1 N HCl, and 2.7-fold faster dissolution than that of pure CAR, which resulted in improved bioavailability. The co-crystals also showed stability for ninety days (Eesam et al. 2020). Table 22.1 provides a comprehensive overview of patent co-crystals.

TABLE 22.1
Current Patented Co-Crystals

Co-Crystals	Method(s)	Improvement	Patent	Patent Number	Reference
Itraconazole with carboxylic acid	Stirring	Solubility, dissolution bioavailability (5-fold)	Transform Pharmaceuticals, Inc.	US7446107 B2	(Remener et al. 2008)
C-Glycoside derivative with L-proline	Stirring	Increase storage stability	Astellas Pharma Inc., Kotobuki Pharmaceutical Co. Ltd.	US8097592 B2	(Imamura et al. 2012)
Ezetimibe with L-proline and imidazole	Wet milling/ grinding or solution crystallization	Rapid dissolution and higher supersaturation	Amplio Pharma, Ltd	WO2012116349 A2	(Childs 2012)
Progesterone with vanillic acid, benzoic acid, salicylic acid, cinnamic acid, and vanillin	Stirring, slow evaporation	Improved stability of API	Amri Ssci Llc	US9982007 B2	(Albert et al. 2018)
Dapagliflozin with dl-proline	Stirring	Improved solubility	Aurobindo Pharma Limited	WO2017191539 A1	(Natrajan et al. 2017)
Ibrutinib with carboxylic acid and fumaric acid	Solvent Evaporation	Improved solubility in the stomach	Ratiopharm Gmbh	WO2016156127 A1	(Albrecht et al. 2016)
Ertugliflozin with L-pyroglutamic acid and L-proline	Stirring	Improved stability of API	Sun Pharmaceutical Industries Limited	WO2016189463 A1	(Santra et al. 2016)
Carfilzomib with maleic acid	Stirring	Preparation of pure API	Dr. Reddy's Laboratories Limited	US20170369528 A1	(Nadgoud et al. 2019)
Agomelatine with adipic acid and saccharin	Dry grinding	Improved solubility	Leiutis Pharmaceuticals Pvt, Ltd.	WO2017115284 A1	(Kocherlakota et al. 2017)
Progesterone with vanillic acid, benzoic acid, salicylic acid and cinnamic acid	Stirring, partial evaporation	Increased stability with respect to humidity and improved bioavailability of API	Aptuit Inc	US20140235595 A1	(Albert et al. 2016)
Nicotinamide with P-coumaric acid	Slow evaporation	Improved solubility and dissolution rate	Aptuit Inc	US20140073674 A1	(Bevill and Schultheiss 2018)
Sildenafil with acetylsalicylic acid	Stirring	poor solubility, stability, hygroscopicity, dissolution rate	Pliva Hrvatska D.O.O., Mcleish, Nicholas, Alastair, Maxwell	WO2007080362 A1	(Miroslav et al. 2007)
Temozolomide with succinic acid, oxalic acid, salicylic acid, malic acid, anthranilic acid, & tartaric acid	Dry grinding	Improved stability and dissolution rate	Ashwini Nangia, Jagadeesh Babu, Nanubolu, Palashsanphui	WO2011036676 A2	(Ashwini et al. 2011)

22.4.1 Case Studies

22.4.1.1 Ibrutinib-Fumaric Acid Co-Crystal

The pharmaceutical compound described in WO 2008/039218, ibrutinib, is a clinically approved selective inhibitor of Bruton's tyrosine kinase (BTK) in the United States. BTK plays a crucial role in regulating several pathways that promote the survival of B-cells, such as apoptosis, cell adhesion and lymphocyte migration and homing. Ibrutinib interferes with these pathways by blocking BTK activity, leading to B-cell death and/or disruption of cell migration and adherence to tumor-protective microenvironments and alters an effective treatment for B-cell malignancies, such as chronic lymphocytic leukemia, small lymphocytic lymphoma and mantle cell lymphoma (Davids and Brown 2014).

Ibrutinib is a BCS Class II drug with reported aqueous solubility of only 0.013 mg/mL at pH 8 for its form A, as disclosed in WO 2013/184572; its solubility varies with pH, which is of course different in every patient (Shi et al. 2021). To develop an improved version of ibrutinib, Israeli pharmaceutical company industry Teva API explored and discovered a novel crystal form: Crystallizing one molecule of fumaric acid with two molecules of ibrutinib synthesized four different crystal forms of ibrutinib: A, B, C and amorphous, each with different crystal structures and therefore different chemical properties. Form C and amorphous form were much more water soluble than form A but turned into form A on stirring with water. Fumaric acid produced a stable, effective product with comparable solubility to that of ibrutinib and a shelf life of at least 18 months (Albrecht et al. 2016).

22.4.2 Ipragliflozin L-Proline Co-Crystal

A class of drug that block the reabsorption of glucose in the kidneys, called sodium-glucose cotransporter-2 (SGLT-2) inhibitors, has emerged as a promising treatment for type 2 diabetes (Kurosaki and Ogasawara 2013). The first SGLT-2 inhibitor to receive approval from the FDA was canagliflozin in 2013, and since then, nine more drugs in this category have been approved worldwide. Ipragliflozin (ASP-1941), was a product of collaboration between two Japanese companies, Astellas Pharma Inc. and Kotobuki Pharmaceutical, marketed as Suglat®, the first SGLT-2 inhibitor in Japan on January 17, 2014.

Ipragliflozin is a small organic compound with the chemical formula (1S)-1,5-anhydro-1-C-{3-[(1-benzothiophen-2-yl) methyl]-4-fluorophenyl}-D-glucitol and molecular weight of 404.5 g/mol. It dissolves well in organic solvents such as ethanol (30 mg/mL) but is poorly soluble in water (0.0299 mg/mL). A novel crystalline form of ipragliflozin with L-proline described in Korean Patent Publication No. 10–1243039 exhibits superior thermal stability and moisture absorption to its parent molecule in addition to high adhesiveness and uniformity and low electrostatic force, demonstrating its suitability for use in oral agents (Hoon 2020).

The authors analyzed the new crystal form of ipragliflozin through powder X-ray diffraction and temperature differential scanning calorimetry and compared the morphologies of solid ipragliflozin and ipragliflozin L-proline co-crystals using a particle size analyzer that measured the uniformity and distribution of particles. The achieved thermal stability of the newly crystalline form was due to polymorphism, the ability of a substance to form different crystal structures with varying thermodynamic energies. During crystallization, the substance initially adopted a less stable polymorph with higher energy and then undergoes a phase transition to a more stable polymorph with lower energy in the presence of a solvent or in the solid state. This process involves rearranging molecules to achieve a more thermodynamically favorable configuration (Hoon 2020).

22.4.2.1 Ibuprofen–Nicotinamide Co-Crystal

Ibuprofen, or (2RS)-1[4–(2-methyl propyl) phenyl] propionic acid, is a highly common NSAID that acts by inhibiting cyclo-oxygenase and blocking the synthesis of prostaglandins, biomolecules with a crucial role in pain, inflammation, and fever. The pain relief effect of ibuprofen depends on how much of it is in the blood, but the BCS classifies it as having low solubility (pKa = 5.3) and high permeability (Bushra and Aslam 2010). To achieve higher solubility, Yuliandra et al. (2018) formulated co-crystals of ibuprofen using nicotinamide as a co-former and characterized them using different techniques, and differential thermal analysis showed a distinct endothermic transition at 96.24°C, confirming its thermal robustness. The co-crystal exhibited a unique diffraction pattern that implied a new co-crystal phase. The co-crystal demonstrated analgesic efficacy in in Swiss-Webster mice with acetic acid (0.5%)-induced pain by decreasing the number of writhes relative to the pure ibuprofen and its physical mixture (Junbo et al. 2015).

22.4.2.2 Isoniazid–Quercetin Co-Crystal

Liver injury caused by anti-tuberculosis drugs is an increasing concern. Among the first-line anti-tuberculosis drugs, isoniazid (INH) has a significant role because of its excellent physicochemical properties and efficacy, but its primary challenge was severe hepatotoxicity, which was mitigated by the rational use of antioxidants. Quercetin (QCT), a well-known natural antioxidant, has a complex structure of three rings and five phenolic hydroxyl groups that protect against the oxidative stress induced by reactive oxygen species. It is a natural compound with reported antioxidant potential found in many fruits and vegetables, including elderberry, apples, and onions, but it cannot be supported in clinical applications due to its low bioavailability because it is highly insoluble in water (Jhun and Koh 2020).

Liu et al. employed a co-crystal design leveraging the strengths of INH with QCT as a co-former. Using vaporization and solvent-assisted methods, they successfully co-crystallized the two into a single crystal lattice that dissolved 51.67 times faster than the pure form. Moreover, the improved *in vitro* dissolution of INH-QCT was consistent with its increased *in vivo* bioavailability and pharmacokinetic properties, with AUC 28.91 times higher and C_{max}

AUC 13.24 times higher than those for the pure compound. The cocrystal also significantly reduced severe INH hepatotoxicity in rats (Alexey et al. 2017).

22.5 FUTURE PERSPECTIVES

Many promising drug candidates suffer from limited solubility, which hampers their absorption and therapeutic efficacy Co-crystal engineering holds tremendous promise for revolutionizing drug development and delivery, particularly via oral administration, ultimately leading to more effective and patient-friendly treatments. Co-crystallization leads to new crystalline structures that often exhibit improved solubility and bioavailability, ensuring that more of the administered dose reaches systemic circulation and thereby increasing the therapeutic effect. Their availability also enables alternative dosage forms such as oral suspensions, dispersible tablets, or even orally disintegrating tablets that would enhance patient compliance and ease of administration, especially in populations with difficulty swallowing. The uncertainty around patenting co-crystals with various co-formers persists, even though USFDA and EMA regulations permit distinct patents for each co-crystal. Notably, the WIPO patent database reflects a notable recent increase in co-crystal patent applications. Collaborative research efforts between pharmaceutical scientists, chemists, and material engineers are likely to lead the development of innovative cocrystal-based formulations, ushering in a new period of precision medicine and improved healthcare outcomes.

REFERENCES

Ahirrao SP, Sonawane MP, Bhambere DS, Udavant PB, Ahire ED, Kanade R, Kube D (2022) Cocrystal formulation: A novel approach to enhance solubility and dissolution of etodolac. Biosci Biotechnol Res Asia 19:111–119

Aitipamula S, Banerjee R, Bansal AK, Biradha K, Cheney ML, Choudhury AR (2012) Polymorphs, salts, and cocrystals: What's in a name? Cryst Growth Des 2:2147–2152

Albert E, Andres P, Bevill MJ, US Patent No. 9,982,007, 29 May 2018

Albert E, Andres P, Lafayette W, US Patent No. 2014/0,235,595, 12 June 2016

Albrecht, Wolfgang, Weg AM, WO Patent No. 2016/1,561,27, 10 June 2016

Alexey P, Drozd VK, Leonidovich G, RU Patent No. 2,630,957, 17 September 2017

Almansa C, Frampton CS, Vela JM, Whitelock S, Plata-Salamán CR (2019) Co-crystals as a new approach to multimodal analgesia and the treatment of pain. J Pain Res 12:2679–2789

Alqahtani MS, Kazi M, Alsenaidy MA, Ahmad MZ (2021) Advances in oral drug delivery. Front Pharmacol 6:1–21

Ashwini, Nangia, Nanubolu, WO Patent No. 2011/0,366,76, 31 March 2011

Barbari GR, Dorkoosh FA, Amini M, Sharifzadeh M, Atyabi F, Balalaie S, Tehrani NR, Tehrani MR (2017) A novel nanoemulsion-based method to produce ultrasmall, water-dispersible nanoparticles from chitosan, surface modified with cell-penetrating peptide for oral delivery of proteins and peptides. Int J Nanomed 11:3471–3483

Benoit DS, Overby CT, Sims KR Jr, Ackun-Farmmer MA (2014) Drug delivery systems history of DDS development. Saltzman:1237–1266

Bevill JM, Schultheiss N, US Patent No. 2014/0,073,764, 16 October 2018

Bhalani DV, Nutan B, Kumar A, Singh Chandel AK (2022) Bioavailability enhancement techniques for poorly aqueous soluble drugs and therapeutics. Biomed 8:1–33

Bhutani U, Basu T, Majumdar S (2021) Oral drug delivery: Conventional to long acting new-age designs. Eur J Pharm Biopharm 162:23–42

Biradha K, Santra R (2013) Crystal engineering of topochemical solid state reactions. Chem Soc Rev 42:950–967

Bushra R, Aslam N (2010) An overview of clinical pharmacology of ibuprofen. Oman Med J 125:155–161

Childs SL, WO Patent No. 2012/1,163,49, 30 August 2012

Childs SL, Hardcastle KI (2007) Cocrystals of piroxicam with carboxylic acids. Cryst Growth Des 7:1291–1304

Coffey JW, Das Gaiha G, Traverso G (2021) Oral biologic delivery: advances toward oral subunit, DNA, and mRNA vaccines and the potential for mass vaccination during pandemics. Annu Rev Pharmacol Toxicol 132:517–540

Davids MS, Brown JR (2014) Ibrutinib: A first in class covalent inhibitor of Bruton's tyrosine kinase. Future Oncol 10:957–967

Drucker DJ (2020) Advances in oral peptide therapeutics. Nat Rev Drug Discov 8:277–289

Duggirala NK, Lacasse SM, Zaworotko MJ, Krzyzaniak JF, Arora KK (2020) Pharmaceutical cocrystals: Formulation approaches to develop robust drug products. Cryst Growth Des 20:617–626

Eesam S, Bhandaru JS, Naliganti C, Bobbala RK, Akkinepally RR (2020) Solubility enhancement of carvedilol using drug–drug cocrystallization with hydrochlorothiazide. Future J Pharm Sci 6:1–9

Ensign LM, Cone R, Hanes J (2012) Oral drug delivery with polymeric nanoparticles: The gastrointestinal mucus barriers. Adv Drug Deliv Rev 15:557–570

Erxleben A (2020) Cocrystal applications in drug delivery. Pharmaceutics 2:1–3

Ezike TC, Okpala US, Onoja UL, Nwike CP, Ezeako EC (2023) Advances in drug delivery systems, challenges and future directions. Heliyon 9:174–188

Fox CB, Kim J, Le LV, Nemeth CL, Chirra HD, Desai TA (2015) Micro/nanofabricated platforms for oral drug delivery. J Controlled Release 22:431–444

Garg U, Azim Y (2021) Challenges and opportunities of pharmaceutical cocrystals: A focused review on non-steroidal anti-inflammatory drugs. RSC Med Chem 54:705–721

Ghadi R (2014) Co-crystals: Emerging approach in pharmaceutical design antimicrobial activity of clove essential oil and extracts of Psidium guajava against dental pathogens view project. Indo Am J Pharm Sci 11:3881–3892

Gokhale MY, Mantri RV (2017) API solid-form screening and selection. In: Developing solid oral dosage forms: Pharmaceutical theory and practice (2nd ed.). Elsevier Inc 13:85–112

Greenwood-Van Meerveld B, Johnson AC, Grundy D (2017) Gastrointestinal physiology and function. Handb Exp Pharmacol 1:239–241

Guo M, Sun X, Chen J, Cai T (2021) Pharmaceutical cocrystals: A review of preparations, physicochemical properties and applications. Acta Pharm Sin B 3:2537–2564

Healthcare (2022) Oral care market size, share & trends analysis report by product (toothbrush, toothpaste, mouthwash/rinse, denture products, dental accessories), by region, and segment forecasts, 2022 – 2030. https://www.grandviewresearch.com/industry-analysis/oral-care-market. Accessed 20 June 2022

Henry R (2023) Case report open access clinical pharmacology & biopharmaceutics advances in drug delivery systems: enhancing therapeutic efficacy and patient compliance. Clin Pharmacol Biopharm 12:6–8

Homayun B, Lin X, Choi HJ (2019) Challenges and recent progress in oral drug delivery systems for biopharmaceuticals. Pharmaceutics 11:1–29

Hoon AJ, KR Patent No. 1,020,972,50, 04 April 2020

Hua S, Marks E, Schneider JJ, Keely S (2015) Advances in oral nano-delivery systems for colon targeted drug delivery in inflammatory bowel disease: Selective targeting to diseased versus healthy tissue. Nanomedicine 5:1117–1132

Imamura M, Nakanishi K, Shiraki P, US Patent No. 8,097,592, 17 January 2012

Jain KK (2020) An overview of drug delivery systems. In: Methods in molecular biology. Humana Press Inc 4:1–54

Jayasankar A, Good DJ, Rodríguez-Hornedo N (2007) Mechanisms by which moisture generates cocrystals. Mol Pharm 4:360–372

Jhun BW, Koh WJ (2020) Treatment of isoniazid-resistant pulmonary tuberculosis. Tuberc Respir Dis 1:20–30

Jones ECL, Bimbo LM (2020) Crystallisation behaviour of pharmaceutical compounds confined within mesoporous silicon. Pharmaceutics 12:1–24

Joshi V, Raut P, Bhosale N, Varda J (2023) Review on co-crystals new approach to modify the physicochemical characteristics of API. Asian J Pharm Res Dev 11:103–111

Junbo ZG, Shen QY, Wang M, Zhang HB, CN Patent No. 1,033,044,76, 1 March 2015

Kale DP, Zode SS, Bansal AK (2017) Challenges in translational development of pharmaceutical cocrystals. J Pharm Sci 106:457–470

Karimi-Jafari M, Padrela L, Walker GM, Croker DM (2018) Creating cocrystals: A review of pharmaceutical cocrystal preparation routes and applications. Cryst Growth Des 18:6370–6387

Kim J, De Jesus O (2025) Medication routes of administration. In: StatPearls [Internet]. Treasure Island (FL): StatPearls Publishing. PMID: 33760436.

Kocherlakota, Chandrashekhar, WO Patent No. 2017/1,152,84, 06 September 2017

Kohane DS, Langer R (2010) Biocompatibility and drug delivery systems. Chem Sci 1:441–446

Kumar S, Kaur R, Rajput R, Singh M (2018) Bio pharmaceutics classification system (BCS) class IV drug nanoparticles: Quantum leap to improve their therapeutic index. Adv Pharm Bull 8:617–625

Kurosaki E, Ogasawara H (2013) Ipragliflozin and other sodium-glucose cotransporter-2 (SGLT2) inhibitors in the treatment of type 2 diabetes: Preclinical and clinical data. Pharmacol Ther 139:51–59

Leyssens T, Tumanova N, Robeyns K, Candoni N, Veesler S (2014) Solution cocrystallization, an effective tool to explore the variety of cocrystal systems: Caffeine/dicarboxylic acid cocrystals. Cryst Eng Comm 16:9603–9611

Lou J, Duan H, Qin Q, Teng Z, Gan F, Zhou X, Zhou X (2023) Advances in oral drug delivery systems: Challenges and opportunities. Pharmaceutics 15:484–595

Miroslav, Zegarac, WO Patent No. 2007/0,803,62, 19 September 2007

Nadgoud RK, Vasam S, Makireddy SR, US Patent No. 2017/0,369,528, 28 May 2019

Nangia AK, Desiraju GR (2019) Crystal engineering: An Outlook for the future. Angew Chem Int 58:4100–4107

Natrajan, Senthilkumar, WO Patent No. 2017/1,915,39, 09 November 2017

Nugrahani I, Jessica MA (2021) Amino acids as the potential co-former for co-crystal development: A review. Molecules 26:1–24

O'Sullivan A, Long B, Verma V, Ryan KM, Padrela L (2022) Solid-state and particle size control of pharmaceutical cocrystals using atomization-based techniques. Int J Pharm 621:121–179

Papich MG, Martinez MN (2015) Applying biopharmaceutical classification system (BCS) criteria to predict oral absorption of drugs in dogs: Challenges and pitfalls. AAPS J 17:948–964

Patrick Stahly G (2009) A survey of cocrystals reported prior to 2000. Cryst Growth Des 9:4212–4229

Pidcock E, Chisholm JA, Wood PA, Galek PTA, Fábián L, Korb O, Cruz-Cabeza AJ, Liebeschuetz JW, Groom CR, Allen FH (2012) The Cambridge structural database system and its applications in supramolecular chemistry and materials design. In: Supramolecular chemistry. John Wiley & Sons, Ltd 9:2927–2945

Prasad V, De Jesús K, Mailankody S (2017) The high price of anti-cancer drugs: Origins, implications, barriers, solutions. Nat Rev Clin Oncol 14:381–390

Remener J, MachPhee M, Peterson M, US Patent No. 7,446,107, 11 April 2008

Rennane S, Baker L, Mulcahy A (2021) Estimating the cost of industry investment in drug research and development: A review of methods and results. Inquiry 58:1–11

Renukuntla J, Vadlapudi AD, Patel A, Boddu SHS, Mitra AK (2013) Approaches for enhancing oral bioavailability of peptides and proteins. Int J Pharm 447:75–93

Saha S, Desiraju GR (2018) Acid Amide supramolecular synthon in cocrystals: From spectroscopic detection to property engineering. J Am Chem Soc 140:6361–6373

Santra, Ramkinkar, Bhogala, Reddy BK, WO Patent No. 2016/1,894,63, 12 January 2016

Sareen S, Joseph L, Mathew G (2012) Improvement in solubility of poor water-soluble drugs by solid dispersion. Int J Pharm Investig 2:12–17

Savjani KT, Gajjar AK, Savjani JK (2012) Drug solubility: Importance and enhancement techniques. ISRN Pharm 2:1–10

Shi X, Wang C, Chen Q, Shen S, Song S, Zhou X (2021) Improving physicochemical properties of Ibrutinib with cocrystal strategy based on structures and natures of the carboxylic acid co-formers. J Drug Deliv Sci Technol 63:102–554

Shishkin OV, Zubatyuk RI, Shishkina SV, Dyakonenko VV, Medviediev VV (2014) Role of supramolecular synthons in the formation of the supramolecular architecture of molecular crystals revisited from an energetic viewpoint. Phys Chem 16:6773–6786

Srikanth P, Raju N, Wasim Raja S, Brito Raj S (2013) A review on oral controlled drug delivery Int J App Pharm 2:51–58

Targhotra M, Chauhan MK (2020) An overview on various approaches and recent patents on buccal drug delivery systems. Curr Pharm Des 26:5030–5039

Tiwari G, Tiwari R, Bannerjee S, Bhati L, Pandey S, Pandey P, Sriwastawa B (2012) Drug delivery systems: An updated review. Int J Pharm Investig 2:2–8

Trask AV (2007) An overview of pharmaceutical cocrystals as intellectual property. Mol Pharm 4:301–309

Tsume Y, Mudie DM, Langguth P, Amidon GE, Amidon GL (2014) The biopharmaceutics classification system: Subclasses for in vivo predictive dissolution (IPD) methodology and IVIVC. Eur J Pharm Sci 57:152–163

Upadhyay RK (2014) Drug delivery systems, CNS protection, and the blood brain barrier. Biomed Res Int 8:1–37

Vargason AM, Anselmo AC, Mitragotri S (2021) The evolution of commercial drug delivery technologies. Nat Biomed Eng 5:951–967

Wang X, Du S, Zhang R, Jia X, Yang T, Zhang X (2021) Drug-drug cocrystals: Opportunities and challenges. Asian J Pharm Sci 16:307–317

Yadav AV, Shete AS, Dabke AP, Kulkarni PV, Sakhare SS (2009) Solid form discovery and design depends on co-crystals: A novel approach to modify physicochemical properties of active pharmaceutical ingredients. Indian J Pharm Sci 6:359–370

Yoshida T, Lai TC, Kwon GS, Sako K (2013) PH-and ion-sensitive polymers for drug delivery. Expert Opin Drug Deliv 10:1497–1513

Yuliandra Y, Zaini E, Syofyan S, Pratiwi W, Putri LN, Pratiwi YS, Arifin H (2018) Cocrystal of ibuprofen–nicotinamide: Solid-state characterization and in vivo analgesic activity evaluation. Sci Pharm 86:2–11

23 Chewable Gels

Pritiman Pothal, *Krishna Dhar* Dwivedi, *Anurag* Kuhad, *and Ranjana* Bhandari*
University Institute of Pharmaceutical Sciences, UGC-Centre of Advanced Study, Panjab, India

23.1 INTRODUCTION

Oral delivery is considered the safest and most practical route of administration for drugs [1]. However, many people, especially among the pediatric and geriatric segments, experience discomfort/dysphagia when taking traditional oral delivery forms, such as tablets or capsules [2, 3].

People often break, chew, or crush hard tablets and mix them with food or water to ease swallowing. However, modifying tablets may alter the rate or extent of drug dissolution and absorption of the active pharmaceutical ingredient (API) [3, 4]. Crushing tablets also enhances the off-taste of some APIs and increases the risk of inaccurate dosing [3], which can lead to increased hospitalization, healthcare costs, and deaths [4]. Alternative formulations such as syrups, elixirs, oro-dispersible tablets, and liquid suspensions have been developed to ease the administration of oral drugs. Although these formulations may reduce dysphagia-related issues, they have certain disadvantages.

For example, the API in an elixir needs to be dissolved, which usually is not optimal for long-term chemical stability [5]. In addition, solubilized APIs are known for their distinct off taste because they can freely interact with taste buds, and oro-dispersible tablets can cause an uncomfortable drying/powdery sensation in the mouth upon disintegrating, a problem especially for people suffering from reduced saliva production, which is common among the elderly [6]. Other issues with these drug delivery systems include homogeneity and accurate dosing. The many challenges with existing formulations encourage the development of innovative oral drug delivery system development [7].

Chewable drug formulations were developed to overcome dysphagia while simultaneously avoiding significant off-taste. Dysphagia, difficulty swallowing, can occur in childhood due to either behavioral or skill-based disorder, and chewable gels, commonly known as gummies, have been establishes as appropriate[8] dosage forms for patients with dysphagia. The primary advantages of gummy drug administration are that its viscous, thick fluid or solid avoids first-pass metabolism, it is rapidly absorbed from the buccal cavity, it is accepted by pediatric populations, it is accessible to handling and packaging in unit-dose or multiple-dose containers, it is suitable for patients with dysphagia since it is easily chewed and converted to liquid that is quickly swallowed [9], and finally, chewable gels allow for a broader range of flavour options than coated tablets [10]. The pleasant flavors of gummy medicines can in particular encourage patients' medication compliance during their therapy. Chewable gels improve taste and texture and enhance patient compliance by reducing the resistance or reluctance to take medication [11].

Liquid dosage forms have limitations of shorter shelf-life, less stability, a requirement of special storage conditions (refrigeration), and are more prone to microbial proliferation and bulky volumes, which in turn increase the shipping cost [12]. The FDA, WHO [13], and European Medicines Agency (EMA), categorize humans into neonates (newborns to 1 month), infants (1 month to 2 years), children (2 to 12 years), adolescents (12 to 18 years [dependent upon on region]), adults, and geriatrics (65 years or older), and the EMA found that different age groups tolerated tablets of different sizes: Respectively, 3 to 5 years, 6 to 11 years, 12 to 17 years, and older accepted tablets of 3–5 mm, 5–10 mm, 10–15 mm, and >15 mm [14]. Table 23.1 provides an overview of the appropriateness of various traditional oral dose forms for various age groups based on evaluations from international drug regulatory bodies.

The disclosed demerits of conventional oral solid and liquid dosage forms (Table 23.2) stress the need for age-appropriate novel solid dosage forms with the flexibility to cover persons with swallowing difficulty and dosage administration with acceptable stability and palatability [16].

23.1.1 Chewable Gels

Chewable gels consist of some or all of gelling agents, sugars, water, sweeteners, and flavoring agents. The sweeteners and flavoring are intended to enhance patient acceptance and mask the taste of the delivered labeled drug substance. Chewable gels maintain their moulded shape, are elastic, and yield to mastication. They are intended to be chewed before swallowing. Chewable gels are also known as "gummies" in the confectionary and dietary supplement industries, Because of the potential for accidental overdose and chewable gels' enticing candy-like look and flavor, chewable gels are not currently advised as therapeutic dosage forms [8]. Chewable gels are often used with children or individuals who have difficulty swallowing pills or capsules. They are valued for their convenience, ease of administration, and palatability, in contrast with forms such as suspensions

* Corresponding author: akb10in@yahoo.co.uk

DOI: 10.1201/9781032660592-28

TABLE 23.1
Conventional Oral Dose Forms Are Suitable for All Ages

Serial No.	Dosage Form	Age Group					Adult	Geriatric (≥65 years)
		Full-Term Newborn Infants (0–28 days)	Infants & Toddlers (1 month to 2 years)	Children Preschool (2–5 years)	Children School Age (6–11 years)	Adolescents (12 to 16/18 years)		
1	Solutions[¥]/Drops	√ (drops)	√ (drops/solutions)	√ (solutions)	√ (solutions)	√ (solutions)	√ # ¥ (solutions)	√ # ¥ (solutions)
2	Emulsions[¥]/Syrups[¥]/ Suspensions	√	√	√	√	√	√ ¥	√¥
3	Powders/ Granules	¥	√	√	√	√	√¥	√¥
4	Tablets	¥	¥	√*	√*	√*		
5	Capsules	¥	√ * **	√*	√*	√*	√*	√*

Notes: √ Suitable for/acceptable to the intended age groups. # Reported causing aspiration in case of swallowing difficulty (dysphagia) in contrast with suspension. * Acceptable in the absence of (dysphagia); size and shape had significant impacts on swallowability. ** From 12 months with body surface area of 0.39 m^2 and above with capsules sized 3 and 4. ¥ Acceptability/suitability not well established [15].

TABLE 23.2
Conventional Oral Dosage Forms and Their Advantages and Disadvantages

Dose Form	Advantages	Disadvantages
	Liquids	
Suspensions	• Acceptability across age groups • Dose flexibility	• Shorter shelf life (solutions & drops) • More prone to microbial proliferation (solutions & drops) • Possibility of spillage while dosing • Special storage conditions (at 2–8 °C) for reconstituted suspension and less stable liquid dosage forms at room temperature • Limited control of dose intake (drops) • Bulky volumes and special storage conditions that in turn increase shipping costs
Solutions, syrups, drops	• Acceptability across age groups (solutions and syrups) • Dose flexibility (solutions and syrups)	
Powders and granules for reconstitution	• Acceptability across age groups • Dose flexibility • Long-term stability (before reconstitution)	
	Solid Preparations	
Tablets Capsules	• More stable than liquid preparations	• Higher doses (i.e., larger materials) can be difficult to swallow • Risk of choking • Dose non-flexibility when there is no divided dose option
Powders and granules	• More stable than liquid preparations	• Dose-measuring device critical (for lower, subdivided doses) • Sprinkles are incompatible with food and drink • Limited dose control with multidose container • Risk of choking on non-disintegrating sprinkles or unwetted powders/ granules in water

and liquid syrups that have high chances of medication error, physical stability, sedimentation and compaction.

Oral delivery is generally considered the safest and the most practical route of administration for drugs. However, many people, especially among the pediatric and geriatric segments, and people taking cancer therapy experience discomfort/dysphagia due to losing gum. When they take traditional oral delivery forms, such as tablets or capsules, people often break, chew, or crush hard tablets and mix them with food or water to ease swallowing. Pediatric patients present problems with accurate dosing and hence lead to medication errors and lack of proper administration. Problems associated with pediatric dosing include palatability issues, inability to swallow large dosage forms like pills or capsules, and inaccuracies in dosing. As such, there is a need for developing tailored dosage forms for such populations which can make treatment more accessible.

Chewable gels are used in a variety of common products. Many vitamins and dietary supplements are available in chewable gel, and some pain relievers and analgesics are formulated as chewable gels for rapid absorption and relief. Pediatric medications are also often made in chewable gel form to make them more appealing to children [17]. Gummy medications have multiple advantages, primarily including ease of consumption: Chewable gels provide a convenient alternative for children and adults alike. Some nutrients are absorbed into the body more easily when delivered in gummy form than in traditional forms. Chewable gels are also available in a wide range of flavors, making them more palatable for those who dislike the taste of pills or capsules, making them particularly popular with children, and the pleasant taste and ease of consumption have demonstrably improved patient medication compliance [18].

There are also some disadvantages of delivery medications in the gummy form, for instance, dosage accuracy. Chewable gels do not always provide the same dosage accuracy as tablets or capsules. They also can still pose a risk if ingested by young children without adult supervision even as they are often designed to be more appealing to children. Interestingly, despite their evident popularity, gummy-textured medications and supplements have a taste and texture that some people find unpleasant, and they also have shorter shelf lives than other forms of medication; they are susceptible to moisture, temperature fluctuations, and degradation over time that can affect their effectiveness, and they also contribute to constipation, diarrhea, and upset stomach in some patients [19]. Figure 23.1 shows some of the current applications of versatile chewable gels [20].

23.1.2 Chewable Formulations

Chewable formulations are designed to be mechanically processed in the mouth to aid API disintegration and/or dissolution. These products offer advantages for their administration because swallowing is aided, and water is not required. In addition, chewable dosage forms are preferred by some patients over other formulations due to their aesthetic properties [21]. However, as in the case of ODTs, chewable products do not offer the same dose flexibility as that with conventional tablets [22]. Some chewable gel formulations are more easily absorbed and therefore more bioavailable than tablets [23], for instance chewable gummy multivitamins. Vitamins are important building blocks of the body and help keep you in good health [24].

23.1.3 Chewable Gels in Cancer Treatment

Emerging technologies in the field of healthcare are continuously advancing to improve the diagnosis, treatment,

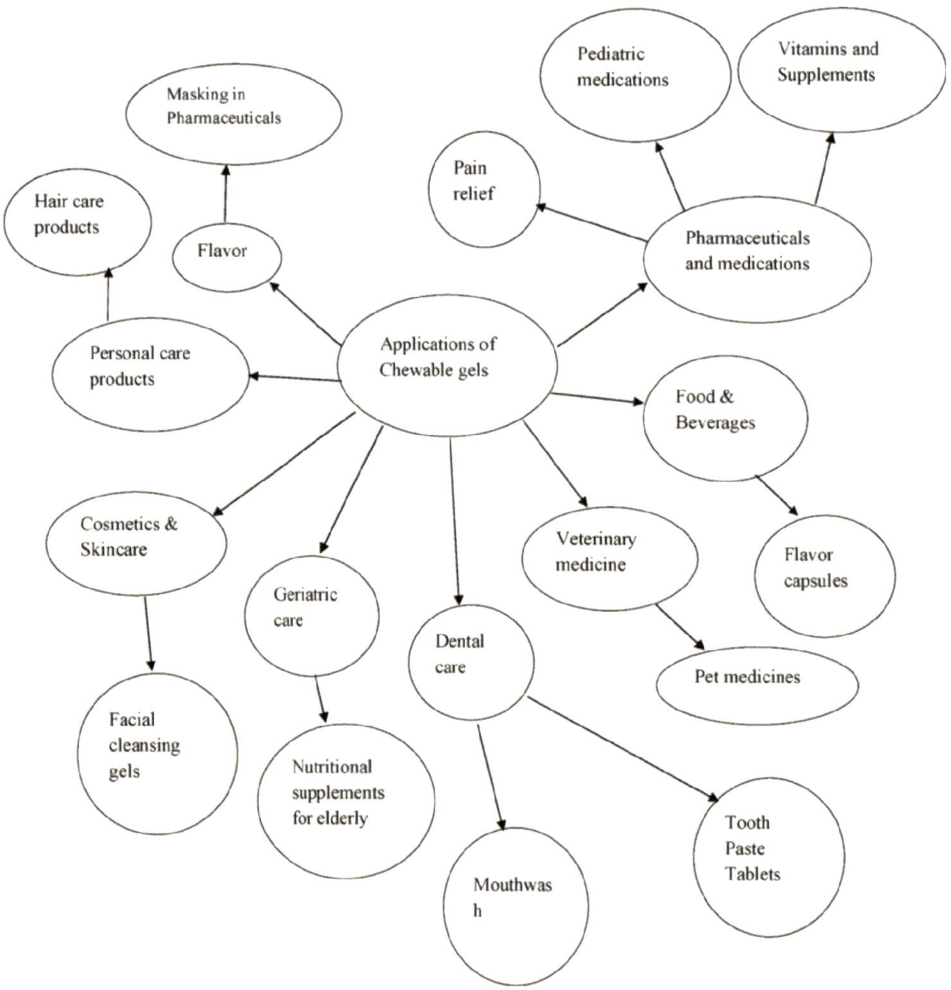

FIGURE 23.1 Applications of chewable gels.

and overall care for patients of all ages, including pediatric and geriatric patients and in particular for cancer treatment [25, 26]. It is important to note, however, that these technologies must be implemented with careful consideration of ethical, privacy, and regulatory issues, particularly when dealing with sensitive patient data and vulnerable populations [27]. These emerging technologies hold the potential to revolutionize cancer care by improving diagnosis, treatment, and overall quality of life [28].

23.1.4 Chewable Gels in Dysmenorrhea Treatment

Menstruation is a regular part of a woman's life, but dysmenorrhea, painful periods, affects many women, most frequently in the form of throbbing, cramping pain in the lower abdomen; other symptoms are possible as well such as nausea; diarrhea; headache; and pain in the hips, lower back, and inner thighs [29, 30]. Premenstrual syndrome manifests in various ways including weight gain, mood swings, tender breasts, vaginal itching and dryness, acne, bloating, irritability, and fatigue [31], usually beginning one to two weeks before the period. The pain usually starts 24 to 48 hours before the period and goes away within 48 hours of getting the period [32]. Menstrual cramps occur when prostaglandin causes the uterus to contract. Higher prostaglandin levels during menstruation lead to stronger contractions that help remove the uterine lining, resulting in cramping and discomfort. Prostaglandin levels rise before menstruation and drop once the period starts, causing cramps to subside after a few days [33, 34].

23.2 SUMMARY AND GOALS FOR THE FUTURE

Oral delivery is the safest and the most practical route but presents challenges for pediatric, geriatric, and cancer patients with dysphagia. These patients often crush or mix tablets with food to ease swallowing, leading to dosing inaccuracies. Tailored dosage forms are needed to improve accessibility for these populations.

Current pediatric drug formulations, such as syrups and suspensions, face issues such as higher costs, microbial contamination, and instability. Chewable gels (gummies) have gained significant interest because of their ability to mask bitter tastes and overcome swallowing difficulties. Unlike tablets, gummies with added sweeteners and flavors improve palatability and reduce the risk of choking. ["Chewable gels, using the heat-induced gelation method, offer a promising candidate for medication delivery due to their favorable pharmaceutical properties, including taste masking, ease of administration, and stability. These gels are especially beneficial for pediatric, geriatric, and female populations, offering a convenient and effective alternative to traditional oral dosage forms. Chewable gels also help alleviate dysmenorrhea and provide an accessible medication option with minimal side effects."]

Chewable gels made using heat-induced gelation are promising candidates for medication delivery due to their favorable pharmaceutical properties taste masking, ease of administration, and stability. These gels are especially beneficial for pediatric, geriatric, and female populations, offering a convenient and effective alternative to traditional oral dose forms. Chewable gels provide an attractive and accessible medication option with minimal side effects.

REFERENCES

[1] Aulton, M. E., & Taylor, K. (Eds.). (2013). Aulton's pharmaceutics: the design and manufacture of medicines. Elsevier Health Sciences.

[2] Khan A, Carmona R, Traube M. Dysphagia in the elderly. Clin Geriatr Med 2014;30:43–53. https://doi.org/10.1016/J.CGER.2013.10.009.

[3] Liu F, Ranmal S, Batchelor HK, Orlu-Gul M, Ernest TB, Thomas IW, et al. Formulation factors affecting acceptability of oral medicines in children. Int J Pharm 2015; 492(1–2):341–3. https://doi.org/10.1016/J.IJPHARM.2015.05.013.

[4] Mateescu M, Vlase G, Budiul MM, Cernuşcă BD, Vlase T. Preliminary study for preparation and characterization of medicated jelly based on Ibuprofen or Ambroxol. J Therm Anal Calorim 2023;148:4601–14. https://doi.org/10.1007/S10973-023-12052-0/FIGURES/8.

[5] Hattrem MN, Molnes S, Haug IJ, Draget KI. Interfacial and rheological properties of gelatin based solid emulsions prepared with acid or alkali pretreated gelatins. Food Hydrocoll 2015;43:700–7. https://doi.org/10.1016/J.FOODHYD.2014.07.026.

[6] Hahm HA, Augsburger LL. Orally disintegrating tablets and related tablet formulations. Pharmaceutical Dosage Forms: Tablets, Third Edition: Volume 2: Rational Design and Formulation 2008:293–311. https://doi.org/10.3109/9781420020298-13/ORALLY-DISINTEGRATING-TABLETS-RELATED-TABLET-FORMULATIONS-HUIJEONG-ASHLEY-HAHM-LARRY-AUGSBURGER.

[7] Bhuiyan J. Development of dry emulsion formulation of Ibuprofen and its characterization. Dissertation. Brac University 2022.

[8] Davydova N. USP chewable gels monographs. Scientific Liaison, May 15, 2018. https://www.usp.org/sites/default/files/usp/document/stakeholder-forum/chewable-gels.pdf.

[9] Villas-Bôas SG, Delicado DG, Åkesson M, Nielsen J. Simultaneous analysis of amino and nonamino organic acids as methyl chloroformate derivatives using gas chromatography-mass spectrometry. Anal Biochem 2003;322:134–8. https://doi.org/10.1016/j.ab.2003.07.018.

[10] Rømer Rassing M. Chewing gum as a drug delivery system. Adv Drug Deliv Rev 1994;13:89–121. https://doi.org/10.1016/0169-409X(94)90028-0.

[11] Sam T, Ernest TB, Walsh J, Williams JL. A benefit/risk approach towards selecting appropriate pharmaceutical dosage forms – an application for paediatric dosage form selection. Int J Pharm 2012;435:115–23. https://doi.org/10.1016/J.IJPHARM.2012.05.024.

[12] Kristensen HG. WHO guideline development of paediatric medicines: points to consider in pharmaceutical development. Int J Pharm 2012;435:134–5. https://doi.org/10.1016/J.IJPHARM.2012.05.054.

[13] WHO Expert Committee on Specifications for Pharmaceutical Preparations . . . World Health Organization – Google Books. n.d. https://books.google.co.in/books?hl=en&lr=&i

d=a6YsDwAAQBAJ&oi=fnd&pg=PA1&dq=World+Health+Organization+(2008).+Development+of+paediatric+medicines:+Pharmaceutical+Development:+points+to+consider&ots=SVrz5d80LW&sig=Jyp6_LK5WyMH9neD4wjeAfp_wB4#v=onepage&q=World%20Health%20Organization%20(2008).%20Development%20of%20paediatric%20medicines%3A%20Pharmaceutical%20Development%3A%20points%20to%20consider&f=false.

[14] FDA Registration – FDA Certificate – FDA Agent. n.d. https://www.fdahelp.us/?gclid=Cj0KCQjw_O2lBhCFARIsAB0E8B-IzA7BJQOEmD3PlbCMwqMnmZ1PU0oLdnHFZ_SBoMKW8swgPccB7R8aAvyTEALw_wcB (accessed July 22, 2023).

[15] Walsh J, Ranmal SR, Ernest TB, Liu F. Patient acceptability, safety and access: a balancing act for selecting age-appropriate oral dosage forms for paediatric and geriatric populations. Int J Pharm 2018;536:547–62. https://doi.org/10.1016/J.IJPHARM.2017.07.017.

[16] Chapman RA. Biopharmaceutical properties of solid dosage forms. Can Med Assoc J 1967;97:753.

[17] Taranum R, Mittapally S. Soft chewable drug delivery system: oral medicated jelly and soft chew. J Drug Del Therap 2018;8:65–72.

[18] Lopez FL, Ernest TB, Tuleu C, Gul MO. Formulation approaches to pediatric oral drug delivery: benefits and limitations of current platforms. 2015;12:1727–40. https://doi.org/10.1517/17425247.2015.1060218.

[19] Kadhim ZM, Ali WK. Preparation and evaluation of granisetron chewable pediatric oral jelly. Intern J Drug Del Techno 2019;39:347–51. https://doi.org/10.25258/ijddt.v9i3.4.

[20] Gonzalez MA, Ramírez-Rigo MV, Gonzalez Vidal NL. Nanoparticle-based chewable gels of praziquantel. Pharm Dev Technol 2023;28:143–51.

[21] Chaudhary SA, Shahiwala AF. Medicated chewing gum – a potential drug delivery system. Expert Opinion Drug Del 2010;7:871–85. https://doi.org/10.1517/17425247.2010.493554.

[22] Mishra B, Sharma G, Shukla D. Investigation of organoleptic characteristics in the development of soft chews of calcium carbonate as mineral supplement. Yakugaku Zasshi 2009;129:1537–44. https://doi.org/10.1248/YAKUSHI.129.1537.

[23] Hattrem MN, Dille MJ, Seternes T, Ege T, Draget KI. The relative bioavailability of Ibuprofen after administration with a novel soft chewable drug formulation. Clin Pharmacol Drug Dev 2018;7:168–76. https://doi.org/10.1002/CPDD.357.

[24] Fletcher RH, Fairfield KM. Vitamins for chronic disease prevention in adults: clinical applications. JAMA 2002;287:3127–9. https://doi.org/10.1001/JAMA.287.23.3127.

[25] Abdullah, Liu L, Javed HU, Xiao J. Engineering emulsion gels as functional colloids emphasizing food applications: a review. Front Nutr 2022;9:890188. https://doi.org/10.3389/FNUT.2022.890188/BIBTEX.

[26] Jensen RE, Rothrock NE, Dewitt EM, Spiegel B, Tucker CA, Crane HM, et al. The role of technical advances in the adoption and integration of patient-reported outcomes in clinical care. Med Care 2015;53:153. https://doi.org/10.1097/MLR.0000000000000289.

[27] Hee Lee D, Yoon SN. Application of artificial intelligence-based technologies in the healthcare industry: opportunities and challenges. Intern J Environ Res Pub Health 2021;18:271. https://doi.org/10.3390/IJERPH18010271.

[28] Huston C. The impact of emerging technology on nursing care: warp speed ahead. Online J Issues Nurs 2013;18.

[29] Menstrual Cramps (Dysmenorrhea): Symptoms, Causes, Treatment, Prevention. n.d. https://www.webmd.com/women/menstrual-cramps (accessed September 11, 2023).

[30] French L. Dysmenorrhea in adolescents: diagnosis and treatment. Pediatr Drugs 2008;10:1–7.

[31] Yonkers KA, O'Brien S, Eriksson E. Seminar premenstrual syndrome, Lancet 2008 Apr 5;371(9619):1200–10. doi: 10.1016/S0140-6736(08)60527-9. PMID: 18395582; PMCID: PMC3118460.

[32] Dysmenorrhea: Menstrual Cramps, Causes & Treatments. n.d. https://my.clevelandclinic.org/health/diseases/4148-dysmenorrhea (accessed September 11, 2023).

[33] Hansen SO, Knudsen UB. Endometriosis, dysmenorrhoea and diet. Eur J Obstet Gyn Rep Bio 2013;169:162–71. https://doi.org/10.1016/j.ejogrb.2013.03.028.

[34] Period Pain – Dysmenorrhoea – Better Health Channel. n.d. https://www.betterhealth.vic.gov.au/health/conditionsandtreatments/menstruation-pain-dysmenorrhoea (accessed September 11, 2023).

Section VI

Innovative Technologies for Tackling Pandemic Diseases Encompassing Vaccines and Vaccination Strategies

24 Vaccine Development and Technologies

*Deepak Kumar Goyal, Poonam Keshav, and Sukhbir Kaur**
Department of Zoology, Panjab University, Chandigarh, India

24.1 INTRODUCTION

Diseases are treated with numerous drugs, generally chemosynthetic, that can be toxic to patients. Therefore, prevention can be safer and more effective than treatment, and vaccines are a key method of constraining infectious diseases.

24.1.1 History of Vaccination

The first approach to vaccination was inoculation, also called variolation. In this method, the disease-causing live pathogen was inoculated to healthy individuals from the target population, at the time with aim of protecting people against smallpox, a contagious and disfiguring disease. Edward Jenner in 1796 noticed that milkmaids who had encountered cowpox were protected from smallpox and confirmed this by taking samples from cowpox-infected milkmaids and inoculated individuals who had not encountered the infection. Mortality was significantly lower in the inoculated individuals (1%–2%) than in those who had not been inoculated (30%) (1).

24.1.2 The Importance of Vaccination

According to the WHO, vaccines annually prevent 3.5–5 million deaths from measles, influenza, diphtheria, pertussis, and tetanus. Currently, vaccines are available for the prevention of such twenty fatal diseases. These vaccines are improving life for all age groups by helping them to live longer and healthier lives (2).

24.2 VACCINE DEVELOPMENT

Vaccines sensitize the immune system against specific pathogens by training the immune system to identify the pathogens, disease-causing agents, or the components they secrete to protect the vaccine recipients. The pathogen can be a virus, bacteria, or parasites. The standard process of vaccine development involves preclinical research and subsequent clinical trials. After successful clinical trials, the tested vaccine is licensed for human use by the Food and Drug Administration (FDA) in the United States and by the Central Drug Standard Control Organization (CDSCO) in India.

24.2.1 Composition

Vaccines comprise multiple components, an active component, the antigen, with other components that boost efficacy (adjuvants), prevent contamination (antibiotics and preservatives), and increase storage life (stabilizers) (3, 4).

In preventive vaccines, the active ingredient is the antigen, the part or whole of the pathogen against which the vaccine is being developed; antigens stimulate the immune response but do not cause the disease. A vaccine might or might not have an adjuvant, a substance that might or might not be immunogenic. Adjuvants can be used to enhance the immune response generated by the active antigen or to prolong the exposure of the antigen to the immune system for longer immune responses. Adjuvants include alum, liposomes, MF59, and so on. Stabilizers such as gelatin or sugar are also added to vaccines to extend their shelf lives, and antibiotics like neomycin are also added to vaccines to prevent contamination by other pathogens during manufacturing. Preservatives such as thiomersal eliminate contamination by fungal or bacterial colonies (generally in vials containing multiple doses), and trace components are residual inactive ingredients in vaccines such as residual cell culture materials and formaldehyde.

24.2.2 Testing

The next step is testing the vaccine candidate for efficacy and safety in preclinical and clinical trials. In preclinical trials, both *in vitro* and *in vivo* studies are performed to assess the immunogenicity of the purified and tested antigens; established animal models are used when needed. In clinical trials, vaccines are tested in humans to evaluate the potential antigen candidates and the immune responses to the vaccine under development. Figure 24.1 presents the phases of vaccine development.

Phase I trials involve a small number of healthy volunteers (generally less than 100) to check immune responses, safety, and appropriate dosages. Phase II trials involve several hundred volunteers to assess the vaccines in different age groups for efficacy and safety. In Phase III vaccine trials, investigators test the vaccine's efficacy and safety in thousands of patients (4).

24.3 VACCINE TECHNOLOGIES

The main principle of vaccines is sensitizing the immune system to attack pathogens or the diseases caused by the same, but investigators have over time developed a number of effective methods for producing them, using the whole pathogen or parts.

* Corresponding author: sukhbir@pu.ac.in

DOI: 10.1201/9781032660592-30

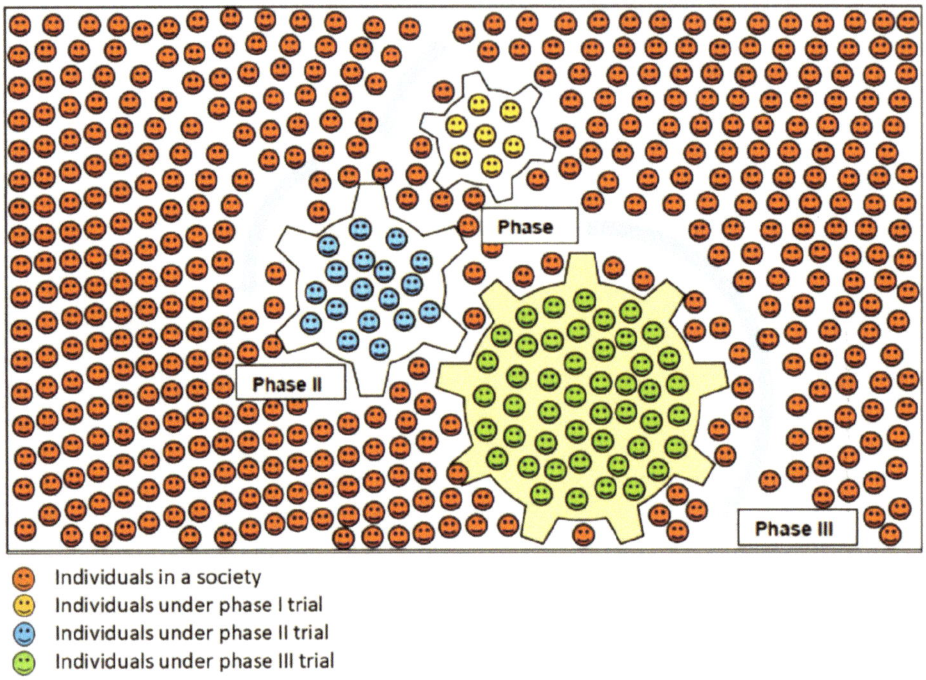

FIGURE 24.1 The phases of human vaccine trials.

24.3.1 Whole-Cell (Whole-Pathogen) Vaccines

With natural infections, whole-pathogenic cells are exposed to the person's immune system; therefore, the earliest vaccines utilized the whole pathogenic cell that caused the disease. Live vaccines are composed of active pathogenic microbes that sensitize the recipient's immune response in the same way as with natural infection. However, these vaccines had high risk of causing active disease that could in turn be spread to others. To preclude such unwelcome situations, modern vaccines are composed of the attenuated or altered form of pathogen that is generally not pathogenic and is safe.

24.3.2 Live-Attenuated Pathogenic Vaccines

In these vaccines, the active antigen is the disease-causing pathogen; its live form is used after it is weakened to sensitize the immune system without triggering the active disease. Advanced cell culture techniques have simplified the development of live-attenuated vaccines, which are generally safer than whole-cell live vaccines. Additionally, one or two shots of live-attenuated vaccines can confer lifetime protection against some diseases, and their development is uncomplicated (5). The measles–mumps–rubella vaccine is the best example of this type of vaccine.

Live-attenuated vaccines do have a few disadvantages, most prominently the risk of microbe reversal to the pathogenic state that can cause active disease. This is a particular threat among the aged, comorbid cases, and people who are immunocompromised. Additionally, virus can cross the placenta, so live-attenuated vaccines can pose risk to a developing fetus, and women are usually advised to delay pregnancy after taking such vaccines (1, 6, 7).

24.3.3 Inactivated Pathogenic Vaccines

These vaccines are comparatively less immunogenic than live or live attenuated vaccines; the microbes are killed or inactivated by either heat or chemical treatment, eliminating the risks associated with live or live-attenuated vaccines. These vaccines are safer for children, the elderly, pregnant women, and persons with comorbidities. However, because they are less immunogenic, more doses are needed to confer the required immune response. These vaccines include rabies, polio, flu, and hepatitis A (7).

24.3.4 Subunit (Acellular) Vaccines

These vaccines are not composed of whole pathogenic cells; rather, they contain only a specific fragment of pathogenic cells and are acellular vaccines; the chemical nature of these fragments of pathogens is polysaccharide and/or protein. Because these vaccines are composed of fragments of pathogen, they pose no risk of disease development and are therefore safe for comorbid, elderly persons, and pregnant females. They are however weaker and require multiple doses for effective protection. These vaccines are utilized against shingles, meningococcal infections, pneumococcal infections, whooping cough, human papillomavirus (HPV), hepatitis B, and *Haemophiles influenzae* type b (Hib) (7).

24.3.5 Viral Vector Vaccines

This technology is based on the use of a modified form of any other virus that is being used as a vector to sensitize the immune response. This is comparatively a novel technology of vaccine development. It utilizes the innocuous virus for the delivery of genetic code of antigen targeted for vaccine development to the body cells of recipient. This results into the synthesis of protein antigen inside the cells of body to sensitize the immune system and generate the immune response. These can be produced swiftly at a large scale using the cell lines. These are comparatively cost-effective than subunit and the nucleic acid vaccines.

24.3.6 Nucleic Acid Vaccines

Nucleic acid vaccines are composed of the pathogen's DNA or RNA. These vaccines do not contain any part of the pathogen and therefore directly do not deliver the antigen to stimulate individuals' immune responses. However, these vaccines direct the synthesis of a particular antigenic protein inside the cells of the body. This technology of vaccine development is easy and swift. Vaccines manufactured by this technology are safer for morbid cases and almost all individuals who need it.

24.3.6.1 DNA Vaccines

DNA is allowed to internalize into the body cells by employing the electroporation technique. After internalization into the cells, this DNA transcribe to mRNA and then translated into a protein that stimulate the immune system to generate the immune response. DNA is quite stable and does not need any protective covering. Currently no DNA vaccine has reached the stage of getting license, but many are under different developmental stages (8).

24.3.6.2 RNA Vaccines

RNA vaccines contain messenger RNA (mRNA) internalized into lipid membranes. This covering of lipid is advantageous for facilitating the entry of vaccine into the body and entering into the body cells by merging with their plasma membrane. After getting into the cells, mRNA is translated into antigenic protein by the cellular machinery. This mRNA is translated into antigenic protein before its natural breakdown and elimination by the system after few days. The resultant protein stimulates the immune response of the recipient. RNA vaccines are not efficient in amalgamating with the human DNA (genome). To date, only two vaccines are licensed for human use in the UK which were developed based on this technology. These two were developed against the COVID-19 infection (8). See Figure 24.2 for the variety of different vaccine types.

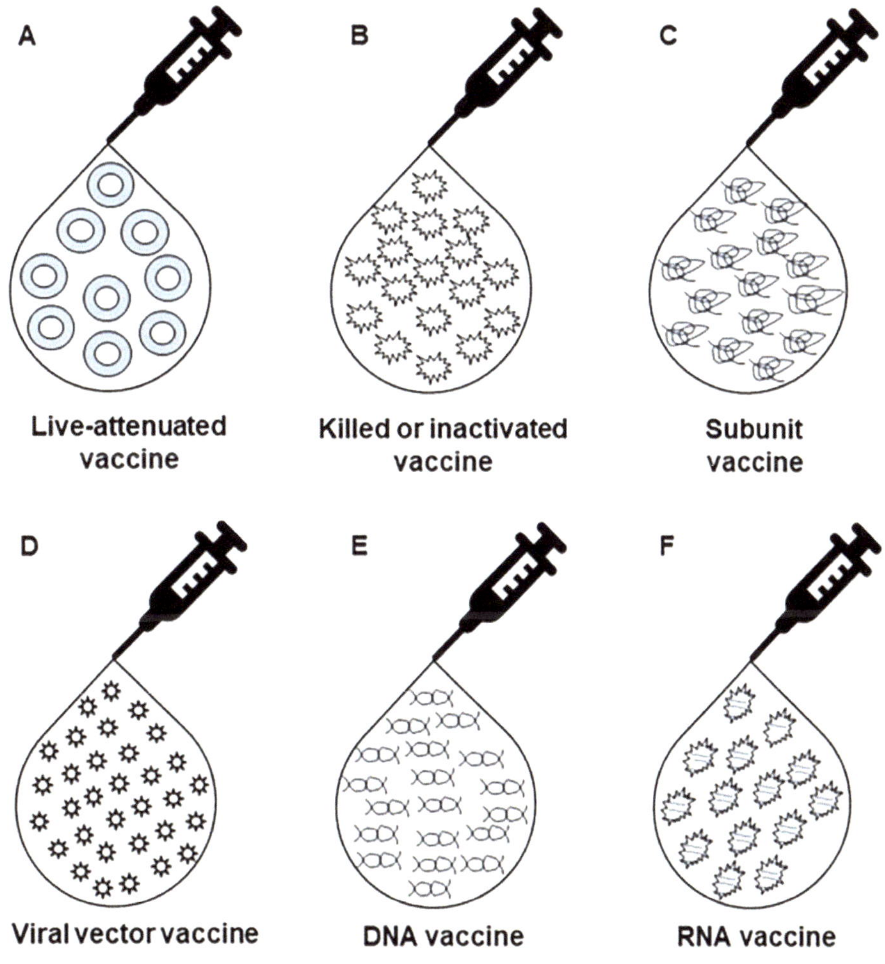

FIGURE 24.2 Different vaccine types.

24.4 EMERGING VACCINE TECHNOLOGIES

24.4.1 Virus-Like Particle Vaccines

Virus-like particles (VLPs) are minute structures that are not able to replicate as the genetic material is absent in them but contain viral proteins and mimic the actions of virus due to the similar form and size. It is a safe technology as there are no chances of disease development after receiving this type of vaccine. Dendritic cells take these VLPs in the body of recipient and process and present them to MHC class I and II molecules thereby activating the immune system by the stimulation of cellular ($CD4^+$ and $CD8^+$ T helper cells) and humoral immune responses (antibody production by B cells). VLP vaccines (Figure 24.3) are significantly immunogenic, and a few are now clinically approved such as HBV (hepatitis B vaccine) and HPV. A few are also in clinical trials against HIV, Zika virus, tuberculosis, rotavirus, influenza, and malaria (9).

24.4.2 Nanoparticle Subunit Vaccines

These vaccines (Figure 24.4) are synthesized after crosslinking the carrier molecule with protein antigens. Crosslinking is done to enhance the immunogenicity of the antigens and simultaneously minimizing the degradation of the same. The carrier molecules used in the production of these vaccines may be metal based (inorganic) or lipids based (organic). NP (nanoparticles) vaccines are less immunogenic than the VLPs but are more stable because VLPs are composed of multiple protein components, and NP vaccines are not; therefore, they are easier and less expensive to produce as well as being safer and more easily reproduced. Their lower immunogenicity can be enhanced by regulating the carrier molecules to the particular antigen. This regulation is based on the hydrophobicity, shape, surface charge and size (9, 10).

24.4.3 Rationally Designed Vaccines

The traditional method of vaccine development involves killing or inactivating the pathogens, but some vaccines can stimulate the immune system as per the requirement to protect from the infection by the active form of the same pathogen and some may not. Therefore, to better sensitize the immune system of the recipient, stimulation of the strong immune response against a pathogen can be done by

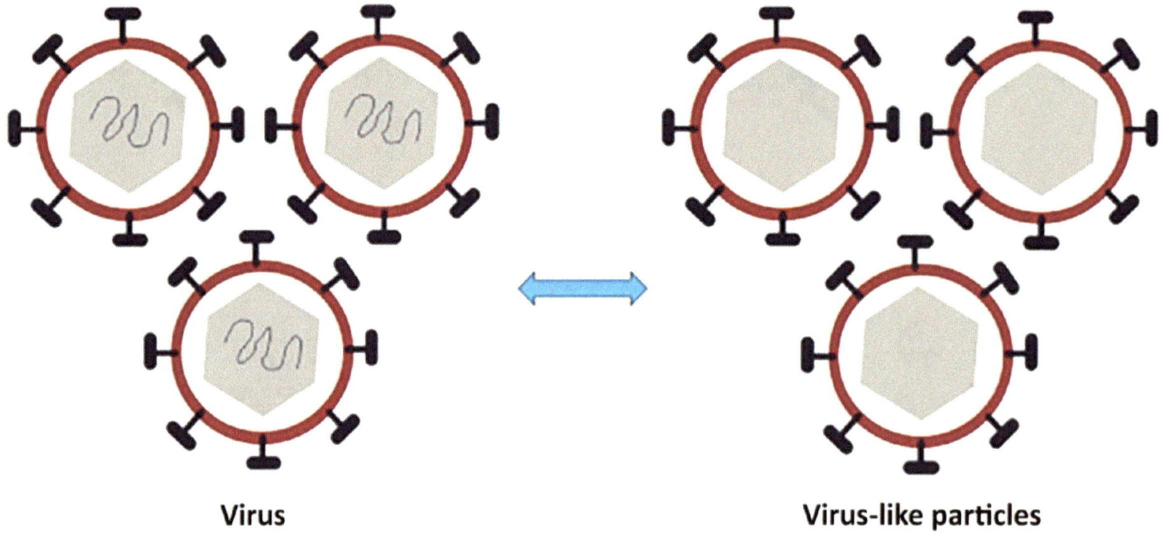

FIGURE 24.3 Virus-like particle vaccine.

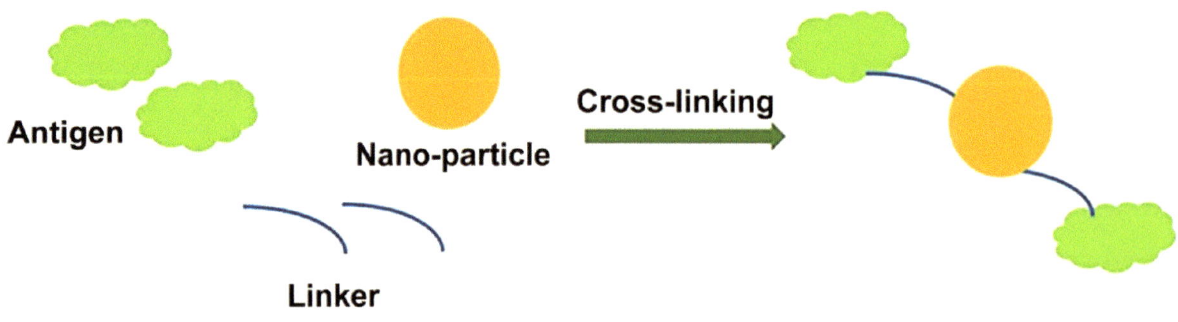

FIGURE 24.4 Nanoparticle subunit vaccine.

developing the vaccine having the antigen of choice. Hence, the rationally designed vaccines (Figure 24.5) constitute the combination vaccines having the selected antigen(s) blend with adjuvant and delivery system. The whole genome is sequenced in this method to screen the entire repertoire of antigens of the pathogen. Computational and bioinformatic tools are also employed to screen the antigens. These vaccines elicit the expectedly strong immune responses to fight the disease-causing pathogen and prevent the development of diseases caused by the particular pathogen. In the absence of a universal approach and technique to design rational vaccines, it wholly relies on the practical and observational data generated from different pre-clinical and clinical trials of vaccine formulations against various diseases (11).

24.4.4 Conjugate Vaccines

These vaccines are composed of lipopolysaccharide antigen from pathogenic bacterial cell which is linked covalently to the carrier proteins. Conjugate vaccines elicit durable immune responses by producing T lymphocyte based memory cells. Cases of disease caused by *Haemophilus influenzae* type B got declined swiftly after the regular immunization programs of infants with conjugate vaccine was initiated. Production of conjugate vaccine is a complicate, multistep and sophisticated procedure. Bacterial strains are grown individually for the production of antigen and carrier protein, and the produced antigen and carrier proteins are purified. The chemical cleavage of lipopolysaccharide (LPS) antigen is performed from the lipid α. The LPS is purified from the lipid and then chemical linking is done to form the conjugate vaccine. The prepared vaccine is subjected to purification process (12). The process of conjugate vaccine production is illustrated (Figure 24.6).

24.5 NONVIRAL PLATFORMS FOR VACCINE DEVELOPMENT

24.5.1 Vaccines for Immunocompromised Individuals

Immunocompromised category includes children, elderly and medically unfit individuals. Children do not possess the fully developed active immune system as that of adults. Old age individuals have weaker immune system due to the age-related conditions. Medically unfit individuals are those having weaker immune system that is suppressed or impaired due to the medical conditions (diseases), for example, cancer, diabetes, HIV infection, etc.

Immune responses by vaccines are generally weaker in the immunocompromised recipients. Live vaccines are not advisable for the immunocompromised individuals. Killed or inactivated and subunit vaccine elicit weaker immune responses and need to be improved by using some immune-modulating substances named adjuvants. The limited repertoire of adjuvants is used in licensed vaccines. But now, this repertoire is being expanded by adding new adjuvants which are oil-in-water emulsion (MF59, an oil-in-water emulsion used in the influenza vaccine) and liposome-based (13). The vaccines amalgamated with adjuvants are more immunogenic and can elicit the profound immune responses in the recipient.

24.5.2 Vaccines with Unconventional Antigens

The conventional antigens are the whole or part of the disease-causing pathogen. But novel approach of using unconventional antigens is getting attention and being employed for the development of vaccines. NP (nanoparticles) and VLP (virus-like particles) are efficient in delivering varied spectrum of biological molecules (proteins, peptides, and small

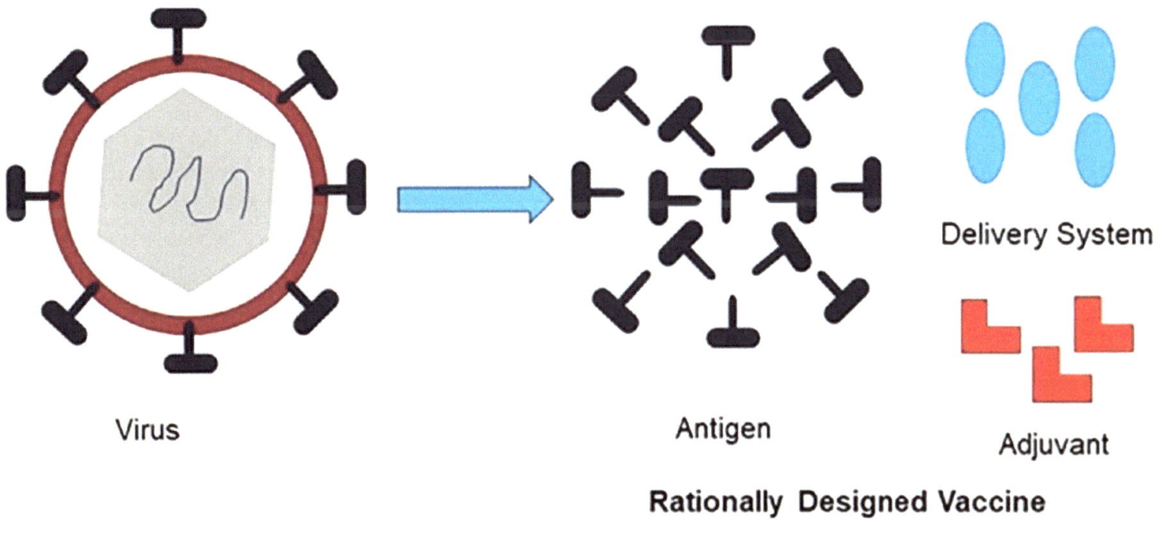

FIGURE 24.5 A rationally designed vaccine.

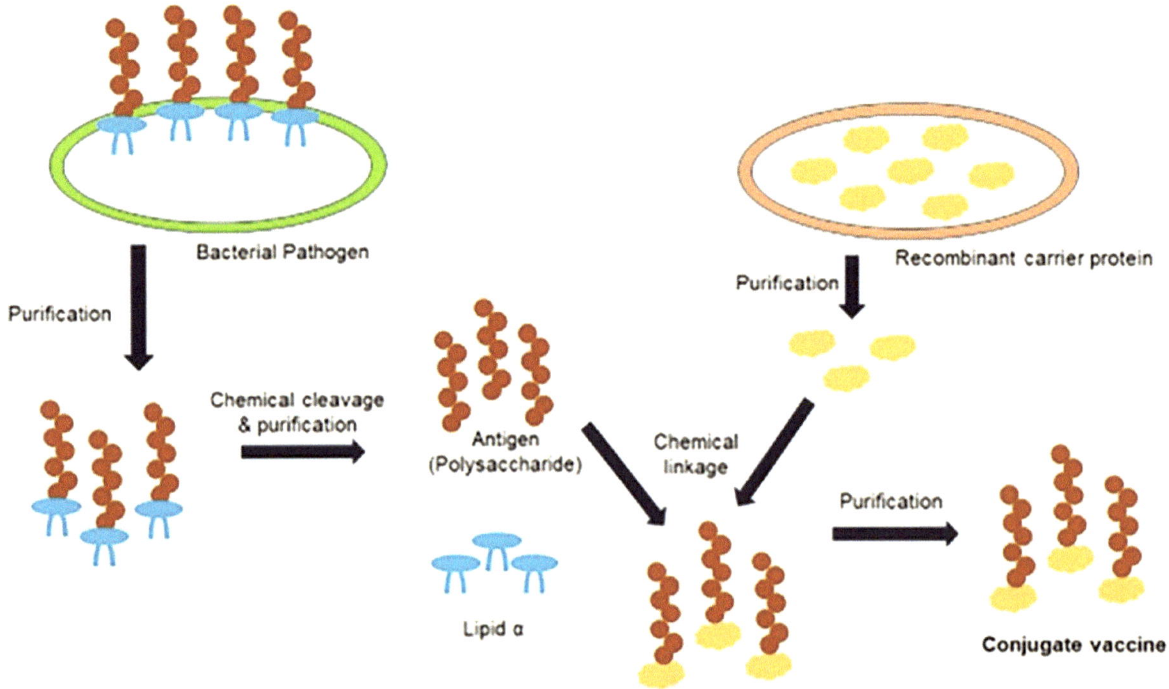

FIGURE 24.6 Developing a conjugate vaccine.

drugs) due to their possession of an internal cavity. These resemble the native virus in structure and can be recognized by the cells hence elicit the immune responses. These are best option for the vaccine development especially for the old age and immunocompromised individuals as they lack the viral genome and incapable of replication. Different VLP-based vaccines have been approved by FDA and are under clinical trials. Mosquirix is an example of one such vaccine against the malaria causing *Plasmodium falciparum* recently developed by Glaxo-SmithKline Inc and has got approval from the FDA. Circumsporozoite protein of *P. falciparum* was fused with surface antigen of hepatitis B. Other VLP-based vaccines which got approval from FDA are developed against human papillomavirus (HPV) and Hepatitis B. Few others are under different phases of clinical trials and are being developed against Influenza Virus (seasonal flu and pandemic flu), Hepatitis E Virus, Respiratory Syncytial Virus (RSV), Norwalk Virus, Ebola Virus, Human Immunodeficiency Virus (HIV) and SARS-CoV-2 virus (COVID-19) (14).

24.6 VACCINES FOR NONCOMMUNICABLE DISEASES

Major NCDs include diabetes, cancer, chronic respiratory and cardiovascular diseases. According to World Health Organization, approximately 74% of global deaths are due to NCDs which account for 41 million deaths annually. Most importantly, the low- and middle-income nations are the major sufferer and 77% of all NCD deaths are happening over here (15). Therefore, this data from WHO indicate the need of some therapeutic interventions especially the vaccines to prevent or minimize the death percentage due these NCDs. Different researchers have employed their efforts to develop the vaccine candidates for the therapy of noncommunicable diseases. Numerous trails have been performed in this direction and some of them are encouraging that give hope for the development of such vaccines. These types of vaccine trials have been done against hypertension, Alzheimer's disease, and diabetes.

24.6.1 Hypertension

Hypertension leads to the death due to its association with cardiovascular diseases. It has been reported that 1.28 billion people around the world aged 30–79 years have hypertension. The highest prevalence of this was reported in middle- and low-income nations. Researchers have started developing therapeutic vaccines against hypertension by following the basic vaccine strategy. Different vaccines tested clinically for their efficacies are ATRQβ-001, PMD3117, AngI-R, CYT006-AngQb, pHAV-4Ang IIs and ATR12181. These have reached phase I and II of clinical trials. PMD3117 is composed of carrier protein KLH (keyhole limpet hemocyanin) conjugated with an analog of angiotensin I and adjuvant (aluminum hydroxide). This vaccine resulted in a significant response in experimental rats and in humans (16, 17).

24.6.2 Alzheimer's Disease

Dementia is a social concern globally with the highly prevalent Alzheimer's disease (AD) being the major cause. More than 50% cases of dementia are due to the Alzheimer's disease. No treatment is available for AD. It is an ailment that grows slowly and starts with the memory loss. But the disease-management is done with symptomatic therapy using drugs that provisionally help in improving the symptoms clinically. In AD, two neuropathological conditions develop namely senile plaque and neurofibrillary tangle. The aggregates of amyloid β are stockpiled in senile plaques whereas precipitates of tau (tubulin associated units) in neuronal cells is observed in neurofibrillary tangles. These two are the basic reasons of AD and are indicated after the observation of piled levels of senile plaque and neurofibrillary tangle in the brain years before the onset of AD. To slack growth of AD in humans, different vaccine trails have been performed and numerous are underway. Vaccine trials have been performed by focusing the aggregated amyloid β. Pathological tau protein has also been employed to assess the immunogenicity and efficacy against AD (18, 19).

24.6.3 Diabetes

Diabetes mellitus (DM) is a metabolic disease and exists in two forms type 1 and type 2. Type 1 DM is autoimmune disease and more severe than type 2 as insulin producing cells get destroyed in this condition. Extrinsic insulin dependency is obvious in this form of disease. Insulin being a polypeptide hormone gets digested upon oral administration and hence, injected to the patients. Up to three injections are needed in severe diabetic cases. Therefore, vaccine can be an option to minimize the invasive procedure of injecting daily insulin.

Researchers have been working in this field for the development of vaccine for diabetes. Few vaccine formulations have been tested and many are underway. Main target antigens are still unknown, however, few antigens from the insulin secreting β-cells are recognized. These include IA-2, GAD and pro-insulin. Different (Oral, nasal and inhaled) routes of administration have been considered for a well-tolerated and effective vaccine. In mouse model, insulin B:9–23 and proinsulin peptide have been assessed for their efficacy via subcutaneous and intraperitoneal routes. Results of such preclinical trials are encouraging for the development of vaccines to control the diabetes. For the vaccine against type 1 DM, GAD is the main target and has shown efficacy in NOD mice. Few vaccine formulations tested are antigen-specific immunotherapies (ASITs). DIAGNODE-3 composed of intra-lymph node rhGAD65 formulated in Alhydrogel resulted in induction of GAD-specific Treg cells but failed to show the similar results in larger studies. DiaPep277 an immunomodulatory protein derived from HSP60 is also another vaccine target for type 1 DM (20, 21).

24.6.4 Cancer

Cancer is perplexing the entire world due to its swift amplification in terms of incidences. Currently, cancer is affecting almost each and every organ of humans. But, the cancers of breast, lung, colon, rectum and prostate are highly prevalent among others. Global death rate due to cancer has been increased (roughly one in six) and in 2020 reached to approximately 10 million. Major causes of cancer are obesity, alcohol, UV radiation, tobacco use, declined physical activity, lower intake of vegetables and fruits, and infections. Although the cases are increasing throughout the world but Asia is alone having a burden of approximately half (48.4%) of the total cases in world. The infection-ascribed cancer is spreading worldwide and the cases are drastically increasing with time. Most common causes of infection-ascribed cancers are hepatitis and HPV, account for prevalence of 30 % cases in low-and lower-middle income nations (22, 23). The causative agent behind the infection-ascribed cancers are certain bacteria, viruses or parasites. *Helicobacter pylori* is the most common bacterium and the viruses are HPV, herpes, HIV, Epstein–Barr virus and hepatitis B and C. An approach of prophylactic vaccines to prevent the infectious diseases can be life-saving from the infection-ascribed cancers. Different types of vaccines such as DNA, antigen-adjuvant, tumor cell, vector-based, anti-idiotype, and dendritic cell vaccines are under development (24).

24.6.5 Viral Diseases

Two categories of infectious diseases are emerging (caused by novel pathogens) and re-emerging (caused due to the pathogens that have affected the society earlier and either were under control or eradicated). These pathogens portray sustained risks to mankind in the health sector and also disturb or disorder the economy thereby having impact on social and political instability. Majority of emerging infections are having their roots from wildlife that causes zoonotic infections. The escalating count of such infections is alarming and thrilling the globe. Consumption of animals from wildlife and livestock quantifies the possibilities of dissemination of the infection. Rising possibilities of vector-borne disease are suggested to have link with global warming and climate change. Many pathogens which fall under this category are Zika virus, dengue virus, Ebola virus, influenza virus, and West Nile virus. These pathogens including the SARS-CoV-2 are posing risk to human health. World Health organization suggested a term 'disease X' to represent an unknown pathogen/disease which could be responsible for future epidemic or pandemic. Therefore, advanced planning is mandatory for making strategy to combat the epidemics or pandemics. Very first step should be to develop prompt resistance against such pathogens which is achievable with the development and dissemination of vaccines. The model vaccine is required for this condition which can be quickly and extensively produced to fulfill the need. The cold chain infrastructure is limited in developing nations hence, the temperature susceptibility must be pondered to develop the heat-resistant vaccines. These vaccines are suitable for the regions where refrigeration and cold chain facilities are wanting. The example of such vaccine is oral pentavalent vaccine for bovine rotavirus (BRV-PV, Rotasil®) produced by Serum Institute of India. If the vaccine is heat sensitive then it is difficult to manage the vaccination programme in the resource limited and affected countries. For example, the vaccine against Ebola virus (rVSV-ZEBOV, ERVEBO®) by Merck and Co., Inc. is heat-sensitive and needs to be stored at -60 °C. The epidemic and pandemic situations necessitate the strategies for quick and massive manufacturing of effective, economical and heat stable vaccines (9, 25). Strategies must include the planning for the dissemination of the vaccine to the underdeveloped nations and on the need basis (giving the equal opportunity) to all.

24.6.6 Shingles

Shingles are caused by the varicella-zoster virus. It is the same virus responsible for the chickenpox. Virus remains dormant in the individuals recovered from the chickenpox. After few years, it can get activated again and may be the reason for shingles in the future. In this condition when virus reactivates, the individuals may suffer from the fluid-filled painful rash or blister on either side of the body or face. A vaccine has been developed which is an adjuvanted recombinant varicella-zoster virus (VZV) glycoprotein E (gE) subunit (non-live) vaccine. Two doses of this vaccines at a gap of 1–2 months in immunocompromised and 2–6 months in immunocompetent individuals are required to prevent the shingles. This registered vaccine named Shingrix® is recommended for the immunocompromised individuals aged 18 years and above. In addition to this, the vaccine is also recommended for the individuals aged 50 years and above (26).

24.6.7 HIV

HIV is an infection that invades and hamper the immune system of infected individuals. HIV attacks and destroys white blood cells or CD4 cells of the infected individuals and pose the risk of opportunistic infections such as severe bacterial, viral and fungal infections including the tuberculosis and also cancers (27). Although, efforts are being put forth to develop a vaccine to prevent the deadly HIV infection. Till date no human vaccine has been developed against this disease. Vaccination of individuals in phase I with eOD-GT8 60mer nanoparticle which was adjuvanted with AS01B resulted into HIV neutralizing antibody precursors in humans (28). A recent study indicated the encouraging results by HIV nanoparticle vaccine targeting first-in-human germline. This vaccine induced helper T cell response (29).

24.6.8 COVID-19

COVID-19 caused by SARS-CoV-2 (severe acute respiratory syndrome coronavirus 2) is extremely transmissible. It was first reported in 2019 hence named COVID-19 in the city of Wuhan, China which was transmitted globally. WHO declared it a pandemic in March 2020 after three years the positive cases crossed the figure of over 762 million. This infectious disease is deadly if not controlled and it has taken a death toll of over 6.8 million worldwide. The globe has experienced a recent pandemic COVID-19, that has given the opportunities to develop an emergency vaccine to combat the situations (30).

This pandemic frightened the world with the severity that was leading to death in the case of negligence of diagnosing the disease and timely therapy to combat the situation to get relief. Earlier there was no drugs or vaccine candidate available to treat the coronavirus disease. The symptomatic therapy was given to patients just to make the situation facile/convenient. The researchers worked hard and their laborious efforts resulted into the development of few vaccine candidates in a short time. The WHO approved these vaccines for use as the emergency vaccines to encounter the difficult circumstances. Initially the vaccination programs were prioritizing the demand and were given to the needy such as immunocompromised individuals, elderly, and first-line workers. After the production was stabilised, the vaccines were available for others too.

The researchers developed few vaccines that were approved in phase 3 or 4 to be used in emergency situations. These include mRNA, inactivated, protein, and viral vector non-replicating vaccines. mRNA vaccine (mRNA-1273 or Spikevax) developed by Moderna and two shots were given at a gap of four weeks via intramuscular route with an efficacy of 93.2%. Another mRNA vaccine (Comirnaty, BNT162b2) was developed by Pfizer and BioNTech which was also administered intramuscularly twice at gap of three weeks with an efficacy of 91%. Gamaleya Research Institute developed a vaccine (viral vector non-replicating vaccines) named Sputnik V (Gam-Covid-Vac) delivered twice, three weeks apart which showed an efficacy rate of 91.6%. Different vaccines that were approved to be used during the pandemic to control the contagious coronavirus diseases are enlisted with their details in Table 24.1 (31).

24.7 VACCINATION CHALLENGES AND SOLUTIONS

Globally, vaccination is a challenge for multiple reasons, with a primary one being vaccine hesitancy, which is still an issue not only in underdeveloped and developing nations but even in developed nations. Many refuse to take vaccines or allow them for their children because they are invasive and because they face stigma or follow religious beliefs that forbid vaccination. Lack of awareness has been an issue as well, although awareness is increasing with social media.

TABLE 24.1
Vaccines Approved for COVID-19

Vaccine Technology	Name	Developed by	Storage	Delivery Route	Dosage	Efficacy
	Sputnik V (Gam-Covid-Vac)	Gamaleya Research Institute	2–8 °C	Intramuscular (3 weeks apart)	Two shots	91.6%
Viral vector nonreplicating	Vaxzevria (AZD1222, or Covishield in India)	Oxford University/ AstraZeneca	2–8 °C	Intramuscular (12 weeks apart)	Two shots	74%
	Ad26.COV2-S	Johnson & Johnson	−20 °C	Intramuscular	Single shot	72%
Protein	Corbevax	Bayer College of Medicine	2–8 °C	Intramuscular (4 weeks apart)	Two shots	90%
	NVX-CoV2373 (Covovax or Nuvaxovid)	Novavax	2–8 °C	Intramuscular (3 weeks apart)	Two shots	90.4%
mRNA	mRNA-1273 or Spikevax	Moderna	−20 °C	Intramuscular (4 weeks apart)	Two shots	93.2%
	Comirnaty (BNT162b2)	Pfizer/BioNTech	−25 °C to −15 °C	Intramuscular (3 weeks apart)	Two shots	91%
Inactivated	CoronaVac (PiCoVacc)	Sinovac	2–8 °C	Intramuscular (2 weeks apart)	Two shots	50.65%
	BBIBP-CorV	Sinopharm	2–8 °C	Intramuscular (3 weeks apart)	Two shots	78.1%
	Covaxin (BBV152)	Bharat Biotech	2–8 °C	Intramuscular (4 weeks apart)	Two shots	77.8%

Additionally, vaccine development is a lengthy process, which hinders swift application.

To conquer some of these challenges, awareness campaigns run by government and nongovernment organizations must disseminate accurate information so the public understands the value of vaccines. Moreover, governments and other stakeholders must develop policies that make vaccine development faster, for instance by incentivizing scientists and expanding laboratories and research facilities.

24.8 FUTURE PROSPECTS

Generally, vaccination is done by giving shot or injection by medical professionals to the recipient. Current invasive procedure of vaccine administration may be updated. New delivery platforms for vaccines will be revolutionary and nasal spray vaccines for influenza are available. Novel platform for vaccine delivery can be MAPs (microarray patches). They have the dimension of a coin and are a dry vaccine coated microneedle that deliver/inoculate the vaccine painlessly. These may have a solvable formula that can infuse in the skin after pressing for 2–5 minutes. Two MAPs for rubella and measles are in market although we need to go much further to develop some more with the required standards.

Inactivated, killed, or subunit antigenic vaccines are less immunogenic than the live vaccines. Therefore, they need some immunostimulants or compounds that can elicit the immune response by enhancing the immunogenicity of the containing vaccines. Non-adjuvanted vaccines evoke a gentle immune response by releasing T helper skewing cytokines, immunoglobulins and T cells (activated). Conversely, adjuvanted vaccines boost the profound immunity by amplifying the mature APCs, escalate the interplay between T cells and APCs, elicit the release of T helper skewing cytokines, diverse and polyfunctional T cells and immunoglobulins (Figure 24.7). Many adjuvants have been assessed preclinically and clinically for the same. Recent preclinical studies tested few adjuvants (MPL-A, alum, saponin, liposome, Montanide ISA 201, imidazoquinolines-gardiquimod, and AddaVax- similar to MF59) in antiparasitic vaccines (against Kala-azar caused by *Leishmania donovani*) in a mouse model. Human compatible adjuvants include alum, emulsion or oil-in-water (MF59 and AS03), TLR agonist (CPG ODN 1018 and AS04) and particulate adjuvant system (AS01- MPLA and QS-21 based). These are approved to be used in broad range of vaccines and elicit humoral and cellular immune responses. Few other adjuvants are under investigation such as synthetic double-stranded RNAs (dsRNA), imidazoquinolines and GLA-derivatives (32–35).

24.9 CONCLUSION

This chapter highlights the different technologies employed for the development of vaccines against infectious diseases. Major platforms or technologies are in use to develop vaccines against diseases such as non-replicating viral vaccines, nucleic acid (mRNA) vaccines, protein vaccines, subunit vaccines, and whole pathogen killed or inactivated vaccines. Many vaccines have been developed and numerous are underway at different phases of preclinical and clinical trials. Adjuvants can be utilized to enhance the efficacy of vaccines for better immune responses, dose minimization and rapid response. Policies and strategies should be made by different stakeholders of the society to combat the challenges of current infectious diseases and newly emerging infections. Governments must encourage the scientific

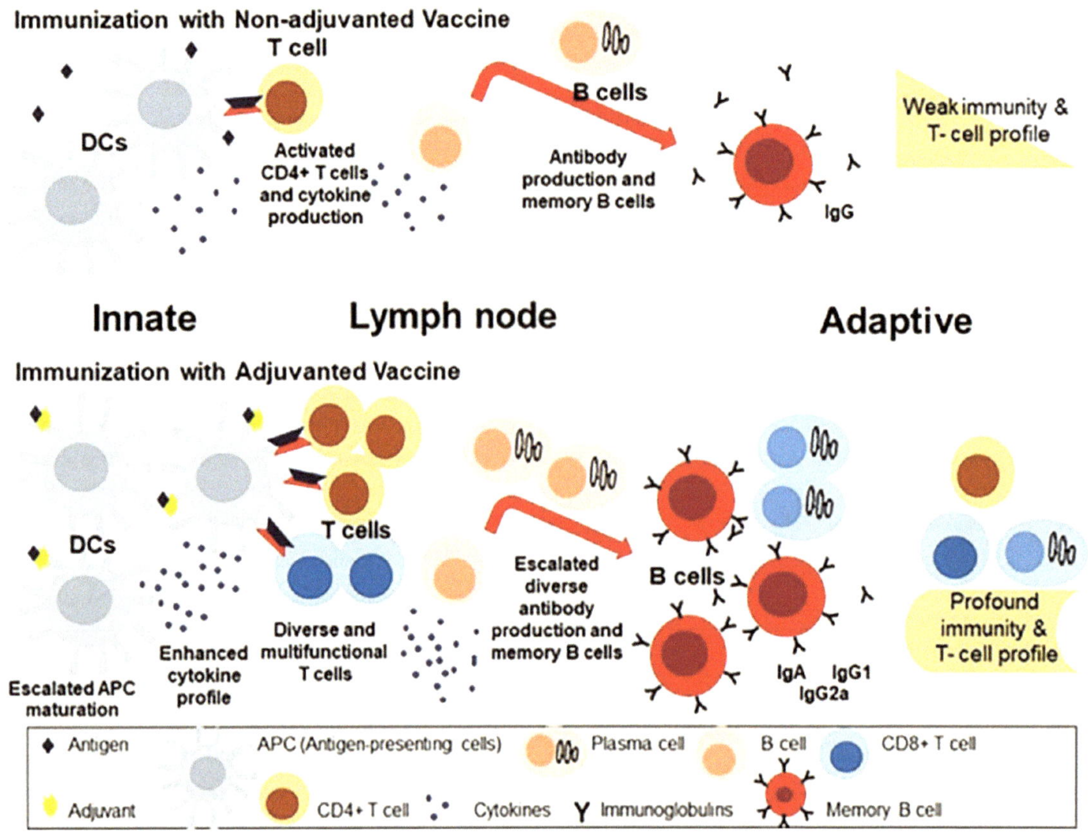

FIGURE 24.7 Immune responses generated by adjuvanted vs non-adjuvanted vaccines.

community by providing excellent infrastructure, man power, fund and research facility to shorten the time to develop vaccine candidates and accelerate the production by escalating the production units and to ensure their justified dissemination (priority based).

REFERENCES

1. Ghosh I, Gandhi MD. Emerging frontiers in vaccine development: A review of changing paradigm. J Biosci Med. 2022;10(7):123–45.
2. World Health Organization. Overview vaccines and immunization [Internet]. World Health Organization. 2024. Accessed on 07.01.2024. Available from: https://www.who.int/health-topics/vaccines-and-immunization#tab=tab_1
3. Ghattas M, Dwivedi G, Lavertu M, Alameh MG. Vaccine technologies and platforms for infectious diseases: Current progress, challenges, and opportunities. Vaccines. 2021; 9(12):1–31.
4. World Health Organization. How are vaccines developed? [Internet]. World Health Organization. 2024. Accessed on 07.01.2024. Available from: https://www.who.int/news-room/feature-stories/detail/how-are-vaccines-developed
5. National Institute of Allergy and Infectious Diseases. Vaccine types [Internet]. National Institute of Allergy and Infectious Diseases. 2019. Accessed on 07.01.2024. Available from: https://www.niaid.nih.gov/research/vaccine-types
6. Shearer WT, Fleisher TA, Buckley RH, Ballas Z, Ballow M, Blaese RM, et al. Recommendations for live viral and bacterial vaccines in immunodeficient patients and their close contacts: Medical Advisory Committee of the Immune Deficiency Foundation. J Allergy Clin Immunol [Internet]. 2014;133(4):961–6. Available from: http://www.ncbi.nlm.nih.gov/pmc/articles/PMC4009347/
7. CDC. ACIP Contraindications Guidelines for Immunization. pdf [Internet]. CDC. 2023. Accessed on 04.11.2023. Available from: https://www.cdc.gov/vaccines/hcp/acip-recs/general-recs/contraindications.html
8. University of Oxford. Types of vaccine. [Internet]. Vaccine Knowledge Project. 2023. accessed on 04.11.2023. Available from: https://vaccineknowledge.ox.ac.uk/types-of-vaccine#viral-vectored-vaccines
9. Brisse M, Vrba SM, Kirk N, Liang Y, Ly H. Emerging concepts and technologies in vaccine development. Front Immunol. 2020;11(September):1–22.
10. Pati R, Shevtsov M, Sonawane A. Nanoparticle vaccines against infectious diseases. Front Immunol. 2018;9:02224.
11. Rueckert C, Guzmán CA. Vaccines: From empirical development to rational design. PLoS Pathog. 2012;8(11).
12. Ihssen J, Kowarik M, Dilettoso S, Tanner C, Wacker M, Thöny-Meyer L. Production of glycoprotein vaccines in Escherichia coli. Microb Cell Fact. 2010;9(August).
13. Pollard AJ, Bijker EM. A guide to vaccinology: From basic principles to new developments. Nat Rev Immunol [Internet]. 2021;21(2):83–100. Available from: http://dx.doi.org/10.1038/s41577-020-00479-7

14. Nooraei S, Bahrulolum H, Hoseini ZS, Katalani C, Hajizade A, Easton AJ, et al. Virus-like particles: Preparation, immunogenicity and their roles as nanovaccines and drug nanocarriers. J Nanobiotechnology [Internet]. 2021;19(1):1–27. Available from: https://doi.org/10.1186/s12951-021-00806-7
15. World Health Organization. Noncommunicable diseases.pdf [Internet]. World Health Organization. 2023. Accessed on 28.12.2023. Available from: https://www.who.int/news-room/fact-sheets/detail/noncommunicable-diseases
16. World Health Organization. Hypertension [Internet]. World Health Organization. 2024. Accessed on 07.01.2024. Available from: https://www.who.int/news-room/fact-sheets/detail/hypertension
17. Tian Y, Hu D, Li Y, Yang L. Development of therapeutic vaccines for the treatment of diseases. Mol Biomed [Internet]. 2022;3(1). Available from: https://doi.org/10.1186/s43556-022-00098-9
18. Takeda S. Tau propagation as a diagnostic and therapeutic target for dementia: Potentials and unanswered questions. Front Neurosci. 2019;13(December):1–8.
19. Thakur A, Bogati S, Pandey S. Attempts to develop vaccines against Alzheimer's disease: A systematic review of ongoing and completed vaccination trials in humans. Cureus. 2023;15(6).
20. Desai S, Buchade S, Chitlange S, Sharma H, Bhombe D, Shewale S, et al. Vaccines for type 1 diabetes: Prevention or reversal? Curr Diabetes Rev. 2020;17(1):30–6.
21. Cook DP, Gysemans C, Mathieu C. Prospects of a type 1 diabetes vaccine. Expert Opin Biol Ther [Internet]. 2017;17(4):403–6. Available from: http://dx.doi.org/10.1080/14712598.2017.1295035
22. World Health Organization. Cancer [Internet]. World Health Organization. 2023. Accessed on 11.11.2023. Available from: https://www.who.int/news-room/fact-sheets/detail/cancer
23. Fan T, Zhang M, Yang J, Zhu Z, Cao W, Dong C. Therapeutic cancer vaccines: Advancements, challenges, and prospects. Signal Transduct Target Ther. 2023;8(1):450.
24. Josefsberg JO, Buckland B. Vaccine process technology. Biotechnol Bioeng. 2012;109(6):1443–60.
25. Trovato M, Sartorius R, D'Apice L, Manco R, De Berardinis P. Viral emerging diseases: Challenges in developing vaccination strategies. Front Immunol. 2020;11(September):1–20.
26. National Centre for Immunzation Research and Surveillance, Australia. Zoster vaccines (Shingrix® [RZV] and Zostavax® [ZVL]) [Internet]. National Centre for Immunzation Research and Surveillance, Australia. 2023. Accessed on 29.12.2023. Available from: https://ncirs.org.au/zoster-shingles/zoster-vaccines-shingrixr-rzv-and-zostavaxr-zvl-frequently-asked-questions-faqs
27. World Health Organization. HIV [Internet]. World Health Organization. 2023. Accessed on 29.12.2023. Available from: https://www.who.int/health-topics/hiv-aids#tab=tab_1
28. Leggat DJ, Cohen KW, Willis JR, Fulp WJ, DeCamp AC, Kalyuzhniy O, et al. Vaccination induces HIV broadly neutralizing antibody precursors in humans. Science (80-). 2022;378(6623):eadff6502.
29. Cohen KW, De Rosa SC, Fulp WJ, DeCamp AC, Fiore-Gartland A, Mahoney CR, et al. A first-in-human germline-targeting HIV nanoparticle vaccine induced broad and publicly targeted helper T cell responses. Sci Transl Med. 2023;15(697):eadf3309.
30. Kushwaha ND, Mohan J, Kushwaha B, Ghazi T, Nwabuife JC, Koorbanally N, et al. A comprehensive review on the global efforts on vaccines and repurposed drugs for combating COVID–19. Eur J Med Chem [Internet]. 2023;260(August): 115719. Available from: https://doi.org/10.1016/j.ejmech.2023.115719
31. Regan JJ, Moulia DL, Link-Gelles R, Godfrey M, Mak J, Najdowski M, et al. Use of Updated COVID-19 vaccines 2023–2024 formula for persons aged ≥6 months: Recommendations of the Advisory Committee on Immunization Practices – United States, September 2023. MMWR Morb Mortal Wkly Rep. 2023;72(42):1140–6.
32. Zhao T, Cai Y, Jiang Y, He X, Wei Y, Yu Y, et al. Vaccine adjuvants: Mechanisms and platforms. Signal Transduct Target Ther. 2023;8(1).
33. Thakur A, Kaur H, Kaur S. Studies on the protective efficacy of freeze thawed promastigote antigen of *Leishmania donovani* along with various adjuvants against visceral leishmaniasis infection in mice. Immunobiology [Internet]. 2015;220(9):1031–8. Available from: http://dx.doi.org/10.1016/j.imbio.2015.05.014
34. Goyal DK, Keshav P, Kaur S. Immune induction by adjuvanted *Leishmania donovani* vaccines against the visceral leishmaniasis in BALB/c mice. Immunobiology [Internet]. 2021;226(2):1–15. Available from: https://doi.org/10.1016/j.imbio.2021.152057
35. Goyal DK, Keshav P, Kaur S. Potential of TLR agonist as an adjuvant in *Leishmania* vaccine against visceral leishmaniasis in BALB/c mice. Microb Pathog [Internet]. 2021;158(June):105021. Available from: https://doi.org/10.1016/j.micpath.2021.105021

25 Enhancing Antibiotic Effectiveness for Sustainable Healthcare

Pallavi Dhawan,[1] *Varsha Gupta,*[1]* *and Tanupriya Bindal*[2]

[1] Department of Microbiology, Government Medical College and Hospital, Chandigarh, India
[2] Department of Pathology, Dr BR Ambedkar State Institute of Medical Sciences, Mohali, India

25.1 INTRODUCTION

The emergence of multidrug-resistant (MDR) strains has led to a significant problem in healthcare, where antimicrobial resistance (AMR) is observed even against last-resort antibiotics (1, 2). Indeed, the WHO in 2019 identified AMR as one of the top ten global public health threats facing humanity (3, 4).

Effectively addressing AMR in a manner consistent with sustainable development will require focusing on two approaches: enhancing the efficacy of current medications and developing new drugs (5, 6). However, the stringent requirements for effective drug design often hinder the process of discovering novel drugs. Therefore, it becomes essential to understand the factors that affect the efficacy of existing antibiotics to improve their effectiveness while considering the long-term health of ecosystems and global health (1, 7).

Clinicians, researchers and microbiologists have long recognized the lack of newer antibiotics and the growing threat of resistance in pathogens. They have emphasized the need for novel antibiotic discovery and development. Despite recent efforts by governments, nonprofit organizations, and public health entities to address resistance and boost investment in antibiotic research, the pharmaceutical industry has not fully returned to antibiotic development, which has delayed introducing new antibiotics to the market (1, 7). However, there is an alternative strategy in combating AMR that resonates with UN Sustainable Development Goal 3 (Good Health and Wellbeing), which involves the use of antibiotic adjuvants (2, 8).

25.2 ANTIBIOTIC ADJUVANTS

Bacterial resistance to antibiotics is a significant challenge in clinical treatment because bacteria have developed various strategies for evading the toxic effects of antibiotics; even slight structural alterations in the target, such as the 30S or 50S subunit of the ribosome, can result in antibiotic resistance, and another resistance mechanism is the active expulsion of antibiotics from the cell through efflux pumps, which are specific transport proteins; efflux pumps can target various classes of antibiotics and confer broad-spectrum resistance, exacerbating the challenge of treating resistant infections (7, 9).

Bacteria also inactivate antibiotics involving enzymes that modify or degrade the key structural elements of the drugs. In particular, β-lactamases hydrolytically cleave the β-lactam ring in penicillin-derived antibiotics. Gram-negative bacteria have an outer membrane that makes them less permeable to antibiotics than Gram-positive bacteria. Porin downregulation reduces antibiotic permeation through the outer membrane and is a prevalent mechanism in Gram-negative bacteria (7, 9). Currently, there is extensive research and clinical utilization of antibiotic adjuvants that specifically target crucial bacterial molecules or cellular structures involved in resistance mechanisms.

25.2.1 β-LACTAMASE INHIBITORS

β-lactam antibiotics such as penicillins, cephalosporins, and carbapenems, are extensively utilized due to their broad-spectrum activity against both Gram-positive and Gram-negative bacteria. However, bacteria can develop resistance through the production of β-lactamases, which are enzymes that render β-lactam antibiotics ineffective by breaking their β-lactam ring (2, 10). In Gram-positive bacteria, resistance initially involved the hydrolysis of penicillins by penicillinases. However, resistance mechanisms in Gram-positive cocci later shifted towards acquiring penicillin-binding proteins (PBPs) with reduced affinities for β-lactams. Conversely, Gram-negative bacteria primarily employ the production of β-lactamases as their common resistance mechanism (1). There are two primary classifications for β-lactamases: the Ambler molecular classification and the Bush-Jacoby-Medeiros functional classification (10).

The Ambler molecular classification categorizes β-lactamases into different molecular classes (A, B, C, and D) based on their amino acid sequences. Classes A, C, and D are known as serine β-lactamases, while class B comprises metallo-β-lactamases (MBLs). On the other hand, the Bush-Jacoby-Medeiros functional classification classifies β-lactamases based on their substrate hydrolysis profiles and their response to inhibitors. This functional classification

* Corresponding author: varshagupta_99@yahoo.com

DOI: 10.1201/9781032660592-31

system further divides β-lactamases into four main groups, along with several subgroups (10).

The antibiotic adjuvant employed nowadays to tackle antibiotic resistance are β-lactamase inhibitors (BLIs), which enhance the activity of β-lactam drugs (1, 2, 10). In the mid-1970s, pharmaceutical companies embarked on research to identify potent inhibitors for β-lactamases, such as TEM-1, staphylococcal penicillinases, and AmpC-related cephalosporinases. Among the notable discoveries was clavulanic acid, derived from *Streptomyces clavuligerus*, which irreversibly inhibited TEM-1 and staphylococcal penicillinase by forming a transiently inhibited intermediate. This breakthrough led to the development of related inhibitors, including sulbactam and tazobactam, primarily targeting class A/group 2 β-lactamases (1). Currently, six β-lactamase inhibitors are approved for clinical use: clavulanic acid, sulbactam, tazobactam, avibactam, varborbactam, and relebactam. Each inhibitor possesses a slightly different chemical structure and activity (2, 10). The next-generation β-lactamase inhibitors, namely zidebactam, taniborbactam, and enmetazobactam, are currently undergoing phase III clinical trials to assess their efficacy when used in combination with cefepime (11).

Various combinations of β-lactamase inhibitors have been introduced into clinical practice, including clavulanic acid with amoxicillin, clavulanic acid with ticarcillin, sulbactam with ampicillin, and tazobactam with piperacillin. These combinations offer broad-spectrum coverage against both Gram-negative and Gram-positive infections. However, the emergence of resistance to BLI combinations remains a concern, attributed to mutated forms of β-lactamases, elevated β-lactamase activity, and the presence of multiple β-lactamases within a bacterial species (1, 2, 10).

With the increasing prevalence of extended-spectrum β-lactamases (ESBLs) and carbapenemases in multidrug-resistant Gram-negative pathogens, newer combinations like tazobactam with ceftolozane and cefepime are being utilized to provide broader spectrum activity. However, these combinations still lack efficacy against strains expressing serine carbapenemases. Researchers continue to pursue innovation in this field, exploring new inhibitors, such as diazabicyclooctanes (DABCOs) like avibactam and investigating compounds that can effectively inhibit metallo-β-lactamases, which are currently unaffected by existing adjuvants (2, 10). Very few effective inhibitors of metallo-β-lactamases are known, but compounds like aspergillomarasmine A show promise in animal models (1). Additionally, a recent study revealed that purine nucleosides can impact MRSA's susceptibility to antibiotics. Guanosine and xanthosine were found to decrease MRSA's antibiotic resistance, while adenosine increased it. As a result, alternative approaches are being investigated, such as using compounds like guanine and xanthosine to reduce cellular levels of c-di-AMP. These approaches aim to restore the susceptibility of organisms like MRSA to β-lactam antibiotics, offering new possibilities for treatment (2, 12).

The use of β-lactamase inhibitors has been crucial in combating antibiotic resistance. These inhibitors enhance the effectiveness of β-lactam antibiotics which are the first-line antibiotics used in community hospitals. However, the constant evolution of resistance mechanisms, including the emergence of ESBLs and metallo-β-lactamases, necessitates ongoing innovation and research in the field. Scientists and healthcare professionals aim to stay ahead in the battle against antibiotic resistance and preserve the efficacy of β-lactam antibiotics in treating infectious diseases by pursuing new inhibitors and alternative mechanisms to inhibit resistance.

25.2.1.1 Efflux Pump Inhibitors

Efflux pumps, situated within bacterial membranes, actively remove substances, including antibiotics, from the interior of the cell. Certain efflux pumps selectively expel specific molecules, whereas others exhibit non-specificity, expelling a wide array of compounds (2). Nonselective pumps are of particular clinical interest as they can render bacterial infections untreatable by available antibiotics, resulting in multidrug resistance (9).

Efflux pump inhibitors (EPIs) are small molecules used in combinatorial antibiotic therapy to overcome multidrug resistance. They work by blocking the efflux pumps, restoring bacterial susceptibility to antibiotics and limiting the emergence of new resistant mutants. Inhibiting efflux pumps can be achieved through various approaches, including gene expression inhibition, prevention of pump component assembly, blockade of membrane outlets or depletion of energy required for pump operation (2, 9).

Developing efflux pump inhibitors requires consideration of the targeted pathogenic bacteria, the specific pump to be inhibited, and the nature of the antibiotic to be potentiated. Additionally, pharmacodynamic and pharmacokinetic characteristics play a crucial role in ensuring the effectiveness of these inhibitors. An ideal inhibitor should be specific to the bacterial target without affecting eukaryotic cells, reach therapeutic concentrations in the body, exhibit maximum specificity and efficacy, avoid antibacterial activity to prevent resistance development, and be nontoxic to humans even at high concentrations (1, 9).

Efflux pumps present challenges due to their diverse physiological functions. Therefore, research focuses on finding compounds that selectively inhibit pumps operating only in prokaryotic cells. Various studies have identified potential substrates and inhibitors for these pumps and explored efflux inhibition therapies. However, to date, no efflux pump inhibitors have been registered for the treatment of bacterial infections in human or veterinary clinics except for MP-601 26 administered as an aerosol in specific cases (2, 7, 9).

Several compounds have shown promising inhibitory activity against efflux pumps. For example:

- Quinoline derivatives inhibited efflux pumps in clinical strains of *Enterobacter aerogenes*.
- The natural alkaloid Conessine inhibited MexAB-OprM or MexB efflux pumps in *Pseudomonas aeruginosa*.

- Compound 4-isopentyloxy-2-naphthamide demonstrated specific inhibition of the AcrB efflux pump in *Escherichia coli*.
- A 3,4-dibromopyrrole-2,5-dione compound exhibited inhibitory effects on efflux pumps in Gram-negative bacterial strains.
- Epigallocatechin-3-gallate (EGCG), a polyphenolic compound found in tea showed antibiotic properties and the ability to inhibit the MexAB-OprM pump in *Pseudomonas aeruginosa*. Furthermore, combining EGCG with the efflux inhibitor PAβN resulted in significantly increased antibiotic potency (1, 9).

These examples highlight the potential of EPIs as adjuvants to enhance the effectiveness of antibiotics. EPIs are potential drugs that can treat bacterial infections by interfering with their efflux pumps. However, unlike β-lactamase inhibitors, none of them are clinically approved yet. To advance EPIs to clinical use, we need more research on their mechanism of action, their bacterial targets, and their safety and efficacy criteria (2, 9).

25.2.2 Membrane Permeabilizers

Another approach to combat antibiotic resistance is to target bacterial cell permeability to enhance the efficacy of existing antibiotics. The chemical composition of the drug molecule and the membrane composition of the bacterium determine how antibiotics penetrate bacterial cells. Resistant strains often exhibit lipid or membrane protein modifications, highlighting the importance of understanding membrane structure in antibiotic development (2, 9).

Various chemosensitizers have been proposed to facilitate antibiotic diffusion and increase intracellular concentration (2, 13). These compounds can alter porins and membrane channels and their use in combination with antibiotics has shown promise in controlling resistant strains. Researchers have also investigated various compounds and derivatives to enhance membrane permeability and potentiate antibiotic activity. For instance, cationic steroidal antibiotics called "Ceragenins" incorporate into membranes and induce dysfunction, leading to cell death (9).

Recently, researchers have developed new compounds called 'Foldamers' that can help antibiotics work better by changing the electrical charge of bacterial cell membranes. These foldamers were designed using a computer and are based on the structure of an antimicrobial peptide called PGLa. They make the membranes more negatively charged, which helps antibiotics get into the bacteria and kill them. This shows how manipulating the membrane potential of bacteria can be a useful strategy for developing new antibiotic adjuvants (14).

Other compounds from plants and sharks can also permeabilize membranes (2). Mefloquine, an anti-malarial drug, can change the membrane fluidity of Staphylococcus aureus, enhancing the activity of oxacillin against methicillin-resistant and susceptible strains. These findings suggest that existing compounds could be repurposed as adjuvants to disrupt membranes and potentiate antibiotics (15).

Addressing bacterial cell permeability can help tackle antibiotic resistance. By understanding membrane composition and creating chemosensitizers that boost permeability, the effectiveness of current antibiotics can be enhanced. Ongoing exploration in this area is crucial for advancing antimicrobial therapy and safeguarding the availability of viable treatment choices for infectious diseases (2, 7, 9, 13).

25.2.3 Emerging Antibiotic Adjuvants under Investigation

Antivirulence agents: The combination of antivirulence agents and antibiotics offers a promising strategy to address antibiotic-induced pathogenicity in *Staphylococcus aureus* infections. A study showcased the efficacy of M21, an antivirulence agent, in conjunction with ampicillin, resulting in the suppression of virulence factors and improved outcomes in mouse models. This innovative approach opens up new avenues for the development of antibiotic adjuvants (16).

Compounds that enhance the activity of the innate immune system: Enhancing the body's defense mechanisms is a target for antibiotic adjuvants. Immunomodulatory peptides, such as LL-37, can enhance the innate immune system and work alongside antibiotics. Exploring small molecules that boost antibiotic action is also promising. Streptazolin, a compound that stimulates macrophage activity, has shown to improve antibiotic effectiveness. However, it is crucial to balance immune system activation to avoid harm to the host (1).

Natural molecules as adjuvants: Topical antibiotics combined with natural adjuvants can minimize the negative impacts of systemic antibiotic use. Phytochemicals and their derivatives were tested as enhancers against *S. aureus* strains, including those from diabetic foot wounds. Farnesol at sub-MIC restored the activity of methicillin and oxacillin on specific strains and MRSA isolates and enhanced the effects of other antibiotics. These findings demonstrate the potential of topical phytochemicals as agents to modify antibiotic resistance in bacterial wound infections (17).

25.3 CONCLUSION

The rise of antibiotic resistance has necessitated the development of innovative strategies to combat this global health challenge. One of the approaches is, antibiotic adjuvants. They offer a distinct and complementary approach to combating this issue. These compounds enhance and sustain the efficacy of existing drugs while mitigating resistance emergence by directly targeting resistance mechanisms or increasing antibiotic potency. Although adjuvants play a vital role in addressing the AMR crisis, yet they alone are insufficient which is their main limitation. They form just one piece of a complex puzzle that underscores the importance of responsible antibiotic usage across agriculture, livestock and human medicine. This emphasis on responsible use aligns with sustainable development goals and highlights the need for the One Health approach – a holistic strategy that encompasses human, animal, and environmental health – to effectively address AMR while safeguarding the wellbeing of present and future generations. Therefore, in future, the search for novel antimicrobials and the expansion of adjuvant diversity remain imperative to ensure sustainable and effective treatment options for bacterial infections.

REFERENCES

1. Wright GD. Antibiotic Adjuvants: Rescuing Antibiotics from Resistance. Trends Microbiol. 2016 Nov 1;24(11):862–71.
2. Barron M. Antibiotic Adjuvants for Combatting Antimicrobial Resistance. ASM.org. [cited 2023 May 2]. Available from: https://asm.org:443/Articles/2023/January/Antibiotic-Adjuvants-for-Combatting-Antimicrobial
3. Antimicrobial Resistance [Internet]. [cited 2023 May 27]. Available from: https://www.who.int/news-room/fact-sheets/detail/antimicrobial-resistance
4. EClinicalMedicine. Antimicrobial Resistance: A Top Ten Global Public Health Threat. eClinicalMedicine [Internet]. 2021 Nov 1 [cited 2023 May 27];41. Available from: https://www.thelancet.com/journals/eclinm/article/PIIS2589–5370(21)00502–2/fulltext
5. Uchil RR, Kohli GS, Katekhaye VM, Swami OC. Strategies to Combat Antimicrobial Resistance. J Clin Diagn Res JCDR. 2014 Jul;8(7):ME01–4.
6. Centers for Disease Control and Prevention [Internet]. 2022 [cited 2023 May 27]. Actions to Fight Antibiotic Resistance. Available from: https://www.cdc.gov/drugresistance/actions-to-fight.html
7. Dhanda G, Acharya Y, Haldar J. Antibiotic Adjuvants: A Versatile Approach to Combat Antibiotic Resistance. ACS Omega. 2023 Mar 28;8(12):10757–83.
8. Jasovský D, Littmann J, Zorzet A, Cars O. Antimicrobial Resistance – A Threat to the World's Sustainable Development. Ups J Med Sci. 2016 Aug;121(3):159–64.
9. Douafer H, Andrieu V, Phanstiel OI, Brunel JM. Antibiotic Adjuvants: Make Antibiotics Great Again! J Med Chem. 2019 Oct 10;62(19):8665–81.
10. Bush K, Bradford PA. Interplay between β-Lactamases and New β-Lactamase Inhibitors. Nat Rev Microbiol. 2019 May;17(5):295–306.
11. Vázquez-Ucha JC, Lasarte-Monterrubio C, Guijarro-Sánchez P, Oviaño M, Álvarez-Fraga L, Alonso-García I, et al. Assessment of Activity and Resistance Mechanisms to Cefepime in Combination with the Novel β-Lactamase Inhibitors Zidebactam, Taniborbactam, and Enmetazobactam against a Multicenter Collection of Carbapenemase-Producing Enterobacterales. Antimicrob Agents Chemother. 2022 Feb 15;66(2):e01676–21.
12. Nolan AC, Zeden MS, Kviatkovski I, Campbell C, Urwin L, Corrigan RM, et al. Purine Nucleosides Interfere with c-di-AMP Levels and Act as Adjuvants to Re-Sensitize MRSA to β-Lactam Antibiotics. mBio. 2022 Dec 12;14(1):e02478–22.
13. Dhanda G, Mukherjee R, Basak D, Haldar J. Small-Molecular Adjuvants with Weak Membrane Perturbation Potentiate Antibiotics against Gram-Negative Superbugs. ACS Infect Dis. 2022 May 13;8(5):1086–97.
14. Nath Bhaumik K, Hetényi A, Olajos G, Martins A, Spohn R, Németh L, et al. Rationally Designed Foldameric Adjuvants Enhance Antibiotic Efficacy via Promoting Membrane Hyperpolarization. Mol Syst Des Eng. 2022;7(1):21–33.
15. Podoll J, Olson J, Wang W, Wang X. A Cell-Free Screen for Bacterial Membrane Disruptors Identifies Mefloquine as a Novel Antibiotic Adjuvant. Antibiotics. 2021 Mar 18;10(3):315.
16. Gao P, Wei Y, Tai SSC, Halebeedu Prakash P, Iu HTV, Li Y, et al. Antivirulence Agent as an Adjuvant of β-Lactam Antibiotics in Treating Staphylococcal Infections. Antibiotics. 2022 Jun;11(6):819.
17. Oliveira D, Borges A, Saavedra MJ, Borges F, Simões M. Screening of Natural Molecules as Adjuvants to Topical Antibiotics to Treat Staphylococcus Aureus from Diabetic Foot Ulcer Infections. Antibiotics. 2022 May;11(5):620.

26 Biotechnological Vector-Borne Disease Control

*Poonam Keshav, Deepak Kumar Goyal, and Sukhbir Kaur**
Department of Zoology, Panjab University, Chandigarh, India

26.1 INTRODUCTION

The majority of the world's underprivileged individuals and nations bear the greatest burden of vector-borne diseases (VBDs). The tropical and subtropical regions of the world are the hub for these VBDs. The no or low-income communities and nations where necessary life-sustaining facilities are lacking are severely affected. The unhygienic housing, surroundings, poor medical amenities and lack of knowledge exacerbate diseases. VBDs have a multifaceted impact on gross domestic product and growth.

Dissemination of VBDs is affected by a range of variables, including land use, investments, trade, mobility of humans and the amount and quality of labor as well as financial sources. According to the WHO report, an estimated 17% of all infectious diseases worldwide are major vector-borne diseases. VBDs are caused by viruses, bacteria and parasites that are transmitted by potent vectors. Vectors are often arthropods like mosquitoes, fleas, lice, bugs and sand flies. Apart from them snails are also usual vectors. Numerous vector-borne illnesses are zoonotic, that is, transmissible between animals and humans either directly or indirectly, and include leishmaniasis, encephalitis, Lyme disease and West Nile virus. A myriad of these vectors are hematophagous, feeding on the blood of the host to transmit the infection.

A strong, effective and sustainable outlook is demanded by VBDs. The majority of VBDs cannot be averted just by vaccination and are instead managed by a comprehensive set of measures. Numerous solutions are proposed to deal with these pressing problems. While there is limited short- or medium-term practical application for some of them, they are nevertheless highly applicable. To genuinely prevent and eliminate the burden of these diseases collective initiatives like vector control, vaccination, diagnosis, population control, climate change, environment management, poverty eradication, education and social equality are crucial (1). The majority of diseases are largely ignored or receive little attention from governing bodies, pharma companies, research and scientific organizations thereby intensifying the plight of infected people and developing countries.

Therefore, the fight against these is possible only by incorporating the biotechnological innovations and advancements to get valid, accurate and resounding proofs. Today when the world is facing a new epidemic or pandemic now and then, it is worth amalgamating the biotechnological, bioengineering and system engineering breakthroughs to mitigate the associated morbidity and mortality (2). The establishment of efficient and swift detection tests is critical in limiting the global dissemination of highly transmissible illnesses. Harnessing the remarkable novel and latest advances in biomedical research may be advantageous to developing nations that are hubs to these VBDs. Applications of biotechnology has revolutionized the diagnostic and therapeutic of the deadliest viral and parasitic infections such as COVID-19 and malaria. The latest biotechnological interventions are answers to global needs as they have the potential to foster the delivery of effective disease-fighting tactics by virtue of monitoring, diagnosis, prevention, or therapeutics. Some of the interventions are discussed below which rightly justify them as the "Technology of hope" for poverty-stricken, infected people grasped in the claws of vector-borne diseases across the world.

26.2 DISCUSSION

26.2.1 Vector-Borne Diseases: A Looming Threat

The global community is severely affected by these VBDs, emerging and re-emerging infections (Table 26.1), and has reignited the discussion among scientific and doctor's communities to prevent their outbreaks and pandemics. A critical approach is required to prevent, detect, and respond to the health emergency propounded by these pathogens. The alarming rate of global warming and escalating anthropogenic emissions of greenhouse gases are imposing human vulnerability to vector-borne diseases. Climate change hugely impacts VBD transmission by altering the ecosystem habitats in which, the host (human/animal), reservoir host and vector reside (Figure 26.2).

Warming of the earth and rise in temperature has caused the mosquitoes and mosquito-borne diseases to spread at higher altitudes and latitudes thereby stretching of transmission season in endemic regions. Climate change-influenced population displacement can also facilitate the transmission and spread of pathogens and vectors to new localities. It has been reported that the vectorial competency of a majority of serious VBDs often spikes at fairly high temperatures. Thus, global warming is posing a grave risk to human health and mounting roadblocks in the vector control measures.

* Corresponding author: sukhbir@pu.ac.in

DOI: 10.1201/9781032660592-32

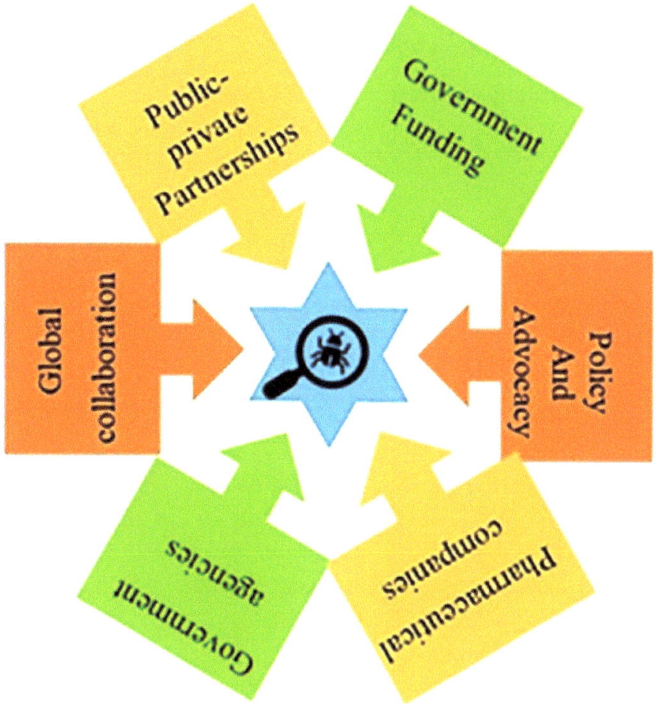

FIGURE 26.1 Developing accessible drugs and diagnostic equipment for VBDs requires a globally collaborative approach.

TABLE 26.1
Disease Vectors and Their Global Distribution

Vectors	Diseases	Causative Organism	Category	Distribution
A. Arthropods				
Sandflies	Leishmaniasis	*Leishmania* spp.	Protozoan parasite	Tropical & subtropical regions of the world
	Sandfly fever/ Pappataci fever	Phlebovirus	Virus	Subtropical regions in the eastern hemisphere
	Carrion's disease	*Bartonella bacilliformis*	Gram-negative bacteria	Endemic to the South American Andes
Mosquito				
Anopheles sp.	Malaria	*Plasmodium falciparum* *P. vivax* *P. ovale* *P. malariae* *P. knowlesi*	Protozoan parasite	Endemic in 84 countries in Southeast Asia, South America, sub-Saharan Africa, and the Western Pacific Islands
	Lymphatic filariasis	Filarial worms *Wuchereria bancrofti, Brugia malayi*, and *B. timori*	Parasitic nematodes	Tropical and subtropical regions of the world
Aedes sp.	Dengue	DENV (Flavivirus)	Virus	Tropical and subtropical regions of the world
	Chikungunya	CHIKV (Alphavirus)	Virus	Asia, Africa, and the Americas
	Yellow fever	Yellow fever virus (Flavivirus)	Virus	Tropical and subtropical areas of Africa and South America
	Rift valley fever	Phlebovirus	Virus	Sub-Saharan Africa
	Zika	ZIKV (Flavivirus)	Virus	Tropical and subtropical regions of the world
	Lymphatic filariasis	Filarial worms *Wuchereria bancrofti, Brugia malayi*, and *B. timori*	Parasitic nematodes	Tropical and subtropical regions of the world

(Continued)

TABLE 26.1
Disease Vectors and Their Global Distribution (Continued)

Vectors	Diseases	Causative Organism	Category	Distribution
Culex	Japanese encephalitis	JEV	Virus	East and Southeast Asia and Western Pacific
	West Nile fever	WNV (Flavivirus)	Virus	
	Lymphatic filariasis	Filarial worms *Wuchereria bancrofti, Brugia malayi*, and *B. timori*	Parasitic nematodes	Tropical and subtropical regions of the world
Fleas	Tungiasis	*Tunga penetrans* (Chigoe flea)	Ectoparasite	Tropical and subtropical regions of the world
	Plague	*Yersinia pestis*	Bacteria	Africa, Asia, and rural areas of western United States
Black flies (*Simulium*)	Onchocerciasis	*Onchocerca volvulus*	Parasitic nematode	99% infection on the African continent, esp. Yemen and Latin America
Tsetse flies	Sleeping sickness	*Trypanosoma brucei*	Protozoan parasite	Sub-Saharan Africa
Triatomine bugs	Chagas disease	*Trypanosoma cruzi*	Protozoan parasite	Latin America
Lice	Typhus	*Rickettsia prowazekii*	Bacteria	North America
Ticks	Lyme disease	*Borrelia*	Bacteria	Tropical and subtropical regions of the world
	Tick-borne encephalitis	TBEV (Flavivirus)	Bacteria	Eurasia
	Rickettsial disease	*Rickettsia*	Bacteria	Tropical and subtropical regions of the world
	Relapsing fever	*Borrelia*	Bacteria	Northern and eastern Africa and the western United States
	Crimean-Congo hemorrhagic fever	CCHF Virus (Nairovirus)	Virus	Africa, the Middle East, and Asia
B. Mollusca				
Freshwater snails	Schistosomiasis	*Schistosomes*	Parasitic trematode	Tropical countries

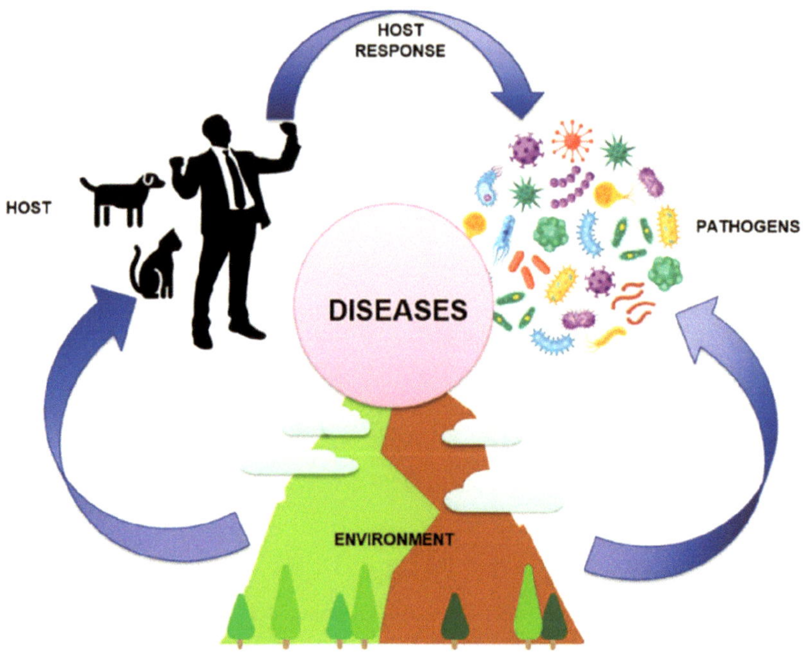

FIGURE 26.2 The host–pathogen relationship depends on the ecosystem, and climate change has shifted global epidemiological trends in vector-borne diseases.

26.2.2 Biotechnological Interventions to Control VBDs

26.2.2.1 Vector Control Strategies

Traditionally and currently, the main technique for managing many VBDs is vector control; chemotherapy drugs show severe side effects, and no effective licensed vaccines have been developed so far. Vector control is the sole technique available for safeguarding communities; its goal is to minimize or completely stop human contact with the vector to control the spread of infections.

Considering the speed at which society and the environment are changing, VBDs provide formidable obstacles to preventative and control measures. They are strikingly instrumental in increasing the global disease burden, and the African continent is maximally affected by the VBDs. The tropical and subtropical regions of the continent exhibit immense suitability for the growth and development of vectors for diseases like malaria, dengue, chikungunya, yellow fever, zika, schistosomiasis, trypanosomiasis and many more. Lack of basic sanitation, medical facilities, poor housing, low socioeconomic status and high vegetation are other factors that contribute to disease transmission in these regions.

Additionally, the rapid expansions and shifts in geographic range driven by climate change have led to the inhabitation of new mosquito species that were not endemic to this continent, which of course is spreading mosquito-borne diseases in the African population. Historically, mosquitoes of *Anopheles* and *Culex* species were predominant in Africa, but now, *Aedes* mosquitoes are also vigorous, prevalent and transmitting diseases across African nations (3). The vast majority of the population lives below the poverty line in peri-urban settings near natural freshwater ecosystems with poor lifestyles and inadequate healthcare, and they are highly prone to freshwater-related VBDs, primarily onchocerciasis, schistosomiasis and lymphatic filariasis. In addition to malaria, these diseases exert severe debilitating health and financial impacts on individuals particularly in Africa and South Asia. Personal prophylactic measures are always the best and the foremost vector control methods and include wearing full-coverage clothing and using vector-repellent creams, lotions, mats and coils as well as bed nets while sleeping.

26.2.2.1.1 Chemical Control

Over the past few decades, chemical exposure and treatment have been proven to be one of the fastest and most effective methods. Indoor spraying and insecticide-treated bed nets are the most promising interventions used in vector elimination programs. Insecticidal nets, bedding and indoor residual spraying are other paramount methods to prevent encounters with mosquitoes and sandflies. The surge in resistant species of mosquitoes, sandflies, horseflies and other vectors pressed the need to search for novel insecticides and larvicide formulations to abate the nuisance of these harmful disease transmitters. Now a days in Africa and India eave tubes are also gaining importance. They are designed to reduce the indoor biting of these insect vectors. Eave tubes are insecticide-laden small PVC tubes placed under the roofs of houses in vector-endemic regions. With the budding biotechnological interventions, the most advanced vector control strategy is to genetically manipulate the insect vectors. There is a rapid surge in the demand for designer strains. This gene-derived approach either aims to stifle the vector population by disseminating sterility or by diminishing the vector competence. The effective and appropriate usage of these tools and approaches alleviates the global disease burden imposed by vector-borne pathogens.

26.2.2.1.2 Biocontrol Strategies

The introduction of biological larvicide and predator species to the vector prevalent sites is the other successful approach. *Bacillus thurigiensis israelensis* (*Bti*) and *Bacillus sphaericus* are biological larvicides and have shown good efficacy against mosquitoes, sand flies (vector of *Leishmania*) and black flies (vector of *Onchocerca*) (4). Some fungal species of the genus *Lagenidium* are also reported to be pathogenic to mosquito larvae and can be used to curb the population of disease transmitting mosquitoes.

A large number of phytocomponents are also being reported to exhibit toxicity, chemosterilant and growth inhibitory activity against both adult and the larval stage of mosquitoes and other dipteran vectors. The bioactive compounds extracted from *Erythrina variegata, Mentha, Azadirachta indica*, *Terminalia chebula, Jatropha curcas, Arachis hypogaea, Sesamum indicum, Artemisia annua* and *Pongamia pinnata* showed good larvicidal activity against different mosquito species (5). Today, they are the most potent weapons in controlling vector-transmitted diseases and to combat surging resistance cases.

The most interesting vector control strategy has been the introduction of *Wolbachia* mosquitoes in the hyperendemic regions. Most of the arthropods (bees, beetles and butterflies) have the endosymbiotic bacterium *Wolbachia. Culex pipiens* naturally inhabits *Wolbachia* and was isolated for the first time in 1924. Experimentally, *Wolbachia* is introduced in male mosquitoes and allowed to mate with females (*Wolbachia* deficient). It causes reproductive manipulation in the females leading to cytoplasmic incompatibility and thereby unviable offspring generation. Moreover, it shortens the lifespan of adult vector, diminishes growth of other pathogens in the vector as well as improves and strengthens the vector's insecticide susceptibility *Wolbachia* is successfully introduced in *Aedes* and *Anopheles* mosquitoes and reported to inhibit dengue, chikungunya virus and malarial parasite replication in them (6).

Wolbachia-Culex mosquitoes have also exhibited diminished receptivity to the West Nile virus (7). Thus *Wolbachia*-based technique is a novel biological and environment-friendly approach to restrict mosquito-borne disease transmission in severely affected populations. *Wolbachia* has been also reported from some sand flies but their effect on vectorial competence is yet to be explored.

Paenibacillus glabratella, an anaerobic bacterial pathogen is used as a biocontrol agent for limiting the population of snails. This microbial pathogen affects both the embryonic and adult stages of freshwater snail, *Biomphalaria glabrata* and causes significant mortality. It appears to be a promising agent to eradicate the schistosomiasis burden. *Chaetogaster limnaei* an oligochaete lives commensally on freshwater snails and is used to confer protection to them from a range of trematodes like *Fasciola hepatica, Fasciolopsis buski*, and *Schistosoma* spp. This predator consumes the free-living miracidium and cercaria larva of trematodes thereby keeping a check on the transmission of dreaded infections such as schistosomiasis and fascioliasis to humans and livestock (8).

Spiders and mites are natural predators of *Phlebotomus argentipes* adults and larvae and thus are biological control agents. These predators are of immense importance as they control the sandfly's population and can avert leishmaniasis transmission. They are good alternative to chemical insecticides (9). The introduction of entomopathogenic nematodes to the breeding site of sandflies is a novel biocontrol approach. A study by El-Sadawy et al (10) revealed the potential of entomopathogenic nematodes, *Heterorhabditis bacteriophoralate* and their symbiotic bacteria against the late third instar larva of *Phlebotomus papatasi.* Similarly, *Steinernema abbasi* and *Steinerenema carpocapsae* has significant larvicidal activity against mosquitoes of *Culex, Aedes* and *Anopheles* spp. *Gambusia holbrooki,* the mosquito fish vigorously feeds on mosquito larvae and thwarts malaria transmission.

26.2.2.2 Surveillance

Social variables such as population density, international trade, tourism and travel, as well as environmental variables like changing climates, agricultural practices and deforestation, are associated with the emergence of novel infections. Timely local and global surveillance of diseases may assist in understanding the disease outcome, treatment, preventive and control measures. The human mobility trends during spread of COVID-19, H1N1 and influenza shed light on better comprehending their epidemic prediction.

Modern surveillance techniques widely rely on computational science and internet-based surveillance. The event-based surveillance is more advanced and swifter in collecting and analyzing raw information/algorithms from multiple sources and thus is at par with traditional surveillance methods. The most extensively used event–based systems are HealthMap, Argus, GPHIN and BioCaster (11). The GPHIN (Global Public Health Intelligence Network) is the one of the best surveillance systems that potently scans for the outbreak and detection of emerging health threats.

Nations possessing proficient early warning and response capabilities could identify hazards in their nascent stages, thereby enabling timely interventions to safeguard public health. Thus, disease transmission can be curbed, and financial and human casualties can be reduced. In 2000 WHO, launched the Global Outbreak Alert and Response Network (GOARN) that utilizes data from GPHIN and integrate it with numerous networks of laboratories and institutions already in place. GOARN was initiated at the WHO organizational level during the H1N1 influenza pandemic in

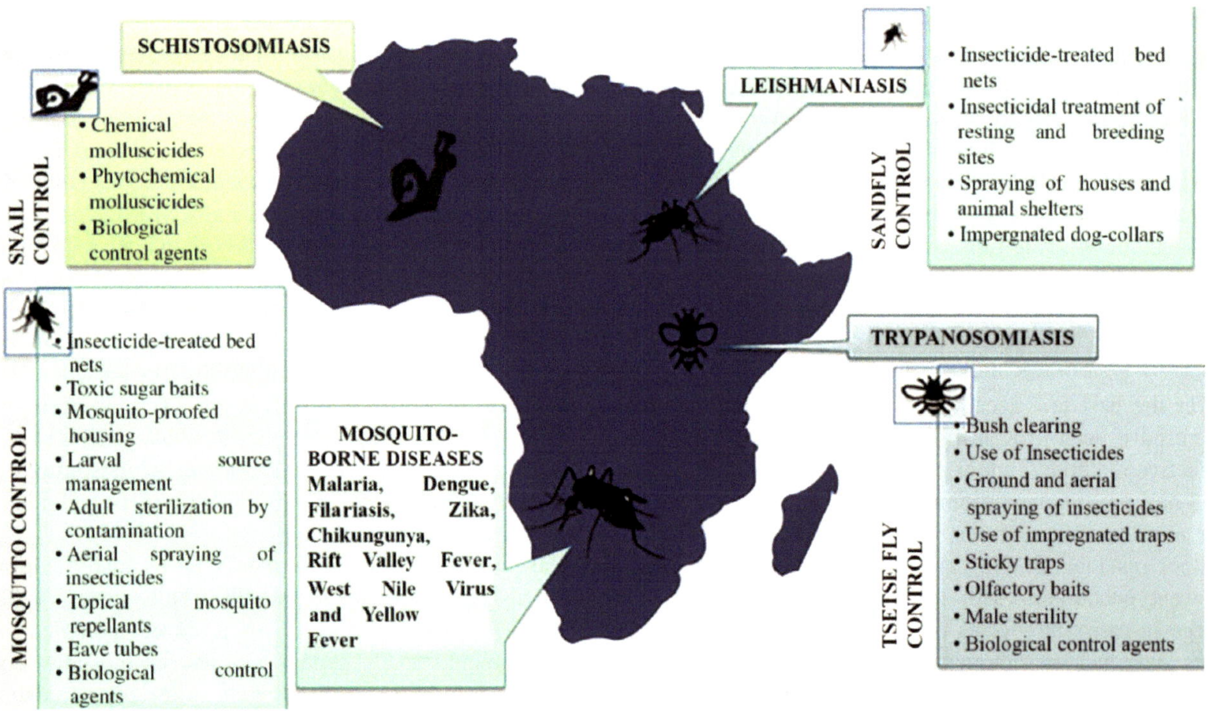

FIGURE 26.3 Vector control strategies employed against major VBDs on the severely affected African continent.

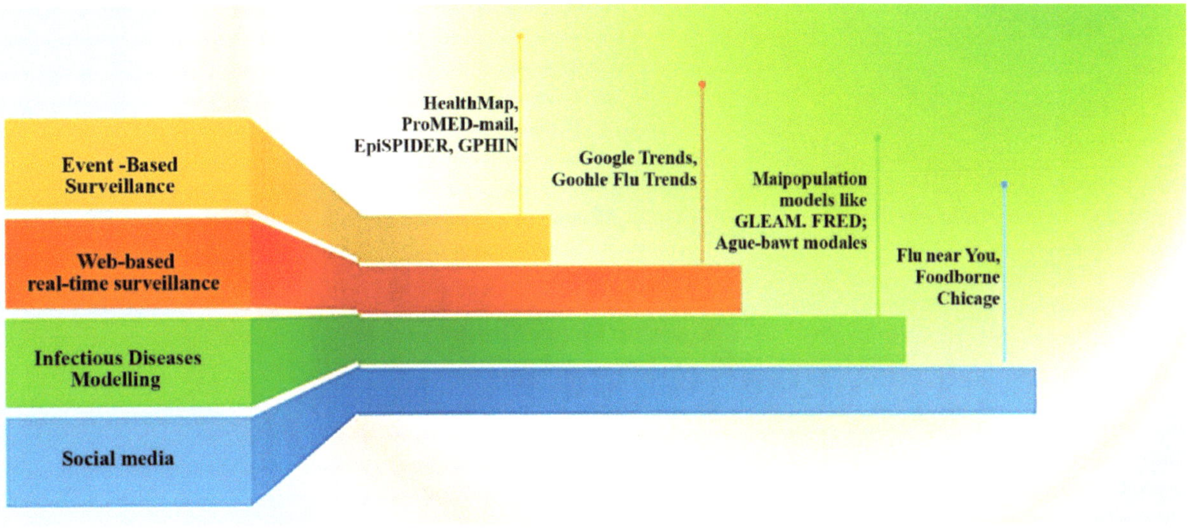

FIGURE 26.4 Potent resources used for managing emerging disease threats.

EpiSPIDER-Semantic Processing and Integration of Distributed Electronic Resources for Epidemics; FRED-Framework for Reconstructing Epidemic Dynamics; GLEAM: Global Epidemic and Mobility; GPHIN: Global Public Health Intelligence Network; ProMED: Program for Monitoring Emerging Diseases.

order to track disease activity, provide technical assistance and execute screening plans and containment tactics in an effort to bring the pandemic to an end. Similarly, Early Warning Alert and Response Networks was launched as it detects variations as well as alerts by combining indicator and event-based methods.

Advanced space-time cluster analysis and visualization are made possible by web-based system in almost real time. Search engine queries on the internet are a useful resource for learning about general health trends. Google Trends and Google Flu Trends are sophisticated web-based tools for near real-time detection of infectious diseases. Google Flu Trends was launched by Google, which helps in detecting the regional influenza outbreaks in the United States (Brilliant, 2009). It alerts healthcare practitioners timely before the outbreak allowing them to quickly implement preventive measures, identify cases and assure appropriate medical accessibility, eventually lowering morbidity and mortality (12). Rising global distress regarding bioterrorism, disease outbreaks, epidemics and pandemics have made real-time surveillance systems increasingly necessary.

Infectious disease modeling is done either by agent-based modeling or metapopulation models. To produce an extensive account of the epidemic, agent-based models necessitate the meticulous entry of data for every member of the target population. While to examine the dynamics of illnesses and forecast the expansion of epidemics, organized metapopulation models incorporate statistics from geographic area censuses as well as inter-population mobility trends, such as aviation and commuting network data. The global relevance of metapopulation models is more in comparison to the agent-based models. Global Epidemic and Mobility (GLEAM) is the most diverse and extensively used metapopulation model as it integrates worldwide statistical population data, short-range mobility and international travel data.

The high-throughput sequencing methods, whole genome sequencing, microarray techniques, and multiplex polymerase chain reaction facilitate early and prompt pathogen identification, syndromic surveillance and disease susceptibility (13). A wide range of viral illnesses are transmitted by mosquitoes including Chikungunya, dengue, Zika virus, Rift Valley virus, yellow fever, and many more. Metagenomics has incorporated lots of information in the field of mosquito-borne viral infections. The metagenomics sequencing has decoded the etiology of yellow fever outbreak in Nigeria in 2020. It provided more accurate risk assessment that resulted in real-time health actions, disease control and vector management (14). Metagenomics and meta-transcrptomics surveillance gives detailed unbiased unique matrices of parasite and viral genomes, although it requires high end equipments, thorough optimization and validation (15).

26.2.2.3 Vaccination

Vaccination has opened fascinating windows into our knowledge of the infectious diseases and its relevance in the context of emerging outbreaks. Vaccination can boost people's likelihood of survival, protect society from rising health hazards and improves socioeconomic productivity. Vaccination is a potent means of disease management that is pertinent to individuals of any age and in every nation, as evident during the COVID-19 pandemic. Recently a major breakthrough milestone is achieved by developing R21/Matrix-M and RTS, S/AS01 vaccine against the deadliest parasitic disease, malaria. It is the most needed innovation

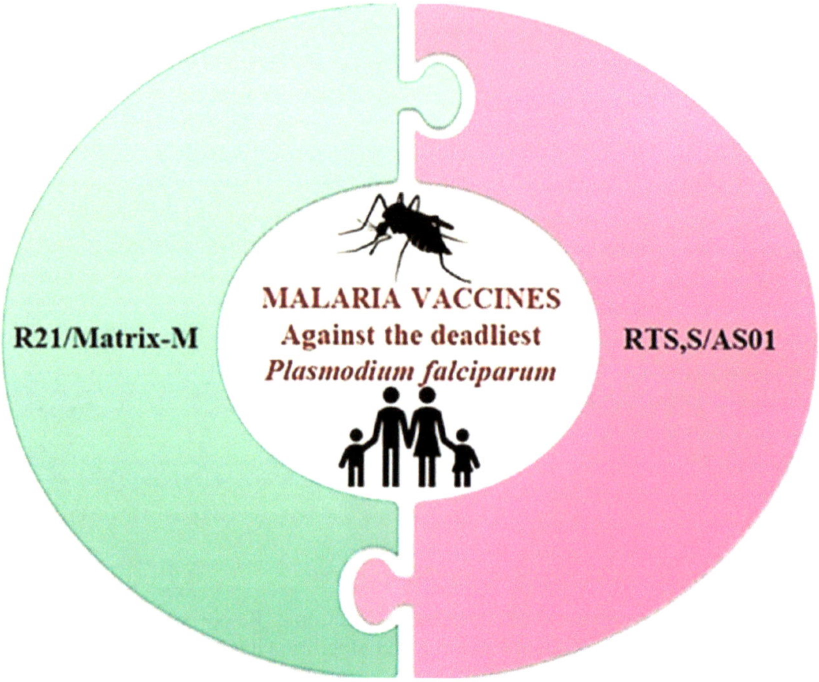

FIGURE 26.5 Current malaria vaccines against the deadly infection.

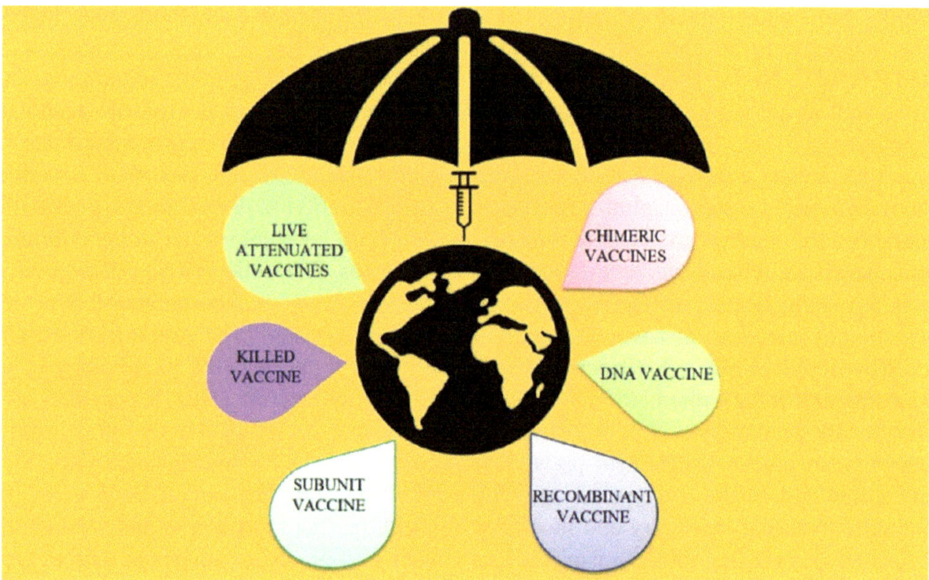

FIGURE 26.6 The umbrella of vaccination: Different vaccines protect the world from deadly VBDs.

as malaria is one of the largest killers of children under the age of five. The development of these two vaccines is a critical step toward protecting *P. falciparum* induced malaria in children living in endemic zones. Since 2019 under the Malaria Vaccine Implementation Programme more than 1.7 million children were administered the malaria vaccine in Ghana, Malawi, and Kenya. It exhibited safety with good efficacy, thereby resulting in decrease in children mortality and enhanced their chances of survival (16). Difficulty in generating and procuring adequate protective antigens is the biggest hurdle in vaccine development against major parasitic infections. So far, a good number of candidate antigens have been singled out (recognized) and are under the process of vaccine development and appraised for potency and applicability.

Developing a vaccine against parasitic infection is highly challenging owing to their complex life cycle rendering them arduous to study *in vitro*. In addition to this the complex multicellular parasites are engaged in complicated interactions with their hosts. They tend to stifle the host immune responses in an effort to sewn up their successful existence. The convention of having secondary, reservoir or partenic hosts also benefits the parasites and rollout to expand the infection (17).

It is not just humans who are affected by parasites. Globally, parasitism has severely impacted livestock health as well and countries' economic health in turn. Additionally, such infections are an important factor in the spread of zoonotic illnesses. To date, the only effective treatments have been antiparasitic and anthelmintic drugs, but rapidly increasing drug resistance is becoming an area of concern. Only a handful of vaccinations are available against helminth infections of livestock, specifically for the cestode *Echinococcus granulosus* in dogs, the nematode *Haemonchus contortus* in sheep, and the nematode *Dictyocaulus viviparus* in cattle (18).

Schistosomiasis is another VBDs caused by trematode *Schistosoma japonicum, S. mansoni* and *S. haematobium* causing a significant morbidity in both humans and livestock. Presently, the only treatment is praziquantel chemotherapy; many vaccine candidates have been identified, but they have shown efficacy and protection in murine models only. Only four vaccine antigens have successfully entered human clinical trial: *S. haematobium* 28 kDa recombinant glutathione S-transferase, *S. mansoni* tetrasapanin, fatty acid binding protein and *S.mansoni* calpain (19).

Scientific communities and researchers ae putting rigorous efforts to control and eliminate leishmaniasis, a neglected tropical disease endemic in more than 100 countries globally. Leishmaniasis exhibits a range of clinical manifestation making it different from other parasitic infections. Cutaneous leishmaniasis caused by *Leishmania braziliensis* and *L. mexicana* complex; mucocutaneous leishmaniasis caused by *L. braziliensis, L. amazonensis* and *L. panamensis*; and visceral leishmaniasis caused by *L. infantum* and *L. donovani*, the most devastating and deadly form. Furthermore, the sequel of visceral leishmaniasis is post kala-azar dermal leishmaniasis characterized by hypopigmented patches, nodular lesions and erythematous papulo-plaques.

According to WHO reports seven hundred thousand to one million new instances are estimated to arise each year. Till now no human vaccine is available. Even though a number of recombinant DNA and protein vaccines, either with or without an adjuvant, demonstrated encouraging outcomes in animal models, their clinical trial results are unsatisfactory. Adjuvants improve the efficacy of vaccines and there is a dire need of finding suitable adjuvants. AddaVax, Gardiquimod and MISA 201 are adjuvants that showed efficacy against *L. donovani* infection in mice model (20, 21). Presently LmCen$^{-/-}$, a live attenuated centrin gene-deleted *Leishmania major* (Gennova Biopharma, India); and mRNA (LEISH F2/F3) a self-amplifying mRNA vaccine (HDT Bio, USA) are in late preclinical stage of development. ChAd63-KH, a third-generation simian adenovirus-vectored vaccine (University of York, Avaxia) is in Phase II of development in Sudan (22).

26.2.2.4 AI

Today is the era of AI and machine learning and we are heading towards building a digital ecosystem. With the rapidly mounting global burden of infectious and vector borne diseases it is the need of the hour to amalgamate the conceptual understanding of the major components of AI like machine learning, neural networks, deep learning and the algorithms to develop a disease free community or ecosystem. AI is the fastest growing field at present. The biggest challenge being faced today by the developing nations and

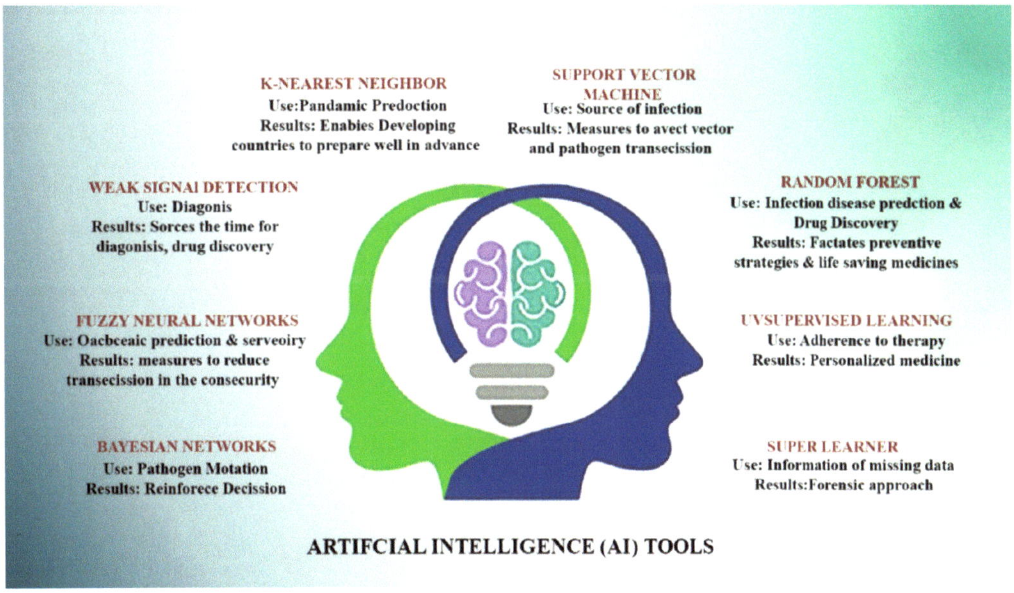

FIGURE 26.7 AI and machine learning tools used in treating vector-borne and infectious diseases.

WHO is monitoring and managing the outbreaks and rampant attack of deadly pathogens such as COVID-19, SARS-Cov-2, H1N1 influenza, Zika virus, and many more.

AI has the potential to enhance the efficiency and precision of disease diagnosis and evaluation, aid in healthcare delivery, fortify medical research and drug discovery and facilitate a range of public health initiatives. AI models are of great relevance in studying imaging techniques, mass spectrometry as well as genetic expression-based diagnostics. Not only this AI is also being utilized in developing vaccines for these troublesome infections, explicating infection biology and drug development. AI has tremendous usage to scientists, doctors and healthcare practitioners to better understand vector-borne and infectious diseases. Like the installment of thermal cameras at international airport terminals to monitor the temperature of individuals travelling is one of the attribute to stop the transmission of infection to other areas. This system proved to be highly beneficial during the COVID-19 pandemic. Other AI tools are Fuzzy cluster, Artificial Immune Recognition System (AIRS), Support Vector Machine (SVM), k-nearest neighbor (k-NN), Artificial neural network (ANN), Decision tree (DT), Autoregressive integrated moving average (ARIMA) model, Seasonal ARIMA (SARIMA), weak signal detection and Bayesian Information criterion (BIC) (23).

These tools have revolutionized the healthcare industry as they are well enabled in epidemic prediction in various cases like Kyasanur Forest Disease, a tick-borne viral infection endemic to South Asia (24), prediction of tuberculosis incidence with changing season evaluated in South Africa, Rift Valley fever (a vector-borne viral zoonotic infection) outbreak in Arabic Peninsula and Africa (25). Recently Zare et al. (26) proposed the Viola–Jones algorithm to diagnose cutaneous leishmaniasis infection resulting in 83% sensitivity and moderate specificity. This *Leishmania* parasite detection method emerged to be a quick, easy to operate and economical. Geographic information system (GIS) based AI approach has been used by Shabanpur et al. (27), using decision tree, linear regression and support vector regression to spatially analyze the prediction of cutaneous leishmaniasis in Iran.

26.2.2.5 Nanobased Approaches

The world is undergoing the threats of emerging and reemerging diseases and outbreaks, and the conventional drugs to treat these infections are dissatisfying owing to their low potency, reduced efficacy, extended therapeutic regimens, short life spans, adverse side effects, rising drug resistance and failures to reach infection sites.

The most demanded anthelmintic drug, praziquantel, has low solubility in aqueous solution as well as low bioavailability; therefore, it rapidly eliminates from the host's body, which necessitates repeated high doses for long durations (28). The best way to overcome these challenges is to develop novel drug carriers for preexisting chemotherapeutics because finding novel drugs is a long process. Nanomedicines have successfully treated many lifestyle diseases like cancer, arthritis, atherosclerosis and diabetes as well as neurodegenerative diseases such as Alzheimers's disease. Nanotechnology has enabled entrapping active drug molecules in specialized nanoscale drug-delivery systems via varied routes including intravenous injections, inhalation and oral and topical applications.

Range of polymers such as PLGA, PCL, poly(anhydrides), poly(cyanoacrylates), poly(amides) and poly(orthoesters) can be used for sustained release of chemotherapeutics. Drugs can also be encapsulated in liposomes, which are lipid vesicles that can hold both hydrophilic and hydrophobic drugs, to achieve sustained release. Nanocarriers have the potential to improve medication efficacy against

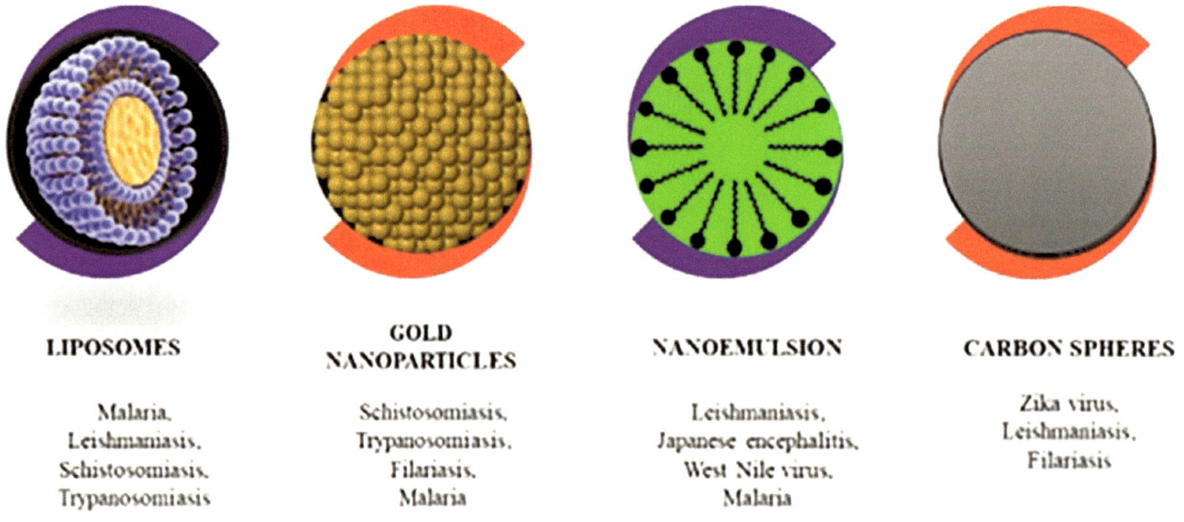

FIGURE 26.8 Nanoparticles in use against some vector-borne diseases.

intracellular pathogens and enhance drug absorption in cells. The encapsulation of unstable drugs such as nucleic acids, in nanocarriers may offer protection against strong physiological triggers like enzymes and low pH concentration. A couple of oral nanoparticles, aerosolized and nebulized nanoparticles, have been developed for treating tuberculosis. Similarly, injectable, immune-liposomes and micro-emulsion against malaria infection are being developed and are under preclinical trials in rodents (29). Liposomal amphotericin B is a great success in treating leishmaniasis in comparison to amphotericin B. Using the biotechnological advancements, the nano-immunoengineering is gaining the attention of scientists and research organizations.

The immune system is an inexorable component of human body that maintains homeostasis throughout (30). To improve therapeutics via biotechnological expertise and nanotechnology, the immune cells can be engineered to understand the intricate processes and correlation between immune cells, drug release, cellular uptake and signal transduction. Nanomedicines are promising means of combating hard to treat viral, bacterial and/or parasitic infections. In the past few decades, a number of inorganic, organic, and hybrid nanoparticle drugs have been developed having antimicrobial and bactericidal efficacy. Not only treatment, nanomaterials are even efficient in diagnosing infections. The nanotechnology associated novel detection methods have promising role in scrutinizing viral infections. Chiral gold nanoparticles-quantum dot, carbon electrodes modified with gold nanoparticles, and Chiral zirconium Quantum dots are some engineered nanomaterials developed for detecting coronaviruses (31).

26.2.2.6 Next-Generation Sequencing

NGS is an incredibly fast and affordable technique. Next-generation sequencing played a vital role in identifying evolution and secondary host of SARS-Cov2 virus. The findings put an end to the speculation of people that it is human made. The NGS data makes it possible to promptly track down the possible source of novel SARS-CoV2, which is essential for preventing the spread of the virus across species. The wide use of NGS was reported to be critical to comprehend the pathophysiologic pathways driving the COVID-19 symptoms and clinical relevance. By employing NGS technology the unknown virus genome sequences can be retrieved for the first time. Vector control efforts have been significantly impacted by the adoption of progressively powerful computer modeling methodologies. The data retrieved by sophisticated gene editing techniques, PCR and NGS has opened the new windows of comprehensive systematic and epidemiologic knowledge that usher the precise and meticulous evaluation of disease risk. It also aids in the remodeling of vector genome sequences to minimize their population and competency. Whole genome sequence and next generation sequencing initiatives have yielded complete genome sequences for the majority of the arbo-vector species. It also benefited us to understand the co-evolutionary links between the vector immunity, heterogeneity and reproduction strategies. Transcriptomic and NGS reported that alteration in the microbiome framework is attributed to the insecticide resistance phenotype and that amalgamated the frequency in resistant mosquitoes. Insecticides resistance polymorphism in *Aedes* mosquitoes was successfully identified by multi-locus target amplicon sequencing. Targeted amplicon based NGS deciphered the mixed trypanosomatid infection in a population. These data enable comprehensive evaluations within and between populations as well as species. Targeted NGS is culture free method with higher sensitivity and less expense than shotgun/deep sequencing, thus making it suitable for diagnosis of vector borne pathogens in humans as well as in live-stock. NGS technology also enabled a rapid detection of microbiome in hematophagous arthropods, ticks. After mosquitoes the ticks are vital vectors of human and zoonotic infections. Understanding of tick microbiome may assist in boosting preventive measures and control strategies to curb tick-borne infections.

26.2.2.7 CRISPR/Cas9

CRISPR/Cas9 is a novel and versatile gene editing technology that targets specific stretches of DNA and acts as molecular scissors that edit genomes at specific locations, and it is bringing new hope. CRISPR/Cas9 is simple and inexpensive, and it can be utilized to perpetually tweak the genome of nearly every organism. Decoding the parasite genome sequences using CRISPR/Cas9 has triggered the scientists and researchers to explore the genetics of host-parasite relationship. To date, scientists have used CRISPR to edit the genomes of trematodes *Schistosoma mansoni* and *Opisthorchis viverrine* (32, 33). By deleting, inserting and modifying specific nucleotides of the genome of an organism, the treatment and diagnosis of major human pathogens can be achieved. CRISPR/Cas is the most promising candidate for molecular diagnosis of parasitic and vector-borne diseases (34). Recently developed CRISPR-d/Cas9 and collateral cleavage-based diagnostics are more robust diagnostic approaches that have opened new horizons for infectious disease diagnosis. CRISPR/Cas12 and Cas13 exhibit collateral activity. They are designed for triggering nonspecific collateral cleavage of genes. Cas13 recognizes and enables the non-targeted cleavage of ssRNA, while Cas12 senses and cleaves the ssDNA (35). Investigators have already used CRISPR/Cas9 DNA detection for the quick and accurate detection of SARS-CoV-2 and diagnosis of protozoan parasites *Leishmania, Trypanosoma, Plasmodium* spp., and *Toxoplasma* (36–39). SHERLOCK, Specific High-sensitivity Enzymatic Reporter unLOCKing, is a CRISPR-based diagnostic technique that entails employing a CRISPR enzyme for collateral recognition with any sort of DNA or RNA pre-amplification. SHERLOCK assays are being used for malaria, leishmaniasis, Zika virus, dengue virus, COVID-19, and human African trypanosomiasis diagnostics (39–41).

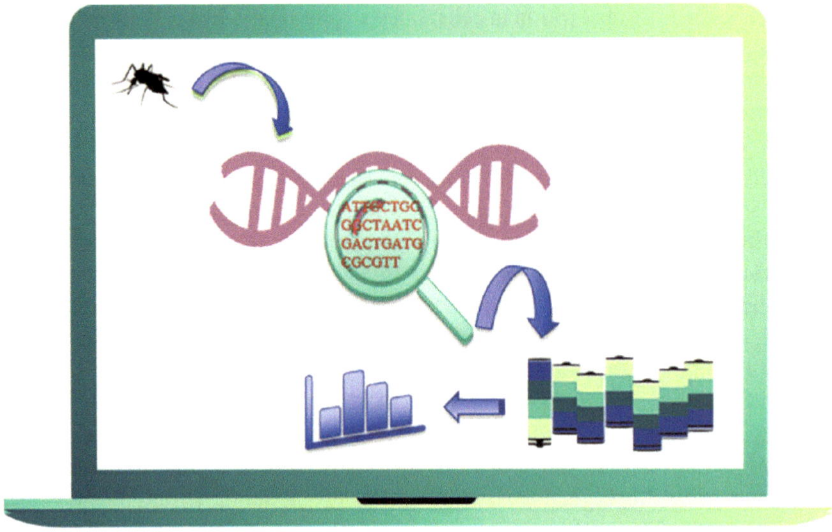

FIGURE 26.9 Next-generation sequencing technology.

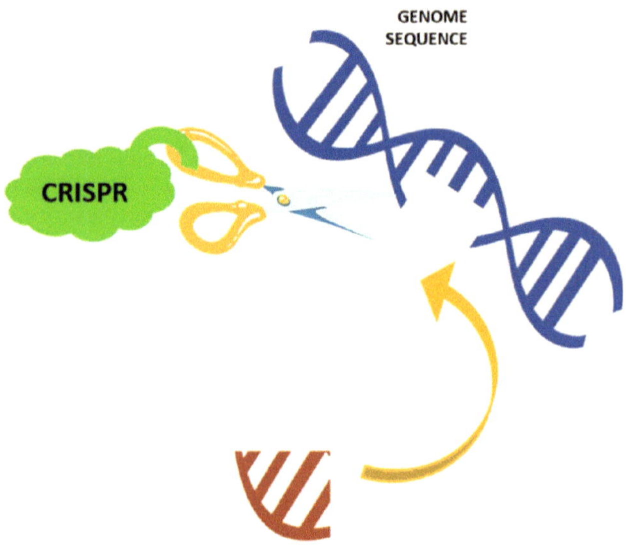

FIGURE 26.10 CRISPR/Cas9 act as molecular scissors and enables gene editing.

26.2.3 Challenges to Leveraging Biotechnology

Emerging biotechnologies and their wide integrative perspective have effectively pitched to control and prevent many infectious diseases. These were of great help during COVID-19 pandemic. Because of these technological advancements the fastest COVID-19 vaccine was developed and deployed to masses worldwide. Identification of viral genome sequences paved the way for designing and developing COVID-19 vaccine. The expanding scope of the biotechnology has created limitless opportunities to further strengthen the healthcare system. But pertinent obstacles in execution as well as associated vulnerabilities are to be corrected for further advancement. The biggest challenge in fighting infectious and vector-borne diseases is inadequacy of a collaborative innovation and research, minimal or little participation of financing bodies and a disparity among industry and academia. Collaborative R&D is paramount to develop rapid and accessible medical technological innovations. NGS has the potential to revolutionize medical diagnoses and therapies, but it also poses complex ethical issues that need to be tackled. A massive data is generated by NGS, so it is critical to identify proper methods for data protection, management and communication. Nano delivery systems and nanomedicines are rapidly developing technology. It also faces several pharmacological, immunological and safety challenges before it is approved for marketing. There are a number of vector control strategies lined up but their proper implementation at the ground level is the key to get rid of vectors and

VBDs. Grossly, a cross-disciplinary collaboration between educationists, entomologists, parasitologists, researchers, and social workers is critical to understand and determine the epidemiological impact of emerging VBDs and disseminate the vector control measures among masses living in endemic or disease prone areas. Biotechnology has the ability to boost health and eliminate VBDs related disparities on an international scale. However sustainable and persistent efforts are pivotal to achieve the vector and disease free ecosystem.

26.3 CONCLUSION

The world is facing unprecedented challenges because of rapidly increasing population, urbanization, deforestation, changing climate at an alarming rate and mushrooming burden of infectious and vector borne diseases. VBDs are major public health problem in tropical and subtropical regions of the globe. Prompt diagnosis, mass screening, treatment of infected, and vector control are the major strategies to curb the menace transmitted by numerous tiny biters. Incorporation of innovative biotechnologies, can reduce the time to reach these goals. Certainly, with the development and advancement in the biotechnological tools and novel gene/genome editing platforms laid the foundations for better diagnostics and therapeutic modalities. The whole genome sequencing and next generation sequencing techniques enabled the alteration of infectious pathogens and aids in vaccine development. The knowledge of computer based modeling and advanced surveillance tools potently scans for the outbreak and detection of emerging health threats. These biotechnological interventions potentially improve the global public health by restraining the vector growth, local and global transmission, timely diagnosis of infection and development of vaccines.

REFERENCES

1. Ellwanger JH, da Veiga ABG, Kaminski V de L, Valverde-Villegas JM, de Freitas AWQ, Chies JAB. Control and prevention of infectious diseases from a one health perspective. Genet Mol Biol. 2021;44(1):1–23.
2. Aileni M, Rohela GK, Jogam P, Soujanya S, Zhang B. Biotechnological perspectives to combat the COVID-19 pandemic: Precise diagnostics and inevitable vaccine paradigms. Cells. 2022;11:1182.
3. Okoro OJ, Deme GG, Okoye CO, Eze SC, Odii EC, Gbadegesin JT, et al. Understanding key vectors and vector-borne diseases associated with freshwater ecosystem across Africa: Implications for public health. Sci Total Environ. 2023 Mar 1;862:160732.
4. Dambach P, Bärnighausen T, Yadouleton A, Dambach M, Traoré I, Korir P, et al. Is biological larviciding against malaria a starting point for integrated multi-disease control? Observations from a cluster randomized trial in rural Burkina Faso. PLoS One. 2021 Jun 1;16(6):e0253597.
5. Baranitharan M, Sawicka B, Gokulakrishnan J. Phytochemical profiling and larval control of *Erythrina variegata* methanol fraction against malarial and filarial vector. Adv Prev Med. 2019 Apr 16;2019:2641959.
6. Minwuyelet A, Petronio GP, Yewhalaw D, Sciarretta A, Magnifico I, Nicolosi D, et al. Symbiotic Wolbachia in mosquitoes and its role in reducing the transmission of mosquito-borne diseases: Updates and prospects. Front Microbiol. 2023;14:1267832.
7. Guruprasad NM, Jalali SK, Puttaraju HP. Wolbachia-a foe for mosquitoes. Asian Pacific J Trop Dis. 2014 Feb;4(1):78–81.
8. Hobart BK, Moss WE, McDevitt-Galles T, Stewart Merrill TE, Johnson PTJ. It's a worm-eat-worm world: Consumption of parasite free-living stages protects hosts and benefits predators. J Anim Ecol. 2022 Jan 1;91(1):35–45.
9. Dinesh DS, Kumar V, Kesari S, Das P. Mites and spiders act as biological control agent to sand flies. Asian Pacific J Trop Dis. 2014;4(S1).
10. El-Sadawy HA, Ramadan MY, Abdel Megeed KN, Ali HH, El Sattar SA, Elakabawy LM. Biological control of *Phlebotomus papatasi* larvae by using entomopathogenic nematodes and its symbiotic bacterial toxins. Trop Biomed. 2020;37(2):288–302.
11. Christaki E. New technologies in predicting, preventing and controlling emerging infectious diseases. Virulence. 2015;6(6):558–65.
12. Carneiro HA, Mylonakis E. Google trends: A web-based tool for real-time surveillance of disease outbreaks. Clin Infect Dis. 2009 Nov;49(10):1557–64.
13. Lipkin WI. The changing face of pathogen discovery and surveillance. Nat Rev Microbiol. 2013;11:133–41.
14. Ajogbasile FV, Oguzie JU, Oluniyi PE, Eromon PE, Uwanibe JN, Mehta SB, et al. Real-time metagenomic analysis of undiagnosed fever cases unveils a yellow fever outbreak in Edo State, Nigeria. Sci Rep. 2020 Feb 21;10(1):3180.
15. Franssen FFJ, Janse I, Janssen D, Caccio SM, Vatta P, van der Giessen JWB, et al. Mining public metagenomes for environmental surveillance of parasites: A proof of principle. Front Microbiol. 2021 Jun 30;12.
16. WHO. 18 million doses of first-ever malaria vaccine allocated to 12 African countries for 2023–2025: Gavi, WHO and UNICEF [Internet]. World Health Organisational and UNICEF; 2023. Available from: https://www.unicef.org/press-releases/18-million-doses-first-ever-malaria-vaccine-allocated-12-african-countries-20232025
17. Mutapi F, Billingsley PF, Secor WE. Infection and treatment immunizations for successful parasite vaccines. Trends Parasitol. 2013;29:135–41.
18. Claerebout E, Geldhof P. Helminth vaccines in ruminants: From development to application. Vet Clin North Am Food Anim Pract. 2020;36:159–71.
19. Al-Naseri A, Al-Absi S, El Ridi R, Mahana N. A comprehensive and critical overview of schistosomiasis vaccine candidates. J Parasit Dis. 2021;45:557–80.
20. Goyal DK, Keshav P, Kaur S. Adjuvant effects of TLR agonist gardiquimod admixed with *Leishmania* vaccine in mice model of visceral leishmaniasis. Infect Genet Evol [Internet]. 2021;93:104947. Available from: https://doi.org/10.1016/j.meegid.2021.104947
21. Goyal DK, Keshav P, Kaur S. Adjuvanted vaccines driven protection against visceral infection in BALB/c mice by *Leishmania donovani*. Microb Pathog. 2021;151:104733.
22. Kaye PM, Matlashewski G, Mohan S, Le Rutte E, Mondal D, Khamesipour A, et al. Vaccine value profile for leishmaniasis. Vaccine. 2023 Nov 3;41(2). Available from: https://doi.org/10.1016/j.vaccine.2023.01.057
23. Agrebi S, Larbi A. Use of artificial intelligence in infectious diseases. In: Artificial Intelligence in Precision Health: From Concept to Applications. Elsevier; 2020, pp. 415–38.

24. Majumdar A, Debnath T, Sood SK, Baishnab KL. Kyasanur forest disease classification framework using novel extremal optimization tuned neural network in fog computing environment. J Med Syst. 2018 Oct 1;42(10).
25. Walsh MG, Willem de Smalen A, Mor SM. Wetlands, wild Bovidae species richness and sheep density delineate risk of Rift Valley fever outbreaks in the African continent and Arabian Peninsula. PLoS Negl Trop Dis. 2017 Jul 25;11(7):e0005756.
26. Zare M, Akbarialiabad H, Parsaei H, Asgari Q, Alinejad A, Bahreini MS, et al. A machine learning-based system for detecting leishmaniasis in microscopic images. BMC Infect Dis. 2022 Dec 1;22(1):48.
27. Shabanpour N, Razavi-Termeh SV, Sadeghi-Niaraki A, Choi SM, Abuhmed T. Integration of machine learning algorithms and GIS-based approaches to cutaneous leishmaniasis prevalence risk mapping. Int J Appl Earth Obs Geoinf. 2022 Aug 1;112.
28. Sun Y, Chen D, Pan Y, Qu W, Hao H, Wang X, et al. Nanoparticles for antiparasitic drug delivery. Drug Deliv. 2019;26:1206–21.
29. Kirtane AR, Verma M, Karandikar P, Furin J, Langer R, Traverso G. Nanotechnology approaches for global infectious diseases. Nat Nanotechnol. 2021;16:369–84.
30. Chuang ST, Conklin B, Stein JB, Pan G, Lee KB. Nanotechnology-enabled immunoengineering approaches to advance therapeutic applications. Nano Converg. 2022;9.
31. Nasrollahzadeh M, Sajjadi M, Soufi GJ, Iravani S, Varma RS. Nanomaterials and nanotechnology-associated innovations against viral infections with a focus on coronaviruses. Nanomaterials. 2020;10:1072.
32. Ittiprasert W, Mann VH, Karinshak SE, Coghlan A, Rinaldi G, Sankaranarayanan G, et al. Programmed genome editing of the omega-1 ribonuclease of the blood fluke, *Schistosoma mansoni*. Elife [Internet]. 2019;8:e41337. Available from: https://doi.org/10.7554/eLife.41337
33. Arunsan P, Ittiprasert W, Smout MJ, Cochran CJ, Mann VH, Chaiyadet S, et al. Programmed knockout mutation of liver fluke granulin attenuates virulence of infection-induced hepatobiliary morbidity. Elife [Internet]. 2019;8:e41463. Available from: https://doi.org/10.7554/eLife.41463.001
34. You H, Gordon CA, MacGregor SR, Cai P, McManus DP. Potential of the CRISPR-Cas system for improved parasite diagnosis: CRISPR-Cas mediated diagnosis in parasitic infections. BioEssays. 2022 Apr 1;44(4).
35. Kaminski MM, Abudayyeh OO, Gootenberg JS, Zhang F, Collins JJ. CRISPR-based diagnostics. Nat Biomed Eng. 2021;5:643–56.
36. Nguyen LT, Rananaware SR, Pizzano BLM, Stone BT, Jain PK. Clinical validation of engineered CRISPR/Cas12a for rapid SARS-CoV-2 detection. Commun Med. 2022 Jan 12;2(7):1–11.
37. Dueñas E, Nakamoto JA, Cabrera-Sosa L, Huaihua P, Cruz M, Arévalo J, et al. Novel CRISPR-based detection of *Leishmania* species. Front Microbiol. 2022 Sep 15;13.
38. Ma QN, Wang M, Zheng LB, Lin ZQ, Ehsan M, Xiao XX, et al. Raa-cas12a-tg: A nucleic acid detection system for toxoplasma gondii based on crispr-cas12a combined with recombinase-aided amplification (raa). Microorganisms. 2021 Aug 1;9(8):1644.
39. Cunningham CH, Hennelly CM, Lin JT, Ubalee R, Boyce RM, Mulogo EM, et al. A novel CRISPR-based malaria diagnostic capable of *Plasmodium* detection, species differentiation, and drug-resistance genotyping. EBioMedicine. 2021 Jun 1;68.
40. Gootenberg JS, Abudayyeh OO, Lee JW, Essletzbichler P, Dy AJ, Joung J, et al. Nucleic acid detection with CRISPR-Cas13a/C2c2. Science (80-). 2017 Apr 28;356(6336):438–42.
41. Sima N, Dujeancourt-Henry A, Perlaza BL, Ungeheuer MN, Rotureau B, Glover L. SHERLOCK4HAT: A CRISPR-based tool kit for diagnosis of Human African Trypanosomiasis. eBioMedicine. 2022 Nov 1;85:104308.

Section VII

Biotechnology and One Health Approach

27 Precision Medicine

Therapy of the Future

Samridhi Kaul,[1] *Kirti* Hooda,[2] *and Mary* Chatterjee[1]*

[1] Biotechnology Branch, University Institute of Engineering and Technology, Panjab University, Chandigarh, India

[2] Center for Nanoscience and Nanotechnology, Panjab University, Chandigarh, India

ABBREVIATIONS

AI	Artificial intelligence
AmCAD	AmCad BioMed Corporation
BRCA	Breast cancer gene
CCP	Cyclic citrullinated peptide
CD	Clusters of differentiation
CFTR	Cystic Fibrosis Transmembrane Conductance Regulator
CML	Chronic myeloid leukemia
CTLA	4-Cytotoxic T-lymphocyte–associated antigen 4
DL	Deep learning
DNA	Deoxyribonucleic acid
EGFR	Epidermal growth factor receptor
FDA	Food and Drug Administration
GWAS	Genome wide association studies
HbA1c	Hemoglobin A1c
HCV	Hepatitis C virus
HDL	High density lipoprotein
HER2	Human epidermal growth factor receptor 2
HR	Homologous recombination
hs-CRP	High-sensitivity C-reactive protein
KRAS	Ki-ras2 Kirsten rat sarcoma viral oncogene homolog
LDL	Low-density lipoprotein
ML	Machine learning
NF1	Neurofibromatosis type 1
PARP	Poly (ADP-ribose) polymerase
pCR	Pathological complete response
RAS	Rat sarcoma virus
RNA	Ribonucleic acid

27.1 INTRODUCTION

Precision medicine refers to therapeutic options such as targeted medication and cell therapy that compare the gene sequences and lifestyles of healthy individuals and patients after analyzing computerized big data (Tan et al., 2023). It's also enumerated as a preclinical, translational, or applied drug conception that considers defined particles, cells, or relations between the networks of particles and cells with the thing of perfecting diagnostics, prognostics, and remedy in all kinds of conditions (Valent et al., 2021). Each patient exhibits unique symptoms, reacts differently to pharmacological effects, and has varying levels of physiological resistance. Adopting a treatment strategy tailored to individual patients can help patients live longer lives while reducing their financial burden (Yu et al., 2021). The planned design of a perfect drug has two crucial factors a near-term emphasis on malice and a longer-term thing of generating knowledge applicable to the entire range of health and illness. Because of developments in introductory wisdom, similar to molecular biology, genetics, and bioinformatics, both factors are now within our grasp (Collins and Varmus, 2015). Still, Wang et al., unveiled the use and development of precision medicine in Chen's paper "The Prospect of Precision Medicine Grounded videlicet the assessment of optimal or near-optimal treatment rules and sweats that directly aid in this (Kosorok and Laber, 2019). It also strives to involve cases as active actors in exploration and remedy, giving birth to the terms" preventative, substantiated, perfection, population, and participatory drug (N. Naithani et al., 2021). Talking about the history of perfection drug, following the successful completion of the Human Genome Project at the turn of the century, proponents of perfection drug principally assimilated pharmacogenetics. Posterior NIH on Modified Antibody or Immunocell Therapy'. It's an immunotherapy that improves targeting and validity by combining standard immunotherapy with gene engineering technologies to change the shape and function of antibodies and vulnerable cells (Wang et al., 2016). It is also concerned with identifying the appropriate dynamic treatment regime, calculating the contrast of an optimal regime with subgroup identification, and assessing the causal influence. It is also curious in how the dynamic treatment regime framework might be generalized, as well as how precision medicine choice concerns tie into precision public health and data-driven decision science. Precision Drug's primary thing is decision support, enterprise under Francis S. Collins aimed to subsidize this new knowledge to ameliorate inheritable drug far beyond the study of well-known mutations and chromosomal rarities (Phillips, 2020). Developments in the field of genomics image early raids into the development of perfection drug from discovery of the double Helix model

* Corresponding author: maryc@pu.ac.in

DOI: 10.1201/9781032660592-34

of DNA in 1953 to the development of Sanger sequencing in 1977 and eventually to the launching of the Human Genome Project in 1990. The Human Genome design took a long 13 years of time for completion, but it was grounded on the premise that all diseases follow Mendelian characteristics (N. Naithani et al., 2021). Exemplifications similar as the use of the monoclonal antibody trastuzumab to target a single gene mutation in the mortal epidermal growth factor receptor 2 (HER2) gene in bone cancer cells give stopgap for this strategy. More optimism for a tailored future has been provided by the Human Genome Project's eventual completion, the molecular subtyping of melanoma based on BRAF, RAS, and NF1 mutations, and numerous other discoveries (Seymour et al., 2017). Integrating precision medicine into healthcare offers the ability to produce more exact diagnoses, anticipate disease risk before symptoms appear, and build individualized treatment strategies that maximize safety and efficiency (Johnson et al., 2021).

27.2 MOLECULAR BASIS OF PRECISION MEDICINE

The molecular basis of precision medicine lies in understanding the genetic, environmental, and behavioral factors that contribute to each person's distinct disease risks and treatment responses (N. Naithani et al., 2021). Precision medicine aims to customize interventions and preventive measures according to these variables, as opposed to employing a universal strategy. Genomic information, customized genomics, environmental and lifestyle factors, pharmacogenomics are some of important components of precision medicine in understanding its molecular basis.

The data gathered from the analysis of an organism's whole DNA, including all of its genes, is referred to as genomic information which facilitates a more thorough comprehension of an individual's genetic composition and its potential effects on health (Brittain et al., 2017). Precision medicine relies heavily on genomic information, which uses a person's genetic profile to drive decisions about illness prevention, diagnosis, and treatment. It uses genomic data in many ways, including population-based techniques, pharmacogenomics, risk stratification, disease diagnosis, treatment selection, and prevention (Strianese et al., 2020). DNA extraction, sequencing, data analysis, interpretation, data sharing, and privacy are the methods used to gather genomic information.

Inheritable variations can impact how specifics are digested, absorbed, and excreted from the body, which can lead to individual difference in response to curatives (Belle and Singh, 2008). Medicine efficacy and toxicity can vary depending on how a medicine interacts with its target, such as an enzyme or receptor, due to genetic variations (Lauschke et al., 2019). Differences in medicine response can also affect from inheritable variations impacting how a medicine interacts with diet, other specifics, or environmental circumstances (Seymour et al., 2017). So, healthcare interpreters can more identify a case's threat for a specific complaint and produce targeted interventions by considering the case's inheritable profile.

Precision medicine could profit from genomic information since it offers a more thorough understanding of a person's inheritable composition and how it affects their health (Roden and Tyndale, 2013). Through better understanding of the underpinning inheritable determinants and remedial guidance, the operation of genomic information in precision medicine can help in the timely and precise discovery of uncommon conditions and malice. A person's threat for a specific complaint can also be determined using genomic information, enabling early discovery and treatment (Khoury and Holt, 2021). The field of substantiated genomics involves the analysis and interpretation of unique inheritable data attained from genome sequencing in order to pinpoint inheritable variants. It makes use of a person's inheritable profile to inform choices about illness prohibition, opinion, and treatment (Brittain et al., 2017). This results in safer tradition medicine recommendations, more accurate judgments, and more successful curatives for a variety of conditions and diseases. This system concentrates on determining the stylish course of action grounded on a person's inheritable profile, enabling customized illness prevention, appreciation of complaint vulnerability, and tradition of suitable specifics or curatives (Strianese et al., 2020). Using substantiated genomics in perfection medicine, multitudinous conditions and affections can be treated more successfully and safely, with judgments and curatives more precisely provisioned to the existent's inheritable composition. It assists medical professionals in relating distinct complaint pitfalls and the most effective curatives for each case, performing in substantiated healthcare. Also, as part of ordinary medical care, precision medicine allows croakers to use cases inheritable and other molecular information, which enhances their capacity to anticipate which drugs will be most effective for particular cases (Goetz and Schork, 2018). Precision drug using customized genomics provides a more accurate, harmonious, and effective system of treating cases by considering their individual inheritable composition as well as other variables that impact their health (Strianese et al., 2020). A newer system of perfection drug called pharmacogenomics focuses on opting and dosing specifics grounded on a case's inheritable characteristics. It might enhance drug safety and efficacity (Cecchin and Stocco, 2020). When pharmacogenomics is applied in clinical practice, losses from medicine failures that affect in significant expenditure can be averted. By precluding negative medicine relations, it can also help aged cases entering multiple conventions (Weinshilboum and Wang, 2017). The right drug and cure can be chosen for the case using the results of pharmacogenomics testing, which can affect in safer, more effective, and more affordable curatives. Customizing medical care for each case is the ultimate thing of individualized drug, and one of the newest uses of genetics in perfection drug is pharmacogenomics. Since substantiated drug is still in its immaturity, there are still numerous obstacles to be addressed. These include the need for fresh exploration on the molecular causes of complaint, the creation of new instruments and styles, and the integration of genomic data with other variables similar as life and terrain. Individualized drug using genomic data has the implicit to transfigure healthcare by furnishing fast and precise opinion, genotype-grounded threat assessment, and acclimatized treatments despite these obstacles (Roden and Tyndale, 2013).

27.3 KEY COMPONENTS OF PRECISION MEDICINE

Complementary diagnostics, targeted medicines, and biomarkers are the mainstays of precision medicine. Treatment decisions can be guided with biomarkers, which identify certain aspects of a patient's disease, such as genetic alterations or protein expression. In drug development, biomarkers are frequently employed to select patients, and certain medications have biomarkers included in their indications (Bakker et al., 2023). The aim of precision medicine is to use disease-specific biomarkers to tailor treatments to specific patients or patient subgroups. Treatments can be more individualized and successful when biomarkers are used to know a patient's response to a given treatment (Slikker, 2018). Early identification and intervention can be done by biomarker testing in assessing cancer risk. Doctors can utilize information from biomarker tests to construct tailored care plans. With targeted therapy, these biomarkers are carefully addressed, leading to more individualized and efficient care. Tests known as companion diagnostics identify particular biomarkers in individuals most likely to benefit from precision medicine treatment, enabling more specialized and individualized care (Mankoff et al., 2016) (Table 27.1).

With targeted therapy, these biomarkers are precisely addressed, leading to further personalized and effective care (Table 27.2). Trastuzumab, frequently known as Herceptin, is a monoclonal antibody that targets the mortal epidermal growth-factor receptor 2(HER2). It's a leading illustration of individualized treatment for bone cancer. Compared with trastuzumab alone, a combination of trastuzumab plus pertuzumab – a humanized monoclonal antibody that binds to a distinct region of HER2 – has been demonstrated to ameliorate complaint-free survival and raise rates of pathological complete response (pCR) (Reynolds et al., 2014). Bone cancer provides a perceptive case study for the advancement and enhancement of targeted treatments in the age of perfection drug. The FDA has authorized trastuzumab as

TABLE 27.1
Biomarkers in Precision Medicine

S. No.	Disease	Biomarker	Application	References
1	Cancer	EGFR (Epidermal growth factor receptor), KRAS, BRAF, HER2 (Human epidermal growth factor receptor 2)	Predicting responsiveness to targeted medicines, detecting possible mutations for tailored treatment	(Chu, 2020; Delcuratolo et al., 2023; Motta et al., 2021)
2	Cardiovascular disease	LDL cholesterol, HDL cholesterol, hs-CRP	Determining cardiovascular risk and directing the use of statins	(Ballantyne et al., 2004; Lee & Lee, 2023; Sweeney et al., 2021)
3	Alzheimer disease	Abeta42, tau	Early diagnosis, monitoring illness progression	(Arafah et al., 2023; Ashton et al., 2022; Tondo & De Marchi, 2022)
4	Diabetes	HbA1c, C-peptide, insulin	Controlling blood sugar and anticipating how the body may react to particular diabetes drugs	(Carr et al., 2022; Galiero et al., 2023; Misra et al., 2023)
5	HIV	CD4 count, viral load	Tracking the course of the illness and directing antiretroviral treatment	(Prasad & Groop, 2019; Saag et al., 1996)
6	Cystic fibrosis	CFTR gene mutations	Tailoring treatment with specific CFTR modulators	(Lopes-Pacheco, 2019; Martiniano et al., 2016)
7	Rheumatoid arthritis	Rheumatoid factor, antibodies to CCP	Patient stratification for individualized treatment	(Bhamidipati & Wei, 2022; Cohen et al., 2023; Lindstrom & Robinson, 2010)
8	Genetic disorders	Various genetic mutations (BRCA1/2 for hereditary breast and ovarian cancer)	Determining who is at risk and directing preventative action	(Neves Rebello Alves et al., 2023; Walsh et al., 2016)
9	Infectious diseases	Viral load, specific pathogen DNA/RNA	Keeping track of antiviral therapy response and directing treatment duration	(Atallah & Mansour, 2022; Saag et al., 1996)
10	Hepatitis C	HCV RNA, HCV genotype	Directing the choice of antiviral medication	(Manns & Maasoumy, 2022; Manso et al., 2020; Tsukiyama-Kohara & Kohara, 2018)

TABLE 27.2
The Role of Targeted Therapy in Precision Medicine

S. No.	Disease	Targeted Therapy	Targeted Molecule	Mechanism of Action	Role in Precision Medicine	References
1	Breast cancer	Herceptin	HER2	Inhibits HER2 signaling	Targets the HER2-positive subtype and adjusts treatment according to particular biomarkers	(Reynolds et al., 2014; Swain et al., 2023)
2	Colorectal cancer	Cetuximab	EGFR	Blocks EGFR signaling	Individualized treatment for individuals with wild-type KRAS	(Landi et al., 2023; Nardeep Naithani et al., 2021; Xie et al., 2020)
3	Melanoma	Ipilimumab	CTLA-4	Blocks CTLA-4 on T cells	Increases the ability of the immune system to attack melanoma cells	(Lipson & Drake, 2011; Nikoo et al., 2023)
4	Chronic Myeloid Leukemia	Imatinib	BCR-ABL	Inhibits BCR-ABL tyrosine kinase	Addresses the BCR-ABL fusion gene, the underlying genetic issue	(Braun et al., 2020; Iqbal & Iqbal, 2014; Min & Lee, 2022)
5	Non-Hodgkin's Lymphoma	Rituximab	CD20	Targets B-cell antigen CD20	Targets particular B-cell markers to treat lymphoma more successfully	(Pierpont et al., 2018; Zhong et al., 2020)
6	Prostate cancer	Enzalutamide	Androgen receptor	Inhibits androgen receptor	Blocks androgen signaling, providing specialized treatment for malignancies driven by androgens	(Ahmed et al., 2014; Borgmann et al., 2018; Sternberg, 2019)
7	Ovarian cancer	Olaparib	PARP	PARP inhibitor	Utilizing DNA repair mechanisms, precision therapy is used to treat ovarian tumors with BRCA mutations.	(O'Malley et al., 2023; Rose et al., 2020)

an adjuvant treatment for HER2 bone cancer when combined with chemotherapy (Subhan et al., 2023). The monoclonal antibody cetuximab targets the epidermal growth factor receptor (EGFR) and has demonstrated pledge in the operation of colorectal cancer among other cancers. Cases with EGFR wild-type status are treated with it (Nardeep Naithani et al., 2021). It has been discovered that cetuximab is useful in adding progression-free survival in cases who don't respond well to single-agent remedy (Xie et al., 2020). It has also been delved in confluence with immunotherapeutic approaches involving natural killer cells, and the results indicate that it has the implicit to modify the tumour immunological terrain. also, studies on a new cetuximab-grounded EGFR targeted immunotoxin have demonstrated enhanced cytotoxicity against cetuximab-resistant colorectal cancer cells (Landi et al., 2023). Targeting cytotoxic T-lymphocyte-associated antigen 4 (CTLA-4), imilimumab is an vulnerable checkpoint asset that has been approved for the treatment of metastatic carcinoma (Lipson and Drake, 2011). Also, it has constantly demonstrated anti-melanoma efficacity in both early- and late-phase clinical trials. The U.S. Food and Drug Administration (FDA) has approved the combination of ipilimumab and nivolumab, ananti-PD-1 antibody, for the treatment of different malice (Nikoo et al., 2023). Imatinib is a new targeted drug that inhibits tyrosine kinases, particularly BCR-ABL. It has shown amazing efficacity in treating gastrointestinal stromal excrescences (GIST) and habitual myeloid leukemia (CML) (Iqbal and Iqbal, 2014). As the gold standard for treating CML, imatinib has demonstrated the capability to evoke deep molecular responses, with utmost cases passing long-lasting responses (Braun et al., 2020). Rituximab is a fantastic monoclonal antibody that specifically targets the B-cell CD20 protein. It's considered a groundbreaking treatment in the perfection drug space. It was the first monoclonal antibody certified for use therapeutically in the treatment of cancer (Pierpont et al., 2018). Rituximab has been used considerably in the treatment of habitual lymphocytic leukemia, B-cell malice, and non-Hodgkin carcinoma. The remedy of castration-resistant prostate cancer (CRPC) has showed pledge with the use of enzalutamide, an oral androgen receptor asset (Sternberg, 2019). Enzalutamide has been demonstrated to ameliorate survival rates in individualities with prostate cancer following chemotherapy and to beget excrescence retrogression in a mice model of castration-resistant prostate cancer (Ahmed et al., 2014). Olaparib is a poly (ADP-ribose) polymerase (PARP) asset that has been approved for the treatment of ovarian cancer (OC) and bone cancer. It specifically targets cancers that have mutations in homologous recombination (HR) genes, like BRCA1 and BRCA2(Rose et al., 2020). PARP impediments like olaparib have demonstrated remarkable success in treating excrescences with high situations of oxidative and replicative stress, anyhow of HR status" PARP Impediments in Ovarian Cancer: A Review," another review paper, provides a thorough summary of data about the PARP impediments rucaparib, niraparib, and olaparib in ovarian cancer cases. The review covers the colorful tolerability characteristics of PARP impediments, their overall well-toleration, and the significance of clinical variables similar platinum perceptivity as pointers of PARP asset response (O'Malley et al., 2023).

Tests known as companion diagnostics identify particular biomarkers in individuals most likely to benefit from precision medicine treatment, enabling more specialized and individualized care (Mankoff et al., 2016). Other essential components of precision medicine, like omics and pharmaco-omics, big data and artificial intelligence, social and environmental aspects, and individualized care and treatment, reinforce these components.

27.4 ROLE OF ARTIFICIAL INTELLIGENCE (AI) IN PRECISION MEDICINE

AI is predicted to aid in the realization of precision medicine's promise in three primary areas: disease prevention, individualized diagnosis, and personalized therapy (Deng et al., 2021). Without AI, there can be no precision medicine (Meskó, 2017). Artificial intelligence and machine literacy are fleetly being employed in healthcare for early outgrowth vaticination (Pammi et al., 2023). Big data-driven artificial intelligence (AI) can uncover retired patterns, critical information, and matching knowledge from massive quantities of data (Liao et al., 2022). AI may be utilized in precision oncology to build a medicine combination based on a patient's individual biopsy and make N-of-1 medication recommendations. AI-based algorithms have already improved diagnostic accuracy and performance in a variety of specialties, including radiology. In this sense, artificial intelligence comprises machine learning (ML), deep learning (DL), and artificial neural networks (ANN). When AI is integrated with high-performance computer technologies, we can develop and forecast illness risk based on individual data (Subramanian et al., 2020). AI-based solutions are used by healthcare officials and the government to control and anticipate illnesses and outbreaks. FINDER, a machine-learned algorithm for real-time identification of foodborne disease utilizing anonymized and aggregated online search and location data, is one example (Johnson et al., 2021).

AmCAD-US (AmCad BioMed Corporation) was introduced in May 2017 as a software device that visualizes and quantifies the statistical distribution of backscattered signals echoed by tissue compositions in the body from ultrasound systems. AmCad BioMed Corporation has also developed and acquired FDA concurrence for AmCAD-UT, which is used largely to diagnose thyroid cancer. The Alternate anthology type CADx(QuantX) from Quantitative perceptivity acquired its first FDA concurrence in July 2017. This technology is intended to target bone glamorous resonance imaging (MRI) and help clinicians make an opinion of bone cancer. Artery Oncology DL (Artery) gained FDA concurrence in January 2018 to help croakers in instantly measuring and tracking lesions and nodes on MRI and reckoned tomography (CT) images. This contrivance was created to prop in the discovery of solid excrescences in general, and it's now in the early stages of clinical testing for liver and lung cancer. The FDA authorized SubtlePET (Subtle Medical), an AI-grounded high-speed positron emigration tomography (PET) imaging fashion, in November 2018. SubtlePet is an AI tool that provides four times quicker imaging, particularly during PET imaging, which is critical in cancer discovery. The FDA authorized cmTriage (CureMetrix) for mammography picture triage in March 2019(Hamamoto et al., 2020). The crossroad of AI with precision medicine has the implicit to transfigure care customization in a variety of ways, including remedy planning, identification of patient phenotypes, and the creation and optimization of pathways for opinion, remedial intervention, and prognostic. AI may help break complicated medical issues by giving croakers with redundant threat information, anticipating threat in early illness stages, and bodying early curatives grounded on individual natural datasets (Schork, 2019).

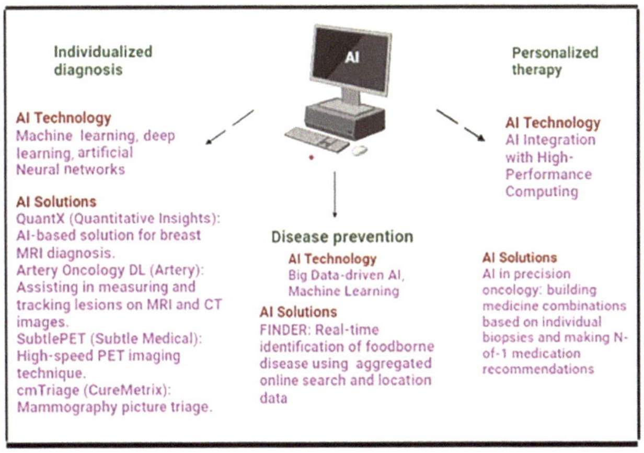

FIGURE 27.1 The role of AI in precision medicine.

(Image created by Biorender.com.)

27.5 APPLICATIONS OF PRECISION MEDICINE

Precision medicine is also important in the development of colorful treatment modalities in oncology, similar as medicines targeting specific mutations and biomarkers in colorful cancers, and has been salutary in the COVID-19 epidemic by using colorful biomarkers to assess complaint inflexibility, develop curatives, and judge vaccine efficacity (N. Naithani et al., 2021). There have been a number of instigative developments in this sphere, ranging from largely individualized advances in ex vivo autologous gene remedy to group position progress in immune oncology curatives for cases who aren't defined by a specific physical mutation or traditional tumour histology (Cardon and Harris, 2016). Seditious brainails, similar as encephalitis, contagious meningitis, vasculitis, T-cell driven seditious brain conditions, and demyelinating conditions, are some of the uncommon conditions that would profit from precision medicine approach (Twilt, 2016). The vacuity of multi-omics data, genotype-phenotype data from genome-wide association studies (GWAS), and literature mining in the

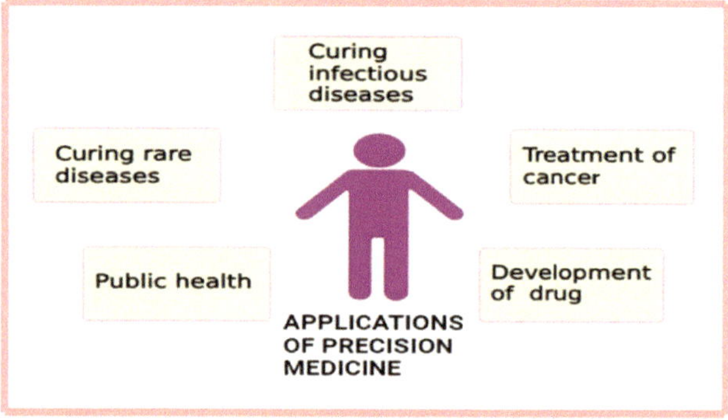

FIGURE 27.2 Applications of precision medicine.

(Image created by Biorender.com.)

field of cancer genomics has fueled the development of advanced AI ways and results that enable medical professionals to give substantiated care through perfection drug (Xu et al., 2019).

Upon its launch in January 2015, the Precision Medicine Initiative committed to investing $70 million to the National Cancer Institute (NCI) with the end of enhancing sweats to uncover the inheritable causes of cancer and exercising this knowledge to develop more effective ways to cancer remedy (Goossens et al., 2015). The perfection drug system allows for the identification of particular treatment in the form of monoclonal antibodies for refractory type 2 high asthmatics (N. Naithani et al., 2021). Cardiovascular drug is an area where AI has had a significant impact, with colorful studies, similar as the one conducted by Li et al, assessing the threat of heart attack using an artificial neuron network (Álvarez-Machancoses et al., 2020).

27.6 CHALLENGES OF PRECISION MEDICINE

Although precision medicine can result in long-term cost savings, the upfront costs may be unaffordable. Certain patients may not be able to afford genetic testing and other customized medicines due to their high cost. This is just the beginning of a lengthy and expensive project. Not all practitioners have the necessary tools to provide individualized care, and precision medicine is not yet generally accessible. As a result, some patients might not be able to get precision medicine, especially those who live in rural areas or have limited financial resources. The application of precision medicine raises additional ethical concerns. Genetic testing, for instance, may reveal private health information about a patient that could be used against them in the job market or when applying for insurance. Precision medicine also requires extensive teamwork. Precision medicine is still lacking in standardization, which makes it challenging to compare patient results and treatments. Because of this, doctors may find it challenging to choose the most appropriate course of treatment for their patients (Janiaud et al., 2019; Khoury and Evans, 2015; Liu et al., 2019).

27.7 CONCLUSION

While the precision medicine initiative will presumably see its greatest benefits years down the road, there ought to be some impressive near-term successes. Apart from the aforementioned cancer research findings, examinations of an extensive study group exposed to an array of pharmaceuticals could provide preliminary understanding of pharmacogenomics, enabling the proper medication at the appropriate dosage to be given to the right patient (Collins and Varmus, 2015). Its main goal is to reduce serious illness-related morbidity, death, and disability while simultaneously enhancing the quality of medical care through technological innovation (Wang et al., 2016). Biomarker screening is another an area that is expanding quickly. Whole genome and transcriptome sequencing methods, in addition to comprehensive genomic panels, are being researched and may be beneficial (Lu et al., 2023). Extensive genomic studies utilizing next-generation sequencing (NGS) and gene expression profiling using DNA microarray are exposing the variety of cancer's genome, epigenome, and expression patterns. These data are used as predictive biomarkers to help choose a particular drug, as well as to monitor, analyze, and anticipate a patient's resistance to and tolerance to a treatment plan (Basu et al., 2018). However, the aim of precision medicine as best clinical practice has always been to provide individualized treatment using the best tools available and understanding of disease biology: "Strive by every method you can think of to gain the greatest certainty attainable," declared Sir William Gowers (Sisodiya, 2021). This is true regardless of how we view the field today or how it will change in the future as new areas of understanding are incorporated into management strategies.

REFERENCES

Ahmed, A., Ali, S., & Sarkar, F. H. (2014). Advances in androgen receptor targeted therapy for prostate cancer. *Journal of Cellular Physiology*, *229*(3), 271–276. https://doi.org/10.1002/jcp.24456

Álvarez-Machancoses, Ó., DeAndrés Galiana, E. J., Cernea, A., Fernández de la Viña, J., & Fernández-Martínez, J. L. (2020). On the role of artificial intelligence in genomics to enhance precision medicine. *Pharmacogenomics and Personalized Medicine*, *13*, 105–119. https://doi.org/10.2147/pgpm.S205082

Arafah, A., Khatoon, S., Rasool, I., Khan, A., Rather, M. A., Abujabal, K. A., Faqih, Y. A. H., Rashid, H., Rashid, S. M., Bilal Ahmad, S., Alexiou, A., & Rehman, M. U. (2023). The future of precision medicine in the cure of Alzheimer's disease. *Biomedicines*, *11*(2). https://doi.org/10.3390/biomedicines11020335

Ashton, N. J., Janelidze, S., Mattsson-Carlgren, N., Binette, A. P., Strandberg, O., Brum, W. S., Karikari, T. K., González-Ortiz, F., Di Molfetta, G., Meda, F. J., Jonaitis, E. M., Koscik, R. L., Cody, K., Betthauser, T. J., Li, Y., Vanmechelen, E., Palmqvist, S., Stomrud, E., Bateman, R. J., Zetterberg, H., Johnson, S. C., Blennow, K., & Hansson, O. (2022). Differential roles of Aβ42/40, p-tau231 and p-tau217 for

Alzheimer's trial selection and disease monitoring. *Nature Medicine*, *28*(12), 2555–2562. https://doi.org/10.1038/s41591-022-02074-w

Atallah, J., & Mansour, M. K. (2022). Implications of using host response-based molecular diagnostics on the management of bacterial and viral infections: A review. *Frontiers in Medicine (Lausanne)*, *9*, 805107. https://doi.org/10.3389/fmed.2022.805107

Bakker, E., Starokozhko, V., Kraaijvanger, J. W. M., Heerspink, H. J. L., & Mol, P. G. M. (2023). Precision medicine in regulatory decision making: Biomarkers used for patient selection in European Public Assessment Reports from 2018 to 2020. *Clinical and Translational Science*, *16*(11), 2394–2412. https://doi.org/10.1111/cts.13641

Ballantyne, C. M., Hoogeveen, R. C., Bang, H., Coresh, J., Folsom, A. R., Heiss, G., & Sharrett, A. R. (2004). Lipoprotein-associated phospholipase A2, high-sensitivity C-reactive protein, and risk for incident coronary heart disease in middle-aged men and women in the Atherosclerosis Risk in Communities (ARIC) study. *Circulation*, *109*(7), 837–842. https://doi.org/10.1161/01.Cir.0000116763.91992.F1

Basu, P., Mukhopadhyay, A., & Konishi, I. (2018). Targeted therapy for gynecologic cancers: Toward the era of precision medicine. *International Journal of Gynecology & Obstetrics*, *143*(Suppl 2), 131–136. https://doi.org/10.1002/ijgo.12620

Belle, D. J., & Singh, H. (2008). Genetic factors in drug metabolism. *American Family Physician*, *77*(11), 1553–1560.

Bhamidipati, K., & Wei, K. (2022). Precision medicine in rheumatoid arthritis. *Best Practice & Research: Clinical Rheumatology*, *36*(1), 101742. https://doi.org/10.1016/j.berh.2022.101742

Borgmann, H., Lallous, N., Ozistanbullu, D., Beraldi, E., Paul, N., Dalal, K., Fazli, L., Haferkamp, A., Lejeune, P., Cherkasov, A., & Gleave, M. E. (2018). Moving towards precision urologic oncology: Targeting enzalutamide-resistant prostate cancer and mutated forms of the androgen receptor using the novel inhibitor darolutamide (ODM-201). *European Urology*, *73*(1), 4–8. https://doi.org/10.1016/j.eururo.2017.08.012

Braun, T. P., Eide, C. A., & Druker, B. J. (2020). Response and resistance to BCR-ABL1-targeted therapies. *Cancer Cell*, *37*(4), 530–542. https://doi.org/10.1016/j.ccell.2020.03.006

Brittain, H. K., Scott, R., & Thomas, E. (2017). The rise of the genome and personalised medicine. *Clinical Medicine (London)*, *17*(6), 545–551. https://doi.org/10.7861/clinmedicine.17-6-545

Cardon, L. R., & Harris, T. (2016). Precision medicine, genomics and drug discovery. *Human Molecular Genetics*, *25*(R2), R166–R172. https://doi.org/10.1093/hmg/ddw246

Carr, A. L. J., Evans-Molina, C., & Oram, R. A. (2022). Precision medicine in type 1 diabetes. *Diabetologia*, *65*(11), 1854–1866. https://doi.org/10.1007/s00125-022-05778-3

Cecchin, E., & Stocco, G. (2020). Pharmacogenomics and personalized medicine. *Genes (Basel)*, *11*(6). https://doi.org/10.3390/genes11060679

Chu, Q. S. (2020). Targeting non-small cell lung cancer: Driver mutation beyond epidermal growth factor mutation and anaplastic lymphoma kinase fusion. *Therapeutic Advances in Medical Oncology*, *12*. https://doi.org/10.1177/1758835919895756

Cohen, S., Curtis, J. R., Mellors, T., Zhang, L., Withers, J. B., Jones, A., Ghiassian, S. D., & Akmaev, V. R. (2023). Commentary on Cohen et al.: Role of clinical factors in precision medicine test to predict nonresponse to TNFi therapies in rheumatoid arthritis. *Rheumatology and Therapy*, *10*(1), 1–6. https://doi.org/10.1007/s40744-022-00506-0

Collins, F. S., & Varmus, H. (2015). A new initiative on precision medicine. *The New England Journal of Medicine*, *372*(9), 793–795. https://doi.org/10.1056/NEJMp1500523

Delcuratolo, M. D., Modrego-Sánchez, A., Bungaro, M., Antón-Pascual, B., Teran, S., Dipace, V., Novello, S., Garcia-Carbonero, R., Passiglia, F., & Graválos-Castro, C. (2023). Liquid biopsy in advanced colorectal cancer: Clinical applications of different analytes. *Journal of Molecular Pathology*, *4*(3), 128–155. https://www.mdpi.com/2673-5261/4/3/13

Deng, J., Hartung, T., Capobianco, E., Chen, J. Y., & Emmert-Streib, F. (2021). Editorial: Artificial intelligence for precision medicine. *Frontiers in Artificial Intelligence*, *4*, 834645. https://doi.org/10.3389/frai.2021.834645

Galiero, R., Caturano, A., Vetrano, E., Monda, M., Marfella, R., Sardu, C., Salvatore, T., Rinaldi, L., & Sasso, F. C. (2023). Precision medicine in type 2 diabetes mellitus: Utility and limitations. *Diabetes, Metabolic Syndrome and Obesity*, *16*, 3669–3689. https://doi.org/10.2147/dmso.S390752

Goetz, L. H., & Schork, N. J. (2018). Personalized medicine: Motivation, challenges, and progress. *Fertility and Sterility*, *109*(6), 952–963. https://doi.org/10.1016/j.fertnstert.2018.05.006

Goossens, N., Nakagawa, S., Sun, X., & Hoshida, Y. (2015). Cancer biomarker discovery and validation. *Translational Cancer Research*, *4*(3), 256–269. https://doi.org/10.3978/j.issn.2218-676X.2015.06.04

Hamamoto, R., Suvarna, K., Yamada, M., Kobayashi, K., Shinkai, N., Miyake, M., Takahashi, M., Jinnai, S., Shimoyama, R., Sakai, A., Takasawa, K., Bolatkan, A., Shozu, K., Dozen, A., Machino, H., Takahashi, S., Asada, K., Komatsu, M., Sese, J., & Kaneko, S. (2020). Application of artificial intelligence technology in oncology: Towards the establishment of precision medicine. *Cancers (Basel)*, *12*(12). https://doi.org/10.3390/cancers12123532

Iqbal, N., & Iqbal, N. (2014). Imatinib: A breakthrough of targeted therapy in cancer. *Chemotherapy Research and Practice*, *2014*, 357027. https://doi.org/10.1155/2014/357027

Janiaud, P., Serghiou, S., & Ioannidis, J. P. A. (2019). New clinical trial designs in the era of precision medicine: An overview of definitions, strengths, weaknesses, and current use in oncology. *Cancer Treatment Reviews*, *73*, 20–30. https://doi.org/10.1016/j.ctrv.2018.12.003

Johnson, K. B., Wei, W. Q., Weeraratne, D., Frisse, M. E., Misulis, K., Rhee, K., Zhao, J., & Snowdon, J. L. (2021). Precision medicine, AI, and the future of personalized healthcare. *Clinical and Translational Science*, *14*(1), 86–93. https://doi.org/10.1111/cts.12884

Khoury, M. J., & Evans, J. P. (2015). A public health perspective on a national precision medicine cohort: Balancing long-term knowledge generation with early health benefit. *JAMA*, *313*(21), 2117–2118. https://doi.org/10.1001/jama.2015.3382

Khoury, M. J., & Holt, K. E. (2021). The impact of genomics on precision public health: Beyond the pandemic. *Genome Medicine*, *13*(1), 67. https://doi.org/10.1186/s13073-021-00886-y

Kosorok, M. R., & Laber, E. B. (2019). Precision medicine. *Annual Review of Statistics and Its Application*, *6*, 263–286. https://doi.org/10.1146/annurev-statistics-030718-105251

Landi, N., Ciaramella, V., Ragucci, S., Chambery, A., Ciardiello, F., Pedone, P. V., Troiani, T., & Di Maro, A. (2023). A novel EGFR targeted immunotoxin based on cetuximab and type 1 RIP quinoin overcomes the cetuximab resistance in colorectal cancer cells. *Toxins (Basel)*, *15*(1). https://doi.org/10.3390/toxins15010057

Lauschke, V. M., Zhou, Y., & Ingelman-Sundberg, M. (2019). Novel genetic and epigenetic factors of importance for inter-individual differences in drug disposition, response and toxicity. *Pharmacology & Therapeutics*, *197*, 122–152. https://doi.org/10.1016/j.pharmthera.2019.01.002

Lee, H. S., & Lee, J.-H. (2023). Early elevation of high-sensitivity C-reactive protein as a predictor for cardiovascular disease incidence and all-cause mortality: A landmark analysis. *Scientific Reports*, *13*(1), 14118. https://doi.org/10.1038/s41598-023-41081-w

Liao, J., Li, X., Gan, Y., Han, S., Rong, P., Wang, W., Li, W., & Zhou, L. (2022). Artificial intelligence assists precision medicine in cancer treatment. *Frontiers in Oncology*, *12*, 998222. https://doi.org/10.3389/fonc.2022.998222

Lindstrom, T. M., & Robinson, W. H. (2010). Biomarkers for rheumatoid arthritis: Making it personal. *Scandinavian Journal of Clinical and Laboratory Investigation*, *242*, 79–84. https://doi.org/10.3109/00365513.2010.493406

Lipson, E. J., & Drake, C. G. (2011). Ipilimumab: An anti-CTLA-4 antibody for metastatic melanoma. *Clinical Cancer Research*, *17*(22), 6958–6962. https://doi.org/10.1158/1078-0432.Ccr-11-1595

Liu, X., Luo, X., Jiang, C., & Zhao, H. (2019). Difficulties and challenges in the development of precision medicine. *Clinical Genetics*, *95*(5), 569–574. https://doi.org/10.1111/cge.13511

Lopes-Pacheco, M. (2019). CFTR modulators: The changing face of cystic fibrosis in the era of precision medicine. *Frontiers in Pharmacology*, *10*, 1662. https://doi.org/10.3389/fphar.2019.01662

Lu, C. Y., Terry, V., & Thomas, D. M. (2023). Precision medicine: Affording the successes of science. *NPJ Precision Oncology*, *7*(1), 3. https://doi.org/10.1038/s41698-022-00343-y

Mankoff, D. A., Edmonds, C. E., Farwell, M. D., & Pryma, D. A. (2016). Development of companion diagnostics. *Seminars in Nuclear Medicine*, *46*(1), 47–56. https://doi.org/10.1053/j.semnuclmed.2015.09.002

Manns, M. P., & Maasoumy, B. (2022). Breakthroughs in hepatitis C research: From discovery to cure. *Nature Reviews Gastroenterology & Hepatology*, *19*(8), 533–550. https://doi.org/10.1038/s41575-022-00608-8

Manso, C. F., Bibby, D. F., Lythgow, K., Mohamed, H., Myers, R., Williams, D., Piorkowska, R., Chan, Y. T., Bowden, R., Ansari, M. A., Ip, C. L. C., Barnes, E., Bradshaw, D., & Mbisa, J. L. (2020). Technical validation of a hepatitis C virus whole genome sequencing assay for detection of genotype and antiviral resistance in the clinical pathway. *Frontiers in Microbiology*, *11*, 576572. https://doi.org/10.3389/fmicb.2020.576572

Martiniano, S. L., Sagel, S. D., & Zemanick, E. T. (2016). Cystic fibrosis: A model system for precision medicine. *Current Opinion in Pediatrics*, *28*(3), 312–317. https://doi.org/10.1097/mop.0000000000000351

Meskó, B. (2017). The role of artificial intelligence in precision medicine. *Expert Review of Precision Medicine and Drug Development*, *2*, 239–241.

Min, H.-Y., & Lee, H.-Y. (2022). Molecular targeted therapy for anticancer treatment. *Experimental & Molecular Medicine*, *54*(10), 1670–1694. https://doi.org/10.1038/s12276-022-00864-3

Misra, S., Wagner, R., Ozkan, B., Schön, M., Sevilla-Gonzalez, M., Prystupa, K., Wang, C. C., Kreienkamp, R. J., Cromer, S. J., Rooney, M. R., Duan, D., Thuesen, A. C. B., Wallace, A. S., Leong, A., Deutsch, A. J., Andersen, M. K., Billings, L. K., Eckel, R. H., Sheu, W. H.-H., Hansen, T., Stefan, N., Goodarzi, M. O., Ray, D., Selvin, E., Florez, J. C. et al. (2023). Precision subclassification of type 2 diabetes: A systematic review. *Communications Medicine*, *3*(1), 138. https://doi.org/10.1038/s43856-023-00360-3

Motta, R., Cabezas-Camarero, S., Torres-Mattos, C., Riquelme, A., Calle, A., Montenegro, P., & Sotelo, M. J. (2021). Personalizing first-line treatment in advanced colorectal cancer: Present status and future perspectives. *Journal of Clinical and Translational Research*, *7*(6), 771–785.

Naithani, N., Atal, A. T., Tilak, T. V. S. V. G. K., Vasudevan, B., Misra, P., & Sinha, S. (2021). Precision medicine: Uses and challenges. *Medical Journal Armed Forces India*, *77*(3), 258–265. https://doi.org/10.1016/j.mjafi.2021.06.020

Naithani, N., Sinha, S., Misra, P., Vasudevan, B., & Sahu, R. (2021). Precision medicine: Concept and tools. *Medical Journal Armed Forces India*, *77*(3), 249–257. https://doi.org/10.1016/j.mjafi.2021.06.021

Neves Rebello Alves, L., Dummer Meira, D., Poppe Merigueti, L., Correia Casotti, M., do Prado Ventorim, D., Ferreira Figueiredo Almeida, J., Pereira de Sousa, V., Cindra Sant'Ana, M., Gonçalves Coutinho da Cruz, R., Santos Louro, L., Mendonça Santana, G., Erik Santos Louro, T., Evangelista Salazar, R., Ribeiro Campos da Silva, D., Stefani Siqueira Zetum, A. (2023). Biomarkers in breast cancer: An old story with a new end. *Genes*, *14*(7), 1364. https://www.mdpi.com/2073-4425/14/7/1364

Nikoo, M., Rabiee, F., Mohebbi, H., Eghbalifard, N., Rajabi, H., Yazdani, Y., Sakhaei, D., Khosravifarsani, M., & Akhavan-Sigari, R. (2023). Nivolumab plus ipilimumab combination therapy in cancer: Current evidence to date. *International Immunopharmacology*, *117*, 109881. https://doi.org/10.1016/j.intimp.2023.109881

O'Malley, D. M., Krivak, T. C., Kabil, N., Munley, J., & Moore, K. N. (2023). PARP inhibitors in ovarian cancer: A review. *Targeted Oncology*, *18*(4), 471–503. https://doi.org/10.1007/s11523-023-00970-w

Pammi, M., Aghaeepour, N., & Neu, J. (2023). Multiomics, artificial intelligence, and precision medicine in perinatology. *Pediatric Research*, *93*(2), 308–315. https://doi.org/10.1038/s41390-022-02181-x

Phillips, C. J. (2020). Precision medicine and its imprecise history. *Harvard Data Science Review*, *2*(1).

Pierpont, T. M., Limper, C. B., & Richards, K. L. (2018). Past, present, and future of rituximab-the world's first oncology monoclonal antibody therapy. *Frontiers in Oncology*, *8*, 163. https://doi.org/10.3389/fonc.2018.00163

Prasad, R. B., & Groop, L. (2019). Precision medicine in type 2 diabetes. *Journal of Internal Medicine*, *285*(1), 40–48. https://doi.org/10.1111/joim.12859

Reynolds, K., Sarangi, S., Bardia, A., & Dizon, D. S. (2014). Precision medicine and personalized breast cancer: Combination pertuzumab therapy. *Pharmacogenomics and Personalized Medicine*, *7*, 95–105. https://doi.org/10.2147/pgpm.S37100

Roden, D. M., & Tyndale, R. F. (2013). Genomic medicine, precision medicine, personalized medicine: What's in a name? *Clinical Pharmacology & Therapeutics*, *94*(2), 169–172. https://doi.org/10.1038/clpt.2013.101

Rose, M., Burgess, J. T., O'Byrne, K., Richard, D. J., & Bolderson, E. (2020). PARP Inhibitors: Clinical relevance, mechanisms of action and tumor resistance. *Frontiers in Cell and Developmental Biology*, *8*, 564601. https://doi.org/10.3389/fcell.2020.564601

Saag, M. S., Holodniy, M., Kuritzkes, D. R., O'Brien, W. A., Coombs, R., Poscher, M. E., Jacobsen, D. M., Shaw, G. M., Richman, D. D., & Volberding, P. A. (1996). HIV viral load markers in clinical practice. *Nature Medicine*, *2*(6), 625–629. https://doi.org/10.1038/nm0696-625

Schork, N. J. (2019). Artificial intelligence and personalized medicine. *Cancer Research and Treatment*, *178*, 265–283. https://doi.org/10.1007/978-3-030-16391-4_11

Seymour, C. W., Gomez, H., Chang, C. H., Clermont, G., Kellum, J. A., Kennedy, J., Yende, S., & Angus, D. C. (2017). Precision medicine for all? Challenges and opportunities for a precision medicine approach to critical illness. *Critical Care*, *21*(1), 257. https://doi.org/10.1186/s13054-017-1836-5

Sisodiya, S. M. (2021). Precision medicine and therapies of the future. *Epilepsia*, *62*(Suppl 2), S90–S105. https://doi.org/10.1111/epi.16539

Slikker, W., Jr. (2018). Biomarkers and their impact on precision medicine. *Experimental Biology and Medicine (Maywood)*, *243*(3), 211–212. https://doi.org/10.1177/1535370217733426

Sternberg, C. N. (2019). Enzalutamide, an oral androgen receptor inhibitor for treatment of castration-resistant prostate cancer. *Future Oncology*, *15*(13), 1437–1457. https://doi.org/10.2217/fon-2018-0940

Strianese, O., Rizzo, F., Ciccarelli, M., Galasso, G., D'Agostino, Y., Salvati, A., Del Giudice, C., Tesorio, P., & Rusciano, M. R. (2020). Precision and personalized medicine: How genomic approach improves the management of cardiovascular and neurodegenerative disease. *Genes (Basel)*, *11*(7). https://doi.org/10.3390/genes11070747

Subhan, M. A., Parveen, F., Shah, H., Yalamarty, S. S. K., Ataide, J. A., & Torchilin, V. P. (2023). Recent advances with precision medicine treatment for breast cancer including triple-negative sub-type. *Cancers*, *15*(8), 2204. https://www.mdpi.com/2072-6694/15/8/2204

Subramanian, M., Wojtusciszyn, A., Favre, L., Boughorbel, S., Shan, J., Letaief, K. B., Pitteloud, N., & Chouchane, L. (2020). Precision medicine in the era of artificial intelligence: Implications in chronic disease management. *Journal of Translational Medicine*, *18*(1), 472. https://doi.org/10.1186/s12967-020-02658-5

Swain, S. M., Shastry, M., & Hamilton, E. (2023). Targeting HER2-positive breast cancer: Advances and future directions. *Nature Reviews Drug Discovery*, *22*(2), 101–126. https://doi.org/10.1038/s41573-022-00579-0

Sweeney, T., Quispe, R., Das, T., Juraschek, S. P., Martin, S. S., & Michos, E. D. (2021). The use of blood biomarkers in precision medicine for the primary prevention of atherosclerotic cardiovascular disease: A review. *Expert Review of Precision Medicine and Drug Development*, *6*(4), 247–258. https://doi.org/10.1080/23808993.2021.1930531

Tan, B. T. N., Khan, M. I., Saleh, M. A., Wangchuk, D., Talukder, M. J. H., & Kinght-Agarwal, C. R. (2023). Empowering healthcare through precision medicine: Unveiling the nexus of social factors and trust. *Healthcare (Basel)*, *11*(24). https://doi.org/10.3390/healthcare11243177

Tondo, G., & De Marchi, F. (2022). From biomarkers to precision medicine in neurodegenerative diseases: Where are we? *Journal of Clinical Medicine*, *11*(15). https://doi.org/10.3390/jcm11154515

Tsukiyama-Kohara, K., & Kohara, M. (2018). Hepatitis C virus: Viral quasispecies and genotypes. *International Journal of Molecular Sciences*, *19*(1), 23. https://www.mdpi.com/1422-0067/19/1/23

Twilt, M. (2016). Precision medicine: The new era in medicine. *EBioMedicine*, *4*, 24–25. https://doi.org/10.1016/j.ebiom.2016.02.009

Valent, P., Orfao, A., Kubicek, S., Staber, P., Haferlach, T., Deininger, M., Kollmann, K., Lion, T., Virgolini, I., Winter, G., Hantschel, O., Kenner, L., Zuber, J., Grebien, F., Moriggl, R., Hoermann, G., Hermine, O., Andreeff, M., Bock, C., Mughal, T., Constantinescu, S. N., Kralovics, R., Sexl, V., Skoda, R., Superti-Furga, G., & Jäger, U. (2021). Precision medicine in hematology 2021: Definitions, tools, perspectives, and open questions. *Hemasphere*, *5*(3), e536. https://doi.org/10.1097/hs9.0000000000000536

Walsh, M. F., Nathanson, K. L., Couch, F. J., & Offit, K. (2016). Genomic biomarkers for breast cancer risk. *Advances in Experimental Medicine and Biology*, *882*, 1–32. https://doi.org/10.1007/978-3-319-22909-6_1

Wang, Z. G., Zhang, L., & Zhao, W. J. (2016). Definition and application of precision medicine. *Chinese Journal of Traumatology*, *19*(5), 249–250. https://doi.org/10.1016/j.cjtee.2016.04.005

Weinshilboum, R. M., & Wang, L. (2017). Pharmacogenomics: Precision medicine and drug response. *Mayo Clinic Proceedings*, *92*(11), 1711–1722. https://doi.org/10.1016/j.mayocp.2017.09.001

Xie, Y.-H., Chen, Y.-X., & Fang, J.-Y. (2020). Comprehensive review of targeted therapy for colorectal cancer. *Signal Transduction and Targeted Therapy*, *5*(1), 22. https://doi.org/10.1038/s41392-020-0116-z

Xu, J., Yang, P., Xue, S., Sharma, B., Sanchez-Martin, M., Wang, F., Beaty, K. A., Dehan, E., & Parikh, B. (2019). Translating cancer genomics into precision medicine with artificial intelligence: Applications, challenges and future perspectives. *Human Genetics*, *138*(2), 109–124. https://doi.org/10.1007/s00439-019-01970-5

Yu, G., Chen, Z., Wu, J., & Tan, Y. (2021). Medical decision support system for cancer treatment in precision medicine in developing countries. *Expert Systems with Applications*, *186*, 115725.

Zhong, M., van der Walt, A., Campagna, M. P., Stankovich, J., Butzkueven, H., & Jokubaitis, V. (2020). The pharmacogenetics of rituximab: Potential implications for anti-CD20 therapies in multiple sclerosis. *Neurotherapeutics*, *17*(4), 1768–1784. https://doi.org/10.1007/s13311-020-00950-2

28 Personalised Healthcare
New Treatment Paradigms in Pharmacotherapy

Ritu Mishra,[1] *Shweta Mishra,*[1] *Shri Krishan,*[2] *Vijay Bhalla,*[1] *and Rakesh Kumar Sharma*[1]*

[1] SGT College of Pharmacy, SGT University, Budhera, Gurugram (Haryana), India
[2] Department of Drug Safety and Pharmacovigilance, Syneos Health Gurgaon, India

28.1 INTRODUCTION

Personalised healthcare is described as the use of genetic and other biotechnologies to provide information about an individual's specific information that may be used to develop a variety of health therapies that are most appropriate for that individual.[1,2] This knowledge can be utilised by healthcare practitioners to adapt therapies for an individual, or it can offer people with knowledge about themselves that they can employ to lower their risk of illness or disease and improves health.[3] Precision medicine emphasises on advancements in healthcare, treatment options, therapies, practices, and products tailored to a specific set of patients based on a knowledge of their genes, environment, and lifestyle.[4] It seeks to achieve optimal therapies while minimising iatrogenic risk and medical costs.[4] The ultimate goal is to deliver rational pharmacotherapy, which means prescribing the right drug to the right patient in the right dose at the right time while minimising or preventing patient harm. The scope of personalised healthcare has grown significantly as a result of recent advances in genetics, molecular biology, and biochemistry.[5] During the COVID-19 pandemic, the importance of making timely and focused decisions was emphasised even more, and the opportunities and challenges for the development of treatments and vaccines against the corona virus established a strong foundation for strengthening personalised healthcare operations.[5] Although healthcare providers have attempted to personalise diagnosis and treatment options based on individual traits throughout long history of healthcare and there have been significant recent changes in this context. The major difference has been improved capacity to recognize, categorise, and treat the patients, bypassing the previously used trial-and-error methods. This shift has had a direct impact on many areas of healthcare system that are currently delivered. Indeed, it has the potential to reign in a revolution in the management of people and populations by reducing uncertainties and boosting the potential of such systems to forecast the effects of various programs and interventions on individuals and the community in particular. However, this promising scenario has also generated challenges that must be addressed as the ongoing evolution of customised healthcare generates more evidence and solutions for its application to both people and populations. A key component of this tailored pharmaceutical formulation is the reliance on, and required participation in, novel biotechnologies, in particular new kinds of genetic analysis. The results of these genetic tests may be supplied and interpreted by a medical professional or by way of a direct-to-consumer service provider. To some extent, healthcare based on personalised biological data has long been used to customise medical treatments for disorders as well as (to a lesser extent) individuals.

28.2 PERSONALISED HEALTHCARE: A MEDICAL ADVANCEMENT

Personalised healthcare encompasses a wide variety of genomic-based goods and services that enable the development of customised methodsfor pharmaceutical care. Thus, health technology comprises pharmaceuticals, equipment, processes, care environments, and diagnostic and prevention programmes, as well as strategies used to enhance health, combat and treat diseases improves rehabilitation and long-term care.[6] The health spectrum, care that is personalised, predictive, preventative, and participatory,[5] comes generally within the scope of health technology assessment (HTA); customised healthcare technologies are primarily used to prevent and detect diseases, and they can be found at various levels of development, ranging from future technology to more established ones.[7,8] Personalised healthcare in oncology chronic disorders like asthma, diabetes, autoimmune connective tissue disorders (AICTD), emergency care infectious disease are discussed briefly as follows.

28.2.1 Personalised Healthcare in Oncology

Surgery followed by traditional chemotherapy and radiation therapy are effective options against cancer cells. This frequently results in a bystander effect on normal fast dividing cells, hair follicles, blood cells, and intestinal tract cells, resulting in severe consequences and toxic effects. Targeted

* Corresponding author: rksharmadrl@yahoo.com

DOI: 10.1201/9781032660592-35

cancer therapy, also known as precision oncology aims at specific sites targeting to reduce the side effects of chemotherapy while retaining the effect on the cancer cell by identifying targets at the extracellular, cell membrane, intracellular, and intranuclear levels. Ligands that bind to receptors on the cell surface and guide cellular activities are examples of extracellular targets. Vascularisation is required for solid tumor growth, and vascular endothelial growth factor-A (VEGF-A) plays a key part in this process. Bevacizumab is a monoclonal antibody that attaches to soluble VEGF-A, blocking it from interacting with the VEGF receptor and preventing neovascularisation pathways. This medicine has been shown to be successful in several malignancies, including colorectal, lung, renal, glioblastoma multiforme, and ovarian tumors, either used in conjunction with chemotherapy or as a standard maintenance treatment.[9–12] The receptors on the cell surface assist the cell survival by activating downstream signalling pathways. Cancer cells with excessive or abnormal receptor expression can be targeted with receptor-directed monoclonal antibodies, lowering the hazard of normal cells. The epidermal growth factor-receptor tyrosine kinase (EGF-RTK) family, also known as the ErbB receptor family, is made up of four members: EGFR (ErbB1, HER1), ErbB2 (HER2, neu in rodents), ErbB3 (HER3), and ErbB4 (HER4). It is one of the most intensively investigated receptors for precision medicine. Among these, overexpression of ErbB2 (Her2/neu) promotes carcinogenesis by homodimerising with certain other HER receptors and stimulating downstream signalling pathways.[13] Approximately 20–25% of breast and gastric tumors have an overexpression of Her-2/neu.[14–16] Her-2/neu overexpression has also been found in a variety of different solid tumor.[17,18] This overexpression has been targeted using monoclonal antibodies such as trastuzumab and pertuzumab, which have proven protective effects in advanced cancers as well as in adjuvant settings. The Her-2 receptor can also be used for targeting by combining a monoclonal antibody with a chemotherapy. This therapeutic method is known as antibody-drug conjugate therapy (ADC). Trastuzumab emtansine is an ADC that combines an antibody with a microtubule targeting agent. When the ADC attaches to the Her-2/neu receptor, it exerts its action via trastuzumab, and the internalisation of the microtubule agent produces cell cytotoxicity, resulting in a dual impact. EGFR-directed therapy in colorectal cancer is based on the principle of mutant receptor or wild-type receptor. In the case of mutant EGFR receptors (rat sarcoma mutations), the downstream signalling pathway is constitutively stimulated, neutralising the effect of surface receptor inhibition. The anti-EGFR therapy (cetuximab, panitumumab) is exclusively used in patients with EGFR wild-type status.[19,20] TKIs are intracellular targets for precision therapy, as previously observed. The RTK initiates downstream signalling pathways by phosphorylating tyrosine residues on protein substrates. RTK mutations, translocations, and amplifications cause aberrant signalling, which promotes carcinogenesis, invasion, and metastasis of malignant cells. Based on the RTK's nature and function, a variety of malignancies have been treated with TKIs that target the inhibition of RTK. Utilising immune checkpoint inhibitors (ICIs) is the other precision therapeutic approach for enhancing host immunity. Along with surgery, chemotherapy, and radiation therapy, the revolutionary work of James Allison and Tasuko Honjo on immunotherapy has been recognised as the fourth paradigm of treatment. Neoantigens are expressed on the cell surface as a result of genetic and epigenetic alterations in cancer cells. This causes the body to produce a T-cell immunological response. Utilising immune checkpoint inhibitors as a second method of precision medicine involves strengthening host immunity (ICIs). However, immunological checkpoints controls the T-cell response as a defence mechanism of the body to stop autoimmunity and an inflammatory state. By expressing the ligands of these immunological checkpoints, the cancer cell successfully sidesteps the host immune system.

In some circumstances, the use of ICIs guides the body's T cell response toward the tumor. The expression of the programmed death ligand (PD-L1), a high tumour mutational burden, and a lack of mismatch repair genes are frequently used to initiate ICI therapy, even if biomarkers for improving the ICI in clinical use are still being researched. As a result, ICI has been approved for use in a wide variety of solid tumors, and its indications are not tumor-specific.[21] Chimeric antigen receptor T-cells (CAR-T) cells are engineered cells with a tumor-associated antigen binding domain, an extracellular hinge domain, a transmembrane domain, and an intracellular domain, as well as a costimulatory molecule.[22] The result is constrained tumor antigen directed T-cells in a non-MHC controlled manner for precision cell killing. CAR-T cell therapy is approved by the FDA for relapsed/refractory acute lymphoblastic lymphoma and diffuse large B cell lymphoma.[23] Over time, molecular knowledge of cancer pathophysiology has resulted in the development of various targeted medicines that are now actively utilised in the clinical practie of precision medicine in oncology.

28.2.2 Personalised Medicine in Chronic Disorders

Chronic diseases, due to their higher prevalence and frequent complications, draw a wide range of pharmacologic therapies, making them an ideal substrate for a precision medicine approach. The multi-omics repertoire is constantly expanding, from disease risk prediction to tailored therapies.

28.2.3 Personalised Medicine and Asthma

It is well known that CD4 T helper cell subsets comprise Th1 and Th2 populations. The production of IL-4 (IgE production), IL-5 (eosinophil activation), and IL-13 (mucus secretion and IgE synthesis) by Th2 cells causes inflammatory processes in asthma. This, along with the recent identification of type 2 innate lymphoid cells as robust makers of the cytokines IL-5 and IL-13, defines a 'type 2 (T2) high' endotype. The 'type-2 (T2) low' endotype is distinguished by the secretion of IL-1 and IL–17. Each endotype is thus defined by a particular biological process that connects clinical features to a molecular pathway. These two distinct groupings are classified as part of the asthma syndrome. The precision medicine method allows for the identification of particular therapy in the form of monoclonal antibodies for refractory type 2 high asthmatics. Omalizumab binds to circulating IgE, while mepolizumab and reslizumab inhibit IL-5; benralizumab inhibits IL-5 receptor activity, and dupilumab, an IL-4RA antagonist, inhibits IL-4 and IL-13 signals. In type 2 mild or 'neutrophilic' asthma, chemokine receptor (CXCR2) antagonism and dual CXCR1/CXCR2 antagonism are treatment options.[24]

28.2.4 Personalised Medicine and Diabetes

With its several subtypes, monogenic diabetes mellitus allows for precision medicine to target therapy options based on discovered gene abnormalities. Maturity-onset diabetes of the young (MODY-2) is characterised by mild non-progressive diabetes with high fasting glucose that responds to diet and exercise. Diabetes is caused by genes coding for hepatic nuclear factor 1a (HNF-1a) in MODY-3 and HNF-4a in MODY-1, which is susceptible to low-dose sulphonylureas. Sulphonylureas are effective in treating neonatal diabetes caused by mutations in the KCNJ11 gene (potassium voltage-gated channel subfamily 11). Several GWAS-based methodologies are identifying novel candidate genes that may influence responsiveness to insulin sensitizers, sulphonylureas, or incretin treatment.[25]

28.2.5 Personalised Medicine and Systemic Autoimmune Connective Tissue Diseases

Systemic autoimmune connective tissue disorders (AICTDs) appear in a variety of ways, with significant overlap, varying course, and remissions and relapses throughout time. There has recently been an effort to categorise them into subgroups based on quantifiable molecular signatures measured utilising genomic technology. This would also allow for targeted therapeutic targeting by applying existing or investigational treatments customised to a certain group. The use of precision medicine to analyse gene expression, methylation, and genetic data to build a molecular taxonomy of AICTDs could be the next therapy paradigm.[26]

28.2.6 Personalised Medicine in Emergency Care

In contrast to chronic illnesses, emergency care settings provide a limited amount of time to make judgements with a tiny margin of error. The urge for action prioritises over the need for analysis. In such circumstance, precision medicine technology combined with clinical medicine would be beneficial. A biomarker score could be used to triage a patient with dyspnea (e.g., Manchester Triage System). To separate causes, a 'urgent' score combined with clinical, epidemiologic, and sociodemographic factors would motivate the use of a diagnostic biomarker panel as shown below:

a. A measure of cardiomyocyte injury i.e., cardiac troponin T levels are raised in acute coronary syndrome, requiring ECG, 2D ECHO, and revascularisation.
b. D-dimer: A value < 500 mg/dL rules out venous thromboembolism, eliminating the need for unessential imaging.
c. Increased levels of brain natriuretic peptide might indicate heart failure, necessitating a 2D ECHO and diuretics.
d. C-reactive protein (CRP) is a universal inflammatory screening test whose levels guide treatment response in infection and inflammation.
e. Elevated levels of procalcitonin would prompt microbiological cultures and guide antibiotic treatment. Contrary to CRP, results are unaffected by concurrent steroid use.
f. Pro-adrenomedullin: a novel biomarker with predictive significance that is thought to be superior to CRP and procalcitonin (PCT), particularly in the setting of community-acquired pneumonia as well as other respiratory tract infections.[27]

Other acute care settings may benefit as well. Vasculopathic alterations associated with the APOE 2 allele have been linked to the intensity and clinical course of intracerebral haemorrhage, according to GWAS. These and other encouraging indicators have inspired the concept of Emergency Precision Medicine, which would include, at its core, an efficient electronic health record system that would receive inputs about prior visit dates, baseline patient data, research updates, and primary care screening data. This, in conjunction with acute omics would help to facilitate complex decision making in an emergency environment. The physician benefits from a large number of inputs, such as scores, clinical symptoms, and biomarker levels, while making judgements and initiating therapies, as does the patient from better medical care. This method of using precision medicine in emergency conditions is also known as the systematic molecular phenotyping technique.[28,29]

28.2.7 Personalised Medicine in Infectious Diseases

In the 1960s, the medical tricorder was designed as a machine with an additional hand-held scanner that could relay life-sign data to the tricorder itself. It has the ability to monitor all important organ activities and detect the presence of dangerous organisms. Its data banks also had information on non-human races, allowing it to treat various life-forms.[30] Today, metagenomics (the study of genomes obtained from environmental samples) strives to realise this seemingly unachievable futuristic concept. MetagenomicNGS (mNGS) is the process of running all nucleic acids in a sample that may contain a mix of microorganism populations. It is possible that in the coming years, metagenomic sequencing may aid in the diagnosis and monitoring of pathogens in both chronic and acute disorders, and may even replace bacterial isolate culture. The identification of resistant microbes is another precision medicine goal.[31]

28.3 THE EMERGENCY MEDICAL GENOME: NEONATAL SCREENING FOR GENETIC DISEASES

Rapid genome sequencing – STATseq – is a ground breaking medical genomics technology that uses a symptom and sign assisted genomic analysis correlation tool to pinpoint hereditary abnormalities. It allows for a two-day genome study of critically ill infants with suspected genetic abnormalities. This technology enables a speedy diagnosis for therapy, if any exists, so avoiding useless intensive care in instances where no treatment exists and instead provide genetic counselling.[32] This tool is also essential for conducting research and developing management standards in a rapidly evolving field. This has also prompted other specialties to create direct-to-consumer diagnostics for a variety of illnesses.

28.4 ADVANTAGES AND DRAWBACKS OF PERSONALISED HEALTH MEDICINE

28.4.1 Safety

The products and services, mostly diagnostics, included within the umbrella of personalised healthcare and improves the safety profile of therapies by delivering more accurate and personalised treatment regimen. This has an impact on both innovative medicines and current research and development of new drugs, as well as existing ones, to the point where personalised diagnosis could offer the platform for better indication of existing treatments.[33] As a result of new research, HTA bodies will be required to examine existing technologies and assist in identifying patient subpopulations where the safety profile of treatments allows their usage. Postmarketing surveillance of health technology, particularly drugs, will be critical in this evaluation process, by providing prospective data on treatment safety. Moreover, HTA bodies will be required to recommend that existing regulatory and reimbursement procedures and will be reviewed in order to approve therapies whose safety and risk-benefit profile is no longer adequate.[34] As a result, advancements in personalised medicine may result in the revision and strengthening of existing legal frameworks and standards for the approval of therapies based on new evidence about their clinical utility. Healthcare systems should ensure that these mechanisms are adopted and implemented as soon as possible to reduce their potential impact on patients safety and delay in the approval of new treatments.

28.4.2 Clinical Effectiveness

It has been suggested that personalised healthcare solutions improve clinical effectiveness in the same way as safety aspects by allowing therapies to be tailored to those patients who are most likely to benefit from them.[35] As a result, the ongoing implementation of personalised healthcare services will reduce the uncertainties that arise while indicating therapies and provide vital information to health professionals for use in making a decision with their patients.[36–38] The effectiveness of personalised healthcare diagnostic tools is closely linked to changes in patient management for their benefit, rather than simply identifying risks or susceptibilities with no corresponding therapies or interventions that could modify the prognosis or course of the disorders. Furthermore, as with other diagnostic tools, efficacy should be judged based on end outcomes on people's health and income, rather than changes in intermediate clinical indicators or outcomes.[39–41]

28.4.3 Cost and Economic Evaluation

According to some authors, personalised healthcare may reduce the fiscal impact of the introduction of new health technologies, primarily drugs, by indicating treatments only for those patients who would benefit the most, as seen with the recently introduced diagnostic tests for the indication of certain novel chemotherapy alternatives. Despite this, other problems remain unanswered, such as who should bear the expenses of diagnostic testing or how new payment methods for pharmaceuticals associated with a more focused indication should effect pricing policies. Regarding the cost-effectiveness of personalised healthcare, it should be noted that other authors have examined this aspect partially.[42] and admitted that personalised healthcare has an impact on the quality adjusted life years calculation, sensitivity analysis, and discounting because of the intrinsic properties of the technology, its accuracy and predictive values and associated clinical utility. As a result, existing pharmacoeconomic regulations should be changed to resolve these difficulties.

28.4.4 Organisational Aspects

Services for personalised healthcare build a new paradigm in the domain of healthcare delivery by emphasising wellness and illness prevention. Thus, whereas present healthcare services are primarily focused on a hospital-centred model of delivery, this new paradigm returns to models in which primary care and public health play a prominent role. To meet these new standards, work- and patient-flow processes will need to be modified. In light of this, extra attention should be paid to staff genomics training, the management of genomic information and its consequences and the development of shared decision-making abilities, and providers should integrate health literacy as part of their obligations.[43]

28.4.5 Social Aspects

Thus, when implementing personalised healthcare, we should explore positive aspects such as patient empowerment and enhancing autonomy, as well as the potential consequences of stigmatising specific subpopulations based on genomic, clinical, and environmental data. Policies aimed to encourage the deployment of personalised healthcare should consider the social implications of these services, including the individuals engaged, the support necessary, and the costs associated, as well as people's reactions to such services. The assumption of value and the empowerment of patients may not always be viewed positively by all groups because they may alter a person's work performance, social interactions and the way they manage with disease.

28.4.6 Legal Aspects

The key legal implications of personalised healthcare are the preservation of citizens' autonomy, legal responsibility, and the preservation of the data produced. Professional ethical guidelines are frequently made into statutes or required norms that must be followed. Healthcare is no exception, and as such, legal problems are an essential component of a comprehensive HTA review. As a result, careful consideration should be given to the legal constraints associated with various reimbursement and pricing laws around the world, as well as their impact on the adoption of customised healthcare services and products as a result of imbalances and disproportionate access to them. Simultaneously, legal liability, marketing authorisation, advertising, and free movement of services should be harmonised to eliminate citizens' misconceptions as well as illegal acts that could lead to abuse and scams.

28.5 CHALLENGES AND FUTURE PERSPECTIVES ON PERSONALISED MEDICINE

Precision medicine, like any new endeavour, will confront numerous challenges. The vast amount of data that must be collected and analysed is not only costly in terms of money, but also in terms of labour and technical know-how. Data anonymisation and security are major concerns. As a result, ethical issues take precedence. The average response time for data processing is 26 hours, which is still too long for judgements in acute care settings. It is also difficult to calculate the validity of such data. Capacity building must proceed at incredible speeds, particularly in terms of healthcare worker training and the availability of high-quality artificial intelligence (AI), machine learning (ML), and laboratory equipment. Population health may be compromised in the pursuit of precision-based treatments.

The Genome India Project was launched in India on January 3, 2020, with the goal of mapping population genomes using genome-wide sequencing in association with 20 scientific institutes. The Indian Cancer Genome Atlas and the Indian Genome Variation Consortium are investigating the differences between Indian and worldwide genomes. The recently announced Unique Methods of Management and Treatment of Inherited Disorders programme and NIDAN (National Inherited Disease Administration Kendra Network) in September 2019 are both advanced initiatives in this field. With these significant advancements, the future of precision medicine in the country is bright.

28.6 CONCLUSION

Given the scope of this discipline and the rate at which it is evolving, it appears just a matter of time before omics technologies and precision medicine assume a prominent position in diagnostic and therapeutic decision making. In terms of public health, it may simply enhance the use of health information and healthcare by wealthy, energetic, and health-conscious persons who are already engaged in their healthcare. However, rather than encouraging autonomy and agency, personalised medicine has the ability to erode conceptions of identity and health Furthermore, by making healthcare the individual's responsibility, personalised medicine may actually relieve the state of responsibility for its citizens' health, aggravating present health imbalances by ignoring social, economic, and environmental determinants of health.

REFERENCES

1. Bentley DR. Genomes for medicine. Nature. 2004;27;429 (6990):440–5.
2. Burke W, Psaty BM. Personalized medicine in the era of genomics. JAMA. 2007;10;298(14):1682–4.
3. Petersen A. The genetic conception of health: Is it as radical as claimed? Health. 2006;10(4):481–500.
4. Wang ZG, Zhang L, Zhao WJ. Definition and application of precision medicine. Chinese Journal of Traumatology. 2016;1;19(5):249–50.
5. Kosorok MR, Laber EB. Precision medicine. Annual Review of Statistics and Its Application. 2019;7;6:263–86.
6. Di Bello V, Tonti G, Barletta G, Di Salvo G, Mancino M, La Carrubba S, Canterin FA, Carerj S, La Canna G. Introduction to health technology assessment. Journal of Cardiovascular Echography. 2012;1;22(3):99–106.
7. Brand A. Public health genomics and personalized healthcare: A pipeline from cell to society. Drug Metabolism and Drug Interactions. 2012;1;27(3):121–3.
8. Tran B, Dancey JE, Kamel-Reid S, McPherson JD, Bedard PL, Brown AM, Zhang T, Shaw P, Onetto N, Stein L, Hudson TJ. Cancer genomics: technology, discovery, and translation. Journal of Clinical Oncology. 2012;20;30(6):647–60.
9. Saltz LB, Clarke S, Díaz-Rubio E, Scheithauer W, Figer A, Wong R, Koski S, Lichinitser M, Yang TS, Rivera F, Couture F. Bevacizumab in combination with oxaliplatin-based chemotherapy as first-line therapy in metastatic colorectal cancer: A randomized phase III study. Journal of Clinical Oncology. 2008;20;26(12):2013–9.
10. Barlesi F, Scherpereel A, Gorbunova V, Gervais R, Vikström A, Chouaid C, Chella A, Kim JH, Ahn MJ, Reck M, Pazzola A. Maintenance bevacizumab–pemetrexed after first-line cisplatin–pemetrexed–bevacizumab for advanced nonsquamous nonsmall-cell lung cancer: Updated survival analysis of the AVAPERL (MO22089) randomized phase III trial. Annals of Oncology. 2014;1;25(5):1044–52.
11. Aghajanian C, Blank SV, Goff BA, Judson PL, Teneriello MG, Husain A, Sovak MA, Yi J, Nycum LR. OCEANS: A randomized, double-blind, placebo-controlled phase III trial of chemotherapy with or without bevacizumab in patients with platinum-sensitive recurrent epithelial ovarian, primary peritoneal, or fallopian tube cancer. Journal of Clinical Oncology. 2012;6;30(17):2039.
12. Kabbinavar F, Hurwitz HI, Fehrenbacher L, Meropol NJ, Novotny WF, Lieberman G, Griffing S, Bergsland E. Phase

II, randomized trial comparing bevacizumab plus fluorouracil (FU)/leucovorin (LV) with FU/LV alone in patients with metastatic colorectal cancer. Journal of Clinical Oncology. 2003;1;21(1):60–5.

13. Harari D, Yarden Y. Molecular mechanisms underlying ErbB2/HER2 action in breast cancer. Oncogene. 2000;19(53): 6102–14.
14. Wieduwilt MJ, Moasser M. The epidermal growth factor receptor family: Biology driving targeted therapeutics. Cellular and Molecular Life Sciences. 2008;65:1566–84.
15. Kreutzfeldt J, Rozeboom B, Dey N, De P. The trastuzumab era: current and upcoming targeted HER2+ breast cancer therapies. American Journal of Cancer Research. 2020;10(4):1045.
16. Bang YJ, Van Cutsem E, Feyereislova A, Chung HC, Shen L, Sawaki A, Lordick F, Ohtsu A, Omuro Y, Satoh T, Aprile G. Trastuzumab in combination with chemotherapy versus chemotherapy alone for treatment of HER2-positive advanced gastric or gastro-oesophageal junction cancer (ToGA): A phase 3, open-label, randomised controlled trial. The Lancet. 2010; 28;376(9742):687–97.
17. Gravalos C, Jimeno A. HER2 in gastric cancer: A new prognostic factor and a novel therapeutic target. Annals of Oncology. 2008;1;19(9):1523–9.
18. Schuell B, Gruenberger T, Scheithauer W, Zielinski C, Wrba F. HER 2/neu protein expression in colorectal cancer. BMC Cancer. 2006;6(1):1–5.
19. Eltze E, Wülfing C, Von Struensee D, Piechota H, Buerger H, Hertle L. Cox-2 and Her2/neu co-expression in invasive bladder cancer. International Journal of Oncology. 2005;1;26(6):1525–31.
20. Hetzel DJ, Wilson TO, Keeney GL, Roche PC, Cha SS, Podratz KC. HER-2/neu expression: A major prognostic factor in endometrial cancer. Gynecologic Oncology. 1992;1;47(2):179–85.
21. Meyers DE, Banerji S. Biomarkers of immune checkpoint inhibitor efficacy in cancer. Current Oncology. 2020;27(s2):106–14.
22. Ma S, Li X, Wang X, Cheng L, Li Z, Zhang C, Ye Z, Qian Q. Current progress in CAR-T cell therapy for solid tumors. International Journal of Biological Sciences. 2019;15(12):2548.
23. Ahmad A, Uddin S, Steinhoff M. CAR-T cell therapies: An overview of clinical studies supporting their approved use against acute lymphoblastic leukemia and large B-cell lymphomas. International Journal of Molecular Sciences. 2020;30;21(11):3906.
24. Kuruvilla ME, Lee FE, Lee GB. Understanding asthma phenotypes, endotypes, and mechanisms of disease. Clinical Reviews in Allergy & Immunology. 2019;15;56:219–33.
25. Gloyn AL, Drucker DJ. Precision medicine in the management of type 2 diabetes. The Lancet Diabetes & Endocrinology. 2018;1;6(11):891–900.
26. Barturen G, Babaei S, Català-Moll F, Martínez-Bueno M, Makowska Z, Martorell-Marugán J, Carmona-Sáez P, Toro-Domínguez D, Carnero-Montoro E, Teruel M, Kerick M. Integrative analysis reveals a molecular stratification of systemic autoimmune diseases. Arthritis & Rheumatology. 2021;73(6):1073–85.
27. Schuetz P, Aujesky D, Mueller C, Mueller B. Biomarker-guided personalised emergency medicine for all-hope for another hype? Swiss Medical Weekly. 2015;145:14079.
28. Anastassopoulou C, Gkizarioti Z, Patrinos GP, Tsakris A. Human genetic factors associated with susceptibility to SARS-CoV-2 infection and COVID-19 disease severity. Human Genomics. 2020;14(1):1–8.
29. Limkakeng AT Jr, Monte AA, Kabrhel C, Puskarich M, Heitsch L, Tsalik EL, Shapiro NI. Systematic molecular phenotyping: A path toward precision emergency medicine? Academic Emergency Medicine. 2016;23(10):1097–106.
30. Lefterova MI, Suarez CJ, Banaei N, Pinsky BA. Next-generation sequencing for infectious disease diagnosis and management: A report of the Association for Molecular Pathology. Journal of Molecular Diagnostics. 2015;1;17(6):623–34.
31. Palacios G, Druce J, Du L, Tran T, Birch C, Briese T, Conlan S, Quan PL, Hui J, Marshall J, Simons JF. A new arenavirus in a cluster of fatal transplant-associated diseases. New England Journal of Medicine. 2008;6;358(10):991–8.
32. Saunders CJ, Miller NA, Soden SE, Dinwiddie DL, Noll A, Alnadi NA, Andraws N, Patterson ML, Krivohlavek LA, Fellis J, Humphray S. Rapid whole-genome sequencing for genetic disease diagnosis in neonatal intensive care units. Science Translational Medicine. 201;3;4(154):154ra135.
33. Seo HJ, Park JJ, Lee SH. A systematic review on current status of health technology reassessment: Insights for South Korea. Health Research Policy and Systems. 2016;14:1–10.
34. Li YY, Jones SJ. Drug repositioning for personalized medicine. Genome Medicine. 2012;4:1–4.
35. Guyatt GH, Oxman AD, Vist GE, Kunz R, Falck-Ytter Y, Alonso-Coello P, Schünemann HJ. GRADE: An emerging consensus on rating quality of evidence and strength of recommendations. BMJ. 2008;24;336(7650):924–6.
36. Guyatt G, Vist G, Falck-Ytter Y, Kunz R, Magrini N, Schunemann H. An emerging consensus on grading recommendations? BMJ Evidence-Based Medicine. 2006;1;11(1):2–4.
37. Ibargoyen-Roteta N, Gutiérrez-Ibarluzea I, Rico-Iturrioz R, López-Argumedo M, Reviriego-Rodrigo E, Cabriada-Nuño JL, Schünemann HJ. The GRADE approach for assessing new technologies as applied to apheresis devices in ulcerative colitis. Implementation Science. 2010;5(1):1–11.
38. Guyatt G, Oxman AD, Akl EA, Kunz R, Vist G, Brozek J, Norris S, Falck-Ytter Y, Glasziou P, DeBeer H, Jaeschke R. GRADE guidelines: 1. Introduction – GRADE evidence profiles and summary of findings tables. Journal of Clinical Epidemiology. 2011;1;64(4):383–94.
39. Ginsburg GS, Willard HF. Genomic and personalized medicine: Foundations and applications. Translational Research. 2009;1;154(6):277–87.
40. Garrido MV, Mangiapane S. Surrogate outcomes in health technology assessment: An international comparison. International Journal of Technology Assessment in Healthcare. 2009;25(3):315–22.
41. Lenoir-Wijnkoop I, Dapoigny M, Dubois D, van Ganse E, Gutiérrez-Ibarluzea I, Hutton J, Jones P, Mittendorf T, Poley MJ, Salminen S, Nuijten MJ. Nutrition economics–characterising the economic and health impact of nutrition. British Journal of Nutrition. 2011;105(1):157–66.
42. Postma MJ, Boersma C, Vandijck D, Vegter S, Le HH, Annemans L. Health technology assessments in personalized medicine: Illustrations for cost–effectiveness analysis. Expert Review of Pharmacoeconomics & Outcomes Research. 2011;1;11(4):367–9.
43. Lewis GH. "Impactibility models": Identifying the subgroup of high-risk patients most amenable to hospital-avoidance programs. The Milbank Quarterly. 2010;88(2):240–55.

Index

D

E

F

G

H

I

K